D0710366

Practical Handbook of

PROCESSING and RECYCLING MUNICIPAL WASTE

A.G.R. Manser
A.A. Keeling

Publisher: Joel Stein
Project Editor: Erica Orloff
Marketing Manager: Greg Daurelle
Direct Marketing Manager: Arline Massey
Cover design: Denise Craig
PrePress: Kevin Luong
Manufacturing: Sheri Schwartz

Library of Congress Cataloging-in-Publication Data

Manser, A. G. R.
Practical handbook of processing and recycling municipal waste / A.G.R. Manser, Alan Keeling
p. cm.
Includes bibliographical references and index.
ISBN 1-56670-164-3
1. Compost plants. 2. Refuse as fuel. I. Keeling, Alan. II. Title.
TD796.5.M328 1996
628.4'458—dc2 95-4649595
CIP

International Standard Book Number 1-56670-164-3
Library of Congress Card Number 95-46495
Printed in the United States of America 1 2 3 4 5 6 7 8 9 0
Printed on acid-free paper

Foreword

The human race is rapidly heading towards the point where it will stifle in its own wastes, and if it continues then sooner or later that will occur. Any waste will damage the environment to some extent; its disposal therefore requires consideration of the best environmental option. Care for the environment and for natural resources is not simply a matter of banning atmospheric emissions or becoming obsessed by land or water pollution in isolation.

This leads to immediate difficulties with the more vociferous members of the community, who would of choice return civilization to the caves. But the reality is that sometimes we must accept a lesser amount of environmental damage to avoid the greater. Nowhere is that position more apparent than in the waste disposal industry.

It is an industry which has long been afflicted by swings of fashion, ill-informed legislation, political interference, and a general lack of understanding among the creators of waste. As a prominent member of the industry in Great Britain once remarked, in response to hostility from a user of the service, "We don't make the waste. You do. We only try to get rid of your problem for you in the cheapest and most appropriate way we know."

Knowledge overcomes dissent. Perhaps one day, if sufficient people, particularly in governments, begin to understand what waste disposal technology is and what its limitations are, then we may begin to make progress. The purpose of this book is to attempt to extend some of the knowledge to a wider audience. It is not aimed at theoretical physicists, so the mathematical modeling techniques described are simple, capable of being conducted on a pocket calculator. They give answers which are acceptable for most practical purposes although they are not perfectly accurate. The design concepts are explained as simply as possible to provide an understanding of how the technology works and how to predict what it will be capable of, not to allow the reader to design a complex plant on his own.

The intention is to give practitioners in local authority environmental health departments, in waste disposal companies, in universities and colleges, a general understanding of how engineers develop recycling and reprocessing designs. Hopefully they will find it of assistance when evaluating proposals and disposal requirements, and in selecting the best environmental option.

Perhaps even the design engineers themselves will read the following chapters, and argue, criticize, question, and above all think. We do not claim that what we have written is engraved in stone. We set out only to explain the state of the art as we know it, based upon a number of hard years of working in the field. A great deal of waste disposal technology is subjective rather than objective. A case of "This is what I have discovered in practice, but I concede that you may have found alternatives."

However the statements and observations in this book are received, it is inescapable that if we do not recycle or reprocess our wastes, then we will destroy not only our environment but that of our children. Let us begin to attack the problem by first understanding it.

A.G.R. Manser
A.A. Keeling
February 1996

The Authors

A.G.R. Manser

Tony Manser trained in heavy mechanical, electrical, and marine engineering with Standard Oil of New Jersey, and spent some 10 years in the oil industry in the Middle East and Venezuela. In 1970, he moved on to the waste disposal industry when he became deputy chief engineer of the largest waste-to-energy plant in Europe.

After 5 years he joined a major British regional authority as its Principal Mechanical Engineer. There he was responsible for the design, development, maintenance, and operation of a number of sophisticated waste treatment facilities, in addition to taking charge of a large transport fleet and workshop complex. During that period he was instrumental in the development and construction of new waste-derived fuel and organic waste composting processes, both fields in which he has achieved some recognition. He also served as operations director of a waste disposal consultancy owned by the authority.

He then joined a major energy corporation as engineering director of a research and development subsidiary, with the task of developing new mechanical processes for the treatment of organic wastes. From there he moved again to become main board engineering director of a substantial British company which built and operated energy-to-waste facilities. He also accepted executive directorships in a process design company and in a waste recycling company, serving as operations director in the latter.

In 1991, with three colleagues, he formed Advanced Recycling Technologies Ltd., a British-registered company operating in the field of waste treatment process design and consultancy. Later those interests expanded with the formation of Advanced Recycling Technologies (Ireland) Ltd., a sister company to the former, based in the Republic of Ireland. The Irish company manufactures plant and machinery to the British company designs. He now lives in Waterford, Ireland, from where he serves as managing director of both companies.

Tony Manser has spent many years exploring the physics of waste and seeking to understand its behavioral characteristics. A significant number of developments occurred in that respect, during the period when Alan Keeling and he worked together upon a major project associated with refuse-derived fuels.

Alan Keeling

After graduating in biochemistry in 1982, Alan Keeling spent 8 years researching into radiopharmaceutical design. During that period he was awarded his Ph.D. as a result of his research, becoming Research Fellow at Birmingham University, Britain.

He then joined a substantial waste disposal company as manager of its biochemistry laboratory, where he conducted research and laboratory analyses into and upon refuse-derived composts and their applications as growing media. His responsibilities included the establishment of growing trials for products, and the establishment of analysis protocols for refuse-derived fuels. Using gas chromatography mass-spectrometry he was instrumental in developing new techniques for the identification of impurities in refuse-derived fuels and composts, and he worked with the company engineering staff to establish treatment processes for their removal.

Later he accepted an appointment as recycling and composting consultant with an international environmental consultancy, providing advisory services to clients in the biochemical and environmental aspects of recycling and composting. During that period he acted as consultant to a British regional waste disposal authority developing yard waste composting facilities.

In 1993 he took up the post of Senior Lecturer at Harper Adams Agricultural College, in Shropshire, Britain. Harper Adams is a world-recognized leader in agricultural research, and there, in addition to his teaching duties, Alan Keeling is heavily involved in researching composting and the biochemistry of soils and plant growth.

Units of Measurement

Throughout this book the following units of measurement are used:

Weight:

1st listed: U.S. "short" tons = 2,000 lb; abbreviation t; 1 t = 907 kg
Imperial pounds; abbreviation lb; 1 lb = 0.4536 kg
2nd listed: Metric tons (tonnes) = 1,000 kg; abbreviation te; 1 te =1.08486 t
Kilograms; abbreviation kg; 1 kg = 2.205 lb

Length/Area/Volume:

1st listed: Inches/feet/yards; abbreviations ft, in., yd, equivalent to:

1 in. = 25.4 mm — 1 ft = 304.8 mm — 1 yd = 914.4 mm
1 in.^{2} = 645.2 mm^2 — 1 ft^2 = 0.092 m^2 — 1 yd^2 = 0.836 m^2
1 in.^{3} = 16.39 cc — 1 ft^3 = 0.028 m^3 — 1 yd^3 = 0.7646 m^3

2nd listed: Millimeters/centimeters/liters/meters; abbreviations mm, cm (cc), m

1 mm = 0.0394 in. — 1 cm = 0.394 in. — 1 m = 3.281 ft (1.094 yd)
1 mm^2 = 0.0016 in.^{2} — 1 cm^2 = 0.16 in.^{2} — 1 m^2 = 10.76 ft^2 (1.195 yd^2)
1 mm^3 = 6×10^{-6} in.^{2} — 1 cc = 0.061 in.^{3} — 1 m^3 = 35.31 ft^3 (1.308 yd^3)

Density:

1st listed: pounds per cubic foot; abbreviation lb/ft^3; 1 lb/ft^3 = 16.02 kg/m^3
2nd listed: kilograms per cubic meter; abbreviation kg/m^3; 1 kg/m^3 = 0.0624 lb/ft^3

Calorific Value:

1st listed: British thermal units per pound weight; abbreviation Btu/lb;
1 Btu/lb = 2.326 kJ/kg
2nd listed: Kilojoules per kilogram; abbreviation kJ/kg; 1 kJ/kg = 0.43 Btu/lb

Pressure:

1st listed: Pounds per square inch, abbreviation lb/in.^{2} or psi
1 lb/in.^{2} = 0.07 kg/cm^2
Or: Inches water gauge, abbreviation in.-wg; 1 in.-wg = 0.433 lb/in.^{2}

Atmospheric pressure = 14.5 lb/in.^{2} (1 atmosphere, or atm)
Gauge pressure (psig) = pressure in excess of 1 atm
Absolute pressure (abs) = atmospheric pressure + gauge pressure

2nd: Kilograms per square centimeter; abbreviation kg/cm^2
1 kg/cm^3 = 14.22 lb/in.^{2} = 32.84 in.-wg.

Atmospheric pressure = 1 kg/cm^2 (1 atmosphere, or "bar")
Gauge pressure (g) = pressure in excess of 1 bar
Absolute pressure (abs) = atmospheric pressure + gauge pressure

Temperature:

1st: Degrees Fahrenheit; abbreviation °F

Temperature °C = (°F – 32) × 0.555

2nd: Degrees Centigrade, abbreviation °C

Temperature °F = (°C × 0.555) + 32

Velocity/Acceleration:

1st: Feet per second, feet per second per second; abbreviations ft/s, ft/s^2

1 ft/s = 0.3048 m/s   1 ft/s^2 = m/s^2

2nd: Meters per second, meters per second per second; abbreviations m/s, m/s^2

1 m/s = 3.280840 ft/s
1 m/s^2 = 3.280840 ft/s^2
Acceleration due to gravity = 32.2 ft/s^2 = 9.8 m/s^2

Force and Force per Length:

1st: Pounds force, pounds force per foot; abbreviations lbf, lbf/ft

1 lbf = 4.448222 N   1 lbf/ft = 14.59390 N/m

2nd: Newtons, kilograms force, Newton-meters; abbreviation N, kgf, N-m.

1 N = 0.1019716 kgf = 0.2248089 lbf
1 N-m = 0.06852178 lbf/ft

Contents

1 INTRODUCTION

When man first started to create significant volumes of wastes, recycling seemed to be an attractive way of disposing of them. After all, among these wastes are materials which can be recovered for reuse, and potentially quite a lot of energy. In the Western societies scavenging was practiced successfully almost within living memory, and in less developed countries it still is. Scavenging is recycling under a less fashionable name, but it is extremely efficient.

Unfortunately as a society becomes more advanced, simple expedients are no longer sufficient. The waste becomes more complex and it grows in volume at an alarming rate, to the point where a few people foraging around on the local dump no longer make much of a contribution to the disposal route. If the potential of the material is to be exploited, technology must begin to play a part. Such an endeavor requires mechanical handling systems, sorting machinery, control systems, process engineering, and skilled operators. It also means the introduction of reprocessing.

The industry of waste recycling and reprocessing has not enjoyed a distinguished history. It is probably the one industry that has been the most bedeviled by the failure of designs. Process plants which either did not work at all or which failed to achieve their production targets are scattered all over Europe and the United States. All too often the cause of the failure has been a lack of understanding of the nature of the feedstock, an oversimplistic view of the design concepts, or even just a bad case of single-mindedness to the exclusion of reality.

1.1 RECYCLING VS. REPROCESSING: THE FUNDAMENTAL DIFFERENCES

Mixed waste is a difficult material to handle, and on occasion almost impossible. For that reason it is important that a structured approach to any treatment policy be adopted, a series of steps that move logically from concept to construction, never sidestepping the difficult questions which appear to argue against the whole idea. And the first stage has to be a clear understanding of what recycling and reprocessing actually are; this is not always as much a statement of the obvious as it sounds.

First the definitions: Recycling is the recovery of items or individual fractions from the waste in such a way that they may be reused. Reprocessing is the reduction of the waste to one or more raw materials from which completely new consumer products are made.

Thus, a materials recovery facility which separates cans, paper, and glass, all of which are sold back to the original material manufacturers, is a recycling operation. So too is the facility where bottles are collected, separated into the three main colors, and destined for the manufacture of new bottles. The facility is not doing anything to the components of the waste to change their characteristics. A can is still a can whether or not it is mixed with household waste or recovered and baled with many others.

A reprocessing facility, however, does change the nature of the waste and of the individual fractions within it. For example, a refuse-derived fuel (dRDF) process removes paper and plastic film, granulates and dries it, and extrudes it into fuel pellets or bales it as flock. The product is then burned in boilers to create energy. It is never again used as paper. Similarly, a composting plant separates organic materials and reduces them biologically to produce a growing medium which bears no relationship to the original waste.

These differences lead to the next; those of cost and capacity. The definitions of the two concepts makes it obvious that reprocessing must be much more sophisticated and therefore much more expensive than recycling, and that difference introduces an order of scale. After all, a reprocessing plant must of necessity have a recycling facility of a sort included in its design. The fractions have to be separated before they can be processed.

Capital investment has to be recovered and a realistic rate of return achieved. Hence a reprocessing facility which cost $8 million may be expected to generate an income of at least $1.5 million a year. The only way in which that can be achieved is in a large plant with a large input of waste.

Conversely, a recycling operation can be anything from a couple of operators picking through the waste delivered to a civic amenity site to a full-scale materials recovery facility. In those cases orders of cost can range from zero to perhaps $3.5 million and waste can be processed from a few hundred tons to a few thousand tons a year. In between are the bottle banks and can banks which are appearing in supermarket parking lots and street corners all over Europe.

Whether recycling or reprocessing is the chosen disposal route, they are commercial undertakings with end products that need marketing. It does not, regrettably, follow that if we make it or retrieve it and put a label saying "recycled" on it, then the public will line up to buy it! It certainly does not follow that simply because *we* think it is a good idea, everyone else will, or that because we can actually be paid to take the waste we cannot fail to make a profit. The marketplace is tough and skeptical.

Moving on to the matter of the market for the products leads to the first difficult question; who is going to buy them? The unpalatable reality of recycling, and one which is often ignored in consequence, is that a tin can is in the waste in the first place because someone didn't want it. Simply taking it out again will not necessarily confer any value upon it. A sweeping generalization, but still a good principle from which to start, is that the products of a recycling operation will all be of low quality and correspondingly low value.

At the moment there is considerable euphoria about recycling throughout the developed world. Manufacturers seem quite happy to buy recovered materials, and some even say they cannot obtain sufficient for their needs. In reality, in Europe alone millions of tons of wastes are disposed of each year, but only thousands of tons are recycled. It does not follow that because there appears to be a market now, there will continue to be one if everyone is recycling. Saturation is an ever-present threat.

Reprocessed products have a potentially much larger market simply because they are new. They do not rely upon accommodation by the original manufacturer or any single buyer. But as stated earlier, the production facilities are of necessity much larger, and so inevitably is the level of output. Marketing is no longer a matter of sending truck-loads of materials at irregular intervals to a manufacturer who will accept them all. Now it is a matter of a number of truck loads every day going to discerning buyers who expect conformance to some specification, and consistency and availability.

And the final important difference between recycling and reprocessing lies in the origins of the waste itself. Any realistic recycling scheme will require some form of prior separation, since any attempt to pick through significant quantities of crude waste for recoverables is both impossible and even potentially hazardous. So the waste producers must be encouraged to keep separate those items destined for recycling, and this in turn leads to separate collections or special vehicles, both expensive activities. The waste collection agencies are not going to pay inflated tipping charges if they are already incurring high collection costs in the name of recycling. Many waste reprocessing designs, however, can accept mixed waste without any presorting. So the collection systems need not change, and tipping charges can be correspondingly higher.

Perhaps the fundamental point is that there is only so much money available to deal with waste, and whether it is spent on collection or disposal is somewhat irrelevant. The more that is spent on the one, the less is available for the other. Recycling tends to place the emphasis upon collection but results in cheaper disposal. Reprocessing is exactly the opposite.

From the financial point of view, when trying to decide whether a waste disposal undertaking should be recycling- or reprocessing-based, the question really becomes one of where the cost can best be tolerated. For example, in a predominantly rural area, a source separation scheme that involves separate collections may be unacceptable. The vehicles already spend more time traveling than collecting, and the cost of increasing the fleet to accommodate recycling will not be welcomed. A city center tower block estate may not even have the facility to keep waste separate at the source, at least not without major structural alterations that would be prohibitively expensive.

If, however, the rural area generates only a few thousand tons of waste a year, then a reprocessing facility to accommodate it would be small, and the capital cost component of the revenue stream would be large. Tipping charges at the plant would have to be high to recover it.

There will be exceptions, since there are other justifications for the concepts that are somewhat more abstract. For example, if the collection service is funded by a public authority that wishes to encourage recycling, then it may be prepared to pay a premium to the collection contractor for source separation. Alternatively, it may be prepared to bear the cost of the initial capital investment in, for example, "wheelie" bins with split compartments.

1.2 HISTORY

It was noted earlier that the recycling and reprocessing industry has had a long and checkered history, mainly due to a lack of understanding of what waste actually is in the realities of the marketplace, or of a failure to grasp the essential design concepts. Such remarks cause outrage in machinery manufacturing and plant design companies, and so it is worth examining that history in a little more depth to illustrate the point. The blame does not, after all, rest with such companies or even attach to them particularly. They can only apply the technology existing at the time and, of course, build what the customer believes he wants.

Perhaps the oldest reprocessing technology is composting, simply because, left to itself, waste will always rot down to a stable material to some extent. That after all is what happens in a landfill, and the restoration of derelict land by controlled tipping could even be said to be itself reprocessing to some extent, in that it makes a new product (usable land) out of waste. From there it was not a long stretch of the imagination to allow it to decompose above ground and to seek a market for the resulting product.

Composting became fashionable, and many process plants were constructed. Most had some form of pretreatment, even if all it consisted of was shredding or milling and magnetic recovery of ferrous metals. The residue could then be allowed to decompose naturally in vessels or bunkers or in heaps until it became dark brown in color and ready for sale to a generally unprepared agricultural industry. Of course, it contained glass shards and metal fragments, and much more sinister components such as cadmium, lead and zinc that were hidden from view, but that seemed unimportant for a comparatively low-grade market.

Unfortunately, nobody spent much time analyzing the resulting products, and so the toxins remained undetected until they built up in the soil to the extent that they actually damaged plants and posed a health hazard. Then there was the matter of how the natural cycle uses organic materials. Stated simply, worms take them down into the soil, thereby enriching it with essential humus and, together with microorganisms, releasing nutrients. They do not take the same accommodating view of glass and metals, and they leave them on the surface, so that in time a field so treated becomes heavily contaminated. The agricultural industry is and has always been a professional and discerning market, and it was not long before it awoke to the dangers of refuse-derived composts. At that point, such materials could not be given away and the technology collapsed.

The lesson in this failure is that the development did not take into account the needs of the market. It only considered the needs of waste disposal. It did not establish the appropriate technology or seek to understand the more subtle characteristics of the raw material. And it failed as a result.

So the fashion moved back a step to recycling and the concept of some form of source separation. Collection vehicles were outfitted with wire-framed trailers intended to recover paper and cardboard separately from the remaining waste. There was a ready market for the material, which could be enhanced by the construction of baling plants to reduce transport costs. It was such a good idea that everyone started to do it, and the limited repulping facilities became overwhelmed. Within a decade or so it again became almost impossible to give the material away, and in fact many recyclers found themselves actually paying to have it removed. No one wanted dirty paper mixed with cardboard and plastic.

Again, we see a failure to evaluate the market for the products. Few, it seems, asked themselves whether saturation was likely, yet the means of exploring that likelihood were always available and easy to apply.

From there the waste disposal industry moved on to incinerators which, in their early form, were indeed dark satanic mills. They created atmospheric pollution on a grand scale and made huge clinkers composed of fused glass and metal and toxic ash, all of which had to be manhandled from the furnaces at regular intervals. But they did create a great deal of heat as well, and it was not long before the idea of recovering that heat became attractive.

So plants were built with waste heat shell boilers installed in their exhaust gas ducts. The intention was generally to supply low-pressure hot water to factory estates and even, in some cases, to housing estates. But the boiler tubes choked up almost immediately after they were put into service, and tube plates and exhaust stacks corroded alarmingly. The reason was simply that the incinerator combustion gases contained a very heavy dust burden made up of sulfur, potassium, calcium, and chlorine compounds with liberal amounts of silicon and metallic vapor. They formed a hard, dense slag on tube plates and blocked the boiler tubes. And those gases that did manage to pass through the boilers were cooled sufficiently to reach the dew-point of hydrochloric and sulfuric acid, which dissolved gas ducts very rapidly indeed.

This situation was brought about by a lack of understanding of the nature of waste when it was burned, and more particularly of the behavior of the high-temperature chemical reactions in the exhaust gases.

So the next stage was steam raising plants designed around water tube boilers, where the gases passed over the tubes rather than through them. That meant that the gas passes could be much larger and more tolerant of dust burdens, but it brought with it new problems. Water tube boilers are associated with high-pressure steam rather than hot water, and the steam was attractive for power generation. Tube metal temperatures and gas velocities were much greater. High temperature corrosion, grit erosion, and flame impingement destroyed water tubes with astonishing rapidity.

As a result, generating incinerators became very expensive to run, since their boilers had to be surveyed annually and retubed regularly. The failure of understanding in this case was again in the behavior of the combustion gases, although it could be excused by the recognition that very little was known about how chemical reactions in complex gases changed at high temperatures.

From these experiences, the industry moved on to a concept that was really a more refined type of incineration: the manufacturing of fuel from a separated and processed fraction. The early incinerators, after all, attempted to burn all of the waste, combustible and noncombustible. It was apparent that if the former were extracted and used as the fuel, and the latter landfilled, then the waste volumes could be reduced dramatically while overcoming boiler problems. But it was not quite as simple as that.

The early refuse-derived fuel plants separated the combustibles in the form of a granulated flock, which was almost unmanageable. It had an initial bulk density of about 3.7 lb/ft^3 (60 kg/m^3), which meant that the introduction of sufficient quantities of it to meet any realistic energy demand in a boiler was virtually impossible. The energy density was far too low, unless the boiler happened to be on the same site and designed for pulverized fuel. So the solution was to pelletize the material and to thereby raise its energy density tenfold.

Pelletizers, however, are very sensitive to moisture contents, and the granulated flock was too moist. It had to be dried, and many ways of doing that were experimented with, some with quite spectacular and explosive results. But still at the end of the process, there remained a fuel with an ash content of 18% and a chlorine content of over 1.5%. The ash level was generally beyond the tolerance of conventional boilers, and the chlorine level led to the formation of hydrochloric acid, which demanded a derating of the boiler plant to avoid damage.

At the same time, the actual processes ran into trouble. All too often they were designed to accommodate a calculated mass balance. This worked upon the principle that if 10 tons (t) of waste are fed into a trommel screen that removes 5 t of glass, metal, and organics, then the discharge conveyor needs to handle 5 t of paper and plastic. Half as much comes out as goes in.

Unfortunately, that is not so. A conveyor is not just a mass-handling device, but is also a volume-handling one. The density of the material leaving the trommel was generally less than half of that of the feedstock, with the result that, extraordinarily enough, a greater volume came out of the screen than went in. The conveyor systems were capable of handling several times the weight, but nowhere near the volume. Thus blockages were prevalent and plant availability was so low as to render many of the early plants almost inoperable.

There were worse problems looming over the horizon in the form of environmental legislation, at least in so far as the European Community was concerned. Refuse-derived fuels were treated with some suspicion in the conviction that they were simply waste incineration under a different name. Nobody was prepared to take into account the fact that the industry was still very much in its development stage and that it needed time to establish its technology. Directives were created to control mass-burn plants, and they included refuse-derived fuel in their classifications. Anyone trying to use it in their boilers would find the activity classed as waste incineration with all of the environmental controls and monitoring regimes which that implied. In the U.K, where the fledgling industry was originally supported by government, there was an attempt to overcome the problem by seeking derogation from the directives, but that loophole was effectively closed by the Environmental Protection Act of 1990.

Here then, the problem started from incomplete understanding of the physical and chemical characteristics of the raw material and the products, and from the refusal of legislators to permit the industry time to establish the technology. That gave refuse-derived fuels a bad name, from which they still have not recovered.

Meanwhile, waste disposal in the developed world was tending back toward recycling rather than reprocessing. The principle seemed to be almost one of desperation. If one cannot make compost out of it, cannot burn it, and cannot make fuel out of it, then perhaps one should revert to picking things out of it. Materials recovery facilities (MRFs) gained popularity.

The concept was embraced with enthusiasm in the United States, where by 1991, 126 plants were in operation and a further 66 were in the development stage. Many were based upon very old-fashioned technology, employing long lines of picking belts and personnel employed to sort through the waste manually. And, of course, they required some form of source-separation before the waste was delivered. There was, it was reported, a ready market for the materials recovered, and the universal panacea for waste disposal problems was at hand.

As suggested earlier, marketing is a difficult business. It is, for example, quite easy to sell a ton of baled cans each week, if for no other reason than this amount is insufficient to cause the purchaser any real problem. Trying to sell 1000 t a week is quite another matter. In the United States this problem was overcome to some extent by the small scale of individual plants; in 1991, some 85 plants out of the total then operational handled less than 100 t of waste a shift. That in the richest and most industrialized nation on earth, where recycling is regarded as vital.

What real contribution, one wonders, does such small-scale recycling really make to the world-wide problem of waste disposal? We repeat, millions of tons are created, thousands of tons are recycled. The waste disposal industry has swung between reprocessing and recycling on average every 20 years or so ever since the turn of the century, looking for an alternative to landfill and the destruction of resources it implies. The technology exists in several forms, each appropriate to some specific arisings. But it has to be understood on a much wider scale than it has been so far. This book is an attempt to assist in that.

2 THE NATURE OF WASTE

2.1 WASTE ANALYSES

Any waste recycling or reprocessing design starts with an analysis of the waste to be handled. Having made that somewhat bland statement, we recognize that there are some who will immediately argue against it on the basis that a design must assume an average quality rather than a specific one. After all, the material will vary hourly, daily, and seasonally, so how much use is detail? Simply this: If the object is to design a reprocessing plant to handle municipal waste, then an analysis taken on one day is useless. There is much more to waste analysis than that. Such an analysis must be wide-ranging enough to give a good overall picture of the arisings over a period, in such a way as to permit the application of experience. It is an exercise in concentrating the mind, not in analytical chemistry.

The strategy for the analysis must be planned, and the first point to be considered is fundamental: any analysis, however it is conducted, will alter some of the characteristics of the material. So, for example, if the purpose is to design a waste storage facility, where knowing the as-received density is important, there is no value in weighing samples in a meter box. Simply separating them and placing them in the box will alter their density quite considerably. The resulting figure will bear little or no relationship to that of the material discharging from the body of the collection vehicle. Conversely, if the purpose is to design a reprocessing plant, the as-received density does not matter much. It is the density and proportions of the individual fractions that are important.

The point of this is that before deciding upon the analysis, we need to ask, What are we doing it for? Then we must tailor the procedure to reveal the information specific to that requirement, so that those characteristics which the very practice of analysis will change do not matter. Now we can move on to the actual procedure.

2.1.1 Waste Analysis for a Recycling Plant

A recycling plant (a materials recovery facility) serves to recover items from the waste for processing elsewhere. Therefore, it is necessary to know three things: (1) the recyclable components of the waste and their proportions, (2) the extent to which those proportions vary over an operating period, and (3) the bulk densities of the individual components.

Note the term "recyclable components." For instance, some part of the waste will be that which is often referred to as the "minus ten millimeter" fraction. It is a mixture of soil, ash, dust, and organics, and a recycling plant will not deal with it, so it is unnecessary to know what it is. Only those items that can be conveniently lifted out, such as cans, bottles, textiles, paper, and plastics, are of interest. Equally, it is unnecessary to know the bulk density of the mixed waste, since a modern recycling plant will not receive it that way in the first place.

Then, bearing in mind the way in which such a plant works, it is unlikely that it will recover every small particle of metal and plastic. So we are really only interested in the larger pieces, and they can be isolated by screening the whole sample initially through a simple vibrating sieve. Following this, a sample exists that contains only intact bottles and cans, etc., and it becomes straightforward enough to pick the individual components to arrive at a fractional analysis.

Now the original sample has been reduced to separate piles of recyclables. That is initially satisfactory, but the recovery lines in the plant will have to be designed to some definite capacity, and it is necessary to know if the analysis obtained will be fairly consistent throughout the operating day and week. After all, the collection service does not visit the same premises or even the same area every day, and there may be differences in the analysis as a result. A random sample from each collection area is needed, so the origin of each sample has to be known and plotted on a map.

Progressively then, over a period of a week a picture starts to build up. It may be, for example, that on Mondays the collection service is preoccupied with clearing the weekend accumulation at commercial premises, and there is a bias toward paper and packaging material. On Tuesday and Wednesday the collectors may be working a predominantly residential area, much of which is middle-class housing. There will be a lot of packaging and plastic bottles again. Thursday collections might, coincidentally, be mainly from an inner-city area of tower blocks and flats, where there is little or no facility for separation at source, and so any recyclables will be heavily contaminated and may not be worth accepting into the plant at all. Fridays might, again, be spent clearing commercial areas ready for the weekend.

From the design point of view, there will be an influx of bulky packaging on Mondays and Fridays, somewhat less packaging but an increase in plastic bottles on Tuesdays and Wednesdays, and not much of anything worthwhile on Thursdays. Had the plant capacity for plastic bottles been designed on the basis of the Monday analysis, then its capacity for some of the time would be too small. But now it is known how much of the week's input of each fraction will occur on which days, and the storage and processing capacity can be designed with some confidence.

Seasonal variations must now be accounted based upon experience. Given the identified collection areas and the proportions of recyclables arising from them, a judgment can be made about what variations might occur. And in the case of a materials recovery facility, the answer is, not much. After all, the fractions which are most likely to change over the year have been eliminated. There is no reason to expect any real change in the amount of packaging from the commercial area, or in the volume of plastic from the residential one except for, perhaps, an increase in beverage containers during the summer months. The organic fraction will change quite considerably in the same period, but it is not a part of the recovery target.

However, to take the examples further, if the plant is being designed to serve a holiday resort, what impact will the influx of tourists have upon each of the areas? It would be reasonable to expect that there will be an increase in the levels of convenience food packaging from the commercial sector, but not much change from any of the others. A judgment of what the change is likely to be can be as simple as obtaining an estimate from the local tourist department of the number of visitors expected. Then it can be assumed that if "*x*" thousand residents generate "*y*" tons of recyclables, each one creates so many kilograms. Multiplying that by the number of visitors provides a good approximation with which to work.

2.1.2 Waste Analysis for a Reprocessing Plant

Adopting a "worst-case scenario," assume that the reprocessing facility will be taking commingled waste, that is, waste that arrives at the plant having been collected without any form of separation at source. Again, it is necessary to decide what sort of process it is to be before we can select the analytical method, but in this case four pieces of information will be needed: the three already discussed for a materials recovery facility, with the addition of the bulk density of the waste as received. That being so, it would be pointless to sieve the material first, because that would alter the as-received density. So the first step in the analysis is to examine a conceptual model of the facility and to decide what is going to happen to the waste at each stage. After all, the whole purpose of the analysis is to be able to design each of those stages.

The process begins with storage of the waste; how much will have to be stored to overcome the difference between plant capacity and waste input in tons per hour. The plant may, for example, start up at seven o'clock every morning, but the first waste may not arrive before ten o'clock. Then between twelve o'clock and two-thirty the vehicles may be out on their second collection, arriving back between three o'clock and three-thirty. All of the waste input arrives in 5 h at a plant which will run for 16.

At very least the plant needs to store sufficient waste to be able to run during the light input periods, but storage space is measured in volume, not mass. So we need to know the bulk density of the waste not as it is in the vehicles but as it is when tipped. That can be estimated by first recording the body dimensions of the vehicle types in service, selecting some which are running at their full gross vehicle weight, and noting their load weights over a weighbridge. Now discharge a series of the vehicles over a period of a week onto a flat area (at the current landfill site will do perfectly well), and record the result by noting the weight, body height and width, and the dimensions of the tipped load.

From the body dimensions and the weight of the load, it is easy to calculate the bulk density inside the vehicle by dividing the weight in tons by the volume cubic feet. Division of the same weight by the volume of the tipped heap predicts the density on the ground. The on-the-ground bulk density arrived at will often exceed 18.7 lb/ft^3 (300 kg/m^3), which is twice as high as the figure most often considered representative, but it is the result of the extremes of compaction of which modern collection vehicles are capable.

Not all vehicles will be fully loaded, which causes some practitioners to doubt the validity of densities arrived at by the means outlined. Obviously, the cargo density of a half-full truck will be less than that of a full one, but it would be a daunting task to go through the exercise for every vehicle in the fleet over even a week. The solution is of course somewhat more straightforward, in that by the end of the period the analysis will provide a series of loads and densities for a reasonable cross section of the collection service. It will also in all probability provide weighbridge records for all deliveries, and by comparison of the two records, it becomes possible to obtain a good indication of how densities vary and to which extreme they tend. It would even be possible to produce a schedule of waste input volumes and densities for every day of the week should one wish, but the purpose is to be able to design a waste reception facility which is big enough to handle the *maximum* amount of waste that may be on site at any time, irrespective of the day.

Let us now assume that the process plant is to manufacture refuse-derived fuel (RDF) and compost as its main products, with baled metals and milled glass as secondary ones. There will be a series of separation stages and then a series of treatment stages, where in each of the former something will be removed from the waste stream, and in the latter each of those fractions will be processed. The first separation stage is likely to be where the two streams, RDF and compost, divide. We need a fractional analysis to discover what will go into each stream, and the simplest way of achieving that is to assess what fractions must report to the RDF line. The remainder can then be assumed to report to composting.

That can be achieved by random sampling from individual collection vehicle loads, and by then picking paper and card, plastic film, plastic containers, cans, and textiles, and weighing each category individually. This yields a mass balance for materials reporting to the RDF line, and the same exercise can be carried out for those going to the composting line. The latter will be glass, smaller heavy plastics, small metal items including batteries, some contaminated paper, and organics. They will include the "minus ten millimeters" referred to earlier; mixed materials each in too small a particle size to be readily identified. The best practical way to carry out the analysis involves picking through random samples by hand and separating each fraction manually, weighing them, and recording them over a period of at least a week.

This is a fairly unpleasant occupation, involving the examination and handling of putrescible materials, and it requires careful attention to hygiene and the strict use of protective clothing, but there is no other practical way of achieving the desired result.

Before introduction to the composting equipment, everything except the organics and paper will be removed, and afterward the product will be refined, leaving a further residue. At this point, the mass balance of the material reporting to the composting line is known, but not what proportion of the minus ten millimeters fraction is compostible. The question is, does it matter? and the answer is, yes it does. If 10% of the original mass of waste is in the minus ten millimeters fraction, then the weight of compost that can be produced from a given input could vary by up to 10% either way, depending upon whether the fraction is all organic, all inert, or somewhere in between. In a plant designed to manufacture 50,000 t per year of compost, that creates a potential variation of several thousand tons, which plays havoc with financial projections if it does not effect engineering ones.

As a rule, it can be assumed that at least 50% of the mass of the organic fraction will be converted to carbon dioxide and water vapor during composting. That is why it is essential to know what part of the minus ten millimeters material is organic, and what that 50% represents in terms of the compost line mass balance.

Now a category analysis exists for the input waste in terms of two process streams, as does a mass balance for each stream. The data is available for collection areas and for a window of time, and we must consider how it will change seasonally. That is where experience comes in, unless there are the resources and time available to continue the analytical work throughout the year — a most unlikely situation.

2.1.3 Regional and Seasonal Variations

It is a common misconception that municipal wastes conform reasonably well to a national average whatever their place of arising; common sense suggests that cannot be true. Table 2.1 in the following section on waste handling and storage (Section 2.2) shows comparative analyses for a rural and an urban region in the U.K. in 1992, and there is actually quite a large difference between the two. If we keep to the urban/rural comparison for the moment, we can give some thought to the reason for such a difference. After all, at first sight there is no reason to believe that a household in a town consumes materials and products differently from a rural one. Yet the wastes are not the same.

To start with, there is much more glass in the rural waste, and that is fairly easily explained. Towns have bottle banks, while rural areas generally do not. It is not cost-effective to collect small amounts of glass from widely scattered sites. There is more paper and card in urban wastes than in rural, since towns tend to be well serviced by supermarkets while small villages and hamlets are

Table 2.1 Waste Analyses from Two Sites

	Proportion % w/w	
	Rural	Urban
Paper and card	24.3	34.1
Plastics	12.7	13.2
Metals	7.4	7.9
Textiles	3.7	1.5
Glass	9.3	6.3
Ash	2.3	3.3
Organics & putrescibles	40.3	33.7

not. The village shop does not go in for packaging to anything like the same extent, and the rural populations are more likely to buy vegetables by the kilogram than in a prepacked carton. That may also be the explanation for the organic fraction in rural waste being larger than that in urban. There is more wastage from raw vegetables than there is from prepacked frozen ones.

Generally, there will also be more ash in rural waste, simply because rural households are more likely to use solid fuels. Wood burning stoves are much more common in villages than they are in town centers, and the fuel is easier to acquire. There may also be more textiles, because people in rural areas tend to buy products in sacks rather than in plastic packets. But the metals fraction may not differ significantly between the two areas, simply because a can is a can whether it comes from the village shop or the supermarket.

A similar comparison could be made between two areas of a large city. One area may be largely commercial, devoted to shops, offices, hotels. The other area may be largely residential, with a preponderance of suburban semi-detached housing. One could expect with some confidence that there will be a significant amount of packaging material in the waste from the first area, and a significant level of metal and plastic in the form of beverage containers. The second area is much more likely to be inclined toward recycling, and so bottle and can banks could reduce the arisings of beverage containers. But households do not produce packaging at anything like the levels that commercial premises do, so the paper and card fraction may be lower. They do create many more food wastes than offices do, so there will almost certainly be more organics.

There is unfortunately no arbitrary rule to follow, and there will be many cases where even the generalized comments made here can be proved false. The point to make is that when planning a waste handling facility it is unwise to assume that all wastes are the same irrespective of their origin. The analyses must themselves be planned carefully with due consideration of the region, or the plant has a very good chance indeed of receiving something unexpected which it was not designed to take.

Then there is the matter of seasonal variations, and the argument that wastes do not change that much from month to month. In fact it is quite common for household wastes to show an increase in moisture content in early spring and in late autumn, and generally this arises from an increase in the proportion of soft organics. In early spring, people are beginning to deal with their gardens, clearing the weeds and the dead plants and starting the first grass cutting. The weather is probably still quite wet and so the organic fraction will reflect that. Then later on, in autumn, there will be another last gardening effort as the residents clear the flower beds of dead annuals and the lawns of leaves from the nearby trees.

In the summer, people spend more time outdoors and enjoying recreation, and they are likely to buy more convenience foods with their associated volumes of packaging materials. Possibly the weather will be hot, and the consumption of beverages will increase as the manufacturers say they do. So one may expect more plastic bottles and aluminum cans. And if the area happens to be in a tourist region, then the population as a whole may increase considerably between, say, March and October.

Once again, there is absolutely no scientific way of predicting any of this. The variations could be insignificant or enormous. Many designers simply adopt the "worst case scenario" and assume that there will be more of everything than any of the analyses suggest. That assumption will mostly ensure that the plant will actually work most of the time, but it also means that there are likely to be wild swings of loading in different parts of the process, and that while some parts never work to capacity, others are frequently overstretched. One thing such an assumption does *not* do is prepare the designer for material that was simply not foreseen, and that lack of foresight has brought more process plants to a halt than anything else. When that happens it is no use deciding to turn that material away in future, since the need for the new process facility may well exist because there are no other disposal routes.

2.1.4 Changing Trends

The final consideration worth applying to waste analyses requires long-term estimation. It is probable that the plant will be in operation for 20 years at least before it is replaced. What will happen to the waste in that time?

Over the last 20 years there have been significant changes. Packaging has increased dramatically as a result of manufacturer's drives for market share and of a growing interest in food hygiene. Plastic materials have taken over much of the role that used to be occupied by glass and cans. In most developed countries there is probably more kitchen waste, as societies become more affluent. In general, while the actual tonnages of waste have remained fairly static or only risen by a few percent a year, volumes have disproportionately increased. The increases in packaging have reduced the density.

One factor which may inhibit that trend from continuing is that society has become more environmentally aware. In most developed countries, there were no waste minimization schemes of any significance 20 years ago. Manufacturers did not design with recycling in mind. Soap powder came in cardboard boxes and drums not in bags, and detergent bottles were not intended to be refillable. The fact that recyclable alternatives exist now and that public demand for them is growing does not mean that the amount of packaging will start to decline, since if it did then many of the developed world's hygiene regulations would have to be abandoned, and sufferers from salmonella poisoning could read of the latest outbreak on recycled paper! It is more likely that the levels of packaging will stabilize and that much more will be capable of reuse. Returnable glass containers will probably become more widely used, and the more difficult materials such as ash will virtually disappear as more and more households move away from the inconvenience of solid fuels. Local recycling centers which are actually just a collection of containers for glass, cans, and paper, etc. may spread, and so recycling and reprocessing plants may find themselves left with the residue after those things that are seen as more desirable are taken away.

Nobody knows what will happen, but the designer may be able to draw some conclusions from observing in the region in which he is interested. Listening to what the politicians are saying about forthcoming legislation, looking at manufacturing trends, considering the long-term strategic planning for the region all supply small pieces of the jigsaw.

2.2 WASTE HANDLING AND STORAGE

2.2.1 Keeping It In One Place

Any recycling or reprocessing facility needs an appropriate capacity to store waste before it can be treated, in part because waste is delivered during periods which are inconvenient for plant operation.

Collection services function during the working day. It may be, for example, that the vehicles start their first rounds at 0730 h, and by 0930 h they have filled and are on their way to the disposal point, to arrive between about 1000 h and 1130 h. Then they are all out again for their second collection round, returning between 1330 h and 1500 h. As a result, all of the waste delivered to the facility has arrived over 3 h, divided into two peak periods. Little waste comes in during the remaining time unless a remote transfer station is used to fill in some of the gaps.

The problem is that nobody in his right mind tries to design a process plant which deals with its total feedstock capacity in 3 h. If the plant has to handle 600 t a day it would be madness to design a 200 t/h flowline. The machinery would be huge, and its energy demands would be prohibitive. It would be much more sensible to plan to operate for 16 h on two shifts. That way the flowlines would only need a capacity of 37.5 t/h, and the energy demand would become manageable.

Unfortunately, implementing that plan means that during the delivery period the plant would process only 3 × 37.5 t, or 112.5 t. It must now store much of the remainder of the 600 t for at least 13 h, and some of it until the next day if the plant is to be able to start up on schedule rather than waiting for the vehicles to arrive. That latter consideration is one of some importance, in fact. It takes time to bring all of the process machines up to speed and to correct any start-up problems that have been encountered. It is not at all sensible to wait for the vehicles before discovering if anything is going to go wrong, because by then it will be too late to make other plans for the waste.

The storage of waste is a science in its own right, and each type of processing facility imposes its own demands upon the design. Specifics are explored in more detail in the relevant sections of this book, but for considering the general characteristics of waste it is worth examining what happens to it when it is kept for any length of time. The first point is that a significant fraction of the material is biologically active, and it will have started to decompose from the time it was first placed in a trash container for collection — that could have been a week ago where collections take place weekly. The only constraint upon it then is that the actual biological mass is small, and it is generally isolated from other materials by trash bags, etc. Once it is in the vehicle, it is mixed with a considerable amount of similar material, and the biological mass increases dramatically. Now there is sufficient mass to begin to retain heat and moisture, which encourages further bacteria growth. And there is the facility for cross-infection of other materials.

By the time the waste is unloaded, the bunker or reception area of the processing facility will have stacked in or upon it one of the more noxious mixtures known to man, and it has to be moved through the process or sorting lines which are going to deal with it. The pile will be warm, and may even be quite hot in localized places. It will probably have unidentifiable liquids draining out of the bottom, and it will already be starting to mat together. The overall bulk density can be almost anywhere between about 9.4 lb/ft^3 (150 kg/m^3) and 31.2 lb/ft^3 (500 kg/m^3), depending upon how many times it has been moved to make way for fresh deliveries.

Even when collected in plastic bags, the waste will tend to create dust emissions every time it is moved, and the more energetic the movement the more apparent becomes the dust. That will certainly have a nuisance value, even if it is not particularly hazardous. It will coat structural steel work with a surprisingly thick layer of light gray fibers which are flammable but not highly so. In the presence of a source of ignition it will smolder slowly, creating considerable volumes of smoke, and it could eventually reach a more sensitive area where a real fire becomes a possibility.

At the same time, the pile will be emitting water vapor, traces of methane, hydrogen sulfide, and ammonia, and probably bacteria and fungal spores as well. Again, there is little published evidence to suggest that any of these emissions are particularly harmful to other than sensitive individuals, and workers in waste disposal plants tend to remain remarkably healthy. Certainly the methane and hydrogen sulfide concentrations will be so low as to be almost undetectable under normal circumstances. The ammonia, which comes from disposable diapers, is often detectable by its pungent smell, but it also is in such small concentrations as to be harmless. Even so, the reception apron is not an area in which visitors should be encouraged, and those employees who must work there should be supplied with dust respirators and safety clothing.

One immediate effect of the biological activity lies in the ratio between free and organic moisture. Free moisture is that water which lies among the particles of the waste — a measure of how wet it is. Organic moisture, however, is the water which is tied up in the cellular structure of the materials. A potato, for example, may feel dry because it has no free moisture attached to it, but it is still 90% water. Once the bacteria start to attack the cellular structure, they release the organic moisture as free moisture, and so the waste becomes wetter. Additionally, some bacteria actually create water as a by-product of their own biochemistry.

In some processes the extra moisture from that source does not matter. For example, a composting plant deals with the organic fraction of the waste, and it is only concerned with the total moisture content. It is irrelevant whether the moisture is all free or all organic or a mixture of both. A waste-derived fuel plant, however, is concerned with the combustible fraction, which has to be dried as a part of the process of making fuel, and so the wetter that material becomes, the more energy is used in the drying stage.

Another aspect of this biological activity and its associated release of moisture is the environmental effect. Increasing "wetness" brings the threat of leachate, with the potential for pollution of drainage systems. It also creates the very real risk of anaerobicity, resulting in foul odors and the release of even more hydrogen sulfide and methane gas. The regulatory authorities in most developed countries will not tolerate any of those problems for long.

Since it is almost inevitable in a process facility that the waste must be stored for some reasonable period, the design must take steps to deal with the problems. Logically, storage should be planned on a first in, first out basis, so that the earliest deliveries are the first ones used. That way the plant is always handling the freshest waste possible. Some years ago it was customary to install deep bunkers that were served by grab cranes, and these of course operated upon exactly the opposite principle. Material could lay in the bottom of the bunker undisturbed for weeks, by which time it would decompose and become completely unsuitable for processing. In fact, some very large plants in recent years have still been designed with deep bunkers. But they are generally big incinerators, where the quantities of waste are simply too huge to store any other way, and where the capacity of the units is such that they can empty one or more bunkers every day.

The first principle of storage is to minimize the biological mass, so that heat retention and moisture build-up are reduced. 600 t of waste is a great deal of material, however one looks at it. Heat retention can be controlled by creating the largest possible surface area, so that radiation and convection deals with it. Given sufficient area, the heat will be released as fast as it is generated, and the material will stay reasonably cool. At the same time, if the material is delivered in plastic bags, it should be left in them as long as possible. Store it in a large number of small biological masses from which air and water is largely excluded, rather than breaking them open any sooner than is absolutely necessary. Also, given such a maximum surface area, any noxious gases that are created can escape easily before they have any chance to build up and become a nuisance.

A flat concrete floor in a ventilated building provides all of those requirements quite well. It allows the incoming waste to be spread out, thus creating a good surface area. It provides for the containment of any liquid effluent in side catchments and interceptors, for treatment before discharge as trade effluent. Planned properly, it allows the collection vehicles to discharge along one face while handling machines work the opposite one, thus permitting the application of the first in, first out principle. It minimizes the drop which the material experiences when it is discharged from the vehicles, so that there will be minimal bag breakage. It also provides a better opportunity for noticing any unwanted contraries which might damage the process machinery, since they are not hidden in a mass of material in the bottom of a hole. The various ways of storing and handling waste from storage are discussed in more detail in Section 3.2.

2.2.2 Types of Deliveries

The source of the waste has an impact upon both the storage system and the process design, and generally there are two possibilities. It may arrive either in collection vehicles direct from its place of arising, or in bulk vehicles from transfer stations. Which of the two applies will make a significant difference to the nature of the material.

What happens to waste from when it goes out for collection to its arrival at the facility in collection vehicles was discussed earlier, but where a transfer station is used the collection vehicles go there instead. Hence the waste from a transfer station will probably be a day or so older than it would be if it came direct. It will be wetter and further decomposed. Then of course, the transfer station facilities will almost certainly not be designed with the same considerations in mind as we have outlined here. The waste will probably have been dumped through a hole in a floor, where it may have been pressed through a hydraulic packer into a compaction container, and all of those activities tend to break bags open and mix all of the fractions. And when it is finally received at the recycling facility, it will be discharged from the bulk vehicle in a huge "sausage," which will have to be spread and leveled by the handling machines. Otherwise, it will coalesce into large lumps which will cause peak loadings in the process lines. The waste will then occupy a much larger floor area than initially expected; a container hardly larger than the body of a collection vehicle may actually hold the equivalent of three vehicle loads if it has been packed under pressure.

A further consideration regarding transfer stations involves a difference in the materials likely to be delivered from them. In many cases a transfer station is built to serve an area where the waste arisings are individually fairly small and are widely spaced, which implies that it services rural collections. After all, collection services outside towns are expensive enough already, without expecting the vehicles to drive over what could be an extended route to a disposal point. The effect is that, in spite of the conviction in some circles that there is no difference between urban and rural wastes, the opposite is more often true and waste from a transfer station may be predominantly rural in nature.

An example of the differences between urban and rural wastes is shown in Table 2.1, which reveals actual data taken from a waste analysis in 1992. The analysis was conducted over a period of a week at two landfill sites, one of which served urban collections and one of which served rural ones. The picture which emerges is that the rural waste contained less paper and card but much more organics and glass. But hidden in the data is a factor which would have a significant impact on a processing facility.

In this case, there is more ash in the urban waste than there is in the rural, which at first sight does not make sense. The urban area is served by natural gas, and it is estimated that 95% of its households use that form of heating. The answer lay in that some of the waste received at the urban site came from a transfer station which served a rural area where there was no natural gas. Ash was actually undetectable in the true urban waste; it all came from the transfer station.

That brings to mind another interesting consideration. The rural waste analysis suggests that one should expect more organics and less paper and card. If that is so, then to what extent is the urban analysis affected by the transfer station material? The urban analysis is misleading, in that the true urban waste is actually higher in paper and lower in organics than the analysis suggests.

Further research showed that all of the transfer station waste arrived at the site each afternoon of four days a week. The rest of the time, the total input came from the adjacent town. Therefore, had a recycling or reprocessing plant been designed using the analysis data alone, without any further investigation, then it would have been exposed to an enormous increase in ash and organics four times a week. In fact, had the analysis only considered the town waste and expected it to be representative, the plant would not have been designed to accept ash at all!

For those reasons, waste from transfer stations is generally undesirable for recycling or reprocessing plants. After all, they all must separate the fractions of the waste to some extent, and the more it has been mixed the more difficult that becomes. The older it is, the greater the environmental risks, and in some plants the greater the energy demand. And however much spreading and leveling is done when it is received, there is every possibility that its density will still be high compared with fresh waste.

Such an analysis does not mean that transfer station waste should not be considered for recycling or reprocessing, only that designers should be aware that it is likely to differ substantially from normal waste. They would be well advised to visit the transfer station and see what happens there so that they can form an opinion upon those differences and then accommodate them in their design. They should consider when the waste from that source will be delivered and to what extent it will alter the design analysis when it arrives. And designers should be prepared to accept that it may so change the process design that it pushes the cost up considerably. It is at that point when they should be asking themselves whether it is worth it, just for the sake of being able to accept a couple of container loads four days a week.

If all of this sounds unnecessarily complicated — and certainly there will be critics who will argue that such obsession with detail is pointless when dealing with waste — then there is one fundamental point to bear in mind. The whole purpose of a recycling or reprocessing facility, as observed earlier, is to make products for commercial markets. That is why the apparent obsession is necessary. No other manufacturing process is indifferent to the quality or specification of its raw materials.

2.3 A MATTER OF DENSITY

2.3.1 Density Concepts

In designing waste handling and processing systems, the first concept which has to be grasped is that of true and apparent density. Far too many designs have gone disastrously wrong in the past because process engineers relied upon mass balances. Waste handling is not that simple. Let us first consider what mass balances are and why they are insufficient.

The traditional concept used to be that to design a waste sorting system, one simply established how much waste would be delivered to the front end. Then if at the first machine the metals were removed, and if those metals made up 8% of the input weight, the next conveyor in the line should now handle input less 8%. If paper and card were then removed, together 25% of the waste, then the remainder of the line should be designed for input less 8% less 25%, and so on.

That principle is sound enough when dealing with dense, homogeneous materials where the weight has more significance than the volume, or where the density does not change. But waste is made up of fractions where the reverse is true, and its handling machinery is as a result partly or entirely volumetric in nature. To design a waste conveying system, for example, it is necessary to know not just the weight it must handle, but also the volume at that stage in the process. It is perfectly possible to build a 20 in. (500-mm)-wide conveyor that will carry several hundred tons of sand an h, but nowhere near the same capacity of wood chip. There is a fundamental difference in density between the two materials, and therefore a vast difference in the volumes.

Unfortunately, dealing with that factor is not just a matter of referring to standard density tables, and it certainly is not one of simply knowing the original waste density. Mechanical separation of any mixture of materials will cause a change in bulk density at every stage in the system. Energy is being applied to it, and depending upon the type of system, is either being absorbed or released by it, and this leads to the concept of true and apparent density.

True density is the **weight per unit volume of a solid block** of material which has no air space within it. *Apparent density* is the density which a material **appears to have as indicated by its behavior in a system.**

Perhaps one way of visualizing the difference between the two densities is to imagine granulated paper falling down a transfer chute between two conveyors. Paper has a true density of between 43.7 lb/ft^3 and 75 lb/ft^3 (700 and 1200 kg/m^3), so each particle will have a density within that range. If, however, one were able to measure that of the whole mass at that point in such a way as to retain the air space between the particles, then it would appear to have a bulk density of about 0.62 lb/ft^3 (10 kg/m^3). It does not fall down the chute in blocks, and it therefore exhibits a true density of around 700 to 1200 kg/m^3 and an apparent density of 10 kg/m^3.

The definitions create something of an impasse, since it follows that the very act of attempting to measure the apparent density will change it. If it is collected in a box, it will compact to some extent under its own weight, and the air spaces between particles will be reduced. There is no way of conducting the test without applying energy to the material in some form, and thereby changing its nature. So if the apparent density cannot be measured reliably, why does it matter and how can it be established for use in the mathematical modeling of a system?

In dealing with the first question, a further conceptual model is useful. Consider a conveyor intended to conduct the granulated paper fraction to a transfer chute and thence to a storage hopper. To design the conveyor it is necessary to know the mass and the volume to be carried. The mass will determine the power characteristics of the machine, and the volume will determine its actual dimensions. So it is necessary to know the apparent density which the material will exhibit as it lies upon the conveyor belt.

Then the material falls down the transfer chute. Now we need to know what the apparent density will be at that point to ensure that the chute can accommodate the full discharge from the conveyor. At this stage we can be sure that the density in the chute will be considerably less than that on the belt simply because it is in free fall, so it is not just a matter of making sure that the cross-sectional area of the chute is at least as large as that of the conveyor.

Now the material reaches the storage hopper, and to design that to a particular capacity we again need to know what density the paper will exhibit as the hopper fills. Even more importantly, it will be necessary to know what the final density will be when the hopper is completely full, because at some point the material will have to be unloaded again, and another effect of bulk density shows up as compaction. The greater the column of material pressing down on the base, the less will be the air spaces between the granules. Paper flock at 37.5 lb/ft^3 (600 kg/m^3) is harder to dig into and needs much more energy to move it than it does at 1.9 lb/ft^3 (30 kg/m^3). And unloaders designed with the wrong density in mind have been known to actually make the situation worse, by further compacting the material to such an extent that it became impossible to remove it at all.

So to the second question, how can the apparent density be established and used? While it is not possible to measure apparent density directly, it is possible to carry out experiments that permit an assessment to be made of what the density must have been. Simplistically, one could, for example, pour a sample of the material onto an existing conveyor of known dimensions and belt speed, and apply a leveling device to establish a chosen bed depth. If the weight of material delivered in a period of time is then measured, it becomes possible to calculate an apparent bulk density for that chosen bed depth.

However, an easier way of acquiring the necessary data is to use a box of known dimensions and weight, preferably 1 m square and 1 m high (or 1 yd square and 12 yd high), with the vertical sides marked off in 50 mm (2 in.) increments. Tip the material into it to a depth of 2 in. (50 mm) and weigh the whole. If we then deduct the weight of the box, and calculate the material volume within it, we can establish a figure for the apparent bulk density of the material at a bed depth of 2 in. If the experiment is then repeated for a series of depths, we obtain a data table of apparent densities at each of those, as follows:

$$\text{Initial volume of material} = 1 \times 1 \times 0.05 \text{ m}^3 \text{ or } (3 \times 3 \times {}^2/_{12})\text{ft}^3$$

$$\text{Weight of material} = 4 \text{ kg or } 8.82 \text{ lb}$$

Therefore, 0.05 m^3 of material weighs 4 kg, and its apparent density at a bed depth of 50 mm is as follows:

$$\text{Apparent density} = 4 \times 1/0.05 \text{ kg/m}^3 \text{ or } (8.82/1.5)$$

$$= 80 \text{ kg/m}^3 = 5.88 \text{ lb/ft}^3$$

Moving on then to the transfer chute, while again the apparent density of the material within it cannot be measured, there are some assumptions that can be made. To begin with, we can be assured that whatever the apparent density is, it will be less than that on the conveyor. After all, there cannot be any compaction influence on a free-falling mass. If anything, air resistance will cause the volume to expand.

Each individual particle will conform to the true density of the material, which brings in the other facet of the difference between the two characteristics. Particles exhibit true density, accumulations of particles exhibit apparent density. And upon that basis there is a limit to how low the apparent density could be, before the mass weighed less than any one of its individual particles. It is impossible to have less than one molecule of material per unit volume.

A further limit is that whatever the material is, it cannot possibly travel down the chute at a speed greater than that imposed by gravity. Therefore if it is assumed that the chute is 6 ft (1.8 m) long, then an individual falling particle will take 0.61 seconds to reach the bottom in the absence of any air resistance. As it does so, air resistance will in reality perform two functions. It will cause the particles to separate and it will slow them down, and the effect of separation will be to reduce the apparent bulk density quite considerably, to an extent constrained by the physical dimensions of the chute. In fact, were the apparent density not allowed to reduce, then some element of compaction must exist within the chute, and it would block very rapidly. The retardation caused by the air resistance will only reduce the rate of acceleration of the particle from rest, and its effect diminishes as the particle gains momentum. The maximum retardation is at the top of the chute, and for that reason badly designed chutes more frequently block at the top than at the bottom.

Let us now assume that the infeed conveyor delivers 10 t/h (9.1 te/h) of paper, granulated to an average particle size of 1 in. (25 mm), into the chute which is 2 ft (610 mm) wide. The volume of each particle of paper (at average sheet thickness) will be approximately 0.005 in.^{3} (8.1×10^{-8} m^3), and as a result of its true density of 43.7 lb/ft^3 (700 kg/m^3) its mass will be 0.00013 lb (5.4×10^{-5} kg). If the effect of air resistance in the chute is to expand the falling mass to the extent that each particle occupies 1 in.^{3} of air space, then the apparent density of the bulk material in the chute becomes 0.00013 lb/in.^{3}, or 0.22 lb/ft^3 (3.5 kg/m^3).

The reduction in apparent density will occur to some extent whatever the materials involved might be, although the higher their true density the less the reduction will be. Air resistance will not cause heavy particles to separate to the same degree as light ones, and the extent of the effect is therefore largely a matter of judgement. However, in the example used the extent of separation is intuitively realistic, and so the apparent density has been reduced from 5.88 lb/ft^3 to 0.22 lb/ft^3 (80 kg/m^3 to 3.5 kg/m^3). Accordingly the volume of feedstock flowing in the transfer chute must increase to $(10 \times 2000/3600)/0.22$ lb/ft^3, or 25.4 ft^3/s (0.72 m^3/s).

Since it has been established that the time taken for a particle to fall to the bottom of the chute is 0.6 seconds, then the enclosed volume of the chute is

$$V = \frac{l \times s_1 \times s_2}{t} \tag{1}$$

where l is the velocity of the material, t is the time taken for the particle to reach the chute bottom, and S_1 and S_2 are the lengths of the sides of the chute. Therefore in this case study, the volume is $25.4 = 6 \times 2 \times s_2/0.6$, and so side s_2 must be 15.5 in. (394 mm) long.

Furthermore, if the retardation effect of air resistance was sufficient to slow the particles by 50% (a not unreasonable conclusion), then side s_2 would need to be doubled in length to 31 in. The overall dimensions of the chute then become 24 in. × 31 in. × 72 in. The essential change in apparent density prevents the chute being designed simply to have the same cross-sectional area as the load of material upon the infeed conveyor.

So we have not defined exactly what the side dimensions of the chute must be, only what they must *not* be, i.e., they must not be less than 24 × 31 in. So, for example, had we built the chute to a cross section of 24 × 24 in., then the material could not fall fast enough to keep in front of that following, and the chute would block.

The model suggested in eq. (1) does not provide a figure for what the actual S_2 dimension should be to permit any given apparent bulk density, and it never can. Apparent density is the density which the material appears to have as a result of its behavior when subjected to handling, and from that it follows that the chute dimensions will effect it. This creates a circular calculation that is insoluble.

All that has been obtained from eq. (1) is a minimum dimension that can be compared with mechanical design constraints to see how it matches. For example, the conveyor must intrude into the chute to some extent unless a considerable amount of spillage is acceptable, and in general a 12 in. (300 mm) diameter head roller conveyor will need to intrude by 12 in. Then, there must be room beyond the chute in the horizontal plane for the material to change direction; if the bed depth on the conveyor is 4 in. (100 mm), the allowance for a direction change cannot be less than 4 in. The conveyor discharge therefore requires side s_2 to be at least 12 in. + 8 in., and the calculation calls for 31 in. There is adequate space for the material change of direction. The calculation is not precise, but it is good enough for most design purposes.

Turning now to the storage hopper, let us imagine that it has an enclosed volume of 10 m^3 and is loaded by the conveyor and chute discussed earlier. If the true density of the material applied, the hopper could contain 6.9 t (7000 kg); while if the apparent density on the conveyor were used instead, it could hold only 0.78 t (800 kg). The actual situation will depend upon the degree of compaction imposed by the height of the material, and the apparent density in the hopper will be somewhere between the two extremes.

In the example earlier it was established experimentally that the apparent density of the material at 50 mm depth on the conveyor was 80 kg/m^3, and that would therefore be so for a 50 mm depth anywhere else. It is a characteristic of the "springiness" of the material, and it will only change if energy is applied to or extracted from it. But, of course, placing another layer on top of it *is* applying energy. Gravity is again the source.

In due course the layer in the bottom will have a column of material on top of it, and the degree of compaction will depend upon the height of that column. The inevitable result is that a tall hopper can hold more weight of material than a short one for the same volumetric capacity. The only thing that can be said about the apparent density at this stage and in this example is that it cannot exceed about 43.7 lb/ft^3 (700 kg/m^3), since that is the true density of a solid block. That is satisfactory so far as it goes, but it is of little help in designing the hopper. The next stage has to be another experiment.

Taking again the 3 ft square and 3 ft high (1 m square and 1 m high) box, fill it loosely with the material in which we are interested — pour it in as if it were falling from a feed chute. Weigh the whole again and deduct the weight of the box. If the box weighed 11 lb (5 kg) and the total weight plus material was 220.5 lb (100 kg), then once again the apparent density of a 1 yd^3 (0.76 m^3) block is 5.9 lb/ft^3 (95 kg/m^3). We can now place a plate 1 m square into the top of the box and apply weights to it until it also weighs 95 kg, simulating another layer of material. The material underneath will compact to fill a correspondingly smaller volume, and that can be calculated by simply measuring its new height. If that height is reduced by, say, 12 in. (300 mm), then we can calculate that the new apparent density has increased to 8.5 lb/ft^3 (135.7 kg/m^3). So now we can estimate that a column of material 5.6 ft (1.7 m) high will have a range of apparent density from 5.9 lb/ft^3 to 8.5 lb/ft^3 (95 kg/m^3 to 135.7 kg/m^3).

If we then double the weight of the compression plate, the material will compact still further, and its apparent density can again be assessed by calculation. Continuing that exercise would yield a series of densities and measurements of compression that relate to the weight applied, until eventually stability would be attained and the material would compact no further. We would have a density figure for the material on the bottom of the hopper, and because the increments of weight placed upon the compression plate equate in each case to a 3 ft column, we have a density figure for each 3 ft level increment upward.

In fact, if the results of such an experiment were plotted graphically, using density for the "Y" axis and increment number for the "X" axis, the resulting curve would probably be almost exactly a simple geometric progression. As the material is compressed, its resistance to compression increases, so that each increment of weight has a correspondingly smaller effect until the true density has been reached.

For this example, we have selected deliberately one of the most difficult materials, where apparent densities are likely to be very low indeed in comparison with true densities, and where changes in them are likely to have a very dramatic effect upon volumes handled at each stage of a process. A conveyor could be substituted for the box to measure the apparent density on a flat belt simply for reasons of analytical variability. The larger the quantity, the less the margin of error. The whole point is that apparent densities are very important indeed to mechanical handling calculations when dealing with waste, and while they cannot be measured directly, they can be assessed by duplicating *as nearly as possible* the environment which the material will experience.

The purpose of the above discussion is to establish the importance of the concept of apparent density. A flowline cannot be designed reliably without it, as has been proved frequently over the years. But the recognition of where to apply it and where to apply the true density is of fundamental importance.

2.3.2 Application of True and Apparent Density

Before committing a pen to a drawing board, it is necessary to visualize what will be going on in each machine or process component. As observed earlier, particles of material exhibit true density, accumulations of particles exhibit apparent density.

Thus it follows that a part of a process that depends upon the behavior of individual particles has to be designed with the true density in mind. An air classifier, for example, is in its simplest form a true density device. The separation of materials depends upon the characteristics of individual particles. So too is a hydraulic separator, where materials are categorized according to their density with respect to a fluid, or a ballistic separator, where it is the mass of the categories that determines their separation. In such cases it is the mechanism delivering material to them that is subject to apparent density, as is that which collects it from them.

In some processes, both true and apparent density apply (to confuse the situation further). For example, a hot-gas dryer in a waste-derived fuel plant is intended to both dry and convey product. The material is introduced into a moving stream of hot gas, which may be passing through a rotating kiln or along a duct. As each particle enters, it is heated and moisture is driven off, so its true density will change. That will effect its aerodynamic and thermodynamic behavior. Loss of moisture will change its thermal conductivity; loss of mass will make it more buoyant in the gas stream. Meanwhile, the delivery rate of material through the machine will be determined by the apparent density of the charge as a whole.

To take this example a little further, imagine such a dryer handling paper with a true density of 43.7 lb/ft^3 (700 kg/m^3), which has been further granulated to a particle size of $^1/_2$ in. (10 mm), and which has a moisture content of 35% by weight. Given the mass of the paper, it is possible to carry out the calculation necessary to establish the amount of heat needed to reduce the moisture content to a selected level. If the infeed rate of the paper were to be 10 t an h, and the final moisture content were required to be 15%, then 3000 lb of water per 10 t of material would need to be released. The mass balance is 10 t in, 8.5 t out, with 1.5 t of emissions.

Now, to turn to the behavior of each particle in the gas stream, it is necessary to know what the stream speed limits are. In a rotary kiln drier, for example, we do not want the particles to simply stay airborne from one end to the other; there would be insufficient dwell time in the machine for the required drying to take place. That stream speed can be calculated (as is explained later in the chapter) on waste-derived fuel processes. Suffice to say for the moment that it depends upon particle mass. So then, given the gas flow speed and a thermodynamic calculation of the volume of gas necessary to conduct sufficient heat energy, we are well on the way to determining what the enclosed volume of the drier must be. To complete that determination we need to know what the apparent density of the product is in those circumstances, since obviously the actual drier volume is the sum of the gas and product volumes.

That, of course, is where it becomes difficult. The apparent density of a mass of free-falling particles in a gas stream obviously cannot be measured directly, since if we capture them and put them in a meter box, they will no longer be in free fall. Also, there is a tendency for some materials, of which granulated paper is one, to simply expand in all directions to fill the available space. The apparent density can therefore be circumscribed to some extent by the chamber in which it is held.

However, from the earlier example in this chapter, it was established that for the plant we have in mind, the apparent density of the material 4 in. (100 mm) deep on a conveyor is 80 kg/m^3. Obviously, in free fall the apparent density cannot be higher than that, only lower. Whatever the apparent density of the paper happens to be, that of the gas will be tiny by comparison. The gas volumes will exceed the paper volumes enormously, so whether we choose 80 kg/m^3 or 5 kg/m^3 for the paper will not matter much. We must avoid assigning a density that is obviously too high, since a low estimate will mean that the machine has a surplus volumetric capacity, while a high one would mean it could be too small for the job.

2.3.3 Mass and Volumetric Balances

When a waste processing line is first conceived, something has to establish the overall design requirement, and that can only be the *mass balance*, which we can define as the changes in the mass of the feedstock in response to the process that will occur at different stages along the line.

This can again be illustrated best by an example. Assume that the waste from the analysis shown in Table 2.2 is delivered to a process line intended to handle 10 t an h and consisting of a trommel screen, a conveyor, a magnetic separator, a picking station for the recovery of paper, and a baling press for the residues. Let us assume that the trommel will remove only the glass, organics, and putrescibles, and the "miscellaneous" fraction. These together make up 51% of the feedstock, so 5.1 t of material an h will leave via the trommel, and 4.9 t will continue onward.

At the magnetic separator, the ferrous metals will be extracted, and these are 6% of the waste, so 1200 lb (544 kg) will be extracted and 4.3 t an h will continue. Now the paper is removed, amounting by the same philosophy to 2.5 t an h, and so at the end of the line 1.8 t of residue comprised of plastics,

Table 2.2 Example Waste Analysis

Fraction	% w/w
Paper & Card	25%
Plastic Film	5%
Dense Plastic	6%
Ferrous Metal	6%
Nonferrous Metal	2%
Glass	12%
Textiles	5%
Organic & Putrescible	30%
Miscellaneous	9%

nonferrous metals, and textiles will remain. In a complex waste treatment design consisting of a number of machines and transfer stages, a mass balance arrived at in such a way would be produced as a large-scale block drawing. There every stage would be represented by a block. The mass leaving the line and the mass continuing on would be marked in at each point, with arrows indicating which of those two alternatives apply. Probably such a plant would have further treatment processes for each of the fractions removed as well, so the resulting mass balance begins to look rather like a tree, with branches spreading along it.

Now the design moves on to the volumetric balance, since as stated waste processing is largely a volumetric business. This is where the two types of bulk density come into the consideration, as shown in the example at the beginning of this chapter. At each stage we know the mass balance to design to, and we need to transform that into a volume to arrive at machinery sizes. The result of all of this, and probably some experimental work as well, is a combination block drawing that shows the mass and the volume applied at each point. Figure 2.1a and 2.1b show an example mass/volume balance for an existing process plant designed to make waste-derived fuel and compost. In summary, waste processing designs start with a mass balance, to which true and apparent densities are applied as appropriate, resulting in a volumetric balance. This concept will be used extensively in later chapters that deal with the mathematical modeling of various types of processes.

2.4 PHYSICAL CHARACTERISTICS OF WASTE FRACTIONS

The term "material" is all too often applied to waste as if it is one consistent, homogeneous mass. It is not. It is a collection of quite dissimilar materials, each with different handling characteristics, all mixed together in an environment where some will interact and others will not. This is another reason why the behavior of waste can be changed completely just by removing one fraction.

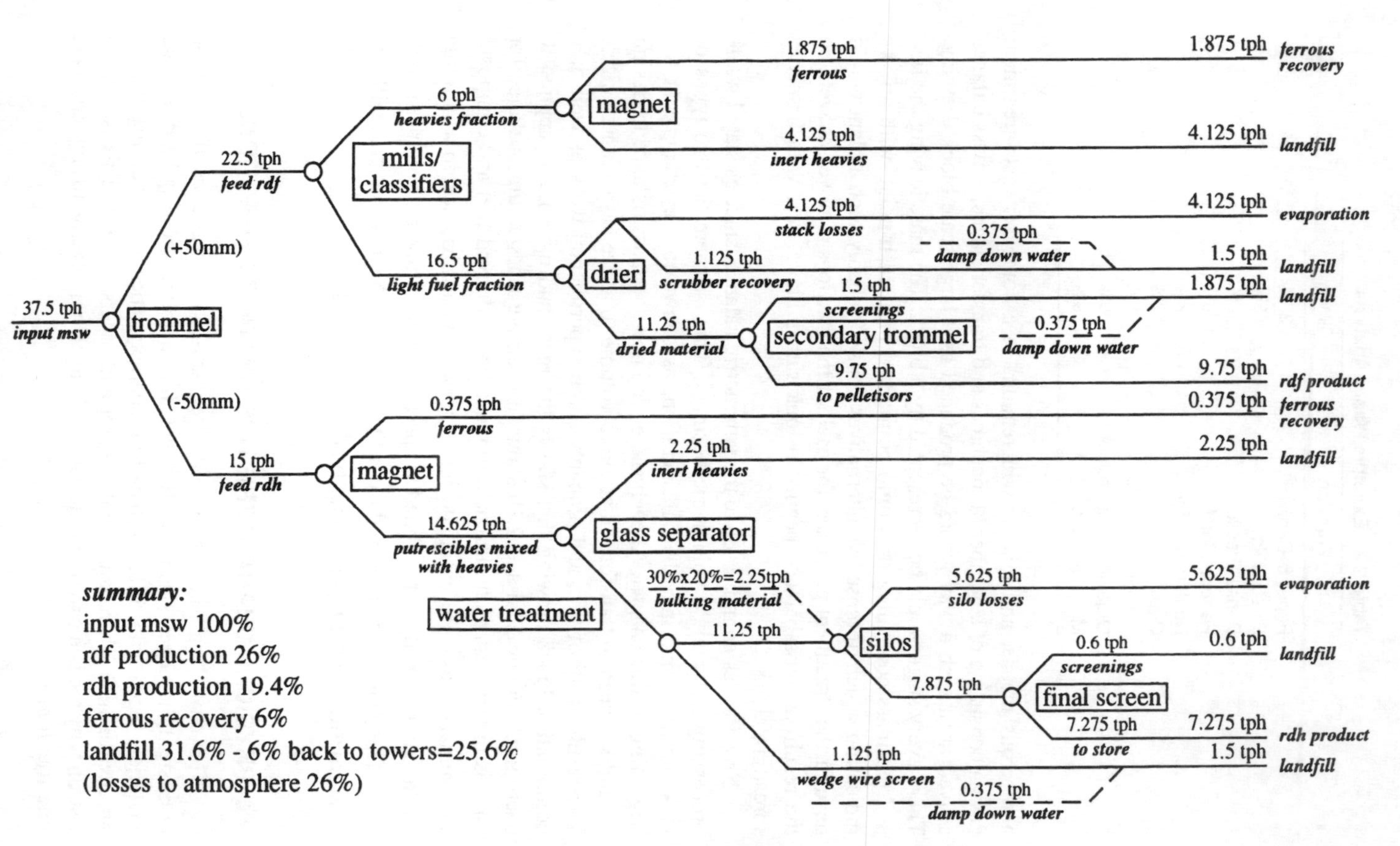

Figure 2.1a 37.5 t/h combined process. a. Mass balance.

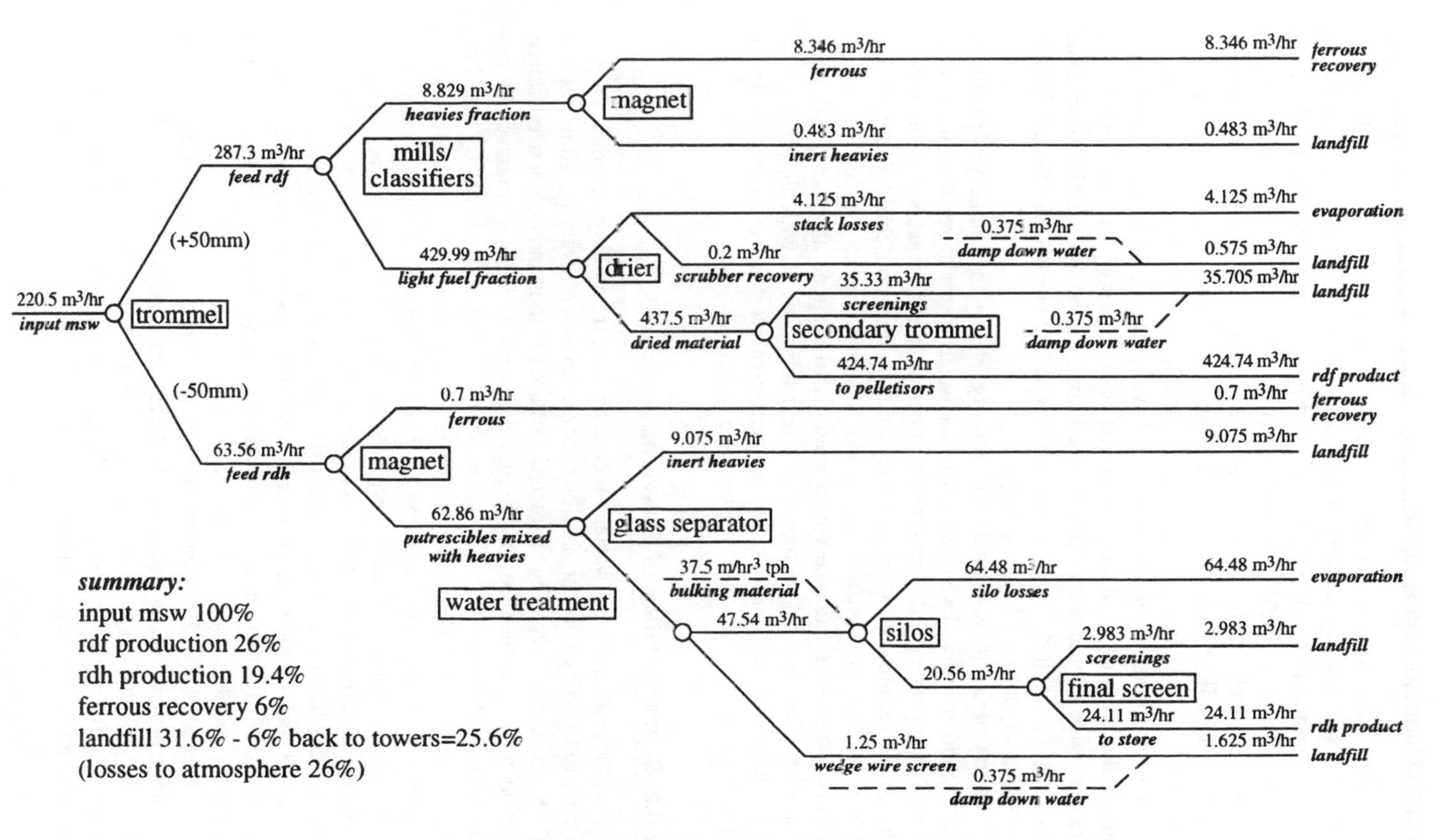

Figure 2.1b 37.5 t/h combined process. b. Volume balance.

2.4.1 Physical Characteristics of Mixed Waste

When it arrives at a disposal facility, municipal waste will be sticky, abrasive, biologically contaminated, and very bulky. It will almost certainly contain animal fats, oils and greases, and protein and vegetable food wastes. It will have hard, brittle materials such as glass and ceramics, and tough, ductile materials in the form of metal containers and metallic components. There will be flexible but tough items made of plastics, and string, rope, and textiles — which often seem designed specifically to wind themselves around the moving parts of machines. It will have a significant level of fine, gritty material displaying all of the characteristics of grinding paste. And in addition, there may be chemical residues, drugs, human excrement, clinical waste, and even explosives. It is indeed a mixture designed by fate to make the mechanical handling engineer's life interesting!

Many of the items listed do not properly fall into the category of municipal waste, and in most developed countries waste disposal regulations actually forbid their presence. It is an offense to permit human wastes, clinical wastes, drugs, and explosives (for example) to enter the household waste stream. However, it is an unwise designer who assumes that as a result such items will not do so. Collection service operatives do not generally look inside every bag they pick up, and members of the public, faced with the much more inconvenient disposal routes for nonpermitted wastes, will put anything in their bags or bins they think will escape notice.

2.4.2 Characteristics of Fractions

Although municipal waste can contain almost anything, it would be ponderous to try to explore the handling characteristics of every possible fraction. Then again, some handling characteristics are obvious and do not need enlarging upon. It is unnecessary to discuss the characteristics of putrescible contamination of cans when all that matters is that it is there. By handling characteristics, we mean the categories into which a mechanical handling engineer would place them as a means of determining the design of a conveyor system, for example. Unfortunately, categorizing the fractions is not as simple a matter as it may seem, for two reasons. First, which category a particular material fits into depends upon what type of handling system is being considered, and second, although there are published lists for a very wide range of materials, waste fractions are not generally among them. Our purpose in this section is to attempt to establish such a list.

There are several parameters that must be known to design a mechanical handling flowline. It is necessary to have some measurement of how free flowing the material is, its tendency to compact, its abrasive and corrosive properties, its average particle size, the maximum angle of incline at which it can be carried, and its natural angle of repose. Before fitting these into standard gradings, it is worth spending a little time establishing what each characteristic means.

The term "free flowing" does not mean that it can be poured out of a hopper or down a spout. It is basically a measure of how "sticky" it is. Does it, for example, tend to stick to conveyors or to the sides and flights of a screw feeder? The less free flowing it is, the more power will be needed to move it, and of course belt conveyor scrapers will need more careful attention.

The tendency to compact is more important for screw feeders and drag link conveyors than it is for belt conveyors. Moist, granulated paper, for example, has an extraordinary ability to compact in screw feeders to the extent that, given any opportunity, the product will compact to a solid mass in the discharge. Compaction is tied to resistance to free flow to some extent, in that materials that do not flow easily usually compact very well indeed.

The abrasive properties are those indicating the ability of the material to damage the handling system. If the material is harder than the carrying mechanism or casing and has sharp edges, then the machinery will wear in preference to the waste. But that is an over-simplification to some extent, in that even materials softer than the machinery components will still cause some wear. Nothing is totally nonabrasive.

The corrosive potential is somewhat obvious, but even here there are some aspects not immediately apparent. If the material is wet, then some of that moisture will be deposited upon handling structures and will permit corrosion. But if it is in addition either acidic or alkaline, then there is an opportunity for electrolytic corrosion as well. That consideration often creates a problem in finding a suitable category, in that some materials are in themselves corrosive but contain other contaminants that tend to reduce the effect. Putrescible kitchen wastes, for example, are wet and generally acidic, but they also contain fats and oils that tend to coat handling machine structures and protect them against corrosion.

The particle size is of importance in determining the dimensions of the handling system, in the sense that if it is on average 150 mm, then there is little purpose in trying to handle it in a 150 mm diameter screw feeder. It simply will not go in. Particle size is a measurement of the "lump" dimensions.

The maximum angle of incline is that angle at which a material can be transported before it starts to slip back, and it is of absolute importance in the design of belt conveyors. Crude waste, for example, has a maximum angle of 27 degrees to the horizontal, so if a flat belt conveyor is inclined more than that, transportation of the material will become increasingly unreliable. If such an angle is inescapable, then the designer has to consider slats or ribs on the belt to hold the material in place, at which point how sticky it is becomes significant. Wet, sticky loads and belt conveyors with slats do not mix.

The angle of repose is not the same as the angle of inclination, but is the angle which the sides of a heap of the material placed upon the ground would adopt. It is important in the design of enclosed systems such as screw feeders, where the action of the flights is to effectively create a series of moving heaps.

With regard to the actual grades, it is necessary to distinguish between the handling systems used. Where belt conveyors are concerned, the grading is that of the conveyor material rather than of the waste, and is an indication of the carrying surface thickness. Three grades are most commonly used: M24 (3.2 mm), N17 (2.5 mm), and N17 (1.0 mm). Enclosed conveyors, however (screw feeders, drag links, etc.), need a number of categories, and those generally applied range from "A" to "D," where "A" and "B" are materials that are nonabrasive, "C" those which are mildly abrasive, and "D" those which are very abrasive.

Each of categories "B," "C," and "D" is then further divided into three subclassifications. So, for example, grade B_1 is for very free flowing material, while B_2 is for material that is still fairly free flowing but less so than B_1. Table 2.3 indicates the categories for the waste fractions on this basis.

Table 2.3 Mechanical Handling Grades

Material	Angle of Repose	Angle of Inclination	Grade
Paper >150 mm	90°	35°	C_3/N17–1.0 mm
Paper <10 mm	90°	35°	C_3/N17–1.0 mm
Dense plastic	<30°	20°	A/N17–1.0 mm
Film plastic	30°	25°	B_2/N17–1.0 mm
Cans	<40°	20°	C_1/N17–2.5 mm
Metals	60°	30°	D_3/M24
Glass, cullet	60°	20°	D_3/M24
Glass, ground	80°	20°	C_2/N17–2.5 mm
Textiles	80°	30°	B_3/N17–1.0 mm
Organics	50°	25°	B_3/N17–1.0 mm
Ash	>45°	23°	C_3/N17–2.5 mm

However, while approximate handling grades for materials that predominate in waste are listed, it should not be assumed that a system designed to that grading will necessarily be able to handle them successfully; there are other factors involved. For example, while textiles are given a grading of B_3, which suggests that they are not particularly abrasive and are not very free flowing, they cannot be delivered reliably through a screw feeder. They would simply wrap round the flights and block the machine.

There is also a fundamental difference between the angle of inclination possible for intact glass bottles, and that for broken glass. The bottles will not convey at much beyond the horizontal for the reason that they will keep rolling back. In the list we have assumed that all glass is either broken or ground, since for all practical purposes glass existing in waste is always one or the other to some extent.

Also, the moisture content of the materials can have a significant effect upon how they handle. Using broken glass again as an example, if it is dry it can be conveyed on rubber belts without many problems, but sharp edges cut wet rubber with remarkable ease. So wet, broken glass on a belt conveyor will be much more aggressive than dry.

These considerations are fundamental to plant design, and they will be used specifically in later chapters.

2.5 HEALTH AND SAFETY

The most prominent health risks associated with municipal waste derive from the organic and putrescible fraction, although there are always physical health risks present from such items as hypodermic needles. Pathogenic bacteria, endotoxins, and human coliforms are always present, and the development of the disposable diaper has significantly increased the concentrations of the latter. Contact with liquors or airborne emissions, or with the waste itself, risks infection by a number of bacteria and possibly viruses, and among the viral infections perceived as the greatest risk is Weil's disease.

Weil's disease, medical name leptospirosis ictero-haemorrhagica, is rare but serious and can be fatal, and it is essential that it is taken into account at the earliest possible stage of illness, particularly where symptoms similar to influenza exist. The disease normally commences as a febrile illness with varying degrees of muscular pain, tenderness, congestion of the conjunctiva, jaundice, and hemorrhages of the mucous membranes and skin. Albumin and casts may be expected in the urine, and polymorphonuclear leucocytosis is usual. Jaundice may not be encountered at all, and is unlikely during the first days.

The occurrences of Weil's disease are becoming more frequent of late in most developed countries, and so its rarity can no longer be assured, particularly in waste disposal operations. In Chapter 6 of this book, dealing with organic waste composting systems, a management policy with respect to the disease is recommended, and a handbook detailing the symptoms and treatment for Weil's disease is illustrated. Issuing the handbook to every employee engaged in a waste disposal process is a worthwhile practice, since the disease is still sufficiently rare for its symptoms to be easily mistaken for a less serious infection. Further health and safety considerations are discussed in each section, where they are specifically related to the handling of waste within each technology.

3 MECHANICAL HANDLING

Engineering science in this context also suffers from thinking that is often too generalistic, and so before proceeding it is worth clarifying the difference between "mechanical handling" and "processing." For the purposes of this book, *mechanical handling* is the technology of moving materials from one place to another in a continuous operation. *Processing* is that of applying energy to materials in such a way as to change their physical or chemical state.

There is a second major difference between the two concepts. Mechanical handling technology is common to all forms of waste treatment, in that a rubber belt conveyor in a materials recovery facility is identical to one in, for example, a composting plant. Processing machinery, however, is almost always quite specific to the particular waste treatment design. Thus, for example, a granulator in a composting plant will be quite different to one in a refuse-derived fuel plant.

3.1 MECHANICAL HANDLING IN AND FROM STORAGE

3.1.1 Storage Capacity

A cost-effective mechanical installation is one where the machinery is operated for the longest possible continuous period, since that minimizes machinery sizes. It is also one that limits the number of times the plant has to be started and stopped, since start-ups demand much more power than normal running. However, it would be unrealistic to expect any waste processing plant to run 24 h a day, 7 days a week. Maintenance and adjustment have to be carried out sometime, and such a plant can never achieve 100% reliability due to the difficult handling characteristics of its feedstock. Therefore, 16 h/day operation with 8 h for maintenance is realistic, whatever the process selected.

Sufficient storage must be provided to support the operating pattern, and the assessment of quantities and the design of the facilities is surprisingly complex. There are many variables to be considered, and many methods that could be used. The purpose of this section is to examine the most common storage systems and assess the quantities of waste they may contain.

3.1.2 Influence of Collection Services

Municipal waste collection services do not normally operate for 16 h at a time. More commonly, they make an early collection, delivering to the plant in the mid-morning, then a further one delivering early to mid-afternoon. Therefore the plant may receive all of its day's quota of waste in two periods, each of about 3 h. It will need to store sufficient to keep it at full load during the remainder of the operating day, less that quantity that it will process during the delivery periods. Returning to the hypothetical plant discussed in Section 2.2, if the plant is rated at 37.5 t (34 te) an h, then it will need 600 t (544 te) of waste a day, or about 300 t (272 te) per shift.

Unfortunately, that is not the whole situation. A full-scale process plant takes some time to start up at the beginning of the operating day, and it is preferable that this task is completed well before the first delivery vehicles begin to arrive. If there are any machinery defects that may prevent or restrict production, it is better that they are encountered before the waste input is committed. It follows, therefore, that there must be a residue of waste from the previous day for that purpose. Table 3.1 suggests the results of carrying over 60 t (54.4 te) from the previous days operations, showing how that and the fresh waste deposited would be processed throughout the day.

The illustration reveals that allowing for an 0700 h start up, the plant would commence full load running by 0800 h, and hence the amount processed between 0700 h and 0800 h would be only half that of each following h. By the time the first new deliveries commence at 0900 h, the plant will have consumed 56.25 t (51 te) of waste, and will have 3.75 t (3.4 te) remaining.

However, from then on until 1200 h, it will be taking in waste at a rate much greater than the throughput of the process lines. Between 1200 h and 1300 h, there will again be a slack period when no waste is received, and the amount in storage will reduce slightly, before increasing again with the afternoon deliveries. At that rate, the maximum amount in storage will be reached by the end of the day shift at 1600 h, amounting to 341 t (309.3 te). With no evening shift deliveries, the store will gradually empty, and allowing half an h for shut down at the end of the second shift, 60 t (54.2 te) of fresh waste will again be carried over to the next day.

If the start-up is delayed for any reason, the surplus carried over will be increased somewhat, as will the peak storage demand. Any well-designed waste processing plant should be able to achieve better than 90% availability, but if 90% were the target chosen, then the waste storage capacity must be further increased by 10%, to 375 t (340 te). In other words, if 10% of the available shift h during the day were lost due to plant failures, there would be sufficient storage to contain the surplus waste.

While the data used in the table are perhaps somewhat generalized, they are reasonably representative of most process plant operations. Most importantly, the assessment of waste storage is not simply a matter of subtracting

Table 3.1 Waste Input Logistics

Time	Waste in	Used	Surplus
	Monday through Thursday		
0800	0	18.75	41.25
0900	0	37.5	3.75
1000	100	37.5	66.25
1100	100	37.5	128.75
1200	100	37.5	191.25
1300	0	37.5	153.75
1400	100	37.5	216.25
1500	100	37.5	278.75
1600	100	37.5	341.25
1700	0	37.5	303.75
1800	0	37.5	266.25
1900	0	37.5	228.75
2000	0	37.5	191.25
2100	0	37.5	153.75
2200	0	37.5	116.25
2300	0	37.5	78.75
2400	0	18.75	60
	Friday		
0800	0	18.75	41.25
0900	0	37.5	3.75
1000	100	37.5	66.25
1100	100	37.5	128.75
1200	100	37.5	191.25
1300	0	37.5	153.75
1400	100	37.5	216.25
1500	100	37.5	278.75
1600	100	37.5	341.25
1700	0	37.5	303.75
1800	0	37.5	266.25
1900	0	37.5	228.75
2000	0	37.5	191.25
2100	0	37.5	153.75
2200	0	37.5	116.25
2300	0	37.5	78.75
2400	0	37.5	41.25
0100	0	37.5	3.75

Note: Initial stock is 60 t.

the amount processed in 6 h from the amount received in the same period. A table of the type shown is indispensable in establishing the true figure, and such a table should be drawn up for the waste delivery logistics that will actually apply in a real situation.

3.1.3 Consequences of Breakdowns

So far, the assessment has dealt with a normal day's operations. However, if a problem occurs that prevents the plant from running at all, rather than just delaying it, and if deliveries continued unabated, then at some point any realistic storage capacity will eventually be overwhelmed. The municipal collection vehicles must be diverted elsewhere, assuming that such a place exists. Since they cannot be diverted at a moment's notice without causing severe and unacceptable disruption to the collection rounds, the plant capacity must be sufficient to contain all of the waste delivered until such time as diversion becomes pòssible. Most collection services require at least 1 working day's notice of diversion, unless there is a suitable alternative facility very close by. Therefore, the potential maximum storage capacity of the plant becomes an average day's input of waste, plus any carry-over from the previous day — a total of 660 t (598.6 te) in the case of the example chosen.

At this point in the design consideration for the plant used as an example, two figures for storage have now been established. There is a normal operational requirement for 375 t capacity, and an exceptional requirement for 660 t. The way in which the two figures are used then depends upon the reception facilities, which are discussed in the following paragraphs.

If the hypothetical plant is to have a 660 t storage capacity, then for design purposes that mass must be converted into a representative volume. It is reasonable to assume that the waste will have a bulk density of about 18.72 lb/ft^3 (300 kg/m^3) in a compacting collection vehicle, but that the bulk density will reduce to about 9.36 lb/ft^3 (150 kg/m^3) upon the load being discharged. However, what it is discharged into or onto will effect that considerably.

3.1.4 Deep Bunker Design

If the load is discharged into a deep bunker, it will compact under its own weight and assume an average bulk density of at least 15.6 lb/ft^3 (250 kg/m^3). Tipped onto a concrete apron, the density would stay at about 9.4 lb/ft^3 (150 kg/m^3). This seems to imply that a deep bunker storage facility is preferable, since it could store more waste in a smaller volume. Against that attraction there are a number of disadvantages.

The most significant disadvantage is that a deep bunker stores waste in such a way that the first waste received is the last to be used. It lies in the bottom of the bunker where it is inaccessible until later waste deliveries have all been processed, whereas ideally the first waste in should be the first processed.

In considering the "first in, first out" principle, the points discussed in Section 2.2 should be remembered. The older the waste is, the more moist and biologically degraded it will have become. Moist combustible fractions require much more energy consumption in the driers of energy-from-waste plants, and are much more likely to cause blockages in earlier stages of any process. More of the light fraction is likely to be lost at the initial screening and separation stages, and while a primary mill may work more efficiently with moist material, an air classifier most certainly will not.

At the bottom of the bunker the wastes may achieve a bulk density as high as 31.2 lb/ft^3 (500 kg/m^3), and settlement and drainage of liquors in that region will cause the combustible fraction there to reach moisture contents in excess of 80% by weight. If stored too long in those conditions, the organic fraction will decompose anaerobically (in the absence of oxygen), potentially generating methane gas, and inevitably creating noxious odors. In spite of that drawback, however, space limitations may make a deep bunker design inescapable. Very large capacity plants may need such a level of storage that any other system would be impractical. If that is so, then the design must at least minimize the disadvantages and maximize the advantages. The basic design principles which achieve this are described in the following section.

3.1.4.1 Bunker Dimensions

When the bunker is filled with waste, the range of bulk densities due to compaction experienced within the mass will range from less than 9.4 lb/ft^3 (150 kg/m^3) at the top (the waste having been "loosened" by the act of tipping), to up to 31.2 lb/ft^3 at the bottom. The extent to which the density increase with depth is linear depends upon the bunker dimensions, but it is reasonable to assume an average bulk density of 15.6 lb/ft^3 (250 kg/m^3). To be able to store the 660 t used in the example, the bunker will need an enclosed volume of 660 × 2000/15.6 ft^3, or 84,615 ft^3 (2,396 m^3).

The deeper the bunker is, the higher the compaction forces are likely to be at the bottom, and so it is preferable to design for the greatest possible area and the minimum depth. Again, a useful rule of thumb is that the depth should not be greater than 1.5 times the length of the shortest side. Therefore, in the hypothetical plant, assuming a bunker with a square cross section, the waste will be contained in a space that is in the ratio 1:1:1.5 in terms of width, length, and depth.

To arrive at the actual dimensions based upon those ratios, the calculation becomes

$$D^2 \times 1.5\, D = V \tag{1}$$

where D is the unit of the ratio and V is the enclosed volume. This equation can then be simplified as

$$1.5\, D^3 = V \tag{2}$$

or

$$D = \sqrt[3]{V/1.5} \tag{3}$$

Substituting the figures for the hypothetical plant in that equation, the dimensions of the bunker become

$$D = \sqrt[3]{\frac{84{,}615}{1.5}} \tag{4}$$

Therefore D = 38.4 ft (12.07 m). The bunker must be 38.4 ft (12.07 m) square and 57.5 ft (17.5 m) deep.

Thus far, the calculation has only yielded the volume to be occupied by the waste, and there must be some "freeboard" to allow for the actual tipping operations. Waste does not conveniently level itself when it is tipped, and so otherwise the last loads in may spill over the sides. What the freeboard should be depends upon the design of the actual tipping arrangements, but a reasonable approach would be to extend the sides to at least half way up the vehicle discharging aperture.

3.1.5 Unloading Cranes

Unloading the bunker implies the use of some form of grabbing crane, which will be operating in a dusty, moist, and demanding environment. The crane will need to be a heavy-duty unit simply to withstand the conditions, and the grab even more so. The installation will be expensive to purchase and costly to run, and may present additional operational problems in that the operator is necessarily somewhat remote from the center of working. He cannot therefore see clearly what he is picking up, and cannot carry out any degree of segregation of unsuitable materials.

If either the crane or the grab breaks down for any reason, and such events are not uncommon due to the mechanical and electrical complexity of the system, then the whole plant will be out of action. In some very large incineration plants that possibility is dealt with by having several cranes and a number of deep bunkers, often with a facility for cranes to be "leapfrogged" over each other in the event of failures. That is, however, a very expensive option indeed, and only really appropriate for plants handling in excess of 500,000 t a year.

3.1.6 Grabs

While grabbing cranes are the most common means of effecting deep bunker unloading, there are alternatives. Most prominent among those are bottom plate conveyors or side wall hydraulic pusher plates, but neither offers sufficient capacity for a large-scale process facility. Grabbing cranes are the only practical mechanisms for unloading, and the grabs they handle must be evaluated.

3.1.6.1 Grab Configurations

There are two basic grab configurations, as shown in Figure 3.1 They are "clamshell," where two opposed, quadrant-shaped "buckets" close together to contain the load, and "cactus," where a number of leaf or finger-shaped arms are mounted in a circle close together on the vertical axis of the machine. In operation, clamshell grabs take an oblong "bite" from the material, while cactus grabs take a circular one. Because of this, clamshells are more effective at clearing waste from bunker corners, but cactus grabs can carry better loads and have much better penetration into the material.

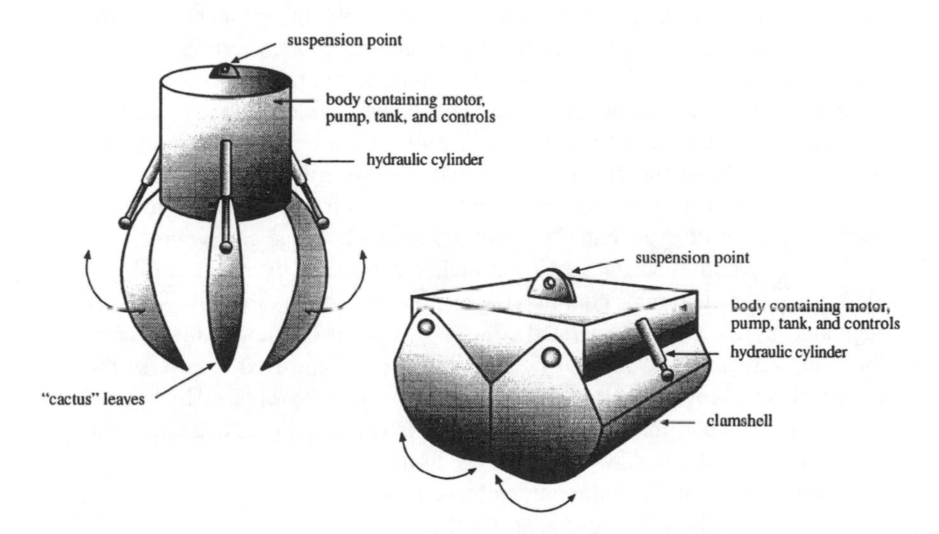

Figure 3.1 Electro-hydraulic grabs.

In waste handling operations, it is the degree of penetration that is most important, and so cactus grabs are most widely used. There is little value in an ability to reach corners if the grab cannot penetrate the waste sufficiently to obtain a realistic payload. The analogy is perhaps that of the difficulty of digging straw with a shovel as compared with a fork. For that reason, while clamshells are not uncommon, they are most often restricted to shallow bunkers with a large surface area, where waste compaction forces do not create such a density that lack of penetration is a problem.

Having chosen the type of grab, the mechanism for its operation must then be determined, and the alternatives are cable-operated or hydraulic. Cable-operated grabs use groups of separate cables to open them, and generally rely upon the hoist wire ropes to close them. They are not very efficient for handling anything other than free-flowing, granular materials, since they do not begin to close until at least some of their weight has been taken by the hoist ropes. Their penetration ability is minimal, and for that reason they are not generally used for waste handling operations.

Hydraulic grabs, meanwhile, use hydraulic cylinders to operate the clamshells or tines, driven by an on-board power pack. Again there are two fundamental designs: for low-capacity operations of around 1 t per load, the hydraulic power pack uses a reversible electric motor and hydraulic pump. Opening is achieved by running the motor in one direction, and closing by running it in reverse. The hydraulic circuit is comparatively simple and robust, and the direction of oil flow is determined by pilot-operated control valves. However, there size limits on machines designed upon this principle, and those limits are determined by the electric motor and its cable. Starting an electric motor requires several times its normal running current, and the current surge each time creates considerable heat. The power cable must be substantial enough to carry the extra current, and the motor must be of small enough mass to be able to dissipate the heat. For motors above about 6 kW (7.5 hp) the cable size becomes such that its own weight, hanging from the crane above, imposes higher stresses than the structural strength of the cable cores.

The alternative design, used for higher capacities over 1 t, employs an electric motor that runs continuously in one direction, the oil flow being regulated by electrically operated control valves. In this application, starting loads are effectively avoided, since the motor does not need to reverse twice in every cycle. However, electrical problems may now be replaced by hydraulic ones, since in such machines hydraulic oil is continually circulating, and in doing so it absorbs heat, which must be dissipated.

Heat dissipation is commonly achieved by providing sufficient on-board oil capacity for the oil to lose heat in the reservoir, and the general principle applied is to install a reservoir capacity equivalent to four times the pump capacity per min. For that reason, a large hydraulic grab will be much more complex than a small one, and increasing size beyond the 6 kW barrier will carry with it a significant weight penalty.

Then, as a further complication, the power cable from the crane to the grab becomes more complex. In addition to the power conductors, it must also contain at least one pair of control conductors for every directional control valve. Since those conductors will be of a much smaller cross-sectional area than the main power conductors, they will exhibit a quite different tensional strength, and will withstand much less stretching. Failures due to control conductor breakage are likely to be common, and simply increasing the physical size of the cable does not overcome the problem. A larger cable weighs more, and so while its tensional strength may increase, so too does its weight.

The tension per unit cross-sectional area remains substantially the same, and in addition the larger diameter may create difficulties at the cable reeling drum on the crane.

3.1.6.2 Grab Suspension

Conventionally, grabs are suspended from their cranes by at least one pair of hoist wire ropes, each of which is wound in separate tracks on the hoist drum. The nature of the hoist ropes is fundamental to the grab operation, in that an incorrect rope selection or installation can quickly render the whole system unusable.

A wire rope is manufactured by winding a number of strands together to make a single "core," after which a number of such cores are wound together to make a rope. The direction of winding, clockwise or counterclockwise, is known as the "lay" of the rope, which may be specified as either left-hand lay or right-hand lay. When placed under tension, there is a tendency for the rope to unwind slightly, and to thereby impose a torsional force over its length. If a grab is suspended from a pair of hoist ropes that are both of the same lay, then the grab will spin. It will usually do so to the extent that the ropes entwine and hoisting becomes impossible, and since the greater the load the greater the torsion, the problem is most likely to occur when the grab is at the bottom of the bunker and loaded with waste. For these reasons it is essential that ropes are fitted in pairs, one of which is left-hand lay, and one right-hand lay. By that means any tendency for one rope to twist is opposed by the tendency of the other to twist in the opposite direction.

That simple remedy addresses the design considerations for the system, but it also imposes an operating discipline that must never be ignored. Hoist wire ropes in waste handling plants work in arduous conditions where acidic moisture, heat, and abrasive dust are all prevalent. The ropes will wear quite rapidly, and may suffer corrosion internally. Weekly inspection is essential if a catastrophic failure is to be avoided, and the wire rope lubricant used should be chosen with care and renewed monthly. And finally, when one or both pairs of ropes have reached the end of their useful lives, they must be replaced as a pair. A new rope of one lay should never be installed alongside an old rope of the opposite lay, since the new rope will tend to unwind more than the old. In that situation the torsional balance will be lost and the grab will again be caused to spin.

The foregoing has considered the various grab options available, but it has not established a formula for determining the actual size of grab to be used. As we will see in the following section, that is not a simple matter.

3.1.6.3 Grab Capacity

Grab design capacity is a function of the process line infeed rate, bunker dimensions, and crane traveling distance per load, all of which are then influenced by the type of grab selected. It is beyond the scope of this book to consider every possible option within those variables. However, in the plant example used so far, a normal infeed rate of 37.5 t/h has been established, and so a grab with a capacity of 1 t would be required to handle 38 loads per h, or one load every 1.6 min. Such a regime would demand a very nimble crane system indeed, but a 2 t grab would increase the turn around time to over 3 min, which is much more practical.

If, as is likely, the plant is built with two process lines, each with a rated capacity of 20 t/h, then two bunkers and two crane–grab rigs would be sensible. There, each process line would be serviced by its own facilities, and provided that the crane tracks spanned both bunkers, one crane could at least maintain half load in the event of a breakdown of the other. There again, a turnaround time of 3 min is established for each crane.

From this broad outline of principles, the route by which crane and grab capacity is established is as follows:

1. Establish bunker dimensions for the amount of waste to be stored.
2. Locate the loading points for the process lines.
3. Calculate hoist, travel, and lowering distances.
4. Assume a grab capacity.
5. Calculate the turn-around time necessary to maintain plant throughput.

The question is, Can the crane–grab rig travel the distances in the turn-around time available, bearing in mind that no crane operator can work at full pace for 8 h?

If the answer to the question is "no," then one or more of the above parameters must be altered. In a good design, the calculated turnaround time should in practice permit the crane driver to relax for a min or two between loads. That may be achieved by considering the plant throughput (#5 above) as being 50% greater than it actually is. Then the driver will be working at full capacity only for two thirds of the time.

Applying this concept to the plant chosen as an example, the considerations become as follows:

1. The bunker is to be 38.4 ft (12.07 m) square and 57.5 ft (17.5 m) deep, holding 660 t each full day.
2. Assume that the process line loading hopper is to be in the center of one side of the bunker.
3. Allow 6 ft (2 m) clearance between the bottom of the grab and the edge of the loading hopper when the grab is fully hoisted. This clearance will be necessary to permit the grab to open without fouling the hopper.
   When recovering waste from the furthest extent of the bunker, the grab must now travel 57.5 ft (18 m) hoisting, about 42 ft (13 m) traveling, and up to 6.5 ft (2 m) lowering. To complete one cycle it must then carry out that operation in reverse, including some time for the grab to open and close.
4. Assume a grab capacity of 2 t.

5. The turnaround time to handle 56.25 t/h (37.5 × 1.5) is such as to require approximately one 2 t load every 2 min continuously. The grab would need to traverse up to 211 ft (66 m) and to open and close during that time. It would need to exceed a combined traveling speed of better than 106 ft/min (33 m/min), which is somewhat strenuous. The answer to the earlier question of whether the crane–grab rig can travel the distances in the turnaround time available, therefore, is "no."

If now two cranes and two grabs are substituted in the consideration, and the bunker design is revised to two bunkers, each 38.4 ft square and 29 ft (9 m) deep, then the crane traveling distance is reduced to 154 ft (48 m) overall, and the loading demand is reduced to 28 t/h. Now a 2 t capacity grab can accommodate that demand in a 4 min cycle, which results in a combined cycle speed of approximately 38.4 ft/min (12 m/min).

That design is much more practical. It should give the driver time to pause during each cycle, and should considerably relieve the load upon the whole system. In this case, the answer to the earlier question: "Can crane–grab rig can travel the distances in the turnaround time available?" is "yes."

3.1.7 Dust Extraction and Suppression

Deep bunkers create more airborne dust, and the dusts are much more difficult to deal with. This characteristic arises from the height from which the wastes are tipped before reaching the floor: the greater the height, the more vigorous the release of dust due to the velocity of impact. The laws of physics then dictate that the airborne particles will be traveling very rapidly indeed, partly due to the "rebound" effect, but mainly due to the high velocity of the air which the tipped load displaces.

The dust particles will generally travel with sufficient velocity to escape most conventional dust extraction systems. For example, a load of 176 ft^3 (5 m^3) tipped into a bunker 57 ft deep may reach a velocity of nearly 62 ft/s (19 m/s) by the time it reaches the floor. At that point it will displace an equivalent volume of air, at the same speed and, due to the constraints of the bunker walls, in the opposite direction. In practice the initial velocity of the air may even be higher than that of the falling waste due to compression and expansion effects.

A conventional air handling unit, based upon a wet scrubber, and mounted in the top edge of the bunker wall, may have an air movement capacity of 20,000 cfm (570 m^3/min). That, at first sight, would appear adequate to deal with the displaced air and its associated dust burden. It is after all capable of exhausting 333 ft^3/s (9.44 m^3/s) when only 176 ft^3 of air are being displaced. However, if the air handling unit ventilates the bunker through a 3.3 ft (1 m) diameter duct, then the air speed it creates will be only about 39.4 ft/s (12 m/s) at the point of entry into the duct, and much less than that only a short distance before it. Since the air speed at the duct is half of that of the contaminated air from the bunker, the momentum of the latter will encourage it to continue in its vertical path, rather than to change direction through 90° into a much slower air stream. In fact, were the air handling unit to be able to deal with the velocities involved, it would need to create an air speed significantly greater than that of the displaced air over the full bunker volume. In other words, the air exhaust duct would have to be of the same cross-sectional area as the bunker itself, or 21.66 ft (6.6 m) diameter in the case of a circular duct, and to generate an air speed of 19 m/s.

A more manageable duct diameter of perhaps 6.6 ft (2 m) would require a corresponding air speed increase at the point of inlet to maintain the desired handling capacity. Unfortunately, due to the inverse square law, every reduction in diameter results in the cross-sectional area decreasing to the *square* of that reduction. Therefore, such a duct, reduced by 6.8 times to 6.6 ft (2 m) diameter, would be only 0.02 of the cross section of the original. The air speed within it would have to increase in inverse proportion, and would accordingly considerably exceed supersonic velocity. No wet scrubber could ever be built to accommodate such conditions.

In many plants, as an alternative solution, spray bars are mounted along the top edges of the bunkers to provide a water mist above the wastes as they are being tipped. Here the principle is one of dust suppression rather than extraction, and in such an application the condition of the mist is important rather than the volume of water. A water-mist dust suppression system relies upon the creation of a "fog" of airborne microdroplets through which the contaminated air must pass. When it does so, the water fog condenses upon the dust particles, rendering them too heavy to remain in the rising air stream. This alternative employs exactly the same principle by which clouds create rain, and the result is that the dust falls back into the bunker.

Water-mist dust suppression systems are dramatically effective, require very little power to operate them, and can be linked to detectors in the tipping hall such that they only operate when a vehicle is actually discharging. They use so little water that they have an insignificant effect upon the moisture content of the waste, but they do have two immediate disadvantages. Perhaps the first and obvious disadvantage is that during the time they are operating, the grabbing crane operator cannot see into the bunker because of the fog. He cannot therefore continue to work another part of the contents while a vehicle is discharging, and during peak delivery periods may not be able to work effectively at all.

The other disadvantage is that not all of the water mist is condensed onto dust particles. A significant proportion may stay in suspension and may rise to the crane and the roof structures due to natural convection currents caused by the heat in the waste below. The result is likely to be corrosion of structures, damage to electrical equipment, and an uncomfortably humid environment in which to work.

3.1.8 Drainage

Further difficulties arise with deep bunkers in that it is necessary to provide some form of drainage to deal with the inevitable liquid effluents. As untreated waste is accumulated in the bunker, compaction forces and biological degradation will cause heavily contaminated water to collect at the bottom, often in considerable quantities. This, as observed earlier, will create unpleasant odors and gaseous emissions due to a process of anaerobicity, and it will certainly render the lower layers of waste unusable as a feedstock. Drainage is, however, extremely difficult to achieve.

It would be impractical to install a pump in the base of the bunker, mainly because it would remain inaccessible for most of the time, and could hardly be expected to operate for long if buried under hundreds of tons of waste. Alternatively, the bunker could be constructed above ground level and the liquors allowed to percolate out to an external catchment. Drainage would then be somewhat simpler, but the design would require a very massive structure indeed to withstand the internal forces.

A second alternative would be to excavate a bunker, but to provide a catchment chamber at basement level, with access from above ground. Such a design makes little difference to the scale of construction, since while the surrounding ground provides most of the resistance to internal stresses, now external stresses from the ground itself must be accommodated. In addition, the provision of service access to the catchment becomes more difficult, and adequate ventilation of the catchment space is essential in order that an accumulation of explosive gases may be avoided. Also, a reliable method for the continuous detection of gases must be installed, and there must be a clear procedure for dealing with any gases in the event of ventilation plant failures.

Irrespective of whether the bunker is above or below ground, drainage is not simple to achieve. As the waste in the base decomposes, as inevitably it will, it settles into a dense mass which tends to block drainage channels. Mesh screens are almost useless, since if they are light enough to have sufficient open surface area, the weight of the waste will collapse them. Heavy-duty screens or grids generally have inadequate openings, and are likely to block. The only effective solution lies in large grid areas over shallow drainage channels all around the bunker base and connected to a number of openings to the catchment.

3.1.9 Effluent Treatment

Once the effluents have been recovered, by whatever means, they must be treated. It is most unusual for direct discharge into main drains to be permitted. Treatment usually involves methods of reducing both suspended solids and biochemical oxygen demand (BOD), and most wastewater utilities have quite specific requirements relating to both. Fortunately, meeting those requirements is not particularly difficult, and the most common method is to store the liquors in settlement chambers, with a method of aerating them without disturbing particulates settling in the bottom. Aeration is achieved by allowing the leachates to flow into the tank over a weir, thereby entraining air, and the only problem then remaining is in the recovery and disposal of settled particulates.

A settlement chamber system is generally designed such that there are two or more entirely independent chambers. At any time only half the chambers are being used, while the remainder are emptied for clearing of sediment. Water from those in service is discharged into main drains through an overflow channel, and sediment from cleaned chambers is landfilled.

3.1.10 Deep Bunker Fire Protection

Fire protection is extremely difficult to achieve in deep bunker installations, mainly because any fires in the waste are almost always some way down in the stored mass. Waste is not particularly flammable, and although spontaneous combustion is a possibility, it is not common in process plants, where the turnover is too fast to permit the necessary level of oxidation and biological activity to become established. More often, fires in bunkers are started by a source of ignition having been delivered with the waste — hot ashes in a collection vehicle load awaiting contact with air to ignite, perhaps. Since there are few precautions that can legislate against the delivery of smoldering material, "fire protection" is almost a misnomer. "Fire fighting" is a more appropriate term.

The first sign of a fire in a deep bunker is usually tendrils of white smoke. It is distinguishable from the more common general emissions of water vapor by being somewhat localized, and once detected or suspected, the usual reaction is to excavate the location with the grab. However, it is just such a reaction that frequently turns a slow smoldering into a raging inferno as oxygen is admitted, and that has been the cause of many destroyed grabs over the years. Lowering a machine filled with flammable hydraulic oil, much of it contained in rubber hoses, into a zone with a temperature of 1500°F (815°C) is not generally a wise undertaking!

A fire needs three conditions to sustain it: a combustible material, oxygen, and temperature. Remove any one of those three sides of the "fire triangle" and no fire can exist. Thus, attempting to grab the burning material clear increases one side of the triangle rather than reducing it. Equally, if the fire is some way down in the stored mass, then removing the combustible material is not an option. Again that cannot be done without increasing the contact with oxygen. A means of reducing the temperature without increasing either oxygen or access to combustible material is required, and the only mechanism for doing that is to introduce copious quantities of water. The storage installation must be fitted with monitors capable of delivering high-velocity, high-volume jets of water to any area of the bunker.

Even then, the results cannot be guaranteed. Often, if a fire is deep enough in the waste, the rapidly increasing temperature gradient approaching it is sufficient to evaporate the water before it can reach the fire. However, something has to be done, and water is the only practical means of doing it. The monitors should be capable of virtually flooding the bunker if necessary, since given enough water, even the most deep-seated fire must be extinguished eventually.

The delivery of large quantities of water presupposes that a secure supply is available, and that to operate high-pressure monitors satisfactorily, it must be pumped. If there is any doubt that the water mains supply is reliable, or if the local water utilities insist upon it, then a fire-water storage capacity becomes essential. Water stored in such a facility should not be available or accessible for any other purpose, and the volumetric capacity should not be less than 25% of the volume of any individual bunker. Pump discharge pressure should not be less than 120 psi (8 bar) to ensure sufficient pressure at the monitors and effective penetration in shallow fires. A make-up supply from local mains should be at least 4 in. (100 mm) bore.

Where the bunkers are substantially above ground level, wet risers will be necessary. These are water mains capable of delivering water from the reserve, through fire pumps, to manifolds at the working levels, and they should not be of less than 6 in. (150 mm) bore. If the riser height is such that fire pump pressure is reduced to less than 60 psi (4 bar) at the working area, then booster pumps at the riser manifolds should be considered.

The number of monitors per bunker depends somewhat upon the surface area involved and the accessibility of the waste within. Commonly, bunkers are only capable of access from three sides, one being taken up by the loading apertures, and an individual member of a group of bunkers may only be accessible from one side. The monitors must also be placed where they do not obstruct plant operation and are unlikely to be damaged.

Fortunately, waste fires in deep bunkers do not develop rapidly. There is ample time to prepare and to instigate action. There is time to mount a monitor in a suitable location, to rig a hose to it, and to commence fire fighting. It is therefore a wise design practice to fix monitor mounting brackets to the top edge of the bunker walls in some numbers, and to support them with nozzles and hose reels in cabinets close to, but not within, the working area. By that means, the amount of fire-fighting equipment is kept to a realistic minimum for effective maintenance, and being normally secure from damage, it is likely to be workable when needed.

In completing these comments about fire fighting, it is also worth observing that in the event of a fire in a deep bunker, perhaps the first decision to make is whether or not to take action at all. There is no guidance upon this decision that would be meaningful in a book such as this, since it is one that has to be made upon the basis of experience. However, if the fire is not very deep-seated or vigorous, simply "tamping" the waste down with a few judicious applications of a closed grab above it may be sufficient to exclude oxygen and extinguish it.

3.1.11 Vehicle Discharge

The final design consideration for a deep bunker is often given less than the priority it deserves. It is that of the means by which waste collection vehicles are to discharge into it.

Collection vehicles are characterized by a low tailgate height, essential to minimize the amount of lifting required of the crew. When the vehicle is reversing to the bunker to discharge, there must be a physical projection against which its rear wheels can engage. Otherwise, the driver will have no means of positioning the vehicle accurately over the aperture. The low tailgate height, however, restricts the maximum possible height of the wheel stop to the extent that vigorous reversing can cause the vehicle to bump over it. There have been many recorded instances of collection vehicles joining their loads in deep bunkers.

Even when reversing is carried out with caution, and the vehicle is apparently safely at rest, the load itself can cause a major accident. Some collection vehicles, and almost all bulk transfer vehicles, discharge by tipping their body to a severe angle and allowing the load to slide out. Unfortunately, mixed municipal waste has a strong tendency to form a homogenous mass, particularly when it has been compacted in the vehicle to increase payload. That mass is very likely to resist movement initially, and then to begin to slide without breaking up at all. In such an event, the mass will reach the rear doors of the vehicle body at a time when much of the total weight of the unit is placed upon the rear axle as a result of the body angle, resulting in severe imbalance. Again, there have been many recorded cases where a vehicle experiencing such an incident has been literally bounced over the wheel stops and into the bunker.

While there is no method by which a careless driver can be prevented from surmounting wheel stops, and none for ensuring that tipped loads do not homogenize, there is a design for deep bunker vehicle discharge that at least ensures that the vehicle cannot fall into the bunker. Fundamentally, the design provides a discharge aperture which is of smaller cross-sectional area than the vehicle body, with a roof section shaped so as to allow a vehicle to tip within it, but not to overbalance backward.

Figure 3.2 shows a way of achieving these parameters. Here the vehicle discharges into an angled chute through an aperture in the bunker wall. The wall aperture is of slightly greater cross-sectional area than the load, but less than the vehicle body. The angled roof structure above it is constructed with a heavy steel beam in its leading edge, to limit the angle a vehicle may reach in the event of it overbalancing. The geometry of the chute forces the load to break up to some extent before it can pass into the bunker, and since the vehicle cannot do that, its potential travel is limited.

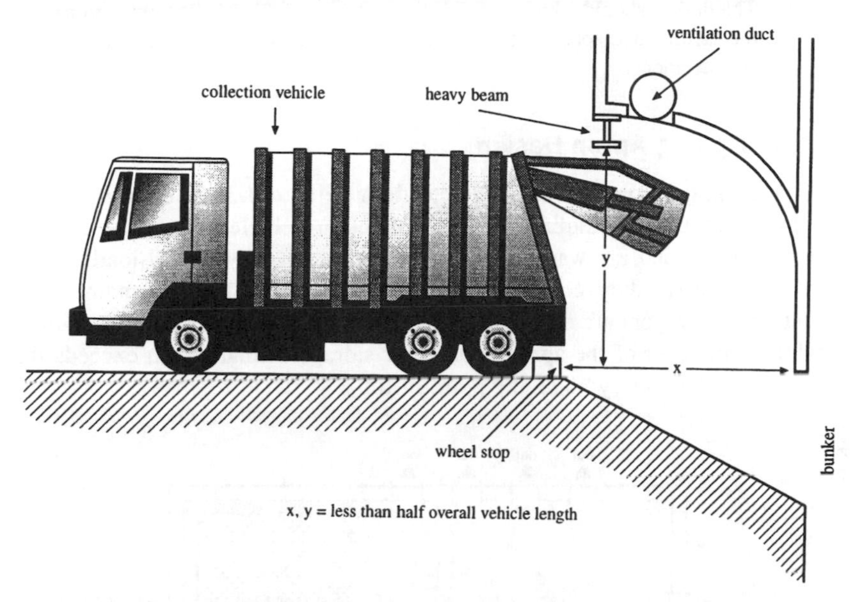

Figure 3.2 Deep bunker vehicle discharging.

The design does not prevent a vehicle from falling into the chute. There is no method by which that can be achieved within the constraints of vehicle design. However, a discharge aperture designed as discussed will at least prevent a serious accident, and where such systems have been used, even vehicle damage in the event of a wheel-stop override has been minimal.

3.1.12 Design Summary

None of the points discussed in the foregoing should be taken to imply that bunker storage systems should be avoided. In fact in a very large process plant design it may be impractical to store the waste any other way. However, best design practice for such a system can be summarized as follows:

1. The bunker should have the maximum cross-sectional area and the minimum depth possible within space constraints. Increasing depth creates increased compaction, liquid effluent generation, excessively wet waste, and handling difficulties.
2. Cranes used for unloading the bunker should be of heavy-duty design, even though the waste infeed rate does not appear to justify it.
3. Dust suppression is better than dust extraction, and if designed properly does not significantly effect the quality of the waste. Dust extraction in deep bunker applications is generally a waste of time.
4. Cactus grabs are preferable to clamshell grabs as a means of keeping below the electrical power demand which requires heavy power cables with control cores. Cactus grabs offer better payloads and greater penetration efficiency.

5. The design should take into account the nature of all vehicles likely to use the facility, and specify the discharging aperture to avoid wheel-stop over-ride accidents.

3.1.13 Tipping Apron Design

A tipping apron is a flat concrete floor (Figure 3.3), usually inside an enclosed, ventilated building. Refuse collection vehicles discharge directly onto the floor, and the waste is subsequently handled by shovel-loaders that may serve heavy-duty conveyors feeding the process lines. In many older plants the conveyors are designed to receive waste directly from the vehicles, and the remainder of the apron is used for storage of that which exceeds the capacity of the conveyors at any point in time.

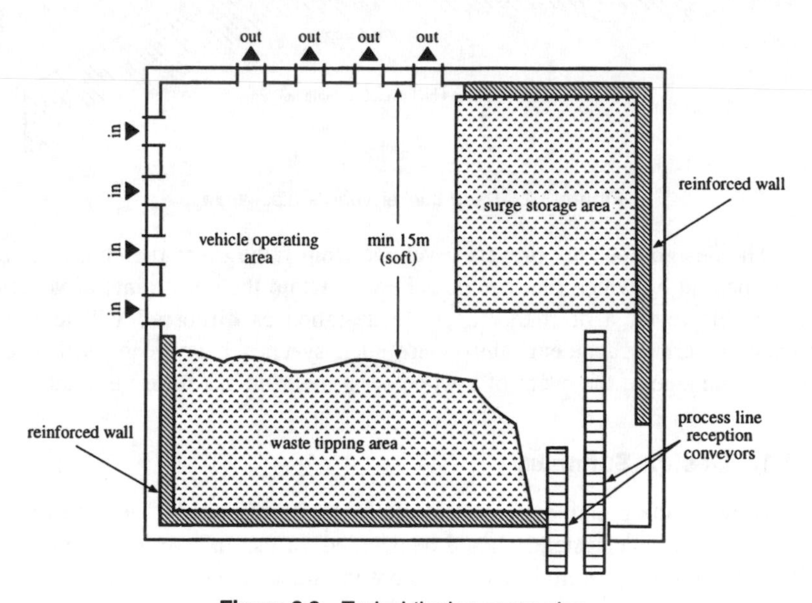

Figure 3.3 Typical tipping apron plan.

Such an apron and building must be large enough to contain the required amount of waste and also to leave sufficient space for vehicle movements and for the handling machines. There is no standard layout ideal for all circumstances, but there are some principles that determine the design to some extent. The site itself will impose some constraints, in that the access road and space for site roads may, for example, all be on one side. It may force the designer to make the building long and narrow or exactly square. But once those constraints have been addressed, the basic principles reenter the equation.

First, the collection vehicles will all be trying to discharge in the same short period of time, and they should not be forced to line up at one door to get in. The size of the collection fleet will give a good idea of how many are likely to arrive all at once, and there should be at least that many doors. Then of course they need to get out again, so the same number of exit doors is needed, and that introduces the geographical relationship between entry and exits.

Excessive vehicle maneuvering is time-wasting, so ideally a vehicle should be able to enter, stop, reverse, and then drive straight out without interfering with any other vehicles in the process. Figure 3.3 shows one way in which that could be done. In this layout, the vehicles enter through one side of the building, reverse once through 90° on the external apron, and then drive straight out again. The points to notice here are (1) since the amount of turning the vehicles have to do is minimized, so too is the space available for that purpose, and the building can be smaller; (2) the waste can be spread out over most of the available floor to maximize its surface area; (3) the process infeed is at the back of the stockpile, and the first in, first out principle can be adopted; (4) the handling machines can approach the stockpile from at least two sides, so they can keep it tidy; (5) from the ventilation point of view, the doors are all in one part of the building, so if the extraction ducts are in the center, outside air is being drawn across the pile and across the vehicles, wherever they choose to be. That way there is no risk of odors and emissions escaping to the outside. And if the highest point of the roof is over the center (as it will have to be if bulk vehicles are to deliver to the plant), then that is where the warm air rising from the waste will go anyway. It will require considerably less fan power to move it away for treatment.

Once again, there are advantages and disadvantages in the principle as a means of handling and storing waste. The main advantages are that tipping aprons permit easy identification of unwanted or hazardous materials, avoid excessive compaction and anaerobicity, are easy to clean and drain, and minimize the creation of airborne dusts. They are also likely to be considerably cheaper to build and operate than deep bunker designs.

Their main disadvantage is that a much greater floor area is required for a given quantity of waste, and to that must be added the space needed for vehicle and shovel-loader maneuvering. In addition, there may be some operational difficulties in that the vehicular traffic moves through the area in which shovel-loaders are working. The design must provide for as much separation as possible, and operational procedures must include a large element of traffic control.

3.1.14 Identification of Unwanted/Hazardous Items

When a collection or transfer vehicle discharges onto a level floor, the waste inevitably expands to some extent, and the vehicle has to move forward slowly to allow all of its load to escape. The result of this is that the waste will form into a long heap, of roughly conical cross section, and of generally not more than 6.0 ft (2 m) in height. It is almost inevitable that the point of discharge is not the same as the place of storage, and so once the vehicle is clear of the working area, the shovel-loader will then move the material to the main stockpile. During this operation, the loader driver will have a reasonably clear view of the nature of the waste, since he will be moving it in comparatively small "bites." He should be trained to identify those types of materials likely to cause blockages or damage inside the process plant, and whenever he sees such material, he must have access to a reject area where he can deposit selected items. For the loader driver to be able to perform those duties, he must be suitably equipped. The design of the apron and of the shovel-loader must be integrated.

3.1.15 Shovel Loader Design

Since the shovel-loader can move more waste by pushing than by lifting and carrying, the concrete apron must be both level and relatively smooth, and the machine needs a dozing blade facility. But pushing the waste requires more power than lifting and carrying, so the floor must not be so smooth that wheel slip becomes a problem. The temptation to achieve an immaculately clean apron by using epoxy surface sealants should be avoided, since when wet and greasy (as they will be most of the time) wheel grip will be impaired.

Although the dozing blade provides the means of shifting the maximum amount of material, it does not provide for the removal of any unwanted items. The driver needs to be able to selectively pick things up, and a 4-in-1 bucket (see Figure 3.4) will do that very well. In such a design, the fore part is hinged to the top of the back plate, and powered by hydraulic cylinders in such a way that it may be swung upward to leave the back plate clear as a dozing blade, or closed downward to provide a conventional loader bucket. This facility to close the fore part also provides the means by which the driver may pick unwanted items. He simply clamps them between the two parts of the bucket and drags them clear of the main pile.

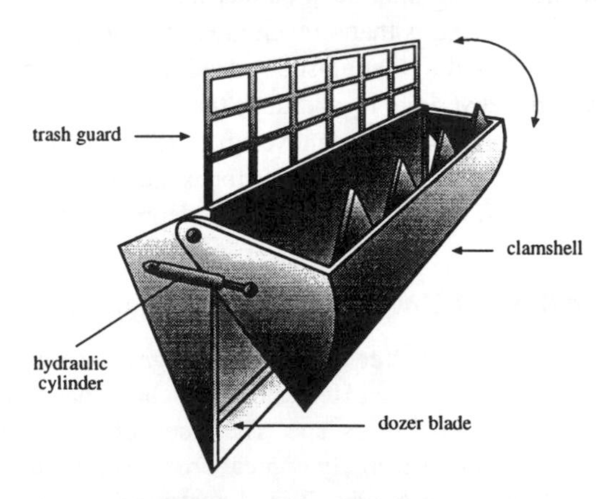

Figure 3.4 Typical 4-in-1 bucket.

4-in-1 buckets are a common and readily available attachment for all shovel loaders, but some care must be taken to specify those to be used to handle waste. Buckets supplied as standard for a particular machine will not be suitable for waste operations, simply because they are too small. Shovel-loaders are primarily designed as earthmoving machines, and their buckets are specified such that the lifting capacity of the machine is not exceeded when handling soil. Waste, however, is of a very much lower bulk density, and so a standard bucket will have insufficient capacity.

In fact, the design imposed by waste disposal operations is not based upon the lifting capacity at all. If it were to be, then the bucket would be larger than the machine upon which it is mounted. For example, a 2 yd bucket (2 yd^3 capacity when filled level) could carry 2 t of soil, but only 620 lb (300 kg) of municipal waste. A bucket large enough to meet the lifting capacity of the machine in those circumstances would need to be more than six times larger, or some 12 yd^3 capacity, and that would be a very large piece of equipment indeed. At least, it would completely obscure the view of the driver, and it would reduce the lifting height of the machine to almost zero.

The critical design parameter for a waste-handling shovel-loader bucket is its width. It is essential that the dozing blade is wider than the wheel track of the machine upon which it is mounted, and for a sound reason: when the machine is pushing waste across the tipping apron, maximum power will be transmitted through the tires to the floor. The tires will distort under load to a greater or lesser extent, and will do so in such a way that they will be particularly prone to penetration by sharp or metallic items. There have been many recorded cases of punctures, even in heavy-duty earthmoving tires, surprisingly caused by even such apparently innocuous items as plastic ball-point pens. If the dozing blade is wider than the wheel track, then the tires will not be running over waste at all, and the threat of punctures is effectively removed.

If the width of the bucket is determined by the wheel track of the machine, then so is its capacity. The geometry of the bucket requires that the width and height are in proportion, since if they were not, then the bucket would have insufficient structural rigidity. The result is that a shovel-loader designed to carry a 2 yd, 4-in-1 bucket may have to be equipped with a $2^1/_2$ yd bucket for waste handling. That requirement suggests that at average municipal waste bulk densities, the shovel-loader will never lift more than about half of its rated capacity, which is why pushing, or dozing, is much more usual.

There will, however, be occasions where the shovel-loader does need to lift the waste: for example, when increasing the height of the storage stockpile to minimize floor area usage. Since the waste is of low bulk density and may be contained in bags, there is a tendency for it to roll over the backplate of the bucket, possibly threatening the safety of the driver. A glass bottle falling over the back of a bucket raised to a height of 10 ft is quite capable of breaking the front screen of the machine.

Such an eventuality may be guarded against by the installation of a "trash guard." This is simply a mesh frame, fabricated from rolled steel angle, and welded to the top of the bucket back plate, extending upward and across the full width (see Figure 3.4). There are no hard and fast rules governing the height of the guard, and a degree of common sense is necessary. However, the guard for even the smallest bucket should be at least 5 ft (1500 mm) high, and one convenient selection method is to make the guard at least as high as the backplate of the bucket.

Having dealt with the bucket design for waste handling, there are some important considerations concerning the design and specification of the shovel-loader. The first of these has already been implied in the foregoing. The fact that the material to be moved is light in weight does not mean that a small machine is suitable. The reverse is true. The machine needs to be large to be able to mount a bucket that is itself large enough to carry or doze a meaningful payload, yet must not be so large as to be difficult to maneuver in the confined space of a tipping room. In most process plants a $2^1/_2$ yd bucket is satisfactory. One machine will be able to handle up to about 50,000 t a year of waste, and greater plant capacities are accommodated by increasing the number of machines rather than their size. A machine designed for earthmoving operations with a 2 yd bucket will adequately accommodate waste handling with a 2 yd bucket.

3.1.15.1 Wheels and Tires

Since the shovel loader will be operating on a firm, level concrete apron, there is a temptation to economize by specifying two-wheel drive (i.e., the engine of the machine transmits power only to the rear wheels), but this should be avoided. If all of the power is transmitted to one pair of wheels only, then in dozing operations tire distortion will be at a maximum, and penetration by sharp objects will become much more likely, particularly through side walls. While a bucket that is wider than the wheel track will minimize the number of occasions when a tire runs over waste, it cannot be relied upon to eliminate them. That is more in the control of the operator, who may make mistakes. A four-wheel drive machine transmits power to all four wheels, and thereby reduces potential tire distortion by at least half. Operationally, the difference between a two-wheel and a four-wheel drive may be punctures every day compared with no punctures at all.

The design of the tires themselves is equally important and worthy of discussion. First, the tread pattern should be an industrial, not an agricultural one. In the former, the rubber treads are wide and closely pitched. There is a considerable amount of rubber in contact with the floor at all times. In the latter, however, the purpose of the tire design is grip on rough and potentially muddy ground. The tread pattern is widely pitched, with deep bars of rubber across the surface, and the amount of rubber in contact with a smooth floor is minimal. Industrial-pattern tires limit distortion under load, while agricultural patterns are designed to distort to some extent to improve grip on unmade ground. Such distortion creates opportunity for penetration by sharp objects.

Many alternative tire designs and treatments have been tried over the years to overcome the puncture problem, and often they created as many problems as they solved. Solid tires were once seen as the solution, but earthmoving machines do not have suspensions. They rely upon the flexibility of the tires, and so a solid tire can transfer unacceptable shock loadings to the axles, leading to reduced life and a potentially increased breakdown rate.

Tire protectors — circular inserts made of hard rubber reinforced by synthetic fibers, and molded in a crescent cross section — have also been used. They are held against the inner face of the tire by the inner tube. Unfortunately, they only protect the tube from penetration through the treads, not the side walls, and it is side wall penetration that is the most likely. They do have the advantage that when a tire is worn out, they can be transferred to a new tire, but if a number of penetrations have occurred a transfer may be more difficult to achieve than one might expect: the protector may have become "riveted" to the casing.

By far the most expensive protection is offered by tire fill. In this application, the tire is filled with expanded synthetic foam before fitting. The idea is that, since there is no inner tube, only foam, there is nothing to puncture. Yet the flexibility of the foam restores the shock-absorbing capacity of the tire, at least to some extent. Unfortunately, it reduces tire distortion to a minimum, and in spite of the correlation between distortion and penetration, some distortion is necessary to maintain grip. It also impairs heat dissipation through the carcass, and may, on a hard-worked machine, lead to early deterioration of the treads. The greatest disadvantage of expanded synthetic foam, however, is its cost, which is usually at least twice that of the tire it is supposed to protect, and it can only be used once. When the tire reaches the end of its service life, the fill cannot be removed and reused.

The conclusion is that by far the best, and cheapest, results can be achieved by using conventional industrial-tread tires upon a machine that carries a bucket wider than the wheel track.

3.1.15.2 Transmission

The transmission of the shovel loader also needs careful attention. The machine will be required to reverse direction repeatedly, during every operation throughout the day. Modern hydrostatic (hydraulic) transmissions can be selected into forward or reverse gear without stopping the machine, and this facility is essential in waste handling. During peak delivery periods, a busy reception hall may be accommodating one collection vehicle every min for up to 3 h at a time, and the shovel loader must be agile to keep the waste moving away from the tipping area. A transmission system that requires the machine to be halted before selecting reverse gear, or one that overheats with repeated use, is a severe handicap.

Since the machine will unavoidably be operating in a fairly crowded space, with vehicles moving in close proximity to it, warning systems are essential if collisions are to be avoided. It must be fitted with a reversing alarm sufficiently loud to be heard over the noise of a truck engine, and since alarms of that nature are necessarily intrusive, it must be tamper-proof and directly connected to the transmission in such a way that it sounds whenever reverse gear is selected.

3.1.15.3 Steering

The necessary agility is also best achieved by specifying an articulated machine rather than a fixed-frame one. In an articulated system, steering is provided by a vertical pivot on the center line, such that the machine effectively bends in the middle. The design provides for a very small turning circle, and an experienced operator can even "crab" the loader sideways if he needs to. Some machines are designed for rear-wheel steer, and are unsuitable for tipping apron waste handling. They are insufficiently agile, and are prone to tire scrubbing when being steered under load.

This discussion of handling machine design leads one, perhaps, to wonder if a tracked loader might be a better option. At least there are no tires to puncture or distort, and such a machine is extremely maneuverable. Unfortunately, concrete floors and steel track plates do not mix. Floor damage and wear would be unavoidable, and track plates would become worn very rapidly indeed.

3.1.15.4 Floor Design

The concrete floor of the reception apron must be strong enough to withstand the wheel loadings of heavy traffic, including plant. It must be impervious, and must have a smooth but not polished surface. The first of those constraints imposes a design standard similar to that of major roads, since the facility may need to accommodate large, articulated bulk trucks. Such a truck, having a gross weight of, say, 38 t, will impose a disproportionate amount of that weight upon the rear wheels during discharging, and if the waste sticks to any extent, there may also be some element of shock loading upon the floor.

Conventional horizontal-discharge collection vehicles impose much less severe design constraints, but if a plant is to be constructed for a 20-year life, it would be imprudent to assume that only this type of vehicle would use it during the whole of that time. Waste collection and disposal is a continually changing industry, where economics often dictate fundamental policy changes, and it would be sensible to design the floor for the largest vehicles likely to use it. Uprating it later would be prohibitively expensive and may even be impractical.

For those reasons it is a wise precaution to construct the concrete slab, on good load-bearing ground, 8 in. (200 mm) thick and double-reinforced, laid upon a bed of at least 8 in. of compacted granular base, which has been blinded with lean-mix concrete and sealed with an impervious membrane. Such a slab can be constructed with a minimum of expansion joints — possibly only every 80 ft (25 m). Other than these general principles, it is difficult in a book such as this to detail the civil engineering design of the reception apron floor. Local geology and topography will dictate the actual specification.

3.1.16 Drainage

Reception aprons do not actually need drainage to any great extent, since the way in which the waste is stored tends to minimize the release of water. However, all waste treatment facilities must have a means of dealing with surface water at least, and environmental pollution controls usually require that all drainage from them is led to interceptors.

An interceptor is a chamber so designed that particulates and oils may be separated from the effluent before discharge into main drains or public sewers, and for a waste treatment facility a three-chamber interceptor is desirable. Such a design will provide reasonably good oil and fuel separation in the event of spills from vehicles, and will protect the main drains from fats and oils from the wastes themselves. Figure 3.5 suggests a suitable design, although the actual dimensions and number of interceptors depends largely upon the size of the process facility and local meteorological conditions. If the interceptor is sized for the anticipated surface water runoff from the buildings and site roads, then there will be sufficient capacity to deal with any drainage from the reception apron.

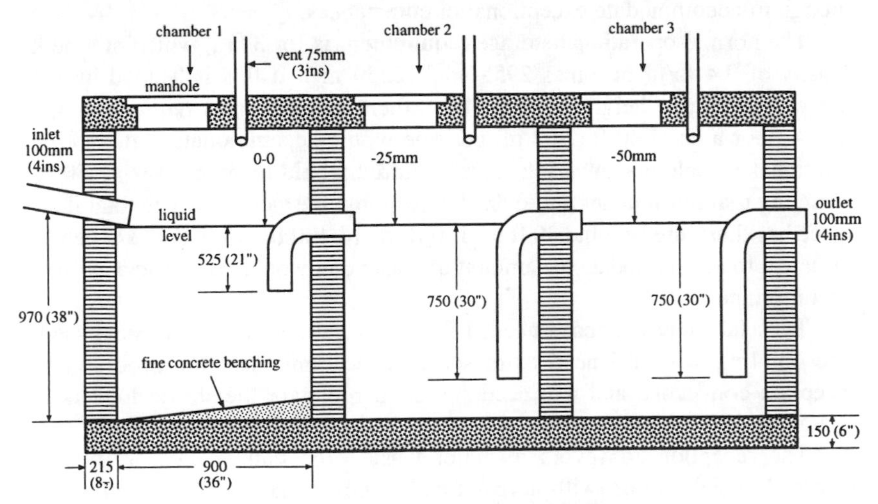

Figure 3.5 Typical oil interceptor.

3.1.17 Building Structure

In order to minimize malodors, dusts, and vermin infestations, the reception apron has to be enclosed within a building, ventilated under negative pressure. The dimensions of the building depend again upon the daily waste arisings, and the amount of storage to be accommodated, but there are some fundamental requirements that apply in all cases.

First, vehicles need a considerable amount of room to maneuver, even when, as in the example, all they have to do is enter the building, reverse, and then drive straight out again. The inside of a reception building is an ever-changing environment, where the tipping area changes as one section becomes filled and another is emptied down by the process lines. There must be a space running either the full length or width of the building, where waste will rarely or never be tipped, and where the vehicles can always find maneuvering space. In order that almost any size of truck may be accommodated, that space should be not less than 50 ft (15 m) deep.

Returning to the hypothetical plant discussed earlier, 660 t of waste at an average bulk density of 9.4 lb/ft^3 (150 kg/m^3) are to be contained. The total volume is then 5200 yd^3 (4400 m^3). Assuming that the shovel loaders serving the plant can heap the waste to a height of 13 ft (4 m), then the floor area needed for waste storage alone becomes 5200/4.3, or 1200 yd^3. That, however, is the maximum area required, to be used in those exceptional circumstances where the plant has become nonoperational for at least 1 complete day. In such an event, waste deliveries should have been stopped temporarily, and should not start again until at least the major part of the waste has been treated. There is little point in continuing to provide the full vehicle maneuvering space when it will not be used, and so the design consideration becomes one of providing sufficient space for normal storage plus vehicle maneuvering, with enough to accommodate exceptional circumstances.

The normal operational storage requirement is for 375 t, which at a bulk density of 9.4 lb/ft^3 becomes 2955 yds^3 (2259 m^3). If this is heaped by the shovel loaders to a height of 13 ft (4 m), then the apron area required is 682 yd^2. A floor area of 80 ft (24.4 m) per side would accommodate that, and the additional vehicle maneuvering space required would then be 80 ft by 50 ft, assuming that the vehicles are to discharge along one side only. The total floor space has therefore become 80 ft by 130 ft, or 10,400 ft^2, which is very nearly sufficient to accommodate the amount of waste that would arise in exceptional circumstances.

Thus far, only the space needed for waste storage has been established. The building will still need extra space to accommodate the process line reception conveyors, and will need sufficient room for the shovel loaders to operate in any circumstances.

The reception conveyors in major waste processing plants are almost invariably steel aprons, with access for the shovel loaders at their ends and possibly along one side. These machines are extremely robust, slow-running, and reliable, and a 6 ft (1.8 m)-wide machine is capable of handling waste infeeds of well over 40 t/h. A plant with a design capacity of 37.5 t/h would normally be built in two flow lines, and it would therefore need two reception conveyors, one to serve each line.

A steel apron conveyor can generally be considered as occupying a space twice its length and twice its width, to allow for maintenance access, cleaning, and shovel loader operation. In the example chosen, we already have a building length of 130 ft and a width of 80 ft. Assuming that the conveyors are to be installed along the long side, and are to project into the building to a distance of 65 ft, then the 130 ft length is adequate. The 80 ft width, however, must be increased by the width required by the two conveyors, installed in parallel. This width is 4 × 6 ft in this case, or 24 ft (7.2 m). The total building floor area, including the operational waste storage, is now 130 × (80 + 24) ft^2: a total of 13,520 ft^2 (1244 m^2).

Industrial buildings are not normally constructed to fractions of a meter or foot, and in this example the calculated dimensions are 130 ft × 104 ft (39.6 × 31.7 m). A convenient clear span would be 115 ft (35 m) wide, and with sufficient sections to enclose 130 ft length, a total floor area of 14,950 ft^2 (1375 m^2) would be achieved. The space would therefore be adequate to contain normal operational storage of 6725 ft^2 (625 m^2) and exceptional storage of 11,836 ft^2 (1100 m^2) resulting from a complete plant breakdown lasting a full day.

From this example, the storage space assessment route is then as follows:

1. Calculate the normal operational storage space requirement using a table similar to Table 3.1, applying known bulk densities and delivery periods.
2. Estimate the space required for 1 day's waste deliveries, assuming none is processed.
3. To the normal operational storage add a vehicle maneuvering space 50 ft (15 m) deep over the full stockpile width.
4. To that space, add conveyor space and access, allowing twice the width and length of each conveyor.
5. Round the result up to the nearest commercially available clear span and length.
6. Compare the resulting floor space, excluding conveyors, with (3) above. If (3) is greater than (2), then the building will be adequate.

In the event that this initial estimation shows (3) is not greater than (2) for some reason, consider extending the length, not the span. Clear spans are limited by structural considerations, lengths are not.

3.1.17.1 Walls

Generally, a steel-framed industrial building clad with corrugated, plastic-coated steel panels is sufficient for a reception hall. Its purpose is to keep the weather separate from the waste, to contain odors and dust, and to keep the waste away from public view. While local planning decisions may determine otherwise in some cases, it is not usually necessary to insulate or sound-proof it. Insulation has no value in a building where large external doors will be open for at least 6 h a day, and sound proofing linings are pointless when vehicle noise inside is, if anything, less than that outside.

Such a building design does not, however, permit waste to be heaped against the wall to a height of 13 ft (4 m), and so an internal retaining wall must be constructed. The retaining wall may be constructed in either mass concrete or steel, and although the former is preferable, the latter has been used with success upon many occasions. Whichever system is selected, the following essential factors must be considered in the design.

- The wall must have sufficient structural strength to withstand the pressure of the waste heaped against it.
- It must also be able to withstand the forces imposed by a shovel-loader pushing against it.
- It must be "lipped" at the top so that a shovel-loader can load waste against it.
- It must isolate the main building structure from any lateral forces.
- It must be rigidly secured to the apron floor.

Figure 3.6 shows a suitable cross section through such a wall. In this case, the height to the underside of the lip is considerably greater than the 13 ft (4 m) discussed so far, because the height is determined by the lift of the shovel-loader, not by the height of the waste pile.

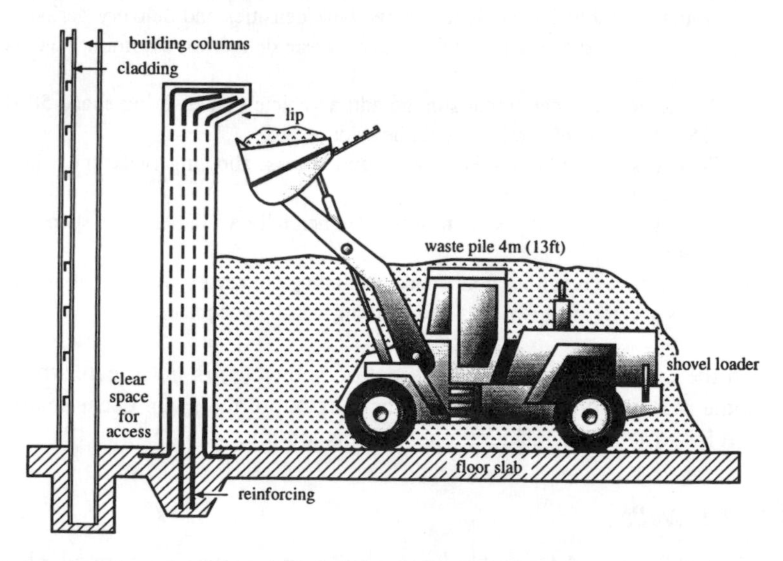

Figure 3.6 Tipping apron retaining wall.

At some point during normal operations, probably every day, the shovel-loader will be recovering residues of waste from against the wall, when dozing would be inappropriate. At such times the operator will be loading the bucket and carrying the waste to the reception conveyors. He will trap the waste between his machine bucket and the wall, will then raise the bucket and "crowd" it (roll it backward to the horizontal). If the wall has no lip at the top, then once the bucket reaches that point, a quantity of waste will fall clear, and will either fall behind the wall if there is space for it to do so, or will lie upon the top and eventually damage the building cladding. Figure 3.6 indicates the importance of the lip in the wall design.

In order that it may withstand forces imposed by both the waste and the shovel-loader, the wall must have considerable structural strength. It is a retaining structure which is unusual in that it must withstand dynamic as well as static loadings. For that reason, mass concrete is particularly appropriate, as its weight provides considerable resistance to lateral forces and impacts.

To achieve the necessary resistance, the wall, whether it is concrete or steel, must be secured to the floor over a strong foundation, but the specification of that foundation is beyond the scope of this book. It will depend upon the amounts of waste to be stored, the storage height, the physical size of the shovel-loader, and local ground conditions. Once the foundation has been designed, the wall should be secured to it by extending reinforcing bars upward through the floor to the full height of the wall, in such a way that foundation, wall, and floor form an integral structure.

If steel is chosen for the walls instead of concrete, and ease of construction and subsequent maintenance often makes it an attractive option, then the panels should not be of less than $^3/_4$ in. (20 mm) thick plate, secured using $^3/_4$ in. (20 mm) countersunk, high-tensile bolts to heavy rolled steel columns pitched every meter length of wall. The columns should be extended down into the foundation at least 6 ft (2 m), and although, as with the mass concrete wall, their actual specification will depend upon local conditions, it would be unwise to use columns of less that 30 × 15 in., 87 lb/ft (356 × 368 mm, 129 kg/m). The columns must be capable of withstanding shock loadings and extended dynamic loadings imposed by the shovel-loader, particularly toward their tops, and steel differs from concrete in that it can be bent beyond the point of recovery. In comparison, the loads imposed by the waste stockpile are insignificant, and it is the dynamic loads that must be accounted for in the design.

One reason often put forward for excluding a steel wall as an option is that corrosion through contact with the waste may be a problem. In fact this is rarely so. Waste contains many fats and oils, all of which coat the wall face very effectively indeed. That, and the frequent scouring effect of the shovel-loader operations, keep steel walls polished and corrosion-free. Protective coatings are unnecessary.

3.1.17.2 Building Height

There are two main considerations applying to building height. The eaves must be sufficiently raised to allow the largest vehicles to tip without any danger of fouling roof structures, and the process block that extends from the reception hall should be of sufficient height to accommodate the optimum machinery installation.

In establishing the tipping height of vehicles, it is dangerous to assume that those which will use the facility initially will be unchanged over its full life. Vehicle design changes rapidly, as do the logistics of waste collection. Raising the roof of the building to accommodate larger vehicles later would be an extremely expensive exercise, even if it were possible. Better by far to design for the largest vehicles, even if there is no immediate intention of using them. A large 38 t articulated bulk vehicle will need a 38 ft (12 m) eaves height.

The process block machinery installation, meanwhile, also needs height. Waste processing separates fractions through a series of machines until clean raw materials have been created, and the amount of mechanical handling between machines can be minimized by using gravity wherever possible. Machines should be installed in "stacks," in such a way that waste is introduced into the first machine in the stack, from which it falls in stages through each downstream machine. An eaves height of 38 ft (12 m) is generally suitable for most purposes, and selecting it permits a single building which is consistent in shape and size throughout. Difficulties that would be encountered in connecting two differently shaped structures are avoided, and construction methods can be simplified, with considerable reductions in cost.

3.1.17.3 Lighting

The best lighting for a reception hall is natural light from outside, but airborne dust may make that difficult to maintain. Windows in side wall must necessarily be at high level to be above the retaining wall and out of reach of the shovel-loaders, otherwise breakages will be common. Also, there will be many times when there is insufficient natural light available, and windows in the building walls are essentially directional. They will only admit significant illumination when they happen to be facing the right way. For those reasons, translucent roof panels are more effective, but they must be supplemented by artificial lighting.

In considering artificial lighting, there are some fundamental rules: The space to be illuminated is large, and potentially humid and dusty, and so simple, exposed tube fluorescent lighting is inadequate. High-intensity halogen lighting in dust-proof, external quality (waterproof) fittings is much better, but it must be placed at a high level in order that its intensity does not blind drivers operating vehicles nearby. Also, it must be placed so as to avoid shadows wherever possible, and that is best achieved by installing fittings around the eaves of the building, and spaced across the roof spans.

Once installed, the lamps will be at a high level, mostly in the less accessible regions of an open span. Maintenance access is essential, and should be provided in such a way that the need to change a few lamp elements does not require the shut-down of the plant. The most effective way of achieving that is to install groups of lamp fittings on carriers that are suspended so that they may be lowered to ground level by hoisting systems located outside the working area. Lighting designs of that nature are common in such high-roofed buildings as theaters and conference halls, and are readily available. The provision of a complete spare lamp group on a carrier allows a bank to be changed in min and the defective unit to be removed for servicing elsewhere.

3.1.18 Ventilation

The ventilation of waste reception buildings is often a somewhat vexed question. The general public, and as a consequence many planning and environmental authorities, view municipal waste as a malodorous, insect- and vermin-infested mixture that must be isolated from the environment in every way possible. That may be true where it is being handled upon a badly run landfill site, but it is not so in a properly designed process plant.

Possibly the most important difference is that when waste is landfilled, it is generally only completely covered at the end of an operating day. The rest of the time it is exposed to the air, and is vulnerable to scavengers. In a process plant, however, the waste is being converted continuously, nearly as fast as it is delivered. It never stays in one place long enough to attract vermin, and is always too fresh to have become excessively malodorous. For those reasons, the ventilation system in a reception building does not have a particularly onerous task to perform. Its main function is fourfold: to remove vehicle exhaust fumes from the working area, to collect airborne dusts, to minimize humidity within the roof structure, and to create a slight negative pressure within the building so that outside air flows in rather than out through the doors.

Structural beams spanning the roof are impossible to avoid, and water-spray dust suppression is impractical in a tipping hall. Dust extraction ducts placed high up in the roof will work quite well, since air velocities are in this case quite modest. Even so, not all dust will be captured and some will lie upon beams. Good housekeeping is the only solution to the problem, since no amount of modification to the shape of the beams will make any significant difference. The beams should be washed down every 6 months with water sprays, hence the reason for waterproof light fittings.

The selection of a suitable ventilation system is based upon the number of air changes per h. That is, the number of times in an h that the extraction fans can exhaust one complete building volume of air. There are many standards applying to air changes for a wide range of applications, and there may also be local building design standards that planning authorities may enforce. Generally, however, a waste reception building would be adequately served by a system providing four air changes per h.

Given that figure as a starting point, the sizing of the ventilation system is simply a matter of calculation the internal volume of the building and multiplying it by four. The result is the fan capacity necessary to achieve four air changes per h. That, however, leads to further design considerations.

First, when the contaminated air is collected, it must be conditioned before being exhausted to the outside atmosphere. If it were not, then there would be little point in enclosing the reception apron in the first place.

There are two ways in which the air may be conditioned: by separators or by filters. Separation involves passing the air over or through a device that encourages particulates to fall out of the air stream, while filtering involves passing the air through a medium that physically holds back particulates. There is a very large range of equipment available within these broad definitions, but the following general descriptions are sufficient to identify the principles involved.

3.1.18.1 Separators — Electrostatic

An electrostatic separator (Figure 3.7) is a device within which an air stream is passed over wires between steel plates. High-voltage electric currents are applied to the wires so as to charge dust particles electrostatically. The particles are attracted to the plates, which are periodically cleaned, usually by the simple expedient of hammering them with mechanical rappers shaft-driven from electric motors. They are highly efficient machines, capable of very good air cleaning, but they are very expensive both to purchase and to operate.

3.1.18.2 Separators — Cyclone

Cyclone separators (Figure 3.8) rely upon centrifugal force to drive dust particles out of air streams, and they operate by causing the air to spin at high velocity, effectively throwing the dust particles out to the side. Particulates fall to a rotary valve in the base under the influence of gravity, from where they are discharged into containers. Cyclone separators are reasonably efficient in removing large particles, but much less so with submicron particles.

3.1.18.3 Filters — Dry

In a dry filter (Figure 3.9), contaminated air is passed through a fabric, ceramic, or chamber filled with a fine, granular material. The porosity of the filter media is such that air can pass through, but particulates cannot. Most commercial types have a means of cleaning, either by reverse air-jet, or by shaking, but all cleaning methods require that the filter be taken out of service at regular intervals for that purpose. Normally that involves two banks of filters at least, such that one bank is in use while the other is being cleaned.

Biological filters (Figure 3.10) are a variety of dry filters often installed in waste treatment plants. Here the air stream is passed through a biologically

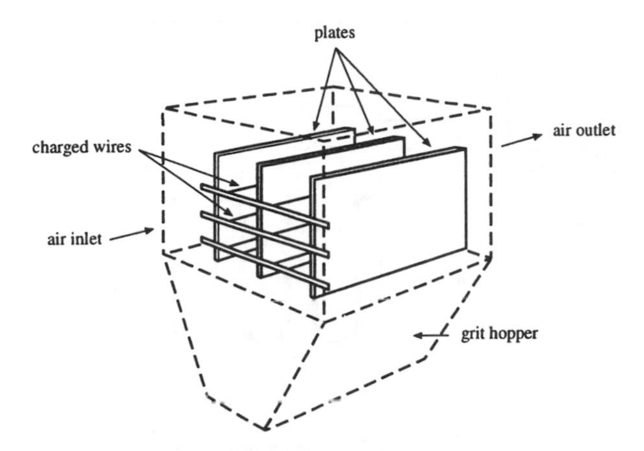

Figure 3.7 Electrostatic separator.

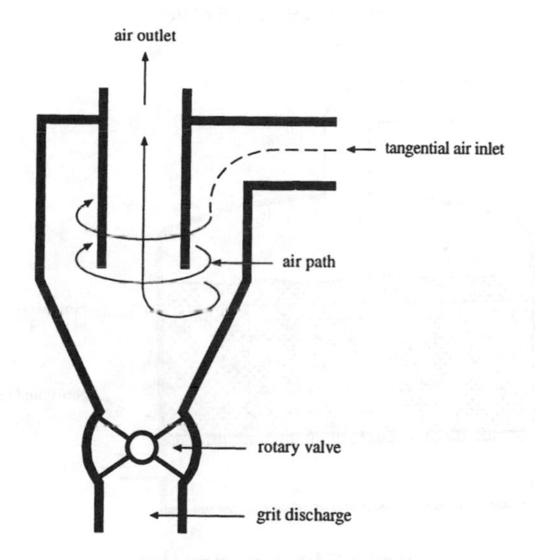

Figure 3.8 Cyclone separator.

active granular medium contained within a chamber, and peat is a commonly used material. Filters of this type are more commonly used to remove odors than particulates, and they rely upon the odor-producing compounds being deposited upon the filter medium, where they are destroyed by aerobic bacteria. Activated carbon filters, where the chamber is filled with charcoal, perform the same function chemically.

Dry filters suffer from the general disadvantage that they have to be cleaned or renewed at regular intervals, and cannot be used while cleaning or refilling is being carried out. They are extremely efficient in removing dust particles, but much less so in dealing with moisture.

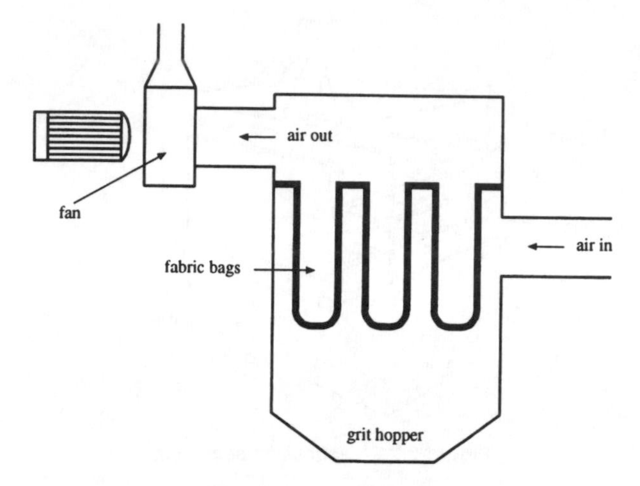

Figure 3.9 Fabric filter.

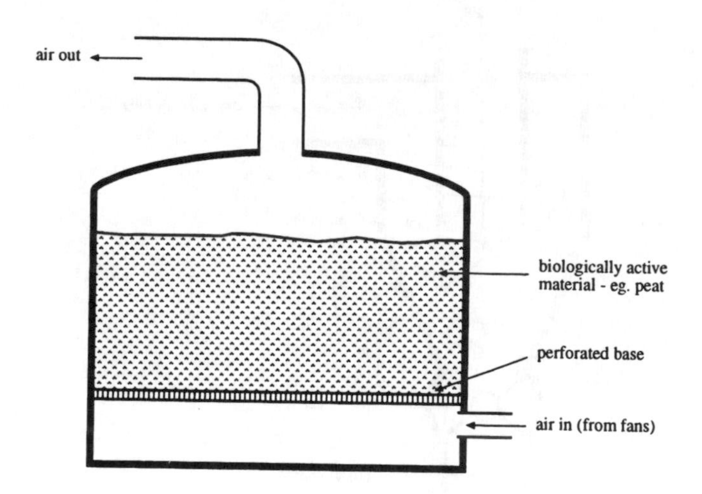

Figure 3.10 Biological filter.

3.1.18.4 Filters — Wet

A wet filter (Figure 3.11) may be no more than a chamber fitted with water sprays, providing a mist through which the contaminated air must pass. Such an oversimplified system is, however, unusual, and not particularly efficient. Wet scrubbers are much more common, and these consist of a water bath in which a series of baffle plates are immersed. Air is drawn over the plates so as to be intimately mixed with the water, effectively washing the contaminants out. Most larger-capacity wet scrubbers have drag conveyors in the bottom of the water tank, allowing the sludges resulting from separation

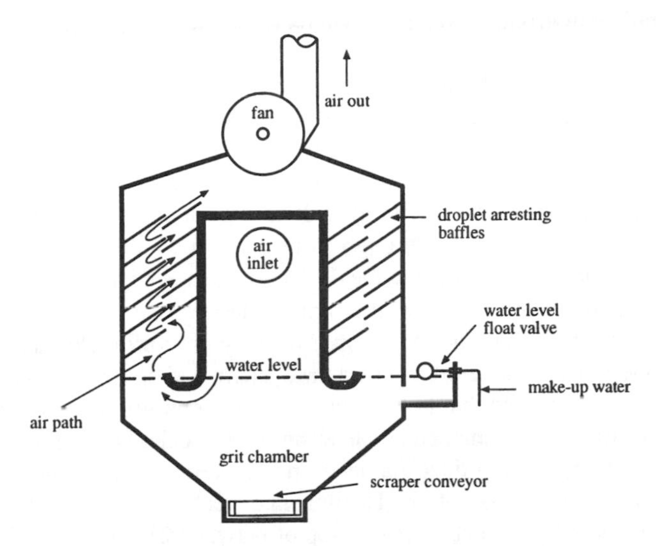

Figure 3.11 Typical wet scrubber.

to be drawn out of the water for subsequent disposal. Devices of that design have the immediate advantage that their operation is continuous, not intermittent. They never need to be taken out of service for cleaning.

Dry filters and wet scrubbers will not eliminate hydrocarbons from vehicle exhaust fumes, but then arguably there is no reason for doing so. The vehicles will emit no more fumes inside the building than they do outside, and no provision is made there for filtering their exhausts. Also, dry mechanical or electrical filters will not have any effect upon odors — only biological or chemical filters will do that reliably. Wet scrubbers, however, will remove dust particles and to some extent moisture efficiently, and will eliminate some of the odors and exhaust gases. They are low-cost, reliable, and continuous in operation, and as such are a suitable compromise for a waste reception building.

3.1.18.5 Air-Handling Ductwork

Having selected a suitable air-conditioning system, the location and sizing of the ductwork must be established. While it would be impossible here to cover every possible eventuality for every conceivable size of plant, there are some fundamental design rules to be followed.

The air-handling ducts must be well out of the way of normal operations, yet be sited where dusts and other pollutants are most likely to be concentrated. Vehicle exhaust fumes would normally be expected to stay at or near ground level, being heavier than air, but this is only true up to a point. They will do so when cold, but they are emitted from the vehicles hot. They will rise until sufficiently cooled, and if they can be captured then, will never have the chance to accumulate at ground level. Meanwhile, since the waste is generally a little warmer than its surrounding environment due to biological degradation, dusts, moisture, and odors will also tend to rise. It therefore follows that the most logical place to put the air exhaust ducts is inside the highest point of the roof structure. There they will be out of the way of the vehicles, positioned where the air will rise naturally as a result of convection currents.

Positioned inside the roof, however, the ducts will not be readily accessible for cleaning. If the air speed through them is too low, then dust will fall out of the air stream and accumulate inside. The presence of moisture will cause the deposits to adhere to the metal casings with surprising tenacity. Unfortunately, noise from a ventilation system is directly proportional to air speed, or more properly to pressure drop across the system, and so a very high velocity may ensure a clean duct but be very noisy. Again a compromise is necessary.

There are various mathematical formulas for calculating pressure drop across a system, given a duct diameter and air velocity, and among them is the Colebrook–White equation. Heating and ventilating engineers use the equation to establish a duct pressure drop of between 250 N/m^2 and 500 N/m^2 for normal residential and commercial ventilation, but even the higher figure is somewhat low for industrial purposes. A more convenient rule of thumb for waste reception area ventilation is an air velocity of between 16 ft/s (5 m/s) and 24 ft/s (7.5 m/s).

Applying this to the hypothetical plant, which has an enclosed volume of 568,100 ft^3 (16,088 m^3), four air changes per h would require an air movement of 2,272,400 ft^3 (64,354 m^3) per h, or 631 ft^3 (17.9 m^3) per s. The duct cross-sectional area necessary to achieve an air speed of 24 ft/s (7.5 m/s) is then calculated by dividing the air movement volume by the desired air speed, in this case 631 divided by 24. The result is a duct of 26.3 ft^2 (2.4 m^2) cross section, and 5.8 ft (1.77 m) in diameter.

These figures may initially appear somewhat alarming, but the air movement volumes are in fact well within the capacity of conventional wet scrubbers. In fact, a suitable scrubber would have capacity in reserve, and it could be used to aspirate at least some of the machinery in the process block. Equally, the duct cross section is not overly large for an industrial installation, and the figure established in the example is only the *total* area. To achieve efficient collection of contaminated air, it is more likely that the wet scrubber would be served by at least two ducts, each of about 5 ft (1.5 m) diameter.

Finally, whatever ventilation system is selected, there will be a residue for disposal. From dry filters it will be a fine, fibrous dust, possibly heavily contaminated with toxic metal salts, needing special handling precautions to ensure operator safety. From wet scrubbers it will be in the form of a slurry, which is much safer and easier to handle, but still operator contact should be avoided. The temptation to simply return any such residues to the waste for treatment within the process should be resisted, since there is no recycling system that can use them. They are process rejects that must be either incinerated or landfilled.

3.1.19 Tipping Apron Fire Protection

Fire protection and dust control are almost two facets of the same problem. While waste tipping upon a flat concrete apron minimizes dust generation, it does not eliminate it completely. Some dust will always be created, and it is the nature of municipal waste that the dust will be fibrous and of low but measurable flammability. The air temperature at around floor level within the building is likely to be somewhat higher than ambient, due to biological activity in the waste and to the operation of vehicles, and so any dust created will tend to rise in convection currents. Roof beams may become coated quite rapidly with a fine, sticky gray layer which, if ignited, will smolder slowly and "track" over considerable distances. It is not unknown in waste process plants for a fire to be started in one area, as a result of ignition of dusts on beams hundreds of yards away.

Dust is not the only potential fire hazard in a tipping hall. Vehicles and mobile plants operate there, carrying highly flammable fuel within their tanks and powered by internal combustion engines that generate considerable heat. They do so in close proximity to a mixture of materials, some of which may be quite flammable, and some of which may already be smoldering when received at the plant.

Returning to the "fire triangle" discussed in relation to deep bunkers, a fire needs oxygen, combustible material, and heat to sustain it. Removal of any of those elements extinguishes the fire, but in the environment of a tipping apron, circumstances may make removal of any one of the elements inappropriate. For example, an engine fire on a vehicle could be extinguished by hosing it with water from a hydrant, thereby removing the heat. The water may, however, simply carry burning fuel to other areas, thereby spreading the fire. Dry powder may exclude the oxygen from the fire, but will leave the metal hot, allowing the fire to reignite as soon as any more combustible material becomes available. Removal of the combustible material is probably impossible if it is the vehicle itself.

Similarly, a fire in the stored waste would be unlikely to respond only to smothering with dry powders or other oxygen excluders. There is likely to be more than enough oxygen contained within the waste to maintain the fire for a long time. Attempts to remove the combustible material may simply admit more oxygen and make the situation worse, and so cooling with copious quantities of water is all that is left.

It follows, therefore, that several means of fighting fires are needed in a tipping hall. Smothering chemicals or possibly foam generators are needed to achieve initial control of vehicle fires. High-pressure water nozzles are necessary to control fires in the waste, and water fog may be necessary to limit smoke and safeguard personnel. All of the systems must be accessible at all times, and must never be obstructed by the waste storage; therefore, the most obvious location for these fire-fighting systems is in that area reserved for vehicle maneuvering. They should be located in color-coded cabinets by the doors, where firefighters can escape if things get out of control and others can cover their exit with spray or fog applicators.

As with deep bunkers, copious supplies of water may be needed, and a fire-water containment should be considered. Main water supplies are not uninterruptible. The containment should not be accessible for any other purpose, and fire hoses should never be used for washing down. In this case, there is no logical definition of adequate water reserve, since local circumstances will dictate what it should be. It may even be that the local fire department will apply bylaws that dictate water reserves, and will certainly be best placed to advise in such matters.

Foam systems are generally based upon the fire hoses, and involve connecting special nozzles that mix foam compound from a storage tank with water at the hose end. They are effective for vehicle fires mainly because they create such large volumes of oxygen-excluding foam that reignition due to residual temperature becomes impossible. They do, however, rely upon the same water reserve as the fire-fighting system as a whole, and if that fails then so does the foam installation.

Dry powder, gas, and liquid systems are all intended to exclude oxygen, and are generally provided in the form of hand-held extinguishers. Powders are usually sodium or potassium bicarbonate, with additives intended to prevent caking and to ensure free-flowing properties. Gas extinguishers rely upon liquified carbon dioxide gas, delivered through nozzles designed to prevent freezing and to minimize air entrainment during discharge. Liquid chemical extinguishers mainly use chlorobromomethane ("CBM" types) or bromochlorodifluoromethane ("BCF" types), both of which are heavy gases at normal temperatures and liquids at about 4° below freezing point.

Other than as a means of effecting rescues of personnel or controlling small fires, hand-held extinguishers are of limited use in waste processing plants. They contain insufficient fire-fighting medium to have any substantial impact upon large-scale fires of the type such plants are likely to experience.

3.1.20 Reception Conveyors

Waste from the reception apron or the bunker must be delivered to the process plant at a consistent rate, not in the individual loads that would arise from a grabbing crane or a shovel-loader. Therefore, a conveyor system capable of the following duties is required.

- It must be capable of leveling and metering individual loads of waste into the process line at a consistent rate.
- It must be able to withstand impact loadings when waste is loaded upon it.
- It must be impervious to damage from sharp, heavy, and abrasive objects and materials.
- It must be reasonably sealed against spillage.

- Since at least a part of its length must necessarily be exposed, it must not run so fast as to present an operator hazard.
- It must be resistant to heat, acids, and alkalis.
- It must be capable of a positive feed of irregular-sized materials.
- It must not be prone to blockages.

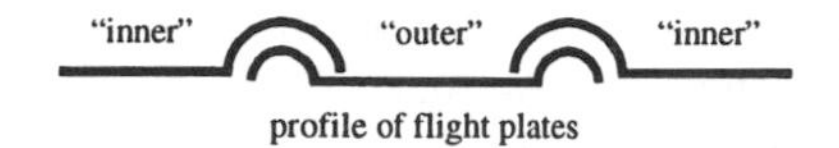

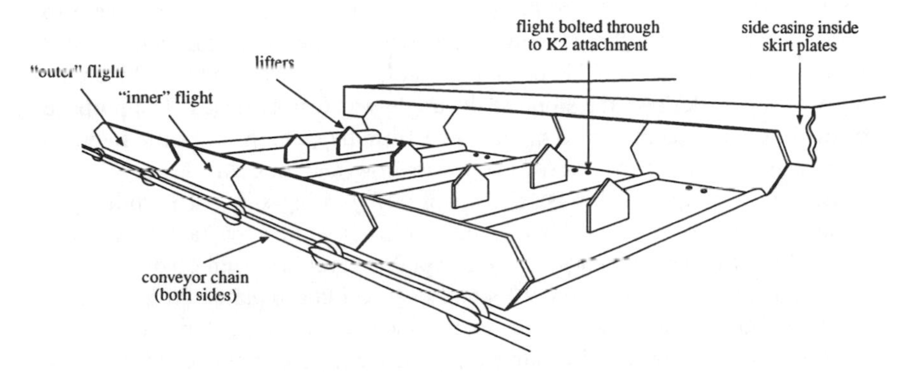

Figure 3.12 Typical plate feeder conveyor.

By far the most common type of conveyor used for the duty is variously referred to as a "steel apron" or a "plate feeder." Basically, it is a series of profiled steel plates ("flights"), which overlap to form a continuous belt, and either mounted upon conveyor chain or linked together and driven directly by sprockets (see Figure 3.12). Steel apron conveyors are extremely robust and slow-running, and they are almost totally impervious to damage from contact with the waste. They can be edge-sealed to prevent spillage, and can provide a positive feed at steep inclinations by being fitted with "lifters" — vertical plates welded to the top faces of the flights.

3.1.20.1 Design Considerations

In a waste processing plant, the machinery installation will almost inevitably be at a higher level than the floor of the reception apron, and so the conveyor must be in three sections. There must be a horizontal loading section long enough to permit access for shovel-loaders, followed by an inclined section rising to the discharge level, and finally a further horizontal section to provide infeed to the first machine. Each section must be the subject of detailed design considerations, almost as if it were a separate conveyor.

The horizontal loading section has to be long enough and wide enough to accommodate the selected loading mechanism, and so the first choice to be made is whether loading should take place from the side, the end, both side and end, or from above in the case of a grabbing crane. The most generally accepted width for those purposes is 6 ft (1.8 m), with a depth of material (the "bed depth") on the belt of about 12 in. (300 mm). The length of the horizontal section should not be less than twice the width of the shovel-loader bucket or the grab.

The flights will have to be fitted with lifters in order that the waste may be elevated up the inclined section, and those lifters will form a part of all three sections of the conveyor. They may add significant weight to the whole, and so their dimensions must be established. Municipal waste has a maximum angle of slip of about 27° to the horizontal. This means that if one were to transfer waste upon a flat, sloping conveyor belt, then once an angle of 27° was reached, transfer would become unreliable. The waste would tend increasingly to slip back down the slope as the angle was further raised. The purpose of the lifters is not entirely to prevent that from happening, since the inclined section is that which will level the waste flow and smooth out any surges caused by loading. They must be capable of limiting slip but permitting a certain amount of "rollback," where the surplus of waste above a certain depth can fall back down the belt until an acceptable level has been obtained.

That function is not achieved simply by welding a plate across the face of the flight and at a right angle to it, since if that were done then sheet materials such as cardboard would simply bridge the space between the lifters, allowing them to pass underneath without any effect upon the waste. Good practice holds that the most effective lifters are in the form of "teeth," welded out of line to successive flights in such a way that they form a series of diagonals across the belt if viewed across their tips (see Figure 3.12).

The vertical projection of the lifters is determined by the angle of inclination of the belt: the steeper the angle, the greater the vertical projection, and here again experience affords a dimension that defies calculation. Since the slip angle of waste is 27°, any angle less than that will not need lifters at all. Meanwhile an inclination of greater than about 50° imposes rapidly increasing loadings upon the conveyor frame at the changes in angle. For most purposes, a convenient inclination is 45°, and in that case lifters 8 in. (200 mm) high, 6 in. (150 mm) wide, and $^3/_8$ in. (8 mm) thick are suitable, pitched as described across successive flights.

On the horizontal sections, the weight of the lifters plays a part in the design calculations, since power is required to move them both upon the loaded and the return strands. On the inclined section, however, their weight is less relevant. There the weight of the loaded strand as a whole is to some extent balanced by that of the return strand. While the drive unit pulls the former, gravity pulls the latter in the opposite direction. The only motive power required is that to move to load and to overcome friction.

Before any calculations of the conveyor specifications can be begun, some further consideration must be given to the construction of the machine. The construction will considerably affect the weight of the moving parts, and therefore the power required to move them. Since the most common plate feeder conveyors consist of steel flights bolted to conveyor chain, we shall restrict this part of the design outline to them.

3.1.20.2 Conveyor Chain Types

A standard conveyor chain is made up of cast iron or cast steel rollers, connected by steel plates in such a way as to form a continuous strand (see Figure 3.13). The rollers may be hollow-bushed or solid-bushed, meaning that the shaft upon which the roller rotates may be either hollow or solid. The former are considerably lighter in weight, and they permit attachments to be bolted directly through the roller. The latter are heavier, but much stronger, and they impose additional weight indirectly in that the linking plates must have attachments that permit the fastening of the flights.

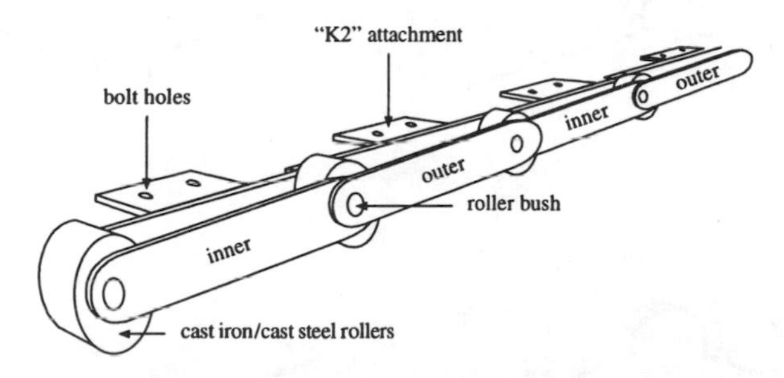

Figure 3.13 Standard conveyor chain with K2 attachments.

Hollow-bushed rollers offer a further advantage in that as the attachments are fixed at the center of the rollers, their turning circle radius at the sprockets is that of the pitch circle of the sprockets themselves. Flights assembled upon the chain in this way do not open out as they change direction, and spillage is therefore easier to control. Where solid-bushed rollers are used, the flights must be fixed to the linking plates, usually by K2 attachments — a right-angled plate projecting from one side, drilled with two mounting bolt holes. The "K2" attachment is welded to the top edge of the link plate, and is therefore higher than the center of the roller. As a result, the turning radius of the flights at the sprockets will be greater than the outside radius of the sprockets, and the flights will move apart from each other as they change direction, leaving a gap through which material may fall. This problem is generally overcome by forming the edges of the flights in such a way that they overlap on the straight sections, and slide against each other over the sprockets (see Figure 3.12).

Forming the edges of the flights for solid-bushed chain has the added disadvantage that the length of the flight is greater than the pitch of the chain, and when that extra metal is added to every pitch on a long belt, it can substantially increase the weight of the whole strand. Furthermore, the nature of the formed overlap must be carefully designed, otherwise the belt will only be able to change direction in one plane (see Figure 3.14).

Waste handling is an arduous duty, and in spite of their attractions, hollow-bushed rollers are of insufficient mechanical strength for most applications. The low bulk density of municipal waste implies that relatively low power is needed to move it, and that as a result the tension upon the conveyor chains should be minimal. However, the material is a mixture of random items, some of which may be bulky and likely to jam against structural framing or conveyor side plates. In those cases, the conveyor must have sufficient power and sufficient chain strength to withstand the significant overloads that may result. Hollow-bushed chain should be reserved for light conveyors, and solid-bushed chain should be used for waste handling.

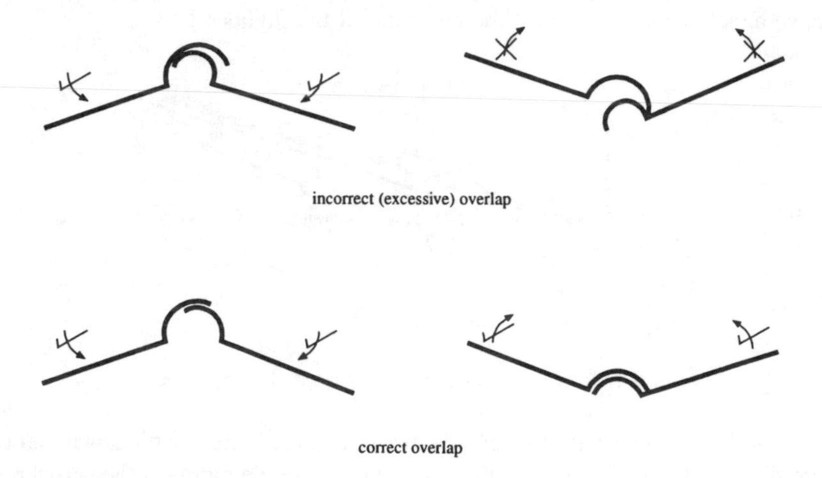

Figure 3.14 Effects of excessive flight overlap.

3.1.20.3 Sealing

When the flights have been edge-formed to provide sealing with solid-bushed carrying chain, they no longer present a flat top surface. Conventional side skirts, of the type used upon rubber belts to prevent side spillage, are ineffective, and side flange plates must be used instead. In such a design, advantage is taken of the relationship of the link plates of the conveyor chain, where the links run in a series of "inners" and "outers" (Figure 3.13). Since the K2 attachments are welded to each link, it follows that the attachments on inner links must be further from the longitudinal center line of the conveyor than are those on the outers. Therefore the flight plates on inner links must be longer than those on outers, by a distance equal to twice the thickness of the link plates. If both types of flights are flanged on their ends, as shown in Figure 3.12, then the flanges of the outers will slide freely inside those of the inners, creating effectively a troughed steel belt. The fixed steel side plating of the conveyor structure can then protrude down below the flange top edge and create an effective seal against spillage.

3.1.20.4 Drive Systems

Having established a general design principle for the reception conveyor, consideration must be given to the location of the drive unit and the configuration of the carrying sprockets. In conveyor terminology, the loading end of the conveyor is always the "tail," and the discharge end is always the "head." Drive units should always be placed at the head end, where they will be pulling the load rather than pushing it as they would if placed at the tail. Pushing is not impossible, but it does make the conveyor framing much more complex, and it certainly makes structural damage in the event of blockages far more likely.

Sprockets for chain-mounted plate feeder conveyors are almost always manufactured from close-grained gray cast iron. Cast steel is available, but not usually necessary. The sprockets must be specified for the type of chain roller to be used, since flanged rollers require a special relieved sprocket tooth. At the tail end of the conveyor, only one of the sprocket pair should be keyed or otherwise fixed to the shaft, in order that the conveyor chain strands may align themselves in the event of unequal stretching.

The choice of the number of teeth on the sprockets is something of a compromise, since if there are too few teeth, the conveyor movement will pulsate due to the polygonal effect. If there are too many, it may become difficult to maintain sufficient engagement with the chain, and the overall depth of the conveyor may become excessive for the space available. For most waste handling operations, 8 teeth is the minimum, and 12 is best.

In the solid-bushed chain plate feeder conveyor, the flight plates will be bolted at their ends to the K2 attachments of two parallel strands of chain, and so not only must the chains be of identical specification but they should also be matched by the manufacturer. Normally the engineering buyer would specify two matched strands of chain, of a stated breaking strain, with steel or cast iron rollers, plain or flanged, solid-bushed, and with K2 attachments one side only on each chain and every link.

The nature of the motive power unit must also be taken into account, even though its rating is not yet known. For waste reception service, the conveyor must be capable of controlled, variable speed. There will be some occasions when lower infeed rates are required, and others where some degree of overfeeding may be necessary. Depending upon the nature of the downstream processes, it may also be necessary to interconnect the controls of the reception conveyors to load sensors on other machines in order to maintain a consistent feed rate.

Plate feeder conveyors operate at very low speeds in mechanical handling terms, but demand a high-torque drive. Variable speed can be delivered by either a mechanical gearbox or by an electrical control system, the former by changing gear ratios, and the latter by varying the alternating current frequency.

Mechanical variable-speed drives are somewhat cumbersome and expensive, and remote control is more complex, but they offer sustained power output over the full speed range and are extremely robust. Electrical controls lend themselves to remote operation and to load sensing, but at lower speed ranges some types are inclined to permit stalling without any indication that they have done so.

The arduous nature of waste reception operations makes mechanical variable-speed drives an option worthy of consideration, and it is the need to link the reception drive to load sensors on downstream machines that makes them attractive in spite of their relative complexity. A reception conveyor that is, for example, related to a granulator, may upon occasion need to reduce speed to as low as 10% of full load. Some particularly difficult material may have been delivered to the granulator, resulting in a temporary reduction in its milling capacity and an associated increase in load current. The three-phase induction motor driving the reception conveyor would normally operate at 1440 rpm, but the conveyor head shaft may only be rotating at 3 rpm even at full load. Therefore, a speed reduction to 10% would imply gearing down from 1440 rpm to 0.3 rpm, which is very close to stalling point. An electrical speed reducer seeks to achieve it by slowing down the drive motor, possibly to half its normal speed in such a case, even allowing for an intermediate gearbox. Where a variable-speed gearbox is used, the drive motor continues to operate at full speed, and therefore full torque, all of the time. It is the gearbox that changes the output speed, and the electric motor never comes close to stalling.

It is the nature of conventional motor running indicators which makes that phenomenon important. Most control systems indicate that a machine is running simply by detecting the current flow to the low-voltage electric motor contactor solenoids in the motor control center. Such systems adopt the logic that if the contactor is closed then the motor must be running, and with a variable-speed gearbox drive that must indeed be the case. If, however, the motor speed has been reduced to zero by changing the supply frequency of the supply current, then the contactor will remain closed and the plant control system will continue to indicate normal running.

In such an event, the first indication of a problem is likely to be a complete cessation of processing. As far as the control system is concerned, every machine is running, but no waste is passing through. When it is discovered that the reception conveyor is in fact stationary, the most obvious action is to switch it to manual control and restart it at full torque. That may be an acceptable solution if the shovel-loader or crane operator has also recognized the failure immediately, but if he has not, the loading end of the conveyor may now have a significant overload upon it. In that case a full-torque restart will simply pass the overload right through the whole process line, and may easily lead to catastrophic blockages downstream.

More sophisticated electrical speed controllers are designed with frequency jogging to permit inching operations, although it is open to question for how long the drive motor will tolerate that, since it is not intended to be a continuous process lasting h. Some have additional built-in stall detection and prevention, but again that is useful as an occasional expedient rather than for continuous operation.

3.1.20.5 Design Calculations

At this point in the design route, the type of conveyor and some detail of its design has been established. A mathematical model must now be constructed to demonstrate that the design chosen is adequate, and for that to be done an assumption must be made. In order that the designer may obtain motive power requirements, from which he may derive chain tensions, he must know the weight of the moving parts. Yet arguably, he cannot know the weight of the moving parts until he has calculated the chain specification necessary to support the motive power. The only solution to this apparent impasse is to assume a chain specification based upon experience, and to show by mathematical models that it is adequate.

The design route which establishes the conveyor specification within the variables is now as follows.

3.1.20.5.1 Capacity Calculation — Before anything can be done about the final design of the conveyor, a handling capacity must be established, and the first step toward this is to consider how many flowlines the process plant is to contain. There will need to be one feeding conveyor for each line, and so the capacity in terms of tons per h will be the total plant capacity divided by the number of lines. At this point it is a wise precaution to increase the calculated capacity by at least 10% to accommodate irregular feeding or increases in waste arisings over the life of the plant.

Having obtained the conveyor capacity in tons per h, the volume to be handled must be established. Conveyors are both volume and mass moving machines, where the volume defines their dimensions and the mass defines the power rating. The volumetric capacity required is arrived at by dividing the mass capacity by the bulk density of the waste; so, for example, an infeed rate of 37.5 t/h at a bulk density of 9.36 lb/ft^3 (150 kg/m^3) in two lines, equates to a volume of 4064.1 ft^3 (115.1 m^3) per h per machine. Increasing this by 10% then yields a design figure of 4470.5 ft^3 (126.6 m^3) per h.

By itself, that figure is somewhat meaningless, since it provides no information on the dimensions of the conveyor. Therefore, the next stage is to determine a realistic depth of material upon the plates, consistent with sufficient width to receive the output of the shovel loader or grab crane.

At the loading end of the conveyor, waste will inevitably be heaped to some degree as a result of the behavior of the loading machinery, and will be intermittent. The inclined section will then level the waste as a result of rollback over the lifters, and that level will be retained on the final discharge section. Since all three sections are part of the one conveyor, it is the bed depth on the inclined section which establishes the running speed, and that bed depth is somewhat arbitrary. With an inclination of 45° and an 6 ft (1800 mm) wide belt, the bed depth is likely to be about 12 in. (300 mm), and this is a reasonably sound figure upon which to design throughout the whole conveyor.

3.1.20.5.2 Belt Speed — At this point, the design route has led to a conveyor 6 ft wide, carrying 4470.5 ft³/h waste at a bed depth of 12 in. The belt speed necessary to achieve this rate is then the result of the volumetric capacity divided by the volume carried per unit of length, that is

$$\begin{aligned}\text{Belt speed} &= \frac{\text{Volume per hour}}{(\text{width} \times \text{bed depth} \times 1)} \\ &= 4470.5/(6 \times 1 \times 1) \text{ ft/h} \qquad (5) \\ &= 745.1 \text{ ft/h } (227.1 \text{ m/h}).\end{aligned}$$

Note that the volumetric capacity is independent of the total length of the conveyor. Conveyors must necessarily discharge at the same rate as they are loaded, and so every foot of the belt must carry the total volume in 1 h.

3.1.20.5.3 Chain Tension — Having established the belt speed, the resulting chain tension must be calculated. This is the sum of tension calculations for each section, each comprising that required to move the top strand, to move the return strand, to overcome frictional resistance, and to move the load. It is usually expressed initially as the "pull" upon the conveyor chain, a figure which can then be translated into a torque at the drive gearbox, and then into horsepower or any other convenient unit of measurement.

Each calculation is then as follows, where

W_m = weight of material handled per unit length of conveyor
W_c = weight per unit length of conveyor components
V = vertical projection of center of head sprocket (or center of radius of change of direction)
L = horizontal projection of center of head sprocket (or center of radius of change of direction)
L_c = length of conveyor between centers
f = coefficient of friction expressed as a percentage of total load carried (1.2)
p_L, p_1, etc. = pull on chain for each component of load
PT = total pull on chain
S = belt speed

Horizontal sections:

$$PT = fL_c(W_m + 2W_c) \qquad (6)$$

Inclined section:

$$p(\text{Load}) = W_m V + W_m fL$$

$$p(\text{Conveyor}) = W_c(V + fL) + W_c(fL - V) \quad (7)$$

$$PT = p(\text{Load}) + p(\text{Conveyor})$$

Conveyor chain manufacturers can provide tables for friction coefficient f, based upon the diameters of the rollers selected, and specified as "from rest" and "in motion." However, waste handling is not such a precise undertaking that such strict accuracy is necessary, and therefore an f of 20% is usually sufficient.

Once the total pull on each strand of chain has been established (PT divided by the number of strands, two in the example), a suitable chain may be selected allowing a safety factor of six to ten against manufacturers listed breaking strains. If the result of the selection implies a heavier chain than originally assumed, then the calculations must be carried out again using the component weights indicated.

3.1.20.5.4 Motive Power — Before calculating the motive power required to drive the conveyor, the total pull upon both chains must be increased to take account of various friction losses in the system. It is customary to increase the calculated chain pull by 10% for head shaft friction, and by 5% for tail shaft friction. A further 5% should be added for each belt or roller chain drive, or gear reduction, and 10% for each cast tooth gear reduction. Another 10% should be added for each change of direction. Finally, because variable speed control is always necessary on reception conveyors to meter the waste into the process lines, 10% should be added to the chain pull to account for the variable-speed drive unit. All of these percentages should be compounded.

When a final total chain pull has been established, one further factor must be considered. So far the calculation has concentrated upon the friction losses within the conveyor alone, but there will be some element of friction caused by the movement of the waste against the conveyor side plating. There can be no arbitrary figure for this, since it is dependent upon the depth of the loading hopper and the height of the plating. However, assuming that the 300 mm (12 in.) material depth is accepted, then a further 20% should be added to the chain pull to accommodate load friction.

Having derived both the conveyor speed and the total chain pull, the motive power requirement can be calculated from the formula

$$\text{Power} = \frac{\text{Chain pull} \times \text{speed}}{\text{Unit of power}} \quad (8)$$

In this case, the power may be metric or imperial horsepower, with chain pull expressed in kilograms or pounds as appropriate. Speed must then be in meters per second or feet per second. The "unit of power" is the standard equivalent to one unit of power, i.e., 75 meter-kilograms per second for metric horsepower or 550 foot-pounds per second for imperial.

The result of the calculation may be converted to the more convenient terms of electrical power demand in kilowatts by dividing it by 1.36 if metric units are used, or by 1.341 for imperial units.

Worked Example

Applying the formulas above to the example conveyor discussed, the design criteria may be tabulated as follows.

Known:

Number of conveyors	2	
Load/conveyor	18.75 t/h	17 te/h
Belt speed	0.206 ft/s	0.063 m/s
Belt width	6.0 ft	(1.8 m)
Cubic capacity	4064.1 ft³/h	115 m³/h
Product density	9.36 lb/ft³	150 kg/m³
Load per unit length	50.33 lb/ft	73.4 kg/m

Assumed:

Loading length	20 ft	6 m
Inclined section		
vertical projection	20 ft	6 m
horizontal projection	20 ft	6 m
Inclination angle	45°	
Discharge length	10 ft	3.05 m
Weight of flights	66.15 lb	30 kg each
Flight width (av)	6 ft	1.8 m
Flight length	12 in.	305 mm inc. overlap
Flight thickness	1/4 in.	6 mm
Flight wt/unit length	67 lb/ft	99.7 kg/m
Chain breaking strain	60,000 lb	27,200 kg
Chain pitch	8 in. standard	203.2 mm
Chain wt/unit length	26 lb/ft	38.68 kg/m
Belt wt/unit length	92 lb/ft	138.68 kg/m

Chain pull

Chain Pull — Loading Section:

$$P_1 = fL_c(W_m + 2W_c)$$

$$= 1.2 \times 20 \times (50.33 + (2 \times 92)) \text{ lbs}$$

$$= 1.2 \times 6 \times (73.4 + (2 \times 138.68)) \text{ kg}$$

$$= 5623.9 \text{ lb}$$

$$= 2525.47 \text{ kg}$$

Chain Pull — Inclined Section:

$$P_1 = W_m V + W_m fL + W_c(V + fL) + W_c(fL - V)$$

$$= 1006.6 + 1207.92 + 4048 + 368 \text{ lb}$$

$$= 456.6 + 547.9 + 1836.2 + 166.9 \text{ kg}$$

$$= 6630.52 \text{ lb}$$

$$= 3007.6 \text{ kg}$$

Chain Pull — Discharge Section:

$$P_d = fL_c(W_m + 2W_c)$$

$$= 1.2 \times 10 \times (50.33 + (2 \times 92)) \text{ lb}$$

$$= 1.2 \times 3.05 \times (73.4 + (2 \times 138.68)) \text{ kg}$$

$$= 2811.96 \text{ lb}$$

$$= 1283.8 \text{ kg}$$

Total Chain Pull

$$PT = (1) + (2) + (3)$$

$$= 15066.38 \text{ lb}$$

$$= 6816.87 \text{ kg}$$

Factor of Safety

$$\text{(11) Factor of safety} = \frac{\text{Chain breaking strain (two chains)}}{\text{Total chain pull}}$$

$$= \frac{120{,}000}{15{,}066.38}$$

$$= 7.96$$

A factor of safety of 7.96 is between the accepted limits of 6 to 10, and is therefore suitable.

Power Requirement

Incremental additions to total chain pull —10% for head shaft, 5% for tail shaft, 5% for roller chain drive, 5% for gear reduction unit, 20% for two direction changes, 10% for variable speed drive, 20% for load friction.

$$\text{Total chain pull} = 15{,}066.38 \times \text{compounded incremental additions}$$

$$PT = 30{,}389.58 \text{ lb } (13{,}701.34 \text{ kg})$$

(12) Horsepower

$$= \frac{PT \times S}{550} \text{ hp} \qquad = \frac{PT \times S}{75} \text{ metric hp (mhp)}$$

$$= \frac{30{,}389.58 \times 0.21}{550} \text{ hp} \qquad = \frac{13{,}701.34 \times 0.063}{75} \text{ mhp}$$

$$= 11.6 \text{ hp} \qquad = 11.51 \text{ mhp}$$

$$= 8.46 \text{ kW}$$

3.1.20.5.5 Design Conclusions — The initial assumptions have now been applied in a mathematical model of a reception conveyor, and together they have specified a machine with a satisfactory chain tension safety factor, capable of handling waste at the required rate at 90% full load power. Chain stretch will be minimal to nonexistent, and the machine will be large, slow moving, and robust. Since its drive unit is designed to operate at near full-load current, it will be electrically efficient.

The conveyor will present a surface area of at least 120 ft^2 (11 m^2) for loading, and will elevate the waste to a height of 20 ft (6 m) while leveling it to approximately 12 in. (300 mm) and producing a relatively consistent infeed to the process lines. Its power consumption may initially seem surprisingly low, but that is the result of low-speed operation and is a characteristic of plate-feeder conveyors. They are generally the largest machines in a process plant in terms of physical size, but among the lowest power consumers.

3.2 HANDLING SYSTEMS FOR VARIOUS MATERIALS

Waste is not a material, even though for convenience it is often referred to as such. It is a mixture of entirely dissimilar materials, each with its own handling characteristics. As a result, a system that is well-suited to handling crude waste may be totally unsuitable for any of the individual fractions. Since all waste recycling and treatment processes begin by separating some or all of the fractions, at some stage in the process the different characteristics become significant. The purpose of the following section is to consider mixed municipal waste, and then to examine each of the fractions within it in terms of mechanical handling.

Section 3.1 defined mechanical handling as the moving of materials from one place to another in a continuous operation. The technology is therefore about conveyors, of which there are very many types indeed. There is, however, a much smaller number of basic designs, upon which all of the derivatives are based, and there are even fewer that can be used in bulk loose material processes. For waste treatment purposes, it is sufficient to consider the five basic designs and 13 appropriate types which conform to them, as listed in Table 3.2.

Table 3.2 Available Conveying Systems

1. Carrying Conveyors
   - Flat/troughed rubber
   - Flat/troughed synthetic rubber
   - Flat/troughed polyurethane
   - Steel plate
2. Dragging Conveyors
   - Drag links
   - Chain conveyors
3. Pushing Conveyors
   - Screw feeders
   - Moving floors
4. Lifting/Elevating Conveyors
   - Sidewall belts
   - Bucket elevators
   - Folding belts
5. Pneumatic conveyors
   - Lean phase — suction systems
   - Dense phase — pressure systems

Before exploring how waste and its various fractions behave upon each conveyor type, it is necessary to define what each type actually is, in order to avoid any confusion with those used in other industries. The following descriptions are not intended to be over-detailed, since the conveyor systems appropriate to each type of process are defined in the relevant sections.

In the power and capacity formulas for each type, the following symbols are used:

W_m = weight of material carried per unit length in kg or lb
W_c = weight per unit length of conveyor components, including all rollers or idlers turned by the strands, in kg or lb
V = vertical projection of conveyor length in m or ft
L = horizontal projection of conveyor length in m or ft
L_c = conveyor length between pulley or sprocket centers in m or ft
f = coefficient of friction of chain on guides
p = pull on chain or belt in kg or lb for each component of load

PT = total pull on chain in kg or lb
S = running speed of belt in m/min or ft/min
C = carrying capacity in kg/min or lb/min
D = material bulk density (apparent) in kg/m^3 or lb/ft^3

3.2.1 Carrying Conveyors

3.2.1.1 Flat/Troughed Rubber

Conveyors using natural rubber are very rarely used any more, and are unlikely to be readily available to the designers of a modern process facility. They are listed here rather as a warning than as an example of suitable devices, since natural rubber should be excluded from all waste treatment process. It deteriorates rapidly in contact with animal, vegetable, and mineral oils and fats, and conveyors made from it blister and develop ply separation within months. Rubber conveyors are listed here because they may still be encountered in some installations, but they should not be used in the design of municipal waste facilities. For that reason, they are not evaluated with respect to individual fractions later in this section.

3.2.1.2 Flat/Troughed Synthetic Rubber

The synthetic compounds most commonly used for conveyor belting are nitrile or butyl rubber, bonded to nylon or terylene fabric. Belts can be designed to almost any length, and there is a wide range of thicknesses. The technical terms used to define the belting are as follows:

1. Weave and weft — The fabric reinforcing to which the synthetic rubber compound is bonded.
2. Ply — The number of layers of synthetic material interspersed with fabric reinforcing.
3. Top and bottom ply — The unreinforced thickness of synthetic material above and below the fabric.
4. Carcass — The combination of fabric and top and bottom ply.
5. Carrying strand — The layer of the belt that carries load; usually the top.
6. Return strand — The unloaded, usually bottom, loop of the conveyor.

A flat belt conveyor is one where the carrying strand lays flat across a series of idler or carrying rollers, presenting a level surface across its width. The rollers support the belt and are placed in series sufficiently close together to prevent sagging. Side skirts are necessary to prevent spillage of loose materials.

In a troughed belt, separate inclined carrier rollers mounted outboard of the flat rollers cause the belt to form a concave transport surface or trough, while the return strand remains flat. Troughed belts retain loose materials better, but may still need side skirts when fully loaded. Troughing angles vary between 20° and 45° depending upon the belt width and the lump size of the load carricd, and most belting manufacturers publish tables for a wide range of combinations.

Both troughed and flat belts can be made to change direction from horizontal to inclined and vice versa, although troughing can only be maintained in the former. Whenever such a change in direction is planned, the radius of curvature of the belt must be such that at no time does it lift clear of the troughing idlers. The minimum radius for that purpose is given by the formula

$$R = WP/(3.278\ B_w)\ \text{ft}$$

$$R = WP/(9.8\ B_w)\ \text{m} \qquad (9)$$

where R is the minimum radius, W is the belt width in mm or feet, P is the belt pull or tension in kN or lbf, and B_w is the belt weight in kg/m or lb/ft run.

Thc same calculation applies to a loaded belt, in which case the weight per m of the load is added to that of the belt. Irrespective of the result of the calculation, the radius of curvature should not be less than 148 ft (45 m), with the result that in many cases it is simpler and less space-consuming to install two belts, one horizontal and one inclined.

Where a belt is to be inclined downward from the horizontal, or to the horizontal from an incline, the minimum radius becomes the minimum bending radius of the belting, since belt tension holds it down upon the idlers. However, the change of direction may also depend to some extent upon the lump size of the load and thc speed of the belt, since very abrupt changes may causc the load to lift off.

3.2.1.3 Flat/Troughed Polyurethane

The polyurethane conveyor belt is chemicals-resistant and totally synthetic, with no rubber in the mix. It is generally both stronger and of thinner carcass than synthetic rubber belts, and much less prone to stretching. The carcass materials are, however, more sensitive to temperature and entirely plastic belting is not widely used in waste processing facilities.

The design parameters for flat and troughed conveyors are calculated as follows:

Weight of Material Per Unit Length —

$$W_m = \text{Belt width} \times \text{depth of material} \times D\ \text{kg (lb)} \qquad (10)$$

Capacity —

$$C = W_m \times S\ \text{kg/min (lb/min)} \qquad (11)$$

Belt Pull —

$$PT = W_m + W_m fL + W_c(V + fL) + W_c(fL - V) \text{ kg (lb)} \quad (12)$$

$$f = 0.045$$

Drive Power (Metric) —

$$\text{Drive power (kW)} = (S \times PT)/6120 \quad (13)$$

$$\text{Drive power (mhp)} = (S \times PT)/4500 \quad (14)$$

where S is measured in m/min and P is measured in kg.

Drive Power (Imperial) —

$$\text{Drive power (kW)} = (S \times PT)/44{,}280 \quad (15)$$

$$\text{Drive power (hp)} = (S \times PT)/33{,}000 \quad (16)$$

where S is measured in ft/min and PT in lb.

3.2.1.4 Steel Plate

This type of conveyor is described in detail in Section 3.1. It is formed from a series of interlinked steel plates, which may be mounted upon a conveyor chain to make a continuous, flexible strand. Steel plate conveyors are widely used for waste disposal, since they are extremely robust and have sufficient power to overcome most blockage problems.

3.2.2 Dragging Conveyors

3.2.2.1 Drag Links

Drag link conveyors are generally but not exclusively made up of two conveyor chains that are spanned at intervals by steel "flights" — bars fixed to the chain links vertically to the longitudinal axis. In operation, the flights slide in a steel, square-section casing, with the chains running in tracks inset into the sides. The flights drag material along inside the casing, and the whole conveyor is generally fully enclosed. The return strand runs inside the same casing, but is separated from the load by a full-length division plate.

Drag links are more prone to wear than belt conveyors because the flights run on or very close to the casing, and because the load slides upon it. The more abrasive the load, the more wear the casing experiences. Maintenance costs are likely to be high, but changes of angle from horizontal to inclined and back again are possible. The conveyors are not limited by the maximum angle of slip of the load.

One very significant attraction of drag links that almost guarantees their use in many waste processes, is their ability to carry loads upon both strands and in opposite directions. As such, they do not have a carrying and a return strand in the conventional sense, since either or both may be used.

3.2.2.2 Chain Conveyors

A chain conveyor is very similar to a drag link in many ways, in that the design uses flights attached to conveyor chain. However, in this type of conveyor, the flights are generally steel bars, formed into any one of a number of shapes, and fixed to a single chain that is usually forged or cast steel. The chain has no roller wheels, but runs instead upon a wear rail along the longitudinal center of the casing. The machines normally run either full or nearly so, depending upon the angle of incline. Vertical conveying is possible, but chain conveyors, having been designed for grain handling originally, are somewhat inclined to blocking if used with fibrous materials.

The design parameters for dragging conveyors are calculated from the following.

Weight of Material Per Unit Length —

$$W_m = F_c \times F_n \times D \text{ kg (lb)} \quad (18)$$

where F_c is the level volumetric capacity between flights and F_n is the number of flights per unit length.

Capacity —

$$C = W_m \times S \text{ kg/min (lb/min)} \quad (19)$$

Belt Pull —

$$PT = L_c(W_m + 2W_c f) + L_{c2f} s \text{ kg (lb)}$$

$$f = 0.045 \quad (20)$$

f_s = coefficient of sliding friction of load on trough (see Table 3.3). May be ignored for material depths of less than 150 mm (6 in.).

Drive Power (Metric) —

$$\text{Drive power (kW)} = (S \times P)/6120 \quad (21)$$

Table 3.3 Coefficients of Sliding Friction for Various Materials

Material	Coefficient "F"	Coefficient "G"
Mixed waste	1.6	0.025
Paper/card	1.4	0.044
Plastics	1.2	0.036
Mixed organics	1.8	0.160
Glass	1.6	0.140
Metals	1.6	0.090
Compost	1.9	0.170
Refuse-derived fuel	1.3	0.038

$$\text{Drive power (mhp)} = (S \times P)/4500 \qquad (22)$$

where S is measured m/min and PT in kg.

Drive Power (Imperial) —

$$\text{Drive power (kW)} = (S \times PT)/44{,}280 \qquad (23)$$

$$\text{Drive power (hp)} = (S \times PT)/33{,}000 \qquad (24)$$

where S measured is ft/min and PT in lb.

3.2.3 Pushing Conveyors

3.2.3.1 Screw Feeders

In a screw feeder conveyor, steel plate is fixed to a central shaft in such a way as to form a helix, rather like a coarse screw thread. The rotating element runs inside a troughed, "U"-shaped casing, and the direction of rotation determines the direction of travel of the load.

The length of a screw feeder is limited by the torsional strength of the center shaft, but longer lengths require hanger bearings to prevent the shaft from also sagging against the trough. Hanger bearings are acceptable with granular, free-flowing materials, but are likely to initiate blockages if the load is wet or fibrous.

Screw profiles are manufactured to a considerable number of designs in addition to the basic design described, some of which are as follows:

Double-helical — two "threads" on one shaft, used where a smoother conveying action is required. Granularity is important, since this type of conveyor cannot accommodate larger lump sizes. Not generally used in waste processing applications.

Double-ended — two "threads" of opposed hand meeting at the longitudinal center of the shaft and used for feeding two ways from the center outward.

Flat-bladed — similar to a series of propeller blades mounted helically along the shaft, used where a degree of mixing is required in addition to conveying. Not appropriate for waste processing.

Ribbon — a helical band standing clear of the shaft on spokes, and most commonly used for mixing materials. Not widely used in waste processing.

Tapered screw — where the outside diameter of the helix reduces or increases toward the discharge end. Used mainly for hopper unloading or for loading from two or more separate points. The carrying capacity varies in direct proportion to the diameter.

Variable-pitch screw — of similar purpose to tapered screws. In this case, the pitch of the helix reduces or increases progressively toward the discharge end.

The design of screw conveyors is extremely sensitive to the nature of the materials to be handled, and is always something of a compromise. It must seek to accommodate abrasiveness, granularity, flow characteristics, wetness, and stickiness, and as a result of those factors and the number of designs available, there is no simple way of calculating capacities and power demands without reference to a specific material.

In principle, the power demand for a standard screw conveyor is expressed as

$$\text{Power (in kW)} = CL_c(K_a + DK_b)/6120 \quad (25)$$

where K_a and K_b are constants that relate to the material, density, handling characteristics, and screw diameter. Most screw conveyor manufacturers publish tables, based upon experience, that provide figures for both constants.

3.2.3.2 Moving Floors

A moving floor conveyor usually consists of a number of segments, each of which is an acute, right-angled triangle in cross section. The segments are connected together, with wide gaps between, by longitudinal bars. Hydraulic cylinders cause the assembly to reciprocate, and load transfer is achieved by pushing the material forward with the flat vertical face of each segment. As the assembly moves through its return stroke, the wedge-shaped end of each segment penetrates to load, filling the voids left between the segments by the forward stroke. The action is similar to that of a ratchet.

Once again, there is no simple way of calculating capacities or power requirements, since the designs are sensitive to material characteristics. Moving floor systems are not widely used in waste processing applications.

3.2.4 Lifting/Elevating Conveyors

3.2.4.1 Sidewall Belts

A sidewall belt conveyor is effectively a flat, synthetic rubber belt across which a series of vertical, flat rubber plates (flights) have been vulcanized. Concertina side walls are then vulcanized along the sides, usually to the full height of the flights. The load upon such a conveyor is constrained between the flights and the side walls, and so steep angles of inclination are possible. Spillage is eliminated by the side walls.

The concertina shape of the side walls permits them to change shape as the belt passes over the head and tail drums, and it also permits changes of direction in the vertical plane along the length. Such changes of direction may be either of increasing or decreasing inclination, or even both along a single conveyor.

The minimum radius of curvature for a directional change in a sidewall belt can be as little as one tenth of that for a conventional flat or troughed conveyor. Since the load upon a sidewall is constrained between the walls and the flights, it is customary for a side strip outboard of the walls to remain clear of material. Tracking wheels, mounted in curved frames and bearing upon the side strips, create radii that are constrained only by the bending radius of the belting and the flexibility of the sidewalls.

The design parameters for a sidewall conveyor are calculated from the following.

Weight of Material Per Unit Length —

$$W_m = F_c \times F_n \times D \text{ kg (lb)} \tag{26}$$

where F_c is the level volumetric capacity between flights, and F_n is the number of flights per unit length.

Capacity —

$$C = W_m \times S \text{ kg/min (lb/min)} \tag{27}$$

Belt Pull —

$$PT = W_m + W_m fL + W_c(V + fL) + W_c(fL - V) \text{ kg (lb)} \tag{28}$$

$$f = 0.1$$

Drive Power (Metric) —

$$\text{Drive power (kW)} = (S \times P)/6120 \tag{29}$$

$$\text{Drive power (mhp)} = (S \times P)/4500 \tag{30}$$

where S is measured in m/min and PT in kg.

Drive Power (Imperial) —

$$\text{Drive power (kW)} = (S \times PT) / 44{,}280 \qquad (31)$$

$$\text{Drive power (hp)} = (S \times PT) / 33{,}000 \qquad (32)$$

where S is measured in ft/min and PT in lb.

3.2.4.2 Bucket Elevators

A bucket elevator, as the name implies, consists of a series of containers or buckets mounted upon an endless belt or chain between pulleys or sprockets, and having a means of loading each bucket in turn. They are most commonly used for elevating materials, often vertically. Changes of direction are possible with some, but not all, bucket elevator designs.

Containment of spillage is an ever-present problem with bucket elevators, particularly with low-bulk-density materials, since it is difficult to match the feeding or transfer arrangements to individual buckets. Some designs in fact do not attempt to do so. Instead, they rely upon the bottom casing of the conveyor being partly filled, and the buckets being required to dig out material as they pass over the tail sprockets. Such a principle, however, leads to the chains and the sprockets being immersed in the product to be handled. Given a granular, free-flowing material (grain, etc.), that design may not be particularly detrimental, but it is unlikely to work with fibrous, sticky materials of irregular granularity.

An alternative approach is employed in "continuous" elevators. Here the buckets are fitted to the chains or belts in such a manner that one bucket forms the inlet (and discharge) chute for that immediately following it. Loading usually takes place at a point where the buckets have started to ascend. Continuous elevators run at speeds between 82 and 130 ft/min (25 to 40 m/min) where the buckets are mounted upon chain, and 100 to 260 ft/min (30 to 80 m/min) where they are mounted upon belting.

Discharging the machine may also be difficult if the material handled is wet or sticky, since it may tend to hang up in the buckets. The problem can be alleviated to some extent by the bucket shape and construction material, and compressed air cleaning is sometimes used successfully. Some designs employ the principle of centrifugal discharge, where the conveyor runs at high speed such that the centrifugal force generated by passing the buckets over the head sprocket or drum throws the material out. Alternatively, a positive discharge system may be used where the buckets are caused to invert completely at the discharge. This is achieved by passing the chain over "snubbing" sprockets which loop the chain inward toward the longitudinal center of the machine.

The choice of chain or belting as the mounting and driving medium for a bucket elevator depends upon the load to be carried and the speed of operation. Chain is suitable for high loads but low speeds, while conveyor belting is appropriate for higher speeds and lighter loadings. Some designs, however, limit the choice further. Centrifugal discharge conveyors, for example, rely upon speed to throw the conveyed material clear, and so they must be assembled using belting. Excavating machines, where the buckets dig material from the bottom of the casing, need power and positive drive at that point. In those designs chain is the only choice.

The design parameters for a bucket elevator conveyor are calculated from the following.

Weight of Material Per Unit Length —

$$W_m = B_c \times B_n \times D \text{ kg (lb)} \tag{33}$$

where B_c is the level volumetric capacity of one bucket, and B_n is the number of buckets per unit length.

Capacity —

$$C = W_m \times S \text{ kg/min (lb/min)} \tag{34}$$

Chain Pull —

$$PT = L_c(W_m + W_c) \text{ kg (lb) (vertical conveyors)} \tag{35}$$

$$PT = W_m + W_m fL + W_c(V + fL) + W_c(fL - V) \text{ kg (lb) (inclined conveyors)} \tag{36}$$

$$f = 0.2$$

Drive Power (Metric) —

$$\text{Drive power (kW)} = (S \times PT)/6120 \tag{37}$$

$$\text{Drive power (mhp)} = (S \times PT)/4500 \tag{38}$$

where S is measured in m/min and PT in kg.

Drive Power (Imperial) —

$$\text{Drive power (kW)} = (S \times PT)/44{,}280 \tag{39}$$

$$\text{Drive power (hp)} = (S \times PT)/33{,}000 \quad (40)$$

where S is measured in ft/min and PT in lb.

3.2.4.3 Folding Belts

A folding belt conveyor is basically a belt conveyor that is so heavily troughed that the two sides meet to form a closed tube along part or all of the length. Generally, the belting used is specially made for the design, being somewhat more flexible across the carcass than that for a conventional flat conveyor. Machines of this type are capable of quite steep angles of elevation by virtue of their enclosure and gripping of the load. They are also capable of gradual changes of direction in any plane.

Folding belts are not widely used in waste processing applications, due largely to a perceived if not fully justified mechanical complexity and susceptibility to damage by the materials conveyed. For some waste fractions, folding belts appear to offer a satisfactory handling method, but there is little operating experience upon which to base that judgment.

Capacity calculations are the same as for flat or troughed belts, with the exception that the unit volume carried is derived from the cross-sectional area of the enclosed, tubular section. Power requirements are unchanged from those for flat or troughed belts, although caution suggests that the coefficient of friction f should be increased to perhaps 0.06 to account for the extra load involved in folding the belt.

3.2.5 Pneumatic Conveyors

A pneumatic conveyor is one where material is transferred through a pipe or duct by a flow of air, and where the air pressure differential between the inlet and outlet may be provided either by suction or blowing. Suction systems are lean phase systems, while blown systems are generally dense phase. In a lean phase system, air is drawn through the pipe or duct by a fan at the discharge end, acting through a cyclone separator and rotary valve. The term "lean phase" derives from the manner in which the material conveyed tends to remain in suspension in the air stream, such that every particle is surrounded by a volume of air. A dense phase system, meanwhile, is one where the material tends to form into "slugs," or dense, homogeneous masses, which travel along the pipe or duct separated from the preceding and following slugs by pockets of high-pressure air.

3.2.5.1 Suction Systems

Suction systems are always lean phase systems because of the principle of air movement which they apply. Air is drawn through the pipe or duct from the discharge end by a fan or rotary blower in such a way that a negative pressure below atmospheric is created at that end. In the event that the conveyed material begins to form a slug, the pressure downstream of it further reduces under the influence of the fan, causing the air trapped between particles to further expand. The expansion of the trapped air forces the particles apart again and maintains their suspension.

When the load reaches the discharge end of the pipe or duct it usually has to be separated from the air stream, and this is commonly achieved by installing a cyclone separator between the end of the duct and the fan. In order to maintain the necessary suction, the cyclone is fitted with a rotary valve at its solids outlet (see Section 3.1.18.2 for a description of cyclone separators). For some special-purpose applications, the cyclone separator is installed downstream of the fan instead, and the material passes through the fan before removal from the air stream. In such applications the purpose is generally to further granulate the material, and the fan used is a "chopper-blower." Fans of this type are centrifugal, with heavy, hardened blades fitted with sharpened edges, arranged in paddle-wheel configuration around the central shaft and running close to a hardened steel casing.

The air speed through a lean phase system is dependent upon the *actual* density of the material conveyed (see Section 2.3 for discussion of actual and apparent densities). The higher the density, the greater the air speed necessary for effective conveying, although lean phase pneumatic systems are usually reserved for low densities.

Mathematical modeling of a pneumatic system is a very complex matter indeed, since there are so many variables. Among those variables which influence any calculation are particle shape, air temperature, duct dimensions and routing, duct material, material moisture content, material infeed rate, and compressibility effects.

Particle shape is significant in that the more streamlined a particle is, the greater the air speed needed to move it. Air temperature is significant because in common with all gases, its density reduces as the temperature rises. The dimensions and routing of the duct, together with the materials from which it is manufactured, determine the frictional resistance to flow. Every bend imposes considerable resistance. Moisture content, at least in the case of low-bulk-density materials, has a considerable effect upon the final density, and therefore upon the mass of each individual particle. The effects of material infeed rate are perhaps more obvious, in that the greater the amount of material to be conveyed, the more energy must be delivered by the air. Compressibility affects the density and therefore the mass of the air, with the result that the lower the pressure the lower the air density, and the lower the quantity of energy it may transmit.

While there are formulas that can deal with some of the variables, the practicalities of waste handling are such that some of the most significant variables are impossible to predict. It may, for example, be possible to establish the granularity of the paper and card fraction after milling, but it will never be possible to determine the shape of each individual particle or even the average shape. Pneumatic conveying is therefore another of the many applications in the waste processing industry where experimental data are the only recourse, unless one happens to be a specialist in that one application.

3.2.5.2 Pressure Systems

Pressure systems are almost always dense phase systems at some point along the pipe or duct. As a particle moves along the duct, a positive air pressure builds up behind it, while atmospheric pressure remains before it. Movement is resisted by duct friction, and to some extent by the atmospheric air downstream, and pressure developed upstream causes compression. Regular-shaped, granular materials can be transferred over short distances by a pressure system in a lean phase form, but eventually frictional resistance causes them to become dense phase.

Most, if not all, of the materials and products encountered in waste processing are highly compressible, and many are abrasive. When a material is compressed in a duct or pipe, the force it imposes upon the walls increases in direct relationship to the degree of compression. Meanwhile, the relative abrasiveness increases the resistance to movement, also in direct relationship to the degree of compression. Since the resistance to movement and the air pressure necessary to overcome it are interrelated through compressibility effects, it is theoretically possible to reach a point where pressure increases simply cause further compression and resistance to movement, without any movement ever taking place.

To some extent the change to dense phase can be postponed or eliminated by employing air at high volumes and velocities, such that the loads are conveyed through the system too rapidly to have the opportunity to densify. Considering an extreme analogy, if there were only one particle of material at any time in any section of duct, and if each were suspended in a high-speed air flow, then the chances of them meeting to form a slug would be remote. Even then, only a minimal effect upon the speed of any one particle could reverse that situation, and its encountering a bend or inclined section could be sufficient.

For those reasons, pressure systems, and particularly, dense phase systems, are very rarely used in waste processing operations. Lean phase suction is operationally more reliable, needing lower power, and having less effect upon the material conveyed.

3.2.6 Material Bulk Densities

Handling systems cannot be designed without accurate data for the *true* and *apparent* bulk densities of the materials they will convey. The true bulk density defines how the materials will behave in suspension in an air stream, while the apparent bulk density establishes volumetric balances upon conveyors. Table 3.4 lists the true and apparent densities most likely to be encountered in municipal waste.

Table 3.4 True and Apparent Bulk Densities of Fractions

Fraction	True density kg/m^3	Apparent density at granularity (kg/m^3)			
		<100 mm	<50 mm	<10 mm	<5 mm
Mixed waste	150–250	ca. 200	ca. 250	ca. 300	ca. 325
Paper/board	700–1100	30	40	55	60
Organics/ putrescibles	800–1050	275	300	350	375
Dense plastics	800–2200	300	325	375	390
Light plastics	800–2200	20	50	80	100
Ferrous metals	7900	200	300	500	1000
Nonferrous metals	2700	100	150	200	300
Textiles	900–1100	150	200	275	300
Timber	400–1300	150	175	225	300
Glass	2400–2600	500	700	1500	1800
Ash[a]	1200				
Mineral, soil[a]	1600				
Dust[a]	300				
Animal fats[a]	940				
Vegetable oils[a]	930				
Mineral oils[a]	900				
Human/animal wastes[a]	500-2000				

[a] Materials always either granulated or liquified in as-received form. Apparent and true densities substantially identical.

3.2.7 HANDLING SYSTEMS

To some extent at least, the suitability of any system for conveying any material is a matter of experience. Rarely, if ever, is it possible to be truly specific, since that which constitutes an unacceptable problem to one operator is not necessarily viewed as such by another with better resources. In the following, experience is used as the basis for judgment, and in order to clarify the opinions expressed, each system is rated in up to 10 points. A rating of 0 is totally unsuitable, and 10 is ideal.

3.2.7.1 Mixed Waste

Mixed waste is moist, sticky, abrasive, biologically active, and very prone to consolidation. It is usually of somewhat higher than ambient temperature, and may reach temperatures of over 158°C (70°F) over 3 to 5 days, given sufficient mass, moisture, and oxygen. Table 3.5 suggests a composition range that is by no means comprehensive, or even representative of any particular

Table 3.5 Average Analysis Range of Percentage of Materials by Weight in Municipal Wastes

Material	% by weight
Paper/board	25–35
Organics/putrescibles	30–40
Dense plastics	6–10
Light plastics	5–8
Ferrous metals	4–8
Nonferrous metals	1–2
Textiles	5–8
Timber, wood products	5–10
Glass	6–12
Ash	2–3
Mineral, soil	2–3
Dust	0.5–1.5
Animal fats	<1
Vegetable oils	<1
Mineral oils	<1
Human/animal wastes	<1

region or country. It does, however, indicate with reasonable accuracy the likely relative proportions of materials likely to occur.

Flat/Troughed Synthetic Rubber — ***Rating 7.*** Flat or troughed synthetic rubber belts can accommodate mixed waste with reasonable reliability provided that certain precautions are observed. The belting must be oil- and fats-resistant, to ensure that it is not damaged by contact with those which inevitably exist within the waste. The carcass top cover must also be abrasion-resistant to withstand the effects of the glass, metals, ash, and soils, and ideally it should be heat-resistant as well. The reinforcing should have "breaker" strips included within it. These are strands laid in the reinforcing in such a way that any penetration or tearing is diverted to the side of the belt, rather than being allowed to split it completely in two.

Belt cleaners must be selected with care if severe belt damage is to be avoided. Hard plastic or metal scrapers should be avoided, since they tend to trap abrasive or sharp contaminants against the belt surface and drag them along it. Wet rubber, whether synthetic or otherwise, is very easy to cut when wet, as a belt handling crude municipal waste is always likely to be. Rotary-brush-type cleaners are not usually effective for long, since they tend to become clogged quite rapidly, after which they may begin to abrade the belt.

Vibratory belt cleaners, where a section of the return strand is caused to vibrate heavily over a short distance, are quite effective in that there is nothing in contact with the carrying surface of the carcass. Often the vibration is created by means of a number of rollers, mounted upon the periphery of rotating discs and placed between the strands of the conveyor so as to strike the inner cover of the return strand as they rotate.

Compressed air jets are a very effective means of cleaning, although they create significant noise and make collection of the contaminants difficult. They may also present some degree of health hazard by driving pathogenic microorganisms and contaminated water droplets out into the surrounding atmosphere.

Water jets are equally effective, and they present fewer health hazards since they do not need high pressures to work efficiently. Against that they inevitably make the belt extremely wet, and may introduce sufficient water to the inner ply to reduce driving grip at the head or tail drums. Also, the recovery of the contaminants that have been removed from the belt is somewhat more difficult, since they must be filtered out from the water before it can be reused.

Faced with those alternatives, the cleaning method most usually adopted is a scraper system that uses a rubber compound softer than the belt carcass, supplied without any reinforcement. It is not an ideal method, since it cannot eliminate belt abrasion completely. It does, however, minimize damage to some extent by allowing the scraper to abrade rather than the belt, and by being mounted in a spring-loaded clamp that maintains a constant pressure.

Concepts similar to cleaning apply to side sealing against spillage. A flat belt, with hard rubber or plastic side skirts rubbing against the top ply edges, is likely to suffer from excessive wear rates. Damage to the edges may weaken the belt and may even create sufficient distortion to make effective tracking difficult. Crude municipal waste is best handled upon a belt that is sufficiently troughed to retain the load, relying upon skirting only to prevent objects from rolling off. The side skirting can then again be softer, unreinforced rubber, since continuous contact with the waste is minimized.

Flat/Troughed Polyurethane — Rating 7. Systems handling crude waste are usually designed for high volumes and are therefore physically large. While plastic belting offers reduced thickness and weight, as well as almost total freedom from stretching, it is expensive. The extra cost is difficult to justify when the alternative is to fit a synthetic rubber belt with a simple tensioner. The flexibility of the material may need more troughing idlers than would a conventional belt, and again there is a cost penalty. For those reasons, plastic belting is not widely used in waste processing.

Steel Plate — Rating 10. The aggressive nature of the mixture is such that only a few conveyor types are suitable for handling it, and the best by far are steel plate feeders. They are powerful, robust, and resistant to corrosion, impact, and heat. They are capable of operating in humid and dusty environments with fairly minimal precautions, and they have a very significant carrying capacity. They are easy to seal against spillage, and wear rates are not generally significant.

The most common use of such machines is at the reception end of process facilities, and Section 3.1 considers them in detail in that environment.

Drag Links — ***Rating 0.*** Drag link conveyors are not suitable at all for handling crude waste before any processing, since their moving parts are prone to binding with the wire, textiles, etc. normally found within the unsorted material.

Chain Conveyors — ***Rating 0.*** As equally unsuitable as drag links, and for the same reason.

Screw feeders — ***Rating 0.*** The rotating action of screw feeders would cause textiles and plastic film to bind around the screws, causing almost immediate blockages.

Moving Floors — ***Rating 0.*** The way in which waste tends to bind together whenever energy is applied to it suggests that moving floors would simply compress the mass and then reciprocate beneath it, without creating movement.

Sidewall Belts — ***Rating 4.*** Sidewalls are a potential if somewhat expensive way of moving crude waste. However, belt cleaning is a potential problem, since the presence of the flights and sidewalls prevents the use of soft rubber scrapers. Vibratory belt cleaners work reasonably well, as do compressed air and water-jet cleaners. Sidewall damage is possible where sharp or abrasive items become trapped by the flexing of the walls as they pass over head and tail drums.

Bucket Elevators — ***Rating 0.*** Bucket elevators are rarely used to handle crude municipal waste because it is difficult to provide the controlled infeed rate they require. Again, binding of moving parts by textiles and wire is likely to create blockages or to cause damage.

Folding Belts — ***Rating 0.*** As a mechanical handling concept, folding belts would appear to have much to offer. They can elevate, change direction in any plane, and completely enclose the load. However, oversized items could be expected to cause serious damage if they are larger than the closed diameter of the belt, and since such items occur in municipal waste with regularity, folding belts are impractical.

Suction Pneumatics — ***Rating 0.*** Not suitable for mixed municipal wastes, where the irregular granularity, abrasiveness, density, etc. would render conveying erratic at best and nonexistent at worst. Heavy materials would not be transported at all, and wear rates in cyclone separators would be untenable.

Pressure Pneumatics — ***Rating 3.*** Dense-phase systems have been used upon occasion with some success, in large-bore systems where very high thrusts have been achieved and transport distances were extended. The principle is not appropriate for waste process plants, however, where alternative and cheaper methods exist and conveying distances are comparatively short.

3.2.7.2 Paper/Card (as received)

Paper and card mainly occur in municipal waste in sheet form, as newsprint and packaging. Such waste is generally damp, having absorbed water from the surrounding waste, and moisture contents commonly range from 20% to 60% by weight. Contact with putrescible wastes may have caused contamination, and the material may be biologically active, although it is not itself capable of significant biological degradation. The material is not particularly prone to compaction under its own weight, but it is inclined to settle in layers on conveyors simply because sheets always fall flat rather than on their edges.

Flat/Troughed Synthetic Rubber — **Rating 3.** Conventional belts of this type can handle paper and card in its as-received, sheet form, but not with great success except at very low loads. A belt carrying sufficient material to cover its full width is unable to provide a sufficiently positive grip, and will simply slide under the load without moving it. Ribs, chevrons, or slats on the carrying ply do not generally improve matters, since the load is inclined to bridge across them, rather than to be gripped by them. Very wide, flat belts are sometimes used, and they rely on the load being so spread upon the carrying strand that the depth is negligible. It is a principle that has more application as picking belts in materials recovery facilities than as pure conveyors.

Flat/Troughed Polyurethane — **Rating 3.** As with flat/troughed synthetic rubber belts, polyurethane belts are less than ideal. The nature of the belt material is irrelevant when the design itself is inadequate.

Steel Plate — **Rating 10.** Lightweight plate feeders are an attractive and widely used way of transporting paper and card, since they can be fitted with lifting teeth at a spacing such that the load cannot easily bridge. This facility for imparting a positive drive to the material permits plate feeders to be fully loaded without significant risk of blockages. A lightweight machine, in this context, is of the same basic design as one suitable for crude waste conveying, but is assembled from thinner plate and mounted upon lower breaking strain chain.

Drag Links — **Rating 2.** While drag links can in theory move unprocessed paper and card, since they provide a positive drive, the very low bulk densities and high volumes legislate against them. They would need to be very large to handle any meaningful quantities, and they would as a result be costly. In addition, since material could be expected to "hang up" on the flights to some extent, return ends would be necessary. A return end on a drag link is a curved plate that matches the outside diameter of the end sprockets, and against which the flights may continue to convey material as they pass over the sprockets. Without return ends, material would become packed into the casing beyond the end station until it finally jammed the conveyor.

Chain Conveyors — ***Rating 0.*** Even more unsuitable than drag links, and largely for the same reasons. Chain conveyors are designed for fine, granular loads rather than for large volume, low density materials.

Screw Feeders — ***Rating 1.*** Screw feeders will handle paper and card to some extent, but do not do so reliably. They also are more suitable for dense materials of a consistent granularity, and are likely to experience difficulties in discharging.

Moving Floors — ***Rating 0.*** As with crude waste, moving floors are unsuitable for this application. Sheet materials will simply bridge across the flights, allowing them to slide underneath.

Sidewall Belts — ***Rating 2.*** Very wide sidewall belts can work up to a point, but with the same difficulties as flat or troughed belts. Sheet material is inclined to bridge across the flights, allowing the conveyor to run underneath.

Bucket Elevators — ***Rating 0.*** The high volumes, large particle sizes, and low densities of untreated paper and card make bucket elevators unsuitable. They are designed for dense, granular products.

Folding Belts — ***Rating 8.*** Potentially very suitable for paper and card, since their method of containment offers minimal spillage. When the belt is fully loaded, the material is gripped strongly, providing positive drive.

Suction Pneumatics — ***Rating 0.*** Inconsistent granularity makes as-received paper and card unsuitable for lean-phase pneumatic systems. Infeed arrangements and cyclone separators essentially involve too many changes in direction, each presenting opportunity for blockages.

Pressure Pneumatics — ***Rating 1.*** Again, irregular granularity is likely to be a problem, although in theory a system, given sufficient pressure, should be able to move mixed paper and card in dense-phase form. There is, however, little operating experience in process plants, as cheaper and less complex alternatives exist.

3.2.7.3 Paper/Card (Granulated)

Once the paper and card fraction has been separated from municipal waste and then granulated, its whole nature changes very dramatically indeed. Compaction under its own weight becomes inevitable and very rapid, and very high densities can be achieved, although initially the bulk density is very low (about 1.8 lb/ft^3, 30 kg/m^3). The material is fibrous, mats together easily, and will adhere to even steeply angled slopes. If allowed to remain in contact for long with unprotected steel platework, its somewhat alkaline moisture content is likely to set up quite rapid electrolytic corrosion, which then "cements" the material to the steel immovably.

Even though it is usually moist, the material has a considerable propensity for the liberation of dusts, and while these may not be highly flammable, they will smolder given a source of ignition. If dried and suspended in an air stream, very finely granulated paper and card (granularities of $<^1/_2$ in.) may become highly flammable and even almost explosive.

Flat/Troughed Synthetic Rubber — ***Rating 8.*** This design is entirely acceptable for handling granulated paper and card, and is widely used for the purpose. Occasionally, improved grip upon the load is sought by installing chevron belts, where the carrying ply is ribbed in a chevron pattern along its length. Since dry paper is surprisingly abrasive, side skirts are usually made from soft rubber without reinforcing. Belt cleaning is not generally necessary unless the load is particularly wet, and then vibrator cleaners are suitable.

Flat/Troughed Polyurethane — ***Rating 9.*** Equally acceptable for granulated paper and card, with some advantage in that a very wide range of surface textures is available.

Steel Plate — ***Rating 10.*** Again, conveyors of this type are widely used for the application, and they are suitable in every sense. Their most common use is to receive the discharge from granulators and flail mills, where heavy contaminants may occasionally be ejected with considerable force, sufficient to damage less robust belting.

Drag Links — ***Rating 5.*** It is not uncommon for drag links to be used in this application, although they need careful design to perform satisfactorily. The volumetric implications of the material's *apparent* density (see Section 2.3) need to be clearly understood, since the machine casing must not be over-filled. Granulated paper and card is very susceptible to compaction, after which it easily forms an unyielding mass. Any restriction or increase in frictional resistance in the conveyor casing, particularly those cased by over-filling, is likely to create compaction and damage to the conveyor parts.

Chain Conveyors — ***Rating 4.*** Not widely used, and being designed for dense materials of fine granularity, not entirely suitable for paper and card. Even more than conventional drag links, they are inclined to compact the load until structural damage occurs. They can be used with some success if the material is densified first.

Screw Feeders — Rating 6. Large-diameter screw feeders work quite well with paper and card that have been granulated to a good degree of particle size consistency, but to achieve high delivery rates, screw diameters of 1 m (3.28 ft) are common. While they offer the attraction of total enclosure, they have the disadvantage that a large-diameter heavy screw needs support at close intervals. The hanger bearings that provide that support also provide initiation points for load compaction and blockages. For that reason, screw feeders are generally only used for paper and card over transfer distances of not more than about 2 m ($6^1/_2$ ft), when hanger bearings are unnecessary. Thus the machines are commonly employed as metering feeders for processing machines rather than as conveyors in their own right.

Moving Floors — Rating 3. The ease with which granulated paper and card can be compacted makes moving floors unsuitable. The material does not flow readily enough to allow it to fill the voids between the flights, having been compressed by the feeding stroke.

Sidewall Belts — Rating 10. Sidewall belts are used extensively for conveying granulated paper and card, particularly where steep elevations are necessary. The positive delivery and load containment make them an option worth considering, particularly in circumstances where the use of one steeply inclined sidewall belt may avoid the need for two or more conventional belts installed at less than the 27° slip angle. Belt cleaning is unlikely to be a problem, as the material is not generally moist enough to adhere strongly to the belt.

Bucket Elevators — Rating 5. Elevators with large, seamless buckets mounted upon chains and loaded at some point on the vertical section are a possibility, provided that spillage into the bottom casing (the "boot") is minimized. Buckets fixed to endless belts should be avoided, as there is a potential for the materials to become trapped between the carrying and return strands. There they will build up upon the drums until they interfere with belt tracking.

"Continuous" elevators are the more logical choice, since there is less opportunity for the load to become dispersed by "windage" — the displacement of air caused by the passage of the buckets within the casing — and possibly by thermal lift if the load is heated.

Folding Belts — Rating 10. Folding belts are in many ways ideal for transporting granulated paper and card, in that their total enclosure eliminates spillage, particularly if any part of the run is outdoors. Their facility to change direction in any plane makes them particularly suitable for long transportation distances. For short distances, however, of the lengths more likely to be encountered inside process plants, troughed belts, sidewall belts, plate feeders, and pneumatics may all be either cheaper or more convenient.

Suction Pneumatics* — *Rating 10. Granulated paper and card handles particularly well in lean-phase systems, which offer the major advantages of vertical elevation, changes in direction in any plane, and conveying over considerable distances, all with only a single power source. The systems are widely used in association with air classifiers and driers, both of which introduce the materials into an air stream as an essential part of their function. Once the materials are in suspension, it is logical to keep them so until they reach their final destination. It is in fact a primary rule that once very low density granulated materials are moving in a handling system, they should be kept so. Allowing them to stop moving, or to change both direction and energy input at the same time as in a transfer chute, invites settlement, compaction, and blockages.

Pressure Pneumatics* — *Rating 7. In theory, dense-phase systems could handle granulated paper and card, but they are not widely used for doing so. They do not attach to classifiers and driers with the convenience of suction systems, and so they offer no advantages over conventional conveyors.

3.2.7.4 Organics/Putrescibles

The organic fraction of municipal waste occurs mainly in the form of vegetable matter, which arises from food preparation and to some extent from gardening. "Soft" organics — vegetable matter high in water content and low in cellulose fiber — predominate, and decompose rapidly.

Putrescibles are largely protein wastes, meat scraps, etc., although the ever-present disposable diapers fit into the same category. Protein wastes also decompose rapidly, although less so than vegetable wastes, and they create odors and attract insects and vermin. Disposable diapers emit copious quantities of ammonia in bulk, and are a major cause of electrolytic corrosion on the internal steelwork of some process machinery.

Vegetable waste liberates vegetable oils, while putrescibles release fats and animal oils. All are potentially damaging to belt conveyors unless the carcasses are specified to be resistant to them.

Flat/Troughed Synthetic Rubber* — *Rating 8. Oils-, fats-, and abrasion-resistant belting is capable of handling the organic and putrescible fraction of municipal waste without major difficulties, although belt cleaning may be less than easy. Soft, unreinforced rubber scrapers work satisfactorily in most applications, and side skirting of similar material is sufficient to contain the load without belt damage.

Flat/Troughed Polyurethane* — *Rating 9. In many ways, plastic belting is a good choice, although there may be some danger from the solvents that occasionally stray into mixed waste. In a process plant, solvents usually report to the organics lines from the initial separation processes. Certainly, plastic belting is more likely to resist oil and fats damage, while remaining easier to clean. Costs alone, however, may make conventional synthetic rubber belting a more attractive option.

Steel Plate — **Rating 9.** Plate feeders are widely used for putrescibles, particularly where the loads are wet. They are used in some processes to recover materials from water baths at the end of scrubbing stages, and are the mechanism for moving organics through some designs of composting systems. Where wet applications are involved, low-grade stainless steel chains are often used, although the oily nature of the fraction tends to limit corrosion. Belt cleaning is not usually a problem since, although deposits build up upon the flights, accretion is limited by the scouring effect of successive loads.

Drag Links — **Rating 9.** Drag links are also quite widely used to handle organics, where their total enclosure makes them environmentally attractive. However, total enclosure can lead to casing corrosion caused by condensation of moisture, and conveyor chain corrosion due to continuous contact with the waste may also occur. Stainless steel chains and casings overcome the problems, but are expensive in relation to belt and plate feeder conveyors. Ventilation of the casing by means of a fan at one end and a breather at the other is usually sufficient to minimize or eliminate condensation.

Chain Conveyors — **Rating 2.** Chain conveyors are not really ideal for organic fractions due to their tendency to compress the material. They can be made to work by careful design of flights, and they offer vertical conveying without the potential spillage problems of a bucket elevator. Solid plate flights are essential, rather than the formed bar flights more often associated with chain conveyors, and they must be close-pitched. Where a conventional machine for grain handling may have a bar flight at every sixth chain link, a conveyor suitable for the organic fraction of municipal waste would need a plate flight at least at every third link.

Screw Feeders — **Rating 6.** Organics can be handled in screw feeders without problems, but they are generally used for metering infeeds to other machines, rather than as conveying systems in their own right. Once again, the need for hanger bearings limits their applications, since much of the organic fraction of municipal waste is fibrous and inclined to compact around obstructions. When designed with ribbon flights, the machines are occasionally used for slurry mixing before anaerobic decomposition.

Moving Floors — **Rating 0.** The compressibility of the fraction makes moving floors unsuitable. They are inclined to simply turn the material into a semi-solid mass that reciprocates with the flights.

Sidewall Belts — ***Rating 3.*** While sidewalls can handle the fraction successfully, belt cleaning is likely to cause problems. The flights and sidewalls prevent the use of scrapers, and some organics are simply too sticky to be liberated by vibrating cleaners. A buildup of congealed dust and fats is to be expected, resulting in further material adhesion and the possibility of carry-over and spillage underneath the return strand.

Bucket Elevators — ***Rating 3.*** Although the concept of vertical elevation is intuitively attractive, bucket discharging can be difficult. Plastic or plastic-lined buckets can alleviate the problems to some extent, but may not overcome them completely. Spillage of organics into the casing boot may present health risks, create odors, and encourage infestation. In general, bucket elevators offer little that cannot be achieved by drag links with fewer problems.

Folding Belts — ***Rating 5.*** Machines of this type are a satisfactory, if costly, way of conveying the organic fraction, provided that the waste is not contaminated by oversized items. They are not ideal, however, and if over-loaded, may permit leakage of liquors that are pressed out as the belt folds.

Suction Pneumatics — ***Rating 2.*** Given sufficient air speed in excess of 32.8 ft/s (10 m/s), lean-phase suction systems can convey the fraction, but the need for elimination of odors from cyclones and the prevention of deposits in rotary valves makes them less than ideal. They do not fit easily into any system likely to be used to process organics, and therefore they offer little more than a conventional belt conveyor offers at lower cost.

Pressure Pneumatics — ***Rating 2.*** Again, given in this case sufficient air pressure, there is no reason why dense-phase systems should not transport the fraction with reasonable reliability. But as with lean-phase, they do not form an inseparable part of any process, and they are a less obvious choice than conventional conveyors.

3.2.7.5 Dense Plastics

In the context of this section, and waste processing in general, the term "dense plastics" is taken to mean the residues of containers and moldings, as distinct from film. It does not imply any specific density compared with other plastic products, and is dealt with separately because a process plant will treat the materials differently. Although there is a very wide range of polymers in common use, not all of them are likely to arise in quantity in municipal waste. The thermoplastics polystyrene, polyvinylchloride, polyethylene, and polypropylene, and derivatives of them, are the most consistent in waste.

Accurate identification of each polymer is difficult, and is generally unnecessary in a process plant unless individual types are to be separated for recycling. However, some indication of the type of plastic may be gained from the use to which it has been put.

General purpose polystyrene is hard, brittle, and very transparent. Its main use is for packaging, toys, and light fittings. In its expanded form, as small white beads or molded blocks, it is widely used for protective packaging of delicate equipment

Toughened polystyrene is tougher and has better heat resistance. Its uses include casings of domestic appliances, vending cups, furniture.

Styrene-based ABS materials are tough and abrasion resistant. They can be metal plated, and they are most often encountered in waste in this form. They are used to make telephone handsets, pipes, and domestic equipment.

Unplasticized polyvinylchloride is both hard and tough, and can be made very transparent. Its main use is for pipes, bottles, and domestic fittings such as curtain rails.

Plasticized polyvinylchloride is very flexible, but otherwise retains many of the characteristics of the unplasticized polymers. It occurs in such items as electric cable insulation, soles of footwear, waterproof footwear, and industrial gloves.

High-density polyethylene occurs in municipal waste mainly in the form of tanks, bottle crates, and garbage cans. It is tough but not particularly hard, and can become fairly brittle at very low temperatures.

Polypropylene is fairly strong and stiff, with a good temperature resistance. It is used mainly for electrical cable insulation, automotive fittings, bottle crates, pipe fittings, etc.

Acrylics are optically transparent, stiff, resistant to ultraviolet light, and do not shatter. They are used for domestic baths and sanitary equipment, for low-cost spectacle lenses, light housings, etc.

Polycarbonates are tough, transparent, and very strong, and among the applications are street lamp covers, baby feeding equipment, glazing, and industrial safety helmets.

All of these materials occur in crude municipal waste as moldings. Most commonly, they are found as bottles, containers, pieces of domestic equipment, and fragments of packaging. Although classified as "dense plastics," some — for example expanded polystyrene — are extremely light. In general, all of the dense plastic items likely to be encountered are quite easy to separate from the waste, if not from each other, and none of them pose any insurmountable mechanical handling difficulties. Their behavior in conveyor systems is broadly as follows.

Flat/Troughed Synthetic Rubber — Rating 10. In many ways, dense plastics are ideal materials for handling on conventional synthetic rubber belt conveyors. They are neither abrasive nor sharp, and they are unlikely to cause any belt deterioration. Containment is simple, either by troughing the belt to the appropriate angle or by fitting side skirts, and cleaning is unlikely to cause difficulties with any conventional cleaner or scraper.

Flat/Troughed Polyurethane — ***Rating 10.*** The above comments apply with equal validity to plastic belts, but there are no immediate advantages over synthetic rubber to justify extra costs.

Steel Plate — ***Rating 10.*** Steel plate conveyors have applications in washing plants where the plastics are to be recovered for recycling, but they have no significant advantages over conventional belts for general handling duties. Dense plastics conveying does not need the power and robustness of such heavy duty machines.

Drag Links and Chain Conveyors — ***Rating 9.*** While both types of machine will convey dense plastics without difficulty, again neither offers significant improvement over a conventional synthetic rubber belt.

Screw Feeders — ***Rating 8.*** Screw feeders will transport plastics, but there is a danger of particles becoming trapped between the screw and the casing. The effects of that may not be immediately obvious, but it may impose shock loadings upon bearings and drive systems. In this application, hanger bearings are unlikely to hinder effective conveying, but even so, other than as metering feeders for other machines, this type has no particular advantages in its favor.

Moving Floors — ***Rating 8.*** Where dense plastics have been granulated and stored in flat-bottomed hoppers, moving floors are an appropriate way of ensuring satisfactory unloading. The materials do not mat together significantly, so that full advantage can be taken of the wide area of discharge that the design offers.

Sidewall Belts — ***Rating 10.*** The design is appropriate where steep inclinations are necessary, and it is capable of transporting the material without difficulty. However, for level conveying applications there are no significant advantages, but there remains a cost penalty.

Bucket Elevators — ***Rating 8.*** "Continuous" elevators are satisfactory for the application, provided that spillage is eliminated. Dense plastics are tough, and are likely to cause damage if they become trapped between carrying chains and sprockets. All other bucket elevators should be treated with caution for that reason.

Folding Belts — ***Rating 9.*** Machines of this type will handle dense plastics, but not sufficiently better than more conventional systems to justify their cost and complexity.

Suction and Pressure Pneumatics — ***Rating 8.*** Again, either type will handle most dense plastics, doing so more effectively if the materials are granulated. Air speeds need to be high, in the order of 49 ft/s (15 m/s), and so fan absorbed power levels are correspondingly high. Moving 2 t/h through a pneumatic system may require 100 hp, while doing so upon a conventional conveyor may require only 5 hp. Thus, there remains the inevitable cost penalty, which limits their application.

3.2.7.6 Light Plastics

Again, the term is not intended to imply any particular *true* density, which is likely to be much the same as that of the polymers listed as dense plastics. In this case, the purpose is to distinguish those materials which occur as films, and which then exhibit a very low *apparent* density in handling systems. The separation stages in a process plant cannot generally distinguish between plastic films and, for example, sheet paper.

The most common films identified in municipal waste are polyvinylchloride, polyethylene, and polypropylene. They may be transparent, as used in packaging, or colored with fillers, and may be printed. Their suitability for conveying is as follows.

Flat/Troughed Synthetic Rubber — ***Rating 7 (granulated film); 2 (as received).*** Films can be conveyed upon flat or troughed belts, but large pieces may be inclined to snag upon side casings, and the containment of spillage may be a problem. At the conveyor discharge, the very light nature of the material makes it prone to windage. Large pieces of film are inclined to wind around rotating shafts, where they bind very tightly indeed and become extremely difficult to remove.

In order to minimize spillage, discharge transfer chutes should be made close-fitting to the conveyor head drum, and should have no internal seams or projections. The bottom openings should be of a greater cross-sectional area than the tops. Platework angles should be greater than 60° to the horizontal, and preferably as near to vertical as possible.

Flat/Troughed Polyurethane — ***Rating 7 (granulated film); 2 (as received).*** Belts of this type offer no particular advantages over conventional synthetic rubber.

Steel Plate — ***Rating 8 (granulated film); 2 (as received).*** Not really suitable for light plastics, since the overlapping side skirts are likely to trap the load. Apart from the snagging risks, a steel plate feeder would be an extremely expensive way of transporting materials that do not form a major fraction of municipal waste.

Drag Links — ***Rating 8 (granulated film); 0 (as received).*** Can work satisfactorily with granulated film, but not with the materials in their as-received form. In the latter case, snagging on moving parts is likely to be a major problem.

Chain Conveyors — ***Rating 8 (granulated film); 0 (as received).*** Again, granulated film is capable of being conveyed, but as-received materials should be avoided due to the risk of snagging upon moving parts.

Screw Feeders — ***Rating 8 (granulated film); 0 (as received).*** Screw feeders are unsuitable for film, which unless it is granulated will bind around the rotating shaft, rapidly clogging the machine completely.

Moving Floors — ***Rating 1 (granulated film); 0 (as received).*** There are no obvious applications in film handling for moving floors.

Sidewall Belts — ***Rating 7 (granulated film); 0 (as received).*** The comments appropriate for flat and troughed belts also apply to sidewall conveyors. They will handle granulated film without difficulty, although discharge arrangements need careful design to avoid excessive spillage. As-received film, however, is even more likely to snag upon sidewall belts than it is upon flat belts, and in that form the flights and sidewalls add nothing to the ease of conveying.

Discharge chutes should be designed as described for flat/troughed belts, although the presence of the sidewalls and flights makes close fitting to the head drums difficult.

Bucket Elevators — ***Rating 5 (granulated film); 0 (as received).*** Large, seamless buckets on a continuous belt are in theory capable of handling granulated film, but spillage into the boot may cause binding upon sprockets or buildup upon drums, and eventually damage the chains or distort the belt. As-received film is difficult or impossible to load into a bucket elevator without overspill, and so machines of this type are unsuitable for plastic film handling in general.

Folding Belts — ***Rating 10 (granulated film); 5 (as received).*** A folding belt is in some ways an ideal conveying mechanism for film, provided that the loading arrangements ensure that the material is fully enclosed when the belt folds. Such a requirement can be met with careful design of the infeed chute work, which should have extended sides and top to direct the material into the center of the belt. The immediate attraction of belts of this type is that as the load is totally enclosed, it is protected against windage. Spillage and snagging problems are eliminated during transport, but may occur at the discharge end, where transfer chute design is important. Again, the transfer chutes should be designed as described for flat/troughed belts.

Suction Pneumatics — ***Rating 9 (granulated film); 0 (as received).*** Lean-phase pneumatic systems are extremely efficient in handling granulated film, and are worthy of consideration. Since the load is enclosed at all times, spillage is nonexistent. Very high conveying speeds are possible, and distance and changes of direction are unimportant. Cyclone separators need to be large to take account of the very low apparent density of the material, and discharge rotary valves must be of large diameter for the same reason. The presence of rotary valves in the systems precludes their being used for as-received film, which would bind around the valve rotors.

Table 3.4 suggests that granulated light plastics may have an *apparent* bulk density of between 1.8 and 6.2 lb/ft^3 (20 kg/m^3 to 100 kg/m^3), and so a system required to deliver, for example, 5 t/h may require a volumetric capacity of 8827.5 ft^3 (5000/20 m^3 or 250 m^3/h). Allowing an air to product ratio of 100:1, then the cyclone drum at the discharge end would have to be at least 6 ft (1800 mm) in diameter, and the whole cyclone could be as high as 23 ft (7 m).

A further difficulty that may be encountered with discharging finely granulated (<$^1/_2$ in.) plastic film is that a single cyclone may not be sufficient to ensure complete separation from the air stream. Depending upon where the air is to be exhausted, a second dust cyclone in series with the first may be necessary.

Pressure Pneumatics — ***Rating 8 (granulated film); 4 (as received).*** Dense-phase pneumatics suggest some advantage over lean-phase systems, in that discharge cyclone separators are not necessary. However, dense-phase systems discharge their contents in slugs, at quite high velocities and with considerable momentum. Granulated plastic film does not mat together, and so slugs are likely to explode apart as they leave the conveying duct, distributing material over a wide area. Wherever the duct discharges, whether it be into a hopper or storage bay, there must be some means of allowing the conveying air to escape, and a considerable amount of film is likely to stay with it. Filters would block almost immediately.

This problem can be overcome provided that the air exhaust is sufficiently far from the duct, and that the receiving space is sufficiently large for the air speed to drop below that which will continue to support the material. However, plastic film will stay suspended in air speeds as low as 10 ft/s (3 m/s), while the air may leave the duct with an expansion velocity of 300 ft/s (91 m/s). Thus, the receiving space would need to be very large indeed to effect satisfactory separation, and would certainly be larger than any realistic receiving hopper for a processing machine.

Some alternative approaches have been attempted with varying success, and among them are the installation of an open cyclone separator or an inclined screen. An open cyclone is one where the exhaust discharges directly into the atmosphere, rather than into a fan, but because of the delivery of the material in slugs, the drum diameter would have to be even larger than that for a lean-phase system.

Inclined screen separators operate by presenting the air flow to a wedge-wire screen inclined at a steep angle to the direction of flow. Here, the principle is that the pressure of the air flow pushes the material down the slope of the screen, leaving a clear mesh for the air to escape. While the concept is simple, the design however is not. The screen must be at precisely the right angle, and it must be positioned at a point in relation to the duct discharge where the air is still more-or-less streamlined, but not so close that it is struck by coherent slugs of material. Such a point is very difficult to locate, and doing so may be impossible if the granularity of the film is inconsistent.

3.2.7.7 Ferrous Metals

Although there is an almost infinite range of ferrous metal products, the variety found in municipal waste is fairly limited. It is generally assumed that the bulk of the metal encountered is in the form of cans, and in volumetric terms this is mainly true. However, considered in terms of mass balances, it is not. Usually municipal waste also contains some quantity of dimensionally small but dense items, among which are likely to be cutlery, automotive parts, cast materials, fastenings, and so on.

In most process plants, cans, because of their relatively low *apparent* bulk density and individual volume, will be separated differently and into different streams from the heavy items. Ferrous metals, therefore, almost fall into two separate categories in terms of mechanical handling, as indicated below.

Flat/Troughed Synthetic Rubber — **Rating 8 (cans); 2 (scrap).** Both cans and heavy items can be handled upon horizontal belt conveyors without any real difficulty, in spite of the sharp edges they may present to the belt. Provided that there are no points along the length where movement could be restricted, belt damage is unlikely to occur. However, if the belt is to be inclined, then precautions must be taken to prevent rollback. Cans rolling on the belt continuously can create a surprising amount of damage to the cover very quickly, and can certainly reduce the life of the conveyor substantially. Chevron belting is usually sufficient to overcome the problem, but even then, inclinations greater than 30° should be avoided.

Heavy metallic items can be handled, again without difficulty, but rather more precautions are necessary. The loading drop should be minimized; otherwise, belt penetration by sharp objects is both possible and likely. The waste processing industry can point to many experiences of belts being split lengthwise neatly in two as a result of penetration by an object that has subsequently jammed against the conveyor structure. Much of the risk can be avoided by designing the loading chute so that the scrap has to slide down a back plate, inclined at not more than 60° to the belt, with the high end toward the tail. By that means, the material is already traveling at some speed in the direction of the belt before it arrives upon it, and penetration is unlikely to occur. The belt should be specified with breaker strips in the reinforcing plies to divert any penetrations and tears to the sides.

The width of the discharge chute is limited by the width of the conveyor, but its projection in the direction of the conveyor is not. That projection should be at least twice the length of the longest items expected, since belt damage caused by lengths of steel tube failing to drop into the discharge chute is not uncommon.

Flat/Troughed Polyurethane — ***Rating 9 (cans); 2 (scrap).*** Polyurethane belting is somewhat more resistant to scuffing from cans, but rollback should still be avoided. Apart from belt life considerations, if excessive rollback is occurring, then the conveyor is operating inefficiently. A polyurethane belt with lifters or chevrons offers no particular advantages over a synthetic rubber belt, other than perhaps some improved resistance to oils and fats. Loading and discharge chute design should be as for synthetic rubber belts, since polyurethane belts can also be split, if only with considerably greater difficulty.

Steel Plate — ***Rating 10 (cans); 9 (scrap).*** Steel plate feeders can accommodate both cans and heavy scrap without difficulty, particularly in large tonnages. Other than ensuring that there is sufficient clearance above the belt to allow large pieces to pass, there are no special design requirements. The belts are very resistant to impact, cannot be penetrated, and cannot be torn. Damage caused by blockages is usually limited to a small number of flights, which are easily replaced.

Drag Links — ***Rating 5 (cans); 0 (scrap).*** Drag links are capable of handling cans, although they are less than ideal for the purpose. Their total enclosure places limitations upon the volume of material that can be conveyed, and there is an ever-present danger of oversized items jamming against the casing. They are entirely unsuitable for heavy scrap, largely for the same reasons.

Chain Conveyors — ***Rating 0 (cans); 0 (scrap).*** Their single, exposed chain design makes chain conveyors quite unsuitable for any form of scrap handling.

Screw Feeders — ***Rating 5 (cans); 0 (scrap).*** Shredded cans can be transported in short screw feeders, perhaps for metered delivery into baling presses. Otherwise, as a mechanism for extended conveying, they have no useful role in dealing with steel scrap.

Moving Floors — ***Rating 9 (cans); 0 (scrap).*** As storage hopper dischargers for shredded cans, moving floors can work well. Otherwise, the irregular shapes of the materials make them unsuitable for scrap handling.

Sidewall Belts — ***Rating 7 (cans); 0 (scrap).*** Sidewall belts are in many ways ideal for transporting cans, shredded or otherwise, although there are some design limitations. One of the attractions of sidewall belts is that they can be designed with a horizontal loading section, before inclining under tracking wheels. In such a configuration they can replace a horizontal belt and an inclined belt with one conveyor, with one drive unit instead of two, and with virtual freedom from spillage. Unfortunately, when used for scrap handling duties, that attraction is lost.

When the belt changes direction into the incline, the concertina profile of the sidewalls closes up, and may trap the load between folds. If that load is in the form of cans or steel scrap, with sharp edges, then sidewall damage will inevitably result.

Belts of this type are unsuitable for general scrap handling due to the risk of large items jamming between the raised flights or against the structure.

Bucket Elevators — ***Rating 4 (cans); 0 (scrap).*** Continuous bucket elevators could handle shredded cans, but as with other materials, spillage into the boot must be avoided. Machines of this type are not the obvious choice for conveying either cans or scrap, and the vertical elevation which is their sole reason for being is rarely needed in the metals recovery sections of process plants.

Folding Belts — ***Rating 0 (cans); 0 (scrap).*** Folding belts should never be used for cans or scrap handling. The pressures exerted upon the load at the folding stage could be sufficient to cause severe belt damage.

Suction Pneumatics, Pressure Pneumatics — ***Rating 0 (cans); 0 (scrap).*** Pneumatic systems in general are difficult to reconcile with scrap handling duties. Although the *apparent* bulk density of the scrap may be low, it is the *true* bulk density which determines the behavior in an air stream. Particles with a density of 490 lb/ft^3 (7900 kg/m^3) would require a very high-speed air flow indeed to move them.

3.2.7.8 Nonferrous Metals

The emergence of can banks for the recovery of aluminum beverage cans has severely limited the amount of aluminum that now occurs in municipal waste, to the extent that it is often difficult to justify recovery of the residues. If the total weight of cans is less than 1% of the municipal waste arisings delivered to a facility, as is often the case, then even the cheapest conveying system may never recover its cost from nonferrous scrap sales.

The arisings of nonferrous scrap are, however, similar to those of ferrous scrap, in that not all of the input is in the form of cans. Again, there are generally two distinct streams, one being almost all cans, and the other being small, dense nonferrous materials such as copper, brass, and coinage. It is this similarity which, in spite of density differences, makes conveying considerations for nonferrous scrap indistinguishable from those for ferrous scrap.

3.2.7.9 Textiles

Textiles present quite complex problems in conveying systems, since the form in which they arise in municipal waste varies considerably. They may be extremely bulky and tough, as in the case of carpets. They may be irregular-shaped and contaminated with pathogenic microorganisms in the case of used clothing. They may be delivered in very long lengths of strong synthetics, in the form of rope, cord, ribbon, and commercial wastes.

Faced with such a broad material specification, it is difficult to conceive of an ideal conveying design. All that can be said is that a few designs may be capable of the task to some extent, but most are unsatisfactory.

For that reason the only mechanisms which have even a remote chance of handling textiles are grouped together for consideration, as are the remainder, which are unsuitable.

Flat/Troughed Synthetic Rubber, Flat/Troughed Polyurethane — ***Rating 4.*** Either type of belting can handle textiles, within the constraint that long strands will inevitably snag upon something, and oversized items such as carpets will cause blockages with equal inevitability. As a result, it is imperative that the conveyor casings and skirts have no projections, and that they present the smoothest face possible to the load. Loading and discharge chutes should be as large as is physically possible in the space available, and the cross section of their outlets should be larger than their inlets.

It is almost an admission of defeat to observe that the design should include easy access to all parts of the conveyor, so that blockages can be cleared as they occur.

Other Conveyor Systems — None of the other systems are capable of handling textiles recovered from municipal waste reliably. Many have too many projections or snagging points, or have moving or rotating parts in contact with the load. Some would be impossible to load reliably, while others could not be discharged.

Some older processing and recycling plants made use of conveyors similar in design to plate feeders, but using wooden slats mounted upon conveyor chain rather than steel flights. The advantage of wooden slats is not immediately obvious in light of modern experience, although since most such conveyors were designed before the advent of modern synthetic textiles, perhaps mechanical handling was somewhat simpler then.

3.2.7.10 Timber, Wood Products

Timber presents similar problems to textiles in that it arises in many forms and sizes, with the added hazard that it is frequently associated with nails and glass. In so far as there is any consistency in the material, it is most often received as sheet or lengths. Fortunately, in some ways, there is little or no market for timber salvaged from municipal waste, and so there is little reason for establishing a method of conveying it by itself. More often, it is delivered with a combustible fraction to a fuel manufacturing process, or with an organic fraction to a composting process. In either case, it is likely to have been shredded with that fraction. Without shredding, timber becomes almost unmanageable in any conveying system, and so the following comments assume a degree of shredding first.

Flat/Troughed Synthetic Rubber — ***Rating 10.*** This system presents no obvious problems, provided that some recognition is given to the likely presence of nails when designing feed and discharge chutes. Feed chutes should be designed in a similar way to those for ferrous metals, and for the same reasons.

Flat/Troughed Polyurethane — ***Rating 10.*** Again, there are no immediate problems with using polyurethane belting, but there are no obvious advantages over synthetic rubber belts, either.

Steel Plate — ***Rating 10.*** Since steel plate feeders are by far the most common way of handling crude waste at the reception stage, they are equally capable of handling timber as it is received and before shredding. That capability, however, is assigned with the caution that it does not apply to loads solely of timber. Were that to be so, then such a machine can still perform a useful duty, but only at the expense of making it very heavy indeed, so that it is capable of breaking any large pieces that become jammed.

Plate feeders are very capable of handling shredded timber, but once the waste has been reduced to that condition, they confer no particular advantages over conventional belt conveyors.

Drag Links — ***Rating 8.*** These systems are capable of handling wood chip provided that the granularity is reasonably consistent and that there are no oversized items. Particle sizes should not be greater than about 4 in. (100 mm).

Chain Conveyors — ***Rating 7.*** As with drag links, a chain conveyor can handle wood chip, although consistent granularity is even more important. In this case, a granularity of greater than 1 in. (25 mm) should be treated with caution.

Screw Feeders — ***Rating 5.*** Given a particle size no greater than half the screw pitch in any dimension, a screw feeder can handle shredded timber. The presence of hanger bearings is unlikely to hinder load transfer, and so extended conveying is possible. There is, however, some danger of pieces becoming trapped between the screw and the casing, and there is little that can be done to prevent it. As a result, screw feeders are not the most obvious conveying mechanisms.

Moving Floors — Rating 9. A moving floor system will handle wood chip extremely well, and the design is widely used where the material is stored in large barns. The flights need to be shaped as very acute wedges, with the acute angle being not greater than 25° to the horizontal. Conversely, such a design will not usually handle unshredded timber at all.

Sidewall Belts — Rating 10. Again, there are no immediate problems in handling wood chip on sidewall belts, and advantage may be taken of their elevating ability. Loading and discharge arrangements should be designed as described for flat synthetic rubber belts. Changes of direction from horizontal to inclined should be treated with caution, due to the potential for splinters to become trapped between the folds of the sidewalls.

Bucket Elevators — Rating 8. There are no insurmountable difficulties with continuous elevators, provided that spillage into the boot is controlled and that the granularity is consistent. Other types should be avoided, since timber is likely to be sufficiently tough to cause damage as it is dredged in the boot.

Folding Belts — Rating 9. Folding belts are a very acceptable way of conveying wood chip, particularly over distances or changes in direction. Given loading arrangements as described above, belt deterioration should be minimal. The folding action of the belt places some requirement for consistent granularity, and prevents the use of conveyors of this type for untreated timber.

Suction Pneumatics, Pressure Pneumatics — Rating 10. (suction); 9 (pressure). Either system may handle wood chip without difficulty, and lean-phase suction systems are often fitted with loading chutes that include a simple classifier to remove nails and other contaminants. No pneumatic system will handle untreated timber.

3.2.7.11 Glass

Municipal waste commonly contains window glass and containers, together with some ceramics, which most processes are unable to isolate. Whatever its origin, glass is of course brittle, extremely sharp, and potentially very abrasive. Broken containers may be biologically contaminated and may present considerable health hazards.

Conventional recycling systems require glass cullet which is separated into colors, and any contamination by ceramics makes the material unusable. A certain amount of experimental work has upon occasion been undertaken to find ways of sorting by color automatically, and some have reached prototype stage. None have confirmed their early promise by reaching reliable production levels of output. For those reasons, it is impossible for a conventional process plant to service the glass recycling industry, other than in producing cullet for glass fiber insulation manufacturing.

A process plant must separate glass, if for no other reason than that it is not wanted in the recyclable fractions, and so mechanical handling systems capable of conveying it are necessary. Unless it is processed in some way, the material has a negative value. It will be costly to dispose of it. Milling it to a granularity approximating that of sand has some attractions, in that the resulting product may then have a positive value, albeit a small one. Where that is done, conveying systems capable of handling broken glass and milled, shard-free glass are required.

Whether the glass is in as-received or milled form, it is likely to be wet, and may also be sticky as a result of contamination by its original contents. As-received glass has a low *apparent* density in mechanical handling terms, and is therefore bulky as well as being sharp and abrasive. Milled glass has a much higher *apparent* density, and its sharpness is reduced substantially. Its abrasiveness is meanwhile increased equally substantially, and since milling processes inevitably reduce a proportion to dust, the presence of moisture may create a material similar to fine grinding paste.

Flat/Troughed Synthetic Rubber — ***Rating 8 (milled glass); 3 (as received).*** Very sharp materials and wet conveyor belting, whether synthetic or natural rubber, do not mix. Both types cut very easily indeed in that condition. Abrasion-resistant covers are not cut-resistant, and loading broken glass onto them via a transfer chute, however shallow, will lead to very rapid deterioration of the cover plies. Therefore, synthetic belt conveyors should not be considered for handling glass in its as-received form.

If the glass has been milled, particularly by ball-milling, then it is likely to be substantially free of shards. Although it becomes more abrasive, it is no longer likely to cut the belting. It assumes properties similar to those of sand, and since that is the origin of glass, it could be argued that the material has been effectively restored to its original form. Belt conveyors are widely used to transport sand, and with some precautions, they do so without difficulty.

When moist, the material does not flow easily. Transfer chutes need to be steeply inclined or the load will simply build up on the sides. Chute slope angles of less than 80° to the horizontal should be avoided. Also, as with wet sand, retention on the return strand of the belt is likely to occur, and if it is allowed to remain, particles will become impacted into the cover ply as it passes over the return idlers. Hard scrapers may cause belt abrasion, and reinforced rubber scrapers tend to retain particles in the reinforcing. Soft rubber scapers work reasonably well, but they need regular inspection, as they may wear rapidly. Vibrating cleaners are reasonably successful, and since the density of the material is sufficient for it to settle out of water rapidly, water jet cleaners are very effective. They should, however, be followed by rubber scrapers to remove water from the belt.

Flat belts should be avoided, since with them, side skirts are essential to prevent spillage. Milled glass will inevitably become trapped between the belt and the skirting, and it will then abrade the belt edges quite rapidly. Troughed belts are preferable, with the loading chute designed to deliver the material into the center of the width.

Flat/Troughed Polyurethane — ***Rating 9 (milled glass); 5 (as received).*** Polyurethane belts are able to transport as-received glass, but while they are more resistant to cuts than synthetic rubber, they are still by no means proof against them. The comments for rubber belts therefore apply equally to polyurethane ones.

Belting of this type is much better able to handle milled glass, since it is more abrasion resistant. Damp materials do not stick to it so easily, but only so as long as the quality of the surface ply is maintained. Therefore, belt cleaning is equally as important as with rubber belting, and similar techniques are appropriate.

Steel Plate — ***Rating 2 (milled glass); 10 (as received).*** Light-weight plate feeders are an excellent way of transporting as-received glass, since they are very resistant to abrasion and impervious to cuts. However, since belt cleaners cannot be fitted due to the profiles of the flights, a certain amount of carry-over must be expected. Little can be done to accommodate that, other than to ensure that there is sufficient space beneath the conveyor for regular cleaning of spillage.

Conversely, machines of this type are less than ideal for handling milled glass, since the particles tend to become trapped between the flight overlaps. There they create rapid wear, and if allowed to fall through onto the carrying chains, may also produce unacceptable wear rates between the rollers and the tracks, and in the roller bushes.

Drag Links — ***Rating 7 (milled glass); 8 (as received).*** Drag links are quite capable of dealing with glass in either form, in spite of the question of abrasion. As-received glass is less likely than milled to cause wear problems, and the steel wearing parts are impervious to cuts. Conveyors for aggressive materials are normally designed such that the platework upon which the load runs is easily replaced as it wears. Being totally enclosed, drag links offer some benefits in terms of safety of personnel.

When used to transport milled glass, the trough in which the load runs should be sufficiently deep to keep the material away from the drag chains. Where this is impossible, the chains should be solid, without rollers, and designed to run upon replaceable wear strips.

Chain Conveyors — ***Rating 6 (milled glass); 8 (as received).*** Again, conveyors of this type can handle glass in either form, and they are somewhat less sensitive to casing wear. However, the single chain and the load are in contact, at least at the feeding point, and in many designs along the length. The chain runs upon a manganese steel wearing rail, and so wear rates even here are somewhat less excessive than might be expected. The rail can be replaced when worn, although doing so is not always an easy task.

Screw Feeders — ***Rating 6 (milled glass); 7 (as received).*** Screw feeders will transport glass in either form, although at the expense of some screw and casing wear. Trapping of particles between the screws and casings is unlikely to cause problems, as glass is too brittle to jam the machines.

Moving Floors — ***Rating 3 (milled glass); 5 (as received).*** Either quality of glass can be moved by these machines, but wear rates are likely to be high. Their ability to transport very large volumes spread over an extensive floor area may make them an acceptable choice for large bulk store unloading.

Sidewall Belts — ***Rating 8 (milled glass); 5 (as received).*** The presence of flights prevents rollback of as-received glass upon inclined sections, and so the gradual accumulation of minor cuts that may eventually ruin a conventional belt is avoided. That benefit may be outweighed by the possibility of sidewall damage where sharp objects become trapped as the walls flex.

Milled glass does not present the same level of difficulties, but a sidewall belt is more difficult to clean if the load is granular, wet, and possibly sticky. Vibrating cleaners provide one solution, and since sidewall conveyors do not have return idlers beneath the flights, a certain amount of spillage from the return strand is tolerable.

Bucket Elevators — ***Rating 5 (continuous elevators); 7 (centrifugal discharge elevators); 8 (positive discharge elevators).*** Continuous elevators will transport glass in either form adequately, although wet milled glass may not be released cleanly by the buckets at the discharge point. Excessive spillage into the boot should be avoided, otherwise chain and sprocket wear may result. Centrifugal discharge elevators are more capable of clean release, and spillage is less significant, since machines of this type are intended to dredge from the boot. Positive discharge machines are best, since they are specifically designed for sticky materials.

Folding Belts — ***Rating 8 (milled glass); 1 (as received).*** Milled glass may be handled without difficulty, but as-received glass should be treated with caution. Sharp edges will inevitably penetrate the cover ply if any pressure is exerted upon the load as the belt folds. Belt cleaning requirements are similar to those for conventional flat and troughed conveyors.

Suction and Pressure Pneumatics — ***Rating 8 (milled glass); 1 (as received).*** Neither system could be recommended for glass transfer. The high *true* bulk density which governs behavior in an air stream would require very high air velocities in a lean-phase system, while load abrasiveness would create unacceptable wear rates in either. Milled or as-received glass being discharged at high speed into a cyclone separator or a hopper would have a "shot blasting" effect upon casings, to the extent that they would be rapidly eroded.

3.2.7.12 Residual Wastes

Material such as ash, minerals and soil, dust, animal fats, vegetable oils, mineral oils, and human and animal wastes are grouped together for convenience, since it is rare that they can be identified separately from each other in municipal waste. Generally, they combine to form the unidentifiable "minus ten millimeter" fraction referred to in many waste analyses. Even so, each has some characteristics that add to the overall behavior of the mixture.

Ash is abrasive and potentially corrosive, and it can liberate inert dust, which contains heavy metals and traces of toxins. Its effect in the mixture is almost entirely abrasive, since its other characteristics are either counteracted or diluted by the other materials.

Minerals and soil occur in municipal waste mainly through the route of the disposal of garden wastes. In this sense, minerals arise in the form of stones, which can create jamming of moving parts in some types of handling systems. Soil, meanwhile, is relatively harmless other than in causing some slight increase in the abrasiveness of the mixture.

Dust, other than that from ash, arises largely from disposable bags from vacuum cleaners, and is characterized by its fibrous nature and high lead content. It is flammable, although not highly so, and is the main source of combustible dust in waste reception areas. In provides the matrix upon which oils and fats accumulate, and it helps to make the mixture sticky.

Animal fats and vegetable oils arise from kitchen wastes, and both are biodegradable, although the latter is much less so than the former. Animal fats particularly, being readily accumulated in handling systems, may create health hazards for maintenance staff as a result of biological activity. They damage natural rubber belting by softening the rubber compound and causing separation from the reinforcing fabric. Vegetable oils adhere to moving parts and under conditions of pressure and temperature, may degrade to form a tough, sticky coating.

Mineral oils do not occur in significant quantities, and should not appear in municipal waste at all. They also are capable of causing severe damage to natural rubber. Their main source is probably from home automotive servicing, where it is convenient to dispose of oil filters etc. in the trash bin. Mineral oils are therefore likely to increase the amounts of toxic heavy metals in the mixture, but otherwise have no serious adverse effects.

Human and animal wastes occur much more frequently in municipal waste than is generally expected. Human excrement, particularly, arises from disposable diapers, and may be mixed throughout the waste in some quantity. The first stages of separation in most treatment processes tend to liberate the contents of diapers and mix them intimately in the minus five millimeter fraction. As a result, the fraction is likely to be contaminated with human coliforms, and is quite strongly alkaline due to the release of ammonia.

Animal wastes arise mainly from cat litter, and therefore are a potentially significant source of parasitic organisms. The litter itself is granular, mildly abrasive, and highly absorbent. It therefore may provide the mixture with the means of retaining both moisture, oils, and fats.

The result of mixing all of these components together in the minus five millimeter fraction is to create a material that is biologically active, contains pathogenic and parasitic organisms, and is abrasive and mildly corrosive. However, the mixture rarely appears in isolation in any part of a waste treatment process, more often reporting to the organic and putrescible stream. There it is diluted to the extent that its mechanical hazards cease to be significant. Its biological hazards unfortunately remain, but do not substantially increase existing contamination of the fraction to which it is added.

For those reasons, there is little to be gained in exploring the mechanical handling characteristics of the minus five millimeter fraction in its as-received form.

3.3 SOME COMMON PROBLEMS IN MECHANICAL HANDLING SYSTEMS

3.3.1 Cranes

Cranes in waste processing plants operate in humid and dusty environments, where regular cleaning is necessary to maintain efficient operation. In these circumstances some of the most common problems are as follows:

Hoist wire rope failures as a result of internal corrosion between strands. Generally, these occur as a result of using the wrong type of wire rope lubricant. Very viscous lubricants should be avoided, since they collect airborne dust very rapidly, and the absorbency of the dust then prevents the lubricant from penetrating through the bundles.

Scuffing and stranding (breaking of outer individual strands) of hoist wire ropes, caused mainly by abrasion of the rope guides of hoist wire reeling drums. Again, the problem is to some extent due to an initial buildup of abrasive dust upon the wire ropes. The rope guides should be regularly inspected and dressed where necessary.

Corrosion in electrical switchgear and motor control centers, caused by the high ambient humidity and alkaline nature of the dusts. Since neither the dust nor the humidity can be entirely eliminated, all switchgear cabinets should be dust- and drip-proof at least, and should be sealed as thoroughly as possible. Each should be fitted with an electrical heating element, which should be left on continuously to prevent condensation.

Wheel slip on gantry rails, generally caused by a buildup of fibrous dust as a result of incautious lubrication of crane wheels. Overlubrication creates many more problems than underlubrication in all mechanical processes, and should be avoided. At the design stage, the bearings of the crane wheels should be specified without automatic lubricators. It is better to introduce into the planned maintenance routines an annual program for opening, cleaning, and repacking the grease casings of wheel bearings. The procedure ensures that they never become overcharged.

3.3.2 Grabs

Waste-handling grabs that have been correctly specified do not generally give too many problems, but here again there are some problems characteristic of the duty unless accounted for at the design stage. These are

Overheating of hydraulic oil, most commonly caused by long periods of over pressure release through spring-loaded relief valves in the hydraulic circuit. As stated in Section 3.2, a waste-handling grab should have a hydraulic oil reserve of at least four times the pump capacity per min. That ensures that in normal circumstances the oil has sufficient residence time in the reservoir to lose the heat it has gained in the pressure circuits. However, space in a grab casing is necessarily limited, and a larger reserve cannot easily be accommodated, however attractive doing so might be. Therefore, in circumstances where the grab is being called upon to deal with particularly compacted waste, the hydraulic loading may still cause overheating. The best defense against that is to specify pilot-operated relief valves, since they are always either fully open or fully closed. They do not throttle the oil flow, and do not generate heat.

Condensation inside hydraulic oil reservoirs, caused by allowing the oil to cool excessively. This happens when grabs are parked for any length of time in the operating area after periods of use. As the air space above the oil also cools, it contracts and draws in humid air, and the solution is twofold. In the plant design, it is advisable to allow for a grab storage bay outside of, and preferably below, the working area. There machines can be allowed to cool away from the humidity of the bunker environment. Second, since the humidity cannot ever be entirely excluded, and because the oil reservoir must "breathe" as the oil temperature varies during service, the tank should be fabricated from stainless steel.

Failure of control conductors in the power cables to larger grabs, commonly the result of stresses induced by the weight of the cable itself. Generally, these failures occur where the hoist distance is too great, and good design avoids them by specifying bunkers with the largest possible surface area and the minimum depth.

3.3.3 Shovel Loaders

Many of the problems that occur with shovel loaders have been dealt with in Section 3.2. However, there are some operational considerations.

Corrosion in fuel tanks is caused by the aspiration of humid air in waste treatment plant reception halls, and the solution lies both in operating procedures and design. Operationally, the fuel tank should be completely refilled at the end of every operating day, and always immediately before the machine is taken out of service. As a result, as the machine cools, there will be insufficient air space in the fuel tank for outside air to be drawn in. At the design stage, meanwhile, it is wise to specify stainless steel fuel tanks and water filters in the fuel lines.

Fires in engine compartments are surprisingly common, and are generally caused by a buildup of spilled oil inside the belly plates beneath the engine. All shovel loaders that operate in waste handling facilities have protective plates underneath the engine compartment, to prevent damage to the engine sump when running over waste. Unfortunately, those belly plates provide a repository for loose paper and dust, which when mixed with oil becomes extremely flammable. Hot exhaust manifolds can provide an excellent source of ignition.

Nothing can be done to prevent material settling upon the plates, and even the most conscientious maintenance routine is unlikely to prevent occasional oil leakage. For those reasons, the machines should be specified with belly plates that are easily removable, so that regular cleaning is possible. Additionally, the provision of a steam cleaner in the machine maintenance workshop is a wise addition to the plant tooling.

3.3.4 Conveyors

3.3.4.1 Volumetric Capacity

One of the most common problems to afflict waste processing plant handling systems is brought about by confusing *mass* balances with *volumetric* balances when establishing conveyor capacity. The process designer should always be aware that, since waste is a mixture of materials and not truly a material in its own right, the removal of any fraction may have a significant effect upon the remainder. That effect may be quite different from anything intuition would suggest.

For example, if crude waste is first treated in a trommel screen, the organic, glass, and minus five millimeter fractions will report to the undersize stream, while the paper, card, plastic, and cans will report to the oversize. In the process, the latter fraction will be violently agitated, and its bulk density will be very considerably reduced. In mass balance terms, every 1 t of waste input has been divided into two fractions, but nothing has been gained or lost. The two still add up to the original.

In volumetric terms, however, the situation is quite different. The considerable reduction in density of the oversize fraction corresponds to an equally significant increase in volume. The undersized fraction meanwhile remains substantially the same as it was. Consequently, it is quite possible for a situation to be reached where a greater volume of material is leaving a machine than is entering, even after a fraction has been removed. For that reason, the concepts of true and apparent bulk density as defined in Section 2.3 are of extreme importance. It is insufficient to assume the original density of waste and to apply it throughout the design, since in that event many of the conveyor systems will be capable of less than half design capacity.

Apparent bulk densities are difficult, and often impossible, to measure. Many factors may affect them, and they are unlikely to remain consistent. For example, the receipt of a day-old load of compacted waste from a transfer station may temporarily double the apparent density of granulated paper and card in a fuel process. A safer design route, therefore, is one where the lowest density of a potential range is always assumed for volumetric conveyor calculations. Use the lowest *apparent* density to establish the belt width, bed depth, and traveling speed, and the mass balance to establish the drive motor power.

3.3.4.2 Housekeeping

Conveyor systems in waste processing plants always suffer from spillage upon occasion. The very low densities of some of the separated fractions make that inevitable, and a good design should accommodate it. Belt conveyors should never be installed below floor level in ducts, however attractive the concept may be from the space-saving point of view. In such locations, cleaning under the return strand is difficult if not impossible, and the risk of fire is considerably enhanced. There should always be a clear space of at least 300 mm (12 in.) below every belt conveyor, and the guards should be designed to allow access for cleaning tools without risk of fouling moving parts.

Even reception plate feeder conveyors will permit a surprising amount of spillage, although in their case it is mainly grit and dust rather than large particles. Again, access for cleaning is essential, even where a part of the conveyor has to be below floor level to facilitate loading. The access space around the below-floor level section should be at least 5 ft (1.5 m) wide along the sides and $6^1/_2$ ft (2 m) at the end. It should extend to at least $1^1/_2$ ft (500 mm) below the return strand of the belt, and should not be of less than 6 ft overall headroom to allow safe working.

The access space will need to be drained, since there is always a possibility of leaching of contaminated water from the waste. If left, any such water will rapidly become foul and highly malodorous, and so the space should be sloped toward one end at an angle of about 1 in 22, at which point a pumpable sump should be installed.

3.3.4.3 Plate Feeder Conveyor Chains

The carrying chains of plate feeder conveyors operate in quite hostile environments and under widely varying load. Some stretching is inevitable, however carefully the machine is designed, and any irregularity in the stretching of the matched strands is likely to impose extra loading and wear upon sprockets. This can be accommodated at the design stage by ensuring that only one of the pair of tail sprockets is keyed to the shaft, the other being allowed to turn freely. By that means, any differences in the strands of chain can be accommodated.

Second, when the conveyor is first installed, and every time new chains are fitted, a set of gauges should be manufactured. The gauges need be no more than plates with two pointed pins fixed to them, where the pins correspond to the pitch of the chains measured at the link bushes. The gauges should then be used to monitor chain stretch during routine maintenance. There the expected result should be some stretching during the first few months of operation, gradually reducing to a stable condition where no further stretching takes place. If that condition is not met, or if it begins to change again at some time in the life of the machine, then either something is preventing the conveyor from running freely, or the chains are reaching the end of their useful life. Action at that point will prevent a breakage, which if allowed to occur may otherwise close the plant for a considerable period.

Lubrication of the chains is another matter for careful design, and the traditional method of providing oil-drip lubricators should be avoided. That method coats all moving parts liberally in oil over a period of time, and causes a buildup of dust and grit upon the roller tracks. Sooner or later individual chain rollers will jam and wear down rapidly, again leading to a potential chain failure.

There are two potential approaches to the problem. One is to specify the chains supplied with the bearing bushes lubricated with molybdenum disulfide during manufacture. The other is to require every link to be fitted with grease nipples, so that the chains may be hand-lubricated at intervals. This second approach is much more labor-intensive than the first, but will provide better lubrication for the life of the conveyor. The slow operating speed of plate feeder conveyors is such that, given the provision of hand-lubricators driven by pressure canisters, grease injection to every link bush is not as onerous a task as might be imagined. One shot of grease to every bush, once every 6 months, is surprisingly easy to achieve.

3.3.4.4 Conveyor Belt Tracking

All belt conveyors need to be accurately tracked in order that they lay centrally upon head and tail drums, and that side skirts bear equally upon both sides. In waste processing plants, however, not only may the density of the material being conveyed vary, but the loading across the belt may not always be consistent. A belt coming out of track, even temporarily, may cause spillage of aggressive, abrasive materials, which may damage the belt or its support structure. It is therefore a sound precaution to design tracking idlers each side just before the head drum on the carrying strand, and just before the tail drum on the return strand. That way, the belt will always remain in track however much the load changes. For this application, a tracking idler is simply a short roller, bearing-mounted, and installed perpendicular to the belt.

3.3.4.5 Pneumatic Conveyors, Casing Wear

Because the product being conveyed is always in high-speed contact with the walls of a pneumatic conveyor, wear will occur. It will mainly be in evidence at bends, and at any changes in diameter of the duct, and virtually nothing can be done to stop it. However, at the design stage, the effect can be accommodated by using swept bends of the largest radius possible within space constraints, and most importantly by making the bends easily replaceable. Even with a circular duct, it is worth considering making the bends of square cross section, and with only the outer plate replaceable, since that is where wear will occur.

In pneumatic systems being used as driers, it is worth considering making the whole duct out of very heavy-gauge material, particularly at the bends. It is one of the stranger aspects of waste processing that a $^1/_2$ in. (10 mm) wall thickness duct lasts three times as long as a $^1/_4$ in. (5 mm) duct.

3.4 HEALTH AND SAFETY CONSIDERATIONS

Operator health and safety is properly and increasingly seen as of supreme importance in industry, and waste processing is a potentially hazardous occupation on both counts. It is now rare for the designers of a facility to be able to ignore such matters or to claim that they are the province of the facility operations management.

All mechanical handling plant and machinery is potentially dangerous, and that which is carrying municipal waste or any of its fractions is doubly so. As well as the more obvious hazards of moving parts and sharp and abrasive loads, there are a number of less apparent ones of which a process designer should be aware. There are an even greater number of potential hazards of which a process *operator* should be aware.

A vast range of legislation has been enacted worldwide in the interests of human health and safety during the current century, and it is beyond the scope of this book to consider even a small part of it. The most that can be done is to establish ground rules for the safe design and operation of plant, restricting ourselves to risks that are to some degree specific to waste processes.

3.4.1 Cranes

A refuse-handling crane is one of the few machines in a waste processing plant where the operator station may be upon or inside the structure. While there are some exceptions, generally in smaller plants, the major health and safety risks occur in the "ride on" designs. Some affect the machine operator, and some affect other personnel in the vicinity.

3.4.1.1 Operator Health

Health questions arise from the location of the operator, in a moist, dusty environment likely to contain a significant concentration of airborne microorganisms. While there is no history of any specific health problem among such personnel, it is perhaps worth recalling that 40 years ago there was no history of any specific illness caused by working with blue asbestos. Therefore, an alert designer or employer should consider how best to insulate the employee from any environment which is unnatural. The crane driver needs to be provided with clean air, and so some form of cab air conditioning is essential.

An acceptable design for external air supply for single-person occupancy of a space of 212 ft^3 (6 m^3), which is about the size of an average gantry crane cab, is 1410 ft^3 (40 m^3) or six air changes per h. However, while the purpose of a cab air conditioning system is to *improve* the operator's environment, it is very easy to make it *worse!* The atmosphere outside the cab will contain suspended particles of between 1 and 100 μm. It may also contain traces of smoke at less than 0.5 μm. Natural ventilation may provide two air changes per h, and so if the conditioning system was capable of removing the contaminants, it would still introduce as many into the cab as would natural ventilation without any filtration at all. Anything less than a separation capacity would actually make the atmosphere in the cab worse.

In assessing the capacity of an air conditioning system, there are four components, three of which are environmental, and one structural. The first two environmental considerations are the amount of air necessary to replace the CO_2 produced by the occupant's respiration, and that necessary to reduce odors and moisture. Of the total demand of 1410 ft^2 (40 m^3/h), only about a tenth is used for replacing CO_2. The third environmental question concerns the amount of air necessary to reduce the temperature to an acceptable level below ambient, and in this case, a temperature drop of no more than 50°F (10°C) is usual. Achieving that would take considerably more air than either the first or second factors.

Finally, there is the structural consideration: simply, the constraints upon the size of the conditioning unit. The unit must be mounted upon the crane, adjacent to the cab, in a position where it will not obstruct access or the operation of the machine. Yet space upon the span of any crane is somewhat limited. There is insufficient for a large filtration and air conditioning system, but a very small installation may not be able to achieve the filtration efficiency needed.

The compromise which addresses all of the requirements is one of recirculation. There is no point in attempting to cool large quantities of external air by 35°F (1°C), when only 140 ft³/h (4 m³/h) of air is required to deal with the products of respiration. Equally, there is no point in admitting large quantities of external air just to reduce odors, when recirculating a proportion of the inside atmosphere through activated carbon filters would remove the odors very successfully. It then follows that if, of the total air demand of 1410 ft³/h (40 m³/h), 80% were recirculated, the cooling requirement could be significantly reduced and the filtration equipment would only deal with 282 ft³/h (8 m³/h). As a result, a very efficient filter could be installed without encountering space problems.

The design of air conditioning units is not the purpose of this section, since that is a matter the process designer should leave to specialist suppliers. However, there is a need to specify accurately what is required so that those suppliers may meet the requirements. From the above, therefore, a satisfactory specification would be for six air changes per h, each of 282 ft³ (8 m³), with 80% recirculation, capable of a temperature reduction of 50°F (10°C) from 90°F (30°C), with a filtration capability down to 0.3 μm and secondary activated carbon filters.

3.4.1.2 Safety

The safety of both the operator and any other personnel who may need to board gantry cranes must be ensured when boarding, leaving, or working upon the unit. Boarding and leaving arrangements must be considered in emergency as well as routine situations. Crane access that is restricted to a parking bay at the end of the gantry may be sufficient for normal shift changes, but will be entirely inadequate in the event that the crane has broken down and there is a fire in the bunker immediately below.

Access must be possible at any point along the gantry, at least at one side. The access must itself be protected in the event of a serious fire, and it must have a number of means of escape away from the bunker area such that personnel can always find an exit by going away from the fire zone. A concrete walkway that is part of the structure of the bunkers is preferable to steel, since it has better fire resistance and can serve a dual role as the housing for fire hydrants, hoses, and monitors. The walkway must have guard rails sufficiently far from the end of the crane span to avoid trapping points as the crane passes. Access gates should be provided at points along the walkway for boarding the crane in normal circumstances. In emergencies, the rails should be sufficiently close to the crane for the operator to be able to climb over them without difficulty.

The provision of access points addresses the question of driver boarding or leaving, but it can increase the risk for other personnel in the same circumstances. If the driver does not know that another is boarding, he may move the machine and trap the newcomer between the guard rails. The risk is best eliminated by a signaling facility between the end of the crane and the cab, and a simple bell-push system is most widely used. In such a system, a bell push is located upon a stand on the crane span, close enough to be reached by personnel upon the walkway without risk of being trapped between guard rails. The push button should operate a bell in the crane cab, where a second push button operates a repeater bell on the span. Operating procedures can then be drawn up making it mandatory that any personnel wishing to board the crane must first signal their intention to the driver by operating the bell. The driver must then move the crane to the nearest access gate, and signal his permission to board by sounding his repeater bell in turn.

Personnel wishing to board must not do so until instructed by the driver, who must not move the crane again until he is once more alone. An exception to the latter condition may upon occasion be necessary, in order that maintenance staff may inspect wheels, driving gears, and winding systems while the machine is in motion. In those circumstances, the only safe procedure is one where the purpose of the visit is made plain to the driver, who must then operate strictly according to instructions from the maintenance personnel.

3.4.2 Grabs

As with all hydraulic machines, grabs need regular hydraulic system maintenance, and it is essential that the work is carried out in a clean environment. The admission of dust to the system while replacing filters, for example, renders the maintenance program ineffective. The maintenance staff also need clean conditions in which to work, if their health and safety is to be protected. Both requirements lead to the provision in the design of a dedicated grab maintenance bay, which is isolated from the bunker area atmosphere, and which contains cleaning and disinfecting equipment for the grab.

The ideal design for such a bay is one where the crane rails extend over a lowering area, where the grab may be dropped to the maintenance level. There, a wheeled trolley with a support frame mounted upon it can receive the machine and transport it through a cleaning bay equipped with steam cleaners, into the plant workshop area. The factors that determine the design of the whole maintenance facility then arise from operator health and safety considerations, since if they are satisfied, then so will be the conditions necessary for hydraulic system maintenance.

3.4.2.1 Health

Refuse handling grabs are in almost continual and intimate contact with waste, and all of their external parts are likely to be contaminated by materials which are at least biologically active, and which may also be toxic. Biological contamination is likely to contain pathogenic organisms and possibly human coliforms. The range of potential toxins is almost endless.

Applying liquid disinfectants to the framework of the machine is likely to be inadequate. Most of the contamination is likely to be an almost totally waterproof mixture of oils, dust, and animal fats. It is for that reason that steam cleaners, which soften the mixture and blast it off, are the more effective way of cleaning the machine. They have the added advantage that they do so without flooding the hydraulics and associated electrical systems. They cannot, however, be guaranteed to sterilize either the contaminates or the machine from which they have been washed. They simply remove them from the immediate vicinity of human operations, and so the cleaning bay must be drained into a dedicated interceptor, with access for a tank-emptying vehicle. The interceptor should be of the design proposed in Section 3.2.

Even when it has been cleaned, the grab may still carry a significant contamination of microorganisms, and so all maintenance workers should be issued with suitable protective clothing, and particularly with nonabsorbent gloves.

3.4.2.2 Safety

When released from the crane, all cactus grabs are top-heavy in the closed position. Therefore, the holding frame on the transporting trolley in the maintenance area should ideally be capable of securing the machine in the open position. When it is fully open, the hydraulic casing will be at a much lower level and the tines will be much easier to clean. A suitable trolley is one made from heavy rolled steel channels, with the flanges downward, arranged in a cross or star shape to match the number of tines on the grab. A heavy, polyurethane-tired castor at each end immediately beneath the tines, provides a suitable means of mobility, and a towing eye at each end permits the rig to be moved in any direction by a forklift truck.

One significant advantage of a skeletal transporting trolley of the type described is in that it allows unobstructed access between the tines, so that an access platform may be moved in close to the upper casing for working upon the hydraulics. Access for such a platform is important, first because ladders, rested against the casing, should never be used. The casings of cactus grabs are cylindrical, and the danger of the ladder slipping sideways is too great. Second, working upon hydraulic systems is always likely to cause oil spills, which would make a ladder an extremely dangerous device indeed.

A further point in favor of the skeletal trolley is that it holds all of the grab tines in the open position. In so doing, it takes up a great deal of space, but it ensures that when a hydraulic circuit is disconnected, workers are not faced with one or more tines suddenly swinging closed under their own weight and forcing high-pressure oil out of the cylinders in the process.

Where clamshell grabs are used, the circumstances are somewhat different. Such a machine is inherently stable when its clams are fully open, and virtually any flatbed trolley large enough to accommodate the clam edges is sufficient. In that condition, the hydraulic casing is much lower than is that of a cactus grab, and access is considerably easier. Even so, it is likely to still be high enough off the floor to be inconvenient, and so again some form of access platform is essential.

Finally, whatever type of grab is used, and however well it has been steam-cleaned, it is still likely to carry sharp pieces of metal or glass in crevices and trapping points. There is little that the process designer can do to overcome that problem, but it is a matter that should be considered and warned against in preparing operating manuals.

3.4.3 Plate Feeder Conveyors

Steel plate feeders are dangerously deceptive machines in that they run very slowly. It almost seems that there is no opportunity for anyone straying near to become entangled in the moving parts. It is tempting to save capital expense by economizing on guards and safety devices. The temptation should be resisted. Plate feeders are also extremely powerful machines, and they do not stop automatically if a member of staff becomes entangled.

3.4.3.1 Health

Any machine that handles crude waste will be heavily contaminated biologically and chemically, and plate feeders are no exception. But in addition they are often, of necessity, installed with at least part of their length below floor level. Liquors that drain out of the waste and through the conveyor structure may lie in the base of the below-floor chamber, and they may be both toxic and biologically hazardous. The chamber may therefore become a place of significant health risk, and there are several design points which should be considered.

It is not unknown for crude municipal waste to contain solvents, and where that occurs, they most often find their way through into the below-floor chamber. As a result, a toxic or even flammable vapor may develop. Even in the absence of solvents, the liquors that may drain from the waste may lead to a buildup of microorganisms in the atmosphere beneath the conveyor. The first design consideration, therefore, is one of ventilation, and an air extraction duct alone is insufficient.

Where air is extracted, external air will be drawn in to take its place. If the plate feeder conveyor is being used for waste reception, then the below-floor chamber will be in the waste reception building, where the low-level air will be charged with vehicle exhaust fumes, and possibly also with fuel vapors. Introducing either component into the conveyor chamber is likely to make the atmosphere there worse rather than better. It is necessary for the chamber to have some means of forced ventilation, but the location of that is important.

If the extraction duct is placed at the closed end of the chamber — the logical location, since that is where stagnant air will otherwise collect — then the forced air inlet should be at the other end. By that means, a flow of air can be encouraged along both sides and beneath the conveyor. However, the inlet duct should be placed toward the extraction, just beyond the open end of the chamber where the conveyor inclines past floor level. Careful balancing of the air inlet and exhaust then ensures that there is no negative pressure zone at the open end of the chamber, and no tendency to draw in air from the tipping apron.

In Section 2.5, the subject of Weill's disease was raised, and nowhere in a waste treatment plant is the threat of infection greater than under and around the reception conveyors. This is particularly so in the drainage liquors — apart from their potential toxicity — and skin contact with them should be avoided. It is the task of the designer to make such contact as unlikely as possible, and sloping and sealing the floor is only the beginning.

At the lowest point of the floor, there must be a sump in which any drainage can collect, and from which it can be pumped. It must be accessible, so that it can be regularly disinfected, particularly at times when any maintenance work on the pump is necessary. It must be sufficiently enclosed by a grid or mesh so that solids cannot gain access and block the pump, and it must be fitted with float switches or other level controls so that the pump keeps the level down.

3.4.3.2 Safety

A common problem with chambers under plate feeder conveyors is one of inadequate lighting. The reasoning appears to be that since the area is rarely visited, and no one works there full time, one or two low-wattage bulbs along the wall will be sufficient. In fact, there is a need for two levels of lighting, one to maintain good operating practice, and one for periods when personnel are active in the chamber.

The purpose of the first level of lighting is to permit plant attendants to see down the chamber, to ensure that all is well with the drainage arrangements and to monitor any occurrence of spillage from the conveyor. To carry out those duties, the operator does not need to enter the chamber, but he does need sufficient lighting to see at a glance that there is nothing needing attention. The second level of lighting, meanwhile, is necessary when any maintenance or cleaning work is being undertaken, and it is usually provided by a separate circuit. Because of its purpose, the second level must not cast dense shadows, and it should ideally illuminate all moving parts of the conveyor. Both lighting systems should use waterproof fittings, but the second level should, in addition, be installed behind diffusers so that the light is glare-free.

In spite of the very low speed of the conveyor, guarding of all of the moving parts is essential. A plate feeder differs from all other conveyors in that it has trapping points at every chain pitch. There are a large number of places where fingers or clothing can become trapped.

3.4.4 Belt Conveyors

3.4.4.1 Health

Belt conveyors are responsible for a considerable number of injuries in an average year, and it is easy to allow this history to conceal the potential health risks that result from their use in waste handling operations. Those risks are not perhaps major, but still they exist.

The most obvious risk is that posed by the creation of dust at transfer chutes, and the more heavily a belt is loaded, the greater the generation of dust. Much of it will occur in submicron particle sizes, and so will not be immediately apparent, and it will always contain bacteria and endotoxins. It is possible that some of the bacteria at least could be of human origin, and that some could be pathogenic. Concentrations are usually high, although surprisingly there is no history of any particular type of illness associated with waste treatment plants. The greatest risk appears to be of respiratory discomfort and infections in predisposed people, although those who are healthy do not appear to be at risk.

Irrespective of the few records of health problems associated with waste handling conveyors, precautions against dust are advisable. They may even be a legal requirement in some countries, and in any case dust presents an ever-present fire and explosion hazard which should be addressed. In transfer chutes, the solution is generally one of total enclosure, such that the dust created there cannot easily escape. Aspiration systems are generally useless in this application, since an air flow sufficient to capture the dust is likely to capture a considerable amount of the load as well. Conversely, a system of sufficiently low velocity to avoid doing so will block rapidly. The dust from wastes is highly fibrous, and the dampness always present in waste handling operations encourages it to form dense deposits upon steelwork. It will therefore settle upon aspiration duct walls and form accretions which eventually block the duct completely.

3.4.4.2 Safety

The most common cause of accidents occurring with belt conveyors is trapping of limbs or clothing between the moving belt and troughing or return idlers. Conveyor belting has a high coefficient of resistance, and it grips readily. It is designed to do so, in order to improve load-carrying performance. Therefore, clothing or fingers entering the pinch point between the belt and an idler inevitably results in the whole limb or even the whole body being drawn in.

The only means of prevention is guarding, and guards should be erected in such a way that any moving part that could be reached, even at full arm's length, is entirely enclosed. The guards should be extended to a height of at least $6^1/_2$ ft (2 m) above any floors or walkways, and higher still if the underside and return idlers are exposed at that level.

Protection in the event of a trapping accident is best provided by spring-loaded trip wire switches. These are simple but effective electrical devices consisting of a switch held closed by a wire that runs along the full length of the conveyor. Ideally, two switches should be used on each side, with the wire strung between them, but a single switch with a spring-tensioned wire is acceptable. The purpose of the two-switch or spring-tension design is to ensure that the system is fail-safe. The trip switches are closed in their mid-travel position, and they disconnect if their operating contact moves in either direction. The provision recommended therefore ensures that if the trip wire breaks or is damaged, the switch will disconnect the supply just as readily as it would if pulled.

In positioning the trip wires, it is important to protect both sides of the conveyor wherever access is possible. The wires should be located along the top edge of the side skirt, such that anyone leaning over or being pulled toward the belt disturbs a wire in the process. Trip wires are the only secure operator protection on belt conveyors. Emergency stop buttons, while still very common, are insufficient, since they rely upon voluntary action by another. There is no guarantee that in the event of an accident, another operator will be nearby and in a position to react.

The other prominent cause of belt-conveyor-related accidents lies with belt fasteners. Often, in a handling system design, it is convenient to join the belt by means of riveted metal strips rather than by heat-splicing. Gradually, as the strips pass over rollers, drums, and scrapers, wear causes them to develop sharp and even hooked points that easily catch clothing or cut unwary fingers. For that reason, belt fasteners should never be used to join manual picking belts, and where they are used on transfer belting, warning signs should be placed upon the sides of the conveyors at regular intervals along the length.

3.4.5 Pneumatic Conveyors

While they are not normally considered dangerous machines, and they do not present any immediate health hazards, pneumatic conveyors are still an occasional cause of minor accidents. The most usual incident involves personnel being unbalanced by the suction of lean-phase systems.

Suction pneumatic conveyors commonly work with air speeds of 10 to 20 m/s, which sounds innocuous but is actually equivalent to a wind speed of 20 to 40 mph, which is equivalent to strong to gale force. It is sufficient to capture limbs or clothing and to cause the victim to fall against the metal casing.

The discharge end of a dense-phase system is clearly dangerous, although it is not normally accessible. However, during routine maintenance it is possible for personnel to be inside the bunkers or hoppers to which such a system delivers. In those situations, it must be impossible for the system to operate, since the consequences of an individual receiving the conveyor discharge, traveling at a velocity of 65 ft/s (20 m/s), could easily be fatal.

3.4.6 Drag Link and Screw Conveyors

Machines of these types are among the safest of mechanical handling devices, since their operation requires them to be totally enclosed. However, they can easily be transformed into the most dangerous machines if covers are left off for any reason. Guarding and safety switches are then of no value, and there is no defense mechanism other than strict operating procedures.

4 COMPOSTING IN WASTE MANAGEMENT

Composting can be defined as the aerobic (oxygen-requiring) biological decomposition of organic waste to yield a stable, hygienic material that is beneficial to soil and plant growth. The products of the decomposition are carbon dioxide, water, and solid matter which is undergoing the process of humification, i.e., being converted into soil organic matter whose major components are humic and fulvic acids. The process has been widely studied but is poorly understood. The organisms involved in the composting process are abundant both in air and soil, so decomposition of this nature will happen without human intervention, as is clear from the generation of odors during the storage of putrescible matter for a short period of time.

Placement of mixed kitchen waste into small compost bins will give rise to material decomposed sufficiently to lose the identity of individual waste components over a few months, but little or no control is exerted over the kind of compost that is ultimately produced, and its chemical structure is very variable.

The key to successful compost production is careful control of the process from beginning to end, not simply relying on nature to take its course. Failure to control the process adequately will result in a number of problems which, if not addressed rapidly, could result in closure of the production plant, or reduce the marketability of the product.

4.1 INCOMING COMPOSTABLE MATERIALS

It is important to ensure that the incoming organic matter from composting is of an adequate quality. Specifically, it needs to be low in contaminating inclusions, particularly glass, plastic, and sources of heavy metals such as batteries, a major source. Ideally, therefore the components of a compost plant feedstock would be raw green waste from parks, gardens, and agricultural processes or well-separated kitchen domestic waste. Organic wastes high in lignin and cellulose (e.g., wood and paper) may be included in mixtures to balance wastes containing high quantities of nitrogen, for example, grass cuttings. Although precise definition of the quantities of each type of incoming waste may not always be possible, sufficient analysis should be carried out to enable a fairly accurate definition of the overall composition of the feedstock. This will allow an estimation of some of the likely properties of the final product.

4.1.1 Compost Production

The aims when undergoing a large-scale compost production must be to control the process in order to minimize emissions to the environment, most particularly to air and water. The emissions to air are mainly carbon dioxide and water vapor, although volatile organic compounds (VOCs) and ammonia may arise. Many VOCs are malodorous and may result in a public nuisance. Water and land pollution could arise from uncontrolled leachates from compost piles.

The most important considerations in satisfactory compost production are

1. Availability of oxygen. In order for decomposition of organic matter to occur at the maximum rate, the microorganisms must have a consistent oxygen supply. Absence of oxygen leads to a preponderance of anaerobic organisms that produce phytotoxins (plant poisons) and malodors, and slows the composting process considerably.
2. Availability of moisture. Water content in a compost pile is important. Above 65% water, the material becomes waterlogged and is difficult to break up and aerate. Below 25% moisture, microbiological activity virtually ceases and the material enters an inert state until rewetted. Active compost piles will lose water at a high rate, since temperatures can rise as high as 80°C. In these cases, it is necessary to ensure that the piles can be sprayed to maintain the required moisture level.
3. Attainment of high temperature. Temperature in large compost piles (with a volume of greater than 1 m^3 or 35 ft^3) can naturally rise to in excess of 80°C (175°F). This heat is generated through the metabolic activity of microorganisms. Different microorganisms live and proliferate at different temperatures; psychrophiles are active up to 35°C (95°F), mesophiles to 50°C (122°F), and thermophiles from 50°C upward. Pathogens may be destroyed by temperatures in excess of 55°C (132°F).

4.2 WASTE MANAGEMENT STRATEGY

Composting can never be a complete waste disposal route by itself. Municipal waste contains inert fractions that are proportionally too large for the technology to treat more than about 55% of the arisings in most developed countries. It does, however, offer a number of attractive contributions to a waste disposal strategy: In achieving volume and mass reductions of 70% and 50%, respectively, it can dramatically reduce the amounts of pollutants going to landfill. It can provide enrichment and humus to soils that are otherwise unproductive. Derelict land can be reclaimed after use for industrial purposes. Waste recycling rates can be increased dramatically. Composting associated with other forms of reprocessing can elevate recycling rates to nearly 90%.

In support of organic waste landfill, it is often argued that any problems caused are only temporary: the wastes soon decompose to a biologically stable residue, and the void space originally occupied is soon restored as a result of the decomposition. The counter argument is that the route to biological stability involves the creation of major pollutants, including methane, hydrogen sulfide, ammonia, volatile organic compounds, and liquid effluents contaminated with heavy metals. Modern landfill sites, where the waste is laid down by heavy compaction machinery, are something of an unknown in terms of their active life, but estimates of 60 years and over have been suggested. The same organic fraction can decompose to near stability in 10 to 15 days in an aerobic process. Irrespective of the biological life of a landfill, it is unarguable that if the organic fraction were first digested aerobically, then on an average western hemisphere waste analysis, the waste inputs would be reduced by at least 15%.

A disposal strategy is not simply confined to municipal waste. In fact, many things fit the general description of waste — either something for which society has no further use, or something society can no longer use. In those terms, derelict industrial land is waste, as is land so lacking in humus and nutrients that it is totally unproductive even as wilderness. A logical strategy does not simply address one problem in isolation, but rather seeks where possible to resolve two by combining one with another. The recovery of derelict land by the use of municipal waste compost is an excellent example of that philosophy.

Human society creates waste on a scale directly related to the standard of living achieved. Rich countries all have a waste disposal problem. Poor ones do not. Rich countries dispose of their waste, while poor ones reuse it. Those statements are largely true whatever definition of waste is applied, and so it is the wealthy nations that have large areas of derelict land as well. Unfortunately, those nations are also generally inclined toward environmental protection and its rewards of comfortable self-satisfaction. Throughout Europe, for example, legislation has been and is being enacted that is intended to protect the soil, by imposing strict limits upon chemical and metal concentrations that can be introduced into it. Many of those controls also exclude waste-derived composts.

The exclusion of waste-derived composts is the result of considering environmental matters in isolation from each other. In some parts of Europe, the materials are prohibited from use on land that has a higher natural heavy metals level than the compost does. It happens because the legislation sets concentrations relating to high-quality topsoil, which rarely exists. It results in the bureaucratic insanity of damaged and polluted land having to stay polluted and useless, while further tracts are used for the disposal of crude waste and rendered equally polluted.

Since the production of waste is inescapable, all human and animal activities are potentially environmentally damaging. Only the human activities create quantities so large and so discrete as to become actually damaging. This being true, there are only a few ways of protecting the environment. One could insist upon best environmental protection, in which case there could be no waste disposal at all. One could require the best available technology, which would make waste disposal unacceptably expensive, or one could consider the best environmental option. Since composting is a controlled process whereas landfill, for all of its improvements in technology, is not, composting must by definition be the best environmental option for dealing with organic wastes.

The strategy, however, is necessarily divorced from environmental ideals, since the production of compost takes the waste disposal operator out of the service sector and firmly into the manufacturing and marketing sector. There he faces all of the constraints of market forces, consumer demand, and product quality. The question is no longer how to dispose of the waste. It is not even how to acquire the technology, since that can be purchased. Instead it is how to run a business that manufactures a very high-volume, low-value product and must seek penetration into a market that is highly competitive, discerning, and subject to wide seasonal fluctuations.

The establishment of a strategy begins with detailed feasibility studies to determine the characteristics of the wastes available. No manufacturer can expect to succeed without a clear understanding of his feedstock. The considerations requiring attention are

- The proportion of the waste that is compostible or can be used in the composting process as a bulking agent.
- What disposal route is available for materials that cannot be processed into compost.
- What seasonal variations in waste analysis are likely to occur, and to what extent they will affect production.
- How the waste will be delivered to and received at the processing facility.
- To what extent environmental pollution controls will affect the capital and running costs of the operation.
- Which technology is most suitable for the wastes available and the product markets to be targeted.

Having worked through such a study, the prospective developer should have a reasonably clear picture of what can be done to manufacture a product or range of products. The next stage is to establish whether it is worth doing so, and a detailed market survey is essential. To achieve that, a further list of considerations must be addressed:

- What products could be made from the wastes available.
- How many similar products are already offered nationally, regionally, and locally.
- How much compost could the proposed facility create, and what level of market penetration would be necessary to dispose of all of it.
- Over what range would the finished product have to be transported. Compost is a low-density material, and transportation is expensive. Market values are such that transportation in bulk loads is rarely economical beyond about 100 miles (161 km).
- To what extent the product would have to be packaged to meet market expectations.

- How accurately and repeatably the product could be specified. There is little point in attempting to sell to professional users and trade suppliers if the analysis cannot be guaranteed.
- What do the markets themselves require, where are the major users, and what products would they use if they were available. Market penetration does not necessarily imply the creation of an identical product to those already available. It is equally likely to require the development of a market for an entirely new product.
- What special attributes would the new product have that would confer advantages, either in terms of price or performance, over those already on the market.
- What is the likelihood of other waste contractors or public authorities moving into the same field, and to what extent market saturation would be a consequence of their doing so.

Following the criteria laid out here provides the information necessary to establish the likely viability of a composting operation, and from it to determine whether the technology could fit into a waste disposal strategy. It is a complex matter which may result in a large financial investment, and may by its acceptance close off other disposal routes. It is, for example, difficult to justify the continuation of landfill, at least politically if not operationally, when a composting facility exists.

[illegible]

[illegible]

5 SIMPLE WINDROW COMPOSTING SYSTEMS

"Windrow" is a term not found in many dictionaries, but a windrow in waste disposal terms is simply a pile of material subjected to aerobic decomposition. The pile must be aerated, and there are two principal ways of achieving this: forced aeration and mechanical agitation.

A few designs actually utilize both techniques, but they are generally reserved for large installations and are discussed in Chapter 6. Most small facilities nowadays will either be forced or mechanical, and the range within those two categories is sufficient for a quite large choice. Therefore, only these will be considered here.

5.1 WHAT CAN BE TREATED?

As with all waste processes, there are some materials that lend themselves to treatment and some that do not. In some cases, it is simply a matter of their being inert in a biological sense, while others may well be capable of decomposition but create unacceptable environmental impacts.

It may be said that anything that is vegetable in origin is capable of biological digestion, while anything that is mineral is not. This is a good enough approximation for normal waste disposal purposes, although it does ignore the fact that wood, for example, needs to be broken down by fungal activity first, and that activity is not normally sustained in composting windrows.

While almost any solid of organic origin can be windrowed, many materials need to be mixed with something else before the process becomes reliable. Others will decompose in a way that produces unwanted by-products. Some are simply not safe to handle that way in the first place, and some can be processed by one windrowing design but not by another. And, recalling that the purpose is to create a compost or soil ameliorant, there are some materials that will decompose quite readily but will not yield the product quality required.

In all of this apparent confusion, some ground rules can be laid. It is possible to first categorize materials by their behavior in a composting system in very general terms, so that later in this section we can look at how individual processes handle them. The first-stage categorization is as follows:

- Soft organics
- Hard organics
- Cellulosic materials
- Protein wastes
- Human and animal excreta
- Municipal wastes

This list and the following more detailed descriptions are not intended to be a scientifically accurate classification in the way that a biologist or botanist would approach it. The intention is to provide a working guide for the waste disposal engineer who neither knows nor wants to know the intricacies of biochemistry.

In the following subsections, we sill look at each of the categories in more detail.

5.1.1 Soft Organics

An organic material is considered "soft" for these purposes when its cellular structure consists mainly of water. Thus most vegetables and fruit wastes fall into this category, as do a large proportion of garden wastes. They all present problems for windrow composting systems.

In Chapter 2 we discussed how moisture contents of waste vary in storage, and how biological activity can change the organic moisture content into free moisture, making the waste appear to be wetter as time goes on. In that sense, windrow composting is very closely allied to storage, so soft organics placed in a windrow for any length of time will simply become wetter as the biological activity increases. Their high initial organic moisture content will be translated into an equally high free moisture content. That in turn leads to anaerobicity, leachate, foul odors, and a disgusting mess that does not resemble the ideal for compost.

Windrow systems are essentially aerobic in nature, that is, they rely upon the activity of oxygen-breathing bacteria. Nature, meanwhile, will happily accept either aerobic or anaerobic bacteria depending upon the environment, and an excess of water can exclude oxygen to such an extent that no amount of windrowing will ever reintroduce it. Therefore, a good working rule of thumb is that soft organics should not be aerobically composted on their own.

Examples of soft organics include root vegetables, leaf vegetables, soft fruits, grass cuttings, soft foliage, and most kitchen wastes.

5.1.2 Hard Organics

These materials have a comparatively low cellular moisture content in relation to their total mass. In a composting system, they will decompose successfully with minimal or even no leachate production, are unlikely ever to go anaerobic, and do not create smells. Often, in fact, their cellular moisture content can be too low to decompose fully without the addition of extra water. A mixture of soft and hard organics is a very successful basis for a windrow composting project.

Examples of hard organics include fresh leaves, most flower foliage, and hedge cuttings.

5.1.3 Cellulosic Materials

These are materials where the cellular structure is largely lignin or cellulose and the moisture content is fairly low. Most either decompose slowly when exposed to bacteria or do not decompose at all.

In general terms, such materials first require the attention of fungi to break them down a form that then is available to bacteria, but most fungi cannot withstand high temperatures. Therefore, it follows that most lignin and cellulosics will not decompose appreciably in windrow composting systems where high temperatures are both the norm and essential to sterilization. That does not mean that they should be excluded from such processes. On the contrary, if the soft organic fraction is unavoidably high, it may be that a proportion of cellulosic material will be necessary to maintain granularity and the corresponding air-filled porosity that aerobicity requires.

Examples of cellulosic materials include chipped wood, straw, dried leaves, bark, and paper.

5.1.4 Protein Wastes

Protein wastes are materials of animal origin and are best avoided in any windrow system. Certainly, they decompose readily enough, and there is an argument that a certain amount of protein makes a very good compost. Chicken offal, for example, has been composted many times with great success, and blood from abattoirs is a potentially excellent source of essential ingredients. But protein decomposition creates a high level of odor and attracts vermin and insect infestation on a significant scale. It is a material for enclosed digester systems only.

Examples of protein wastes include food wastes, animal processing wastes, and dairy wastes.

5.1.5 Human and Animal Excreta

There is a further distinction here between the two potential sources of such materials: Whether they come from carnivores or herbivores. Carnivores, including human beings, excrete wastes that may contain human and animal pathogens and very likely parasitic organisms as well. They can be treated in windrow systems, but not before they have been first processed by some other means to reduce their protein content, and the process control necessary to ensure the safety of the final product makes their use hardly worthwhile.

There are many composting systems in the U.S. where treated sewage sludge is routinely mixed with the organic fraction, and the results seem beneficial. Sludges are a good source of essential nitrogen and carbon, and provided that careful monitoring reduces the risk of heavy and toxic metals contamination, then biologically there is no reason to exclude them.

However, from the process control point of view, one should always approach sewage sludges with healthy skepticism. It is surprising what toxic metals will pass through the human body and be concentrated by it, quite apart from those that may result from industrial processes. And at some stage the product has to be sold to someone, who may well not be attracted by the compost if he knows what it is made from. The days when human wastes were considered good for the garden are all but past.

Herbivore wastes are a different matter. Generally they do not pose a great threat in terms of pathogens, although the possibility of parasitic organisms cannot be ruled out. However, even in the latter case it should be realized that such parasites are generally designed to live on or in creatures of a quite different physiology from human beings.

Again, the materials can be a very good source of nitrogen for the composting process, and they generally do not contain enough protein to create any particular odor problem if the process is maintained aerobic. They are mostly regarded favorably by the buying public, with whom the idealistic concept of horse manure on the roses has by no means been abandoned.

Even so, there may be problems. Many herbivore wastes come in the form of slurries, which are acceptable in reasonable quantities when injected into the windrow. They can, however, be a source of smells if sprayed over it in any quantity, and may create anaerobicity (due to their high biological oxygen demand) if applied to excess.

Examples of excreta sources include sewage sludge, sludge cake, stable litter, agricultural wastes, slurries.

5.1.6 Municipal Wastes

These arisings hardly deserve a category of their own here, since they are effectively a mixture of all of the preceding materials. That fact alone makes them unsuitable for windrow composting, at least outdoors, and the level of contamination from metals and glass, etc. mean that a marketable product cannot be made from them without some form of prior separation.

5.2 TYPES OF WINDROW COMPOSTING SYSTEMS

5.2.1 Forced Aeration Systems — Static Piles

Forced aeration designs used to be known as "aerated static piles" before a range of indoor designs became available and "piles" was no longer an adequate description. In its simplest form, a forced aeration windrow is simply a heap of organic waste placed over a perforated pipe on a prepared base. Air from a fan is blown or sucked down the pipe, with the result that oxygen is delivered to or drawn through the material. Figure 5.1 shows a typical installation.

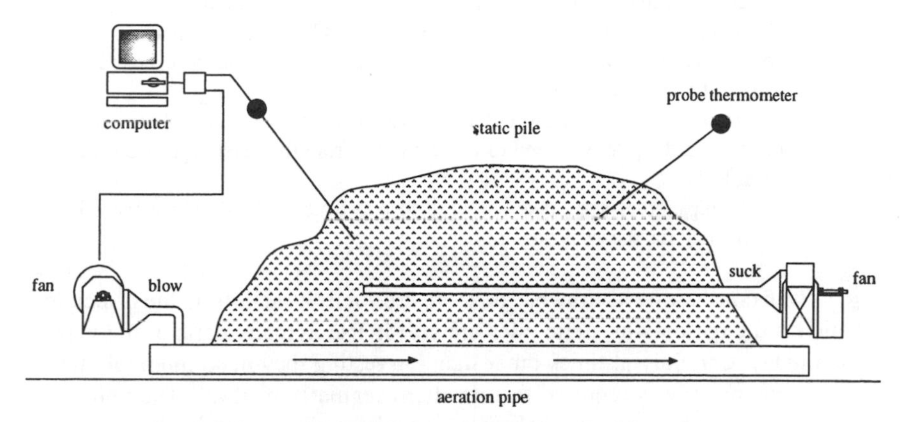

Figure 5.1 Aerated static pile.

Unfortunately, designs of that type suffer from a number of practical disadvantages, although they are still being used even today. The most obvious disadvantage is that they need a power supply for the fans, and a control system to alternate the fans between suction and blowing. But that is only the beginning. The next problem is that when the time comes to harvest the material, the air pipe is in the way of the handling plant, and in the meantime processing is erratic.

It is inevitable that a static pile built over a pipe must be conical in cross section, and not so wide at the base that air cannot possibly be forced out through the mass. Immediately, the initial angle of repose (the angle the sides of the pile will adopt as a result of the nature of the material) comes into play, and it dictates the maximum possible height. Thus, there are practical limits to the cross-sectional area such a pile may have, unless multiple air pipes are employed — not something done often since it simply compounds the obstruction to harvesting.

A conical pile, then, is not the most efficient way of composting for the following reasons.

1. Heat rises, and so is lost by convection and radiation, mainly through horizontal or sloping surfaces. The effect is not as apparent from vertical surfaces, since there is no thermal pressure in those regions. The radiating surface area of a conical pile is significantly greater for a given cross section than is that of a rectangular pile.
2. Since the rising warm air carries moisture with it, the increased radiating surface means that a conical pile dries out more quickly.

3. The cross-sectional area reduces toward the top, limiting the thermal insulating capacity in that region and allowing the material to be colder higher up the pile.
4. Since the material is rarely or never agitated, settlement tends to create voids and fissures, which allow the air to escape without penetrating the material. Inadequate aeration leads to pockets of anaerobicity.
5. The cold outer surfaces are likely to be attacked by fungi, and may also be able to support pathogenic organisms, since they are never exposed to sterilization temperatures.
6. The material nearest the ground may never be aerated at all, simply because the heat in the pile will tend to make the air from the blast pipe rise away from it.
7. The conical sides can make spraying to add water difficult, since there is a tendency for the water to run off.

Even considering its drawbacks, there are some attractions in the principle, which is the reason why static piles are still in use in many parts of the world even today. Certainly at times other than harvesting they need minimal attention, since nobody is required to turn them regularly if at all. They employ fairly simple technology, and where a suitable power supply is available (on a farm, for example), establishing a static pile is straightforward. If there is not too much concern about product quality, where perhaps the material is to be used in-house, then even the prepared base need be no more than a piece of more-or-less level ground. In these circumstances, static piles are probably the cheapest systems available.

It follows, however, that there are some materials for which aerated static piles are less appropriate, and some they will handle without too many problems. They are not particularly good for soft organics alone, since being unturned they tend to settle, becoming waterlogged and anaerobic quite easily. They are also not ideal for entirely hard organics, which tend to dry out too fast in the surplus of air. They are certainly unsuitable for agricultural and domestic effluents, because of the tendency toward runoff and blocking of air paths.

5.2.2 Contained Piles

There are again several proprietary types of contained pile systems, and it is beyond the scope of this book to examine each one in detail. The general principle behind them all, however, is that of containing the material between walls over a perforated floor through which air can be blown, as shown in Figure 5.2. They are effectively basic static piles with improvements, intended usually to address the municipal waste market.

The most obvious advantages of contained piles over more primitive systems are that, being between walls, there is no longer any influence from an angle of repose. The material is contained vertically, and the presence of the walls means that the forced air can only go upward through the pile.

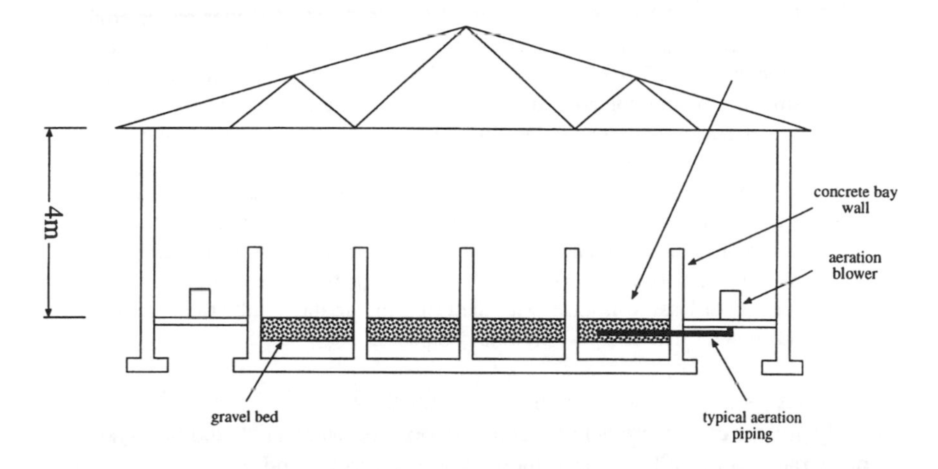

Figure 5.2 Contained pile.

Leachate is also much more easily contained, since it too can only go one way, to the end of the enclosed bay.

Then of course, given a series of bays with concrete walls, erected upon an impervious concrete floor, the next logical step is to simply enclose the whole by extending the outermost walls upward and placing a roof on top. That way, the process becomes enclosed in a building which can then be ventilated through a biological filter. Efficient odor control is now possible, and so kitchen wastes with their inherent protein fraction can be treated without releasing odors into the environment.

As always, however, there are potential problems with the design:

1. It is expensive for what it does: mass concrete walls cost a great deal of money, and if the facility is to have any realistic capacity, there must be a large number of them.
2. It still does not overcome the problems of cold spots in the material. That which is on top of the pile at the beginning stays there.
3. The cold areas encourage fungal growth, and may permit the survival of pathogenic organisms.
4. Again, settlement may result in fissuring, which allows the air to escape without aerating the pile evenly.
5. Since the air pipes are now inset into the floor, the problems caused by having them in the way of harvesting are overcome, but now if a suck–blow regime is employed, the pipes may fill with leachate during the suction cycle.
6. The considerable surplus of air through the pile may cause excessive evaporation, leading to increased drying in local areas and incomplete composting. The evaporation will be difficult to capture and will probably condense upon the roof structures of the building, where it will give rise to severe corrosion problems if the roof is not protected and maintained.

7. Unloading the bays may not be as easy as it sounds. Composted material compacts remarkably well, even without the constraints imposed by mass concrete walls. The effect of a shovel loader pushing into it will cause the structure of the compost to collapse even further, to the extent that the loader bucket may finally contain a material that has been so compressed as to be difficult to refine and screen.
8. Volume loss resulting from the decomposition process may not be easy to make up in a long, comparatively narrow bay. The material laid out first will reduce the most, and so even if the bays are filled from both ends, the middle will soon have less depth. Air can escape there preferentially, taking the line of least resistance, with the result that in the middle the mass will be cooler, drier, and possibly less composted.

In terms of materials suitable for composting, and in spite of the considerably greater cost, there is not much to choose between enclosed bay systems and aerated static piles. The principles are the same and so are the wastes. They do, of course, make the processing of household waste a possibility, but only because they are enclosed within a building. Even then, mainly because of the tendency for the inaccessible center region of the bay to eventually contain less material, they are best for wastes that have a high cellulose content. Straw, for example, will not decompose much in the normal production cycle, and so it will tend to limit volume loss.

5.2.3 Mechanically Turned Bay Systems with Forced Aeration

Many of the problems with enclosed bays can be overcome by mechanically turning the material as well as force-aerating it. Several major waste disposal engineering companies have their own designs, and all are successful to some degree or another. The general principle here is to again build one or more long concrete bays with thick walls, but now to install rails or tracks along the tops and to span them with a turning machine slung from a gantry.

There are several types of machines in use. One is rather like a large paddle wheel which picks up the material and deposits it onto a short conveyor, which delivers it into the next bay along. Once it has reached the end of the track, the machine is switched over to that bay and the process is repeated. Thus, the decomposing waste is gradually transferred across the whole width of the plant, and the concept is that once it has reached the last bay the process is completed.

Obviously, the machine at least addresses the problems of uneven settling of the material. It eliminates the possibility of fissuring, and since water can be added during the turning operation, it overcomes any excessive drying. It ensures that pathogens and fungi are destroyed by making all of the material enter the high-temperature zone at some time or other.

Another system operates in a similar manner, but in this case the machine is rather like a wide, suspended plate conveyor set at an angle from the gantry. As it travels along the tracks, it turns the material and slowly drives it toward the end of the bay by picking it up from the end of the pile and taking it back to drop it behind the direction of travel. The angle of inclination of the cutter belt is such that, in theory at least, material from the top goes on the bottom.

A third mechanical windrow design, the Secit ® system, uses one single, large concrete bay that is floored with slotted metal sheets. The bay is constructed at an angle, with the material infeed at the high point and a collecting conveyor at the low point. Air is blown through the waste from fans along the high side wall, while a large gantry with augers suspended from a traveling carriage runs over the top of the whole bay. As the gantry travels along, with the augers rotating, material is lifted from the bottom of the pile to the top, and the slope of the floor makes it tend to travel sideways toward the low side. The speed at which the material travels can be regulated to some extent by the number of sweeps of the gantry. Obviously, the more often it is used, the faster the waste moves down the slope.

Again, problems with fissuring and uneven drying and heating are overcome, although the principle does not lend itself to a great depth of material. The augers are only supported from the gantry, and there cannot be any bottom bearings. Therefore, there is a limit to the side-thrust they can withstand.

The Buhler Wendelin® system is similar to the Secit system in that again a single large bay is used. In this case, the gantry spanning the bay has a bucket wheel suspended from a traveling carriage, and the wheel picks up material and deposits it onto a conveyor system mounted on the gantry. Facilities exist for the machine to deliver to the final outloading and refining area, or to replace it upon the composting floor so as to maintain the height of the pile.

All of the mechanically larger turned bay systems are intended primarily for the organic fraction of municipal waste, and they deal with it satisfactorily. The atmosphere can become somewhat hostile within the buildings in which they are housed, being saturated with moisture vapor and carbon dioxide gas, and ventilation can become an interesting experience due to the air-movement volume needed to have much effect. But the machinery is mostly automatic in service, and so the area is not a working one in the human operator sense. Maintenance, particularly of the building structures, has to be given careful attention, but it is not beyond the realms of technology.

Suitable materials for such processes cover the range of most organic wastes. The turning devices inevitably introduce a considerable element of shredding, and so even quite hard materials can be treated. It is fairly easy to add water as necessary, and so dry materials need not be a problem, and of course the fact that the whole process is enclosed within a ventilated building means that protein wastes are also suitable, provided there are not too many of them.

5.3 MECHANICALLY TURNED SYSTEMS

Forced aeration technology begins with simple piles over air pipes, and progresses to large facilities that introduce a degree of mechanical agitation.

In some plants, a point is reached where the mechanical agitation equipment costs exceed those of the forced aeration, and they become hybrids. Entirely mechanically aerated facilities, however, tend to adhere to the same principles however large they become. They are, by comparison at least, very primitive. But there is more to producing usable compost than meets the eye even there.

Basically, there is only one way of mechanically aerating a pile of material, and four common ways of achieving it. Ensuring an oxygen supply to the bacteria means that the wastes must be agitated in some way in air so that the air is entrained within the mass, and it follows that since the bacteria will consume the air, agitation must be carried out fairly regularly or they will become anaerobic, resulting in the creation of odors. Generally, the four mechanisms are (1) conventional earthmoving machines, (2) side-cutting windrow turners, (3) straddle turners, and (4) preaeration in a rotating drum.

5.3.1 Earthmoving Machine Operations

The attraction of the earthmoving plant principle is that it is cheap and simple. It uses plant which is probably already on site for another purpose, and it does not need any special understanding of dedicated machinery. It is almost always the starting point for operators wishing to enter the composting industry, which is unfortunate, because it is the mechanism least likely to be successful.

If mechanical turning were only to aerate the mass, it would still not be a very good way of doing it. As organic materials decompose and liberate moisture, they bind together remarkably well. Digging into them with a wide, deep bucket on a tractor or loading shovel will tend to cause even greater compaction. The material will then tip from the bucket in a lump, and it will stay that way to some extent however one tries to overcome it. The lumps will become almost impervious to air, and will become anaerobic (decomposition in the absence of oxygen), liberating traces of methane and hydrogen sulfide gas, both of which are fatal to plants. Pockets of anaerobicity will lead to pockets of acidity in an otherwise somewhat alkaline material, with the result that the final product will be highly variable in all respects.

The purpose of turning, however, is to do more than simply aerate. It is also required to mix the material consistently, to break it down and expose more surface area to biological attack, and to ensure even temperature distribution. Loading shovels will not mix adequately, because they only discharge in lumps, and they will not break the material down at all. They may assist in even temperature distribution, but are unlikely to do so consistently and repeatably.

The question "what are we making compost for?" needs to be posed. Presumably the intention is to sell a product, and if that is so, then it will have to meet some standards the purchaser will consider to be acceptable. Those standards are primarily that it must be consistent in its ability to sustain plant growth, evenly balanced in nutrient levels between batches, and of a regular granularity without too many contaminants. Regular granularity is achieved finally by screening the finished product, and feeding a screen with lumps will result in more oversize reject than compost. That is undesirable when the compost may sell for $100 a cubic yard, but the reject costs $25 a ton to landfill.

While the employment of earthmoving plant is at least cheap, a good argument can be produced against it. One further and major factor that mitigates against it lies in the use of space. On any given area of land laid out to windrows, there must be space between each pile for the machinery to operate. But loading machines attack the material at right angles. So the gap between piles must be sufficient to allow the shovel to advance and reverse — some 30 ft (10 m) even with a small machine in the hands of a competent driver. It follows then that a pile some 13 ft (4 m) wide and $6^1/_2$ ft (2 m) high must be accompanied by a clear space of 30 ft (10 m) down one side at least; in other words, only about 40% of the available width is being used for composting, and that is an expensive proposition where land is costly. Earthmoving plant can make composting an interesting hobby but a commercial disaster.

5.3.2 Windrow Turners

All windrow turners are dedicated machines designed solely for the purposes of agitating and aerating windrows, and there are many manufacturers competing for the market. Each has his own ideas of the right way to do the job, but almost all of the machines available fit into one of two main types. They are either side-cutters or straddle turners. The difference is primarily in the principle of operation: Side-cutting windrow turners cut a slice from the side of the windrow and deliver it via a mixing chamber off to one side where a new one is formed. Straddle turners, as the name implies, straddle the pile and agitate it by simply turning it over. The pile as such stays where it is.

Windrow turner technology originated mainly in the U.S., and to a more limited extent in Germany. In the U.S., of course, things are done on a large scale if they are done at all, and straddle turners are prevalent. Generally, they are very large machines indeed, being in design rather similar to container handling gantries familiar to those living near docks. The technology arrived in Europe in scaled-down versions reflecting the smaller waste arisings there, but the operating principle remained much the same.

In due course, the German company, Willibald GmbH, developed an entirely new approach by creating a side-cutting machine, followed some time later by a British one designed by Advanced Recycling Technologies Ltd (ART). By American standards, the Willibald was small, having a capacity of about 525 yd^3 (400 m^3), against the American demand for several thousand per h. The ART machine was smaller still at about 200 yd^3 (160 m^3), but both were aimed at their local markets. In Germany, sites commonly handle 10,000 to 20,000 t a year of organic wastes, while in the U.K., where the whole idea of composting is still in its infancy, a site capacity of 5000 to 10,000 t is considered adventurous.

So which is best — a straddle turner or a side-cutting turner? As always, each of the machines is to some extent a compromise between that which is possible technically and that which is desirable biologically, with the result that there probably is no "best" as such. It much depends upon what one wants to do and upon the anticipated market for the end product. Even so, it is worth looking at the two principal designs in more detail.

5.3.2.1 Straddle Turners

As noted earlier, straddle turners extend across the compost clamp and turn it over without moving its position in any way (see Figure 5.3). Some achieve this by means of a rotating drum, fitted with welded scrolls or teeth, which spans the side frames of the machine at ground level. Others use a wide steel plate conveyor which again spans the side frames, but which in this case is inclined toward the rear of the machine. The first type works by forcing the drum into the material, which is then picked up, turned over, and redeposited in more or less the same place. The scrolls are generally arranged so as to turn the material inward toward the center of the pile, so that it is effectively rotated in two planes.

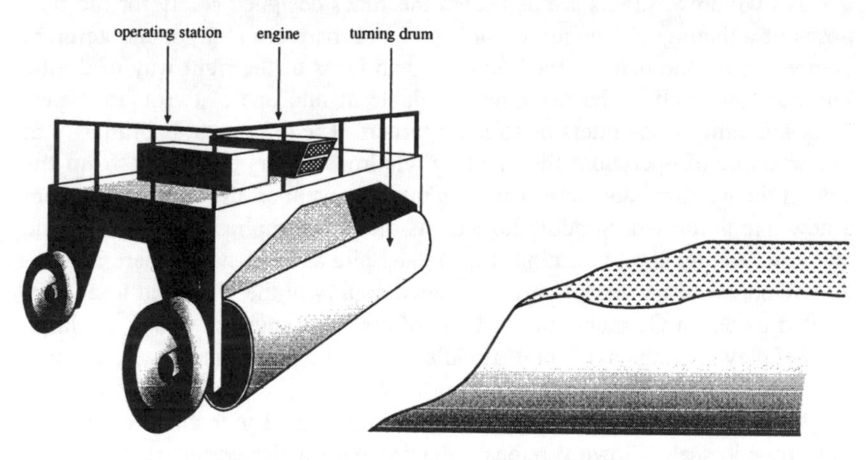

Figure 5.3 Straddle-type windrow turner.

The plate conveyor type of machine works by applying the bottom of the conveyor to the base of the pile. The conveyor belt is usually toothed so that it cuts into the material, which then falls upon the belt, to be transported and dumped again behind the machine. In this case, there is rather more shredding than is likely with a drum-type, but obviously, there is only one plane of rotation. So, in theory at least, the result is better size-reduction but less aeration. In fact, it would probably be difficult to establish any real difference as far as the composting operation is concerned.

With one notable exception, the straddle turners are almost always self-powered. Obviously, they need to be, since it would be impossible to tow them from a central tow-hitch: the windrow would be in the way. That in a sense makes them fairly cumbersome machines, since the power unit needs to be mounted upon the span, together with the control station. The span must be strong enough to withstand the weights and forces imposed upon it, and the side frames must be fairly robust for the same reasons. In theory, at least, the machines can operate on uneven surfaces, but in reality they must not be too uneven or the drum or conveyor will bottom on the ground and the forces on the side frames will become excessive.

The exception to the general design rule is the attachment for the Mercedes-Benz Unimog® all-terrain vehicle. The machine attaches to the rear chassis of the Unimog, and is held by a stay-bar off the midframe so that it extends from the side of the vehicle across the compost windrow. The span is in the shape of a shallow triangle with a rotating drum beneath it, and so in all other respects the machine is a drum-type straddle turner. The immediate advantage of the design is that it is carried by and powered from a vehicle with on–off road capability. A water tank can be installed on the truck chassis for watering the pile as necessary during the turning operations, and the wide range of power takeoffs and secondary processing machinery available from Mercedes-Benz effectively offers a complete composting process in one vehicle. It is often an attractive option to firms already owning Unimogs, which are somewhat remarkable machines by anybody's standards.

The advantage of virtually all straddle turners with the exception of the Unimog is, in principle at least, that they need very little space between windrows: the gap only needs to be wide enough to accommodate the side frames of the machine. Unfortunately, that is not always so, since the turners are fairly unwieldy and may need quite a lot of room at the ends of the piles for maneuvering. Their disadvantage is that they severely limit the dimensions of the windrow that can be constructed, so less waste can be processed in a given pile.

While it is possible to build a wheeled gantry frame to almost any size, mechanical strength considerations indicate that the larger it becomes, the more the cost mounts up and the heavier the structure must be. Deeper, wider windrows mean that the drive to the drum or conveyor must be large to provide sufficient torque, and the drive line on a very large machine may have to be very heavy to prevent too much torsion. Hence, straddle turners in general create windrows that are fairly narrow, shallow, and usually conical in cross section. Thus, the advantage of efficient space utilization is lost in the smaller volume per windrow. The principles of large-volume and small-volume windrows are discussed later in this section, but in general terms it can be argued that the greater the cross section, the better the heat and moisture retention. Small, conical windrows dry out and dissipate heat too rapidly.

5.3.2.2 Side-Cutting Turners

Side-cutters (see Figure 5.4) work by slicing a layer from the sides of the windrow and depositing it off to one side to start a new one, and they do it very aggressively indeed. The Willibald machine is constructed upon a trailer, which is towed by and powered from an agricultural tractor. It consists of a cutter drum mounted at an angle on one side of the trailer, and a collecting and discharge conveyor which extends across the rear. As the machine advances along the side of a windrow the teeth on the fast-rotating drum tear material away from the pile and throw it onto the conveyor, which then deposits it on the other side of the aisle. The design is fairly simple and robust, but the trailer is quite large and does not really lend itself to operating several sites connected by public roads. The trailer is not a road unit in any sense, and it is somewhat wide for normal vehicle transport.

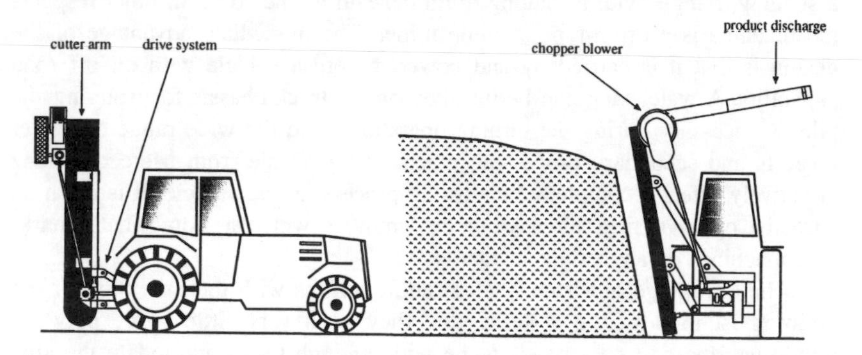

Figure 5.4 Side-cutting windrow turner.

The Advanced Recycling Technologies machine, while still a side-cutter, is quite different in concept. It is more an agricultural implement than a machine in its own right, being mounted upon the rear frame of a tractor and carried by it. It has no wheels, and is driven directly from the tractor power take-off (PTO). In operation, the ART cutter, which is a toothed heavy-duty chain on a frame extending at an angle from the side, slices material from the pile and delivers it to a discharger mounted on the top of the arm, from where it may either be delivered off to one side or back onto the original pile. The cutter, which is best described as a very large chain saw mounted vertically, runs at quite high speed, with the result that very considerable shredding takes place. Since the cutter arm can be retracted hydraulically to a position within the wheel span of the tractor, it is perfectly feasible to transport it from site to site on public roads.

The first obvious advantage of side-cutters is that they cut into the pile very aggressively, so that the amount of preshredding needed to establish the pile in the first place can be reduced. This vigorous shredding action gives very good aeration of the material, and that is further enhanced by the discharge mechanism. Both machines create windrows $9^1/_2$ ft (3 m) in height that can be of any length and width, and such a large block has very good heat and moisture retention capabilities. The two side-cutter designs need much greater space between windrows than do straddle turners, but the windrows themselves are potentially of much greater cross section, so the result is therefore better utilization of space.

Their disadvantages are that they have fast-rotating components, and so they are perhaps more sensitive to contraries within the windrow. They are unlikely to tolerate a piece of concrete fence post or a few house-bricks, but they are lighter than straddle turners and are subject to much lower structural stresses. Wear rates are basically no different in spite of the greater speed, since it is the material that wears the components. Turning capacities are lower, but most European composting operations are not large enough to need high-volume processing. There is little purpose in spending a great deal of money on a big machine that only operates two days a fortnight.

5.3.2.3 Rotating Conditioning Drums

The final plant design in the classification uses a rotating drum to preaerate the waste before windrowing (see Figure 5.5), but the system hardly ever stands alone nowadays. Invariably, it is used only for preconditioning before windrowing by some other method. The principle is one of tumbling the wastes in a large steel vessel for some period of time, often with the addition of water via spray bars, with the intention of reduction of particle size and of intimate mixing.

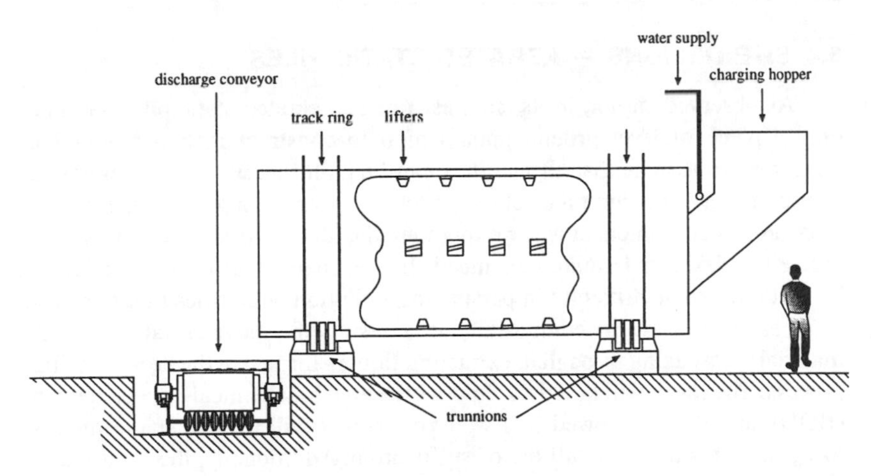

Figure 5.5 Rotating drum aerator.

The technology is actually older than that of most of the simpler straddle and side-cutting machines, and was originally intended for the processing of municipal wastes. Two prominent examples of the concept are the Dano® drum and the Vickers Seerdrum,® although the latter is more often used as a simple breaker than as a preconditioner for composting. In both, it was not unusual to introduce sewage sludge instead of plain water, and this had the effect of increasing the nitrogen levels in the biologically active solids.

Although drum plants are still available, they have been overtaken to some extent by the mechanically turned, forced aerated systems. There are, however, still many of them about the world, working quite successfully, but for the purposes of manufacturing composts for retail or trade markets they have become somewhat outdated. Their high capital cost in what is after all only the first stages of a process is difficult to justify when an acceptable product still needs some other form of treatment, by windrowing or forced aeration or both.

In the United States the most prominent example of the rotating conditioning drum principle is offered by the Bedminster Bioconversion Corporation of New Jersey. In the Bedminster system a mixture of crude municipal waste and sewage sludge is mixed in a drum 180 ft (56 m) long and 12 ft, 6 in. (3.9 m) in diameter for a period of 3 days, followed by screening to separate contraries and then windrow composting for up to 6 weeks. The first plant was constructed in 1972 at Big Sandy, Texas, and was followed by others at Pinetop and at Sevier County, Tennessee.

The Sevier County plant is perhaps the most advertised, serves a population of 51,000, and treats 150 t of municipal waste and 75 t of sewage sludge per day. It reduces volumes by 80 to 85% and is reported to have reduced local landfill demand from $4^1/_2$ acres to $^1/_2$ acre per year. The compost, which conforms to EPA 503 regulations, is distributed through local soil merchants.

5.4 SITE DESIGNS — AERATED STATIC PILES

As observed earlier, in its simplest form an aerated static pile may need only a piece of level ground upon which to construct it, but it would be dangerous to assume that all facilities can be built that way. Any composting operation may at some time release leachates, even though in general all of them are net consumers of water rather than the reverse. Even in circumstances where the designer is quite convinced that leachates will never be released, he will have some difficulty in persuading pollution authorities that this is so.

Leachates arise in composting as a result of rainwater falling on the material and soaking through it, extracting liquors formed in the decomposition process. The result will be a dark fluid with a high biochemical oxygen demand (BOD), and if it is allowed into water courses it will tend to starve them of oxygen; plants and fish will die of suffocation. Additionally, the fluid will be rich in nutrients, and will encourage algae to grow where they are not wanted, making the situation even worse. From the viewpoint of the water supply industry, the nightmare scenario is where the leachates find their way through permeable soil into groundwater reserves. There they may reside for years, doing immense damage and making the water unusable. Those factors combine to make the supplies utilities sensitive about composting operations in general.

The intended use of the product is a fundamental consideration. If the facility is built on unmade ground, then whenever the material is harvested some of the ground will inevitably be recovered with it. Potential buyers of composts will not wish to receive identifiable levels of soil and stones in every bag. And even if the final use is such that quality is immaterial, the base of the pile will have been exposed to soilborne bacteria and fungi, and the degree of decomposition there may be quite different from that further up the pile.

During the harvesting period, loading plant will be running back and forth around the pile, rapidly churning the ground into a quagmire even in fairly dry seasons. During the composting process a considerable amount of organic moisture is converted to free moisture, and much of the water will sink to the bottom. The soil under the pile may well be very wet, even though the surrounds are dry and dusty, and a shovel loader will turn that part to mud in no time at all. It will mash the wet compost into the mud, and will by so doing create a biologically active stew that will become anaerobic sooner or later, making any leachates much more aggressive and creating foul odors in the longer term if not immediately.

Unfortunately, all of the ways of overcoming those conditions are expensive, and probably the most common is the laying of a drained concrete slab. The area covered will need to be about 1.5 times the total planned area occupied by the composting windrows, and the actual civil design will depend to some extent upon the nature of the ground beneath. There are, however, some basic principles.

The slab will need to be capable of withstanding fairly light traffic, in the sense that the shovel loaders used to construct the windrows and harvest the compost do not need to be overlarge, and so their wheel loading should not be excessive. The permeability of the ground beneath it will be of interest to the pollution control authorities, and it is probable that they will require that the slab is sealed against leachate. The surface will need to be rough enough to give reasonable grip to the mobile plant, yet smooth enough to permit good housekeeping. The design arising from those considerations (see Figure 5.6) is a subbase of ground, excavated to a depth of about 12 in. (300 mm), graded and roller-compacted. Upon that about 6 in. (150 mm) of granular material should be laid. The most obvious granular material is hardcore, which should again be coarse-rolled for compaction. On top of that a layer of about 2 in. (50 mm) of dry lean-mix concrete blinding (1:15/24) should be laid and leveled, followed by a 500 gauge polythene membrane with 3 ft (1 m) overlaps at joints. Finally, the concrete surface should be 6 in. (150 mm) of 1:2:4 mix, single-reinforced with not less than 0.5 lb/ft^2 (2.5 kg/m^2) fabric. Expansion joints should be made at a maximum pitch of 80 ft (25 m), and they should be $^3/_4$ in. (20 mm) wide with slightly rounded arises, filled to within about $^1/_2$ in. (15 mm) of the surface with fiberboard, and sealed to the surface with bitumen.

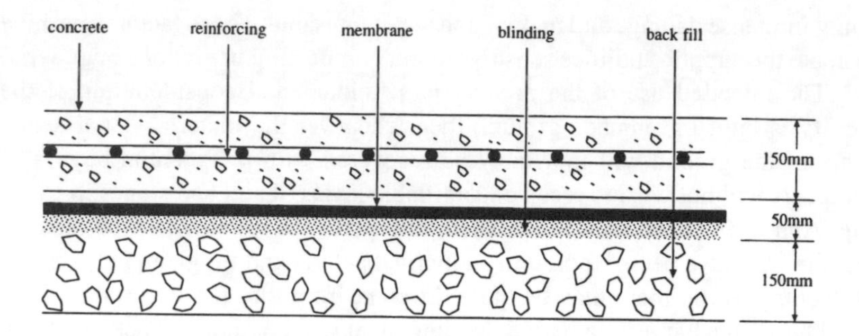

Figure 5.6 Concrete apron for windrows.

The slab should be constructed with a fall from the center to the sides of about 1 in 50. This is considerably less than would normally be used for roads, for example, but then the site is not intended for high-speed traffic, and slow drainage is not a bad thing. One does not want a fall of such severity that surface water runoff carries compost with it to fill and block the drainage. One measure worth taking at this point is to finish the surface with tamping boards parallel to the intended line of the composting windrows, installing the drainage channels across the ends only. That way, the mobile plant will always be working across the grooves in the slab, thus gaining the best grip, while the water will be running down the grooves to the drainage, in parallel to the windrows. There will be a slight tendency for the water to stay within the boundaries of the windrows rather than to spread out around them.

As the slab is laid, precautions must be taken to prevent it from drying out too fast. If it is allowed to do so it will crack, and all of the efforts to avoid penetration by leachates will be wasted. The old-fashioned way of achieving this, and still one of the best, is to cover it with wet burlap as soon as the surface has dried enough to take it without marking, and to leave it there until ready to use the site. An alternative is to cure the concrete using a resinous curing compound, generally at a rate of about 200 ft^2/gal (4 m^2/L).

A more sophisticated alternative to the design outlined is the prepared bay system (see Figure 5.7). It is not widely used, since it reduces the inherent flexibility of the static pile system, but it is worth describing. The principle is to construct working lanes for the mobile plant around bays that have a porous base, and so the bays become effectively long channels with the air pipes set into their bottoms. The plant lanes are effectively independent roadways, and so can be cambered to a fall of about 1 in 35 to give good drainage. They can be coarsely tamped to create a rough surface with good tire grip, since the compost never stands on them.

In this case, the bays are set into the base, with the liner being taken down and across them and the air pipe being laid on top. They are then filled to ground level with a granular material; wood chip or straw are preferable, since

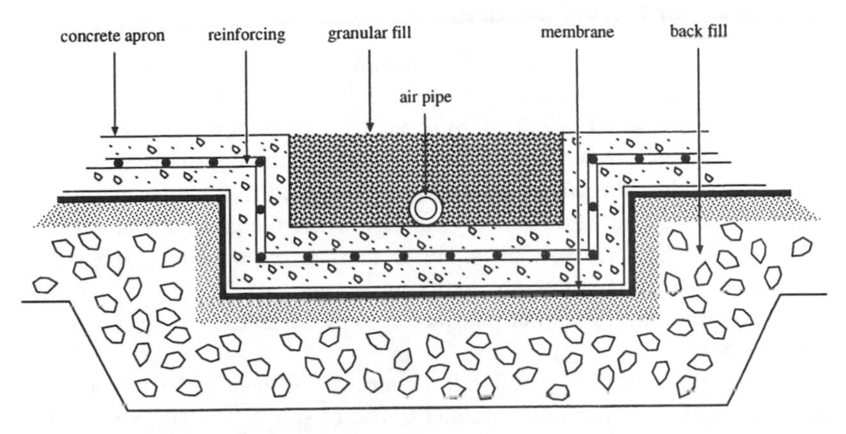

Figure 5.7 Recessed bays for static piles.

it will not matter too much if a small amount gets into the compost. The idea is that the bays should individually be narrow enough so that the bucket of a shovel loader can reach across them while the front wheels of the machine are still on the hardstand.

The immediate advantages of this design are that the air pipes are below the working level, so they do not need to be removed every time the product is harvested. Also, since the base of the pile is granular and porous, air can spread out through it and give more even aeration of the pile, while leachates sink through the base material and can no longer create a wet, sticky mess in the bottom of the compost. In fact, the porosity of the base and its attendant convenient airflow may even contribute to some extent to a reduction of the BOD of the leachate. The disadvantage, of course, is cost.

5.5 SITE DESIGNS — MECHANICALLY TURNED WINDROWS

Mechanically turned windrows without forced aeration are almost exclusively outdoor activities, restricted to parks, garden, and yard landscaping waste. There is no fundamental reason why they should not be constructed inside a building, except that the maneuvering space required for the machinery would tend to make the structure somewhat larger. The absence of any enclosure, ventilation, and odor control system means that they are unsuitable for protein and soft organic wastes, including kitchen wastes. So the designs are not intended for municipal materials.

Designs of this type are extremely plant-dependent; in other words, the civil engineering design depends to a greater or lesser extent upon the type of machinery to be used for processing. Most of the dedicated windrow turners need a level hardstand to work on, and some of them are quite large machines with fairly high wheel-loadings. Four basic designs are in fairly common use.

5.5.1 Sites for Shovel Loaders

The simplest design is that which could be used for primitive facilities operated by loading shovels. Here the base is just coarse aggregate, rolled into the ground and covered to a depth of about 100 mm with wood chip or straw. The windrows are erected upon it, and the whole area is usually surrounded by a drainage ditch to collect any surplus water runoff. Since the base is not sealed in any way, the design is only suitable where the ground is of very low permeability (clay perhaps), or where there is no groundwater. Even then, pollution control authorities are likely to require that the drainage ditches discharge into collectors where surface water and leachates can be treated before release.

The working base of wood chip or straw will need to be restored and made up regularly, since the continual traffic of shovel loaders will cause it to deteriorate quite quickly, and inevitably a certain amount of it will become mixed into the compost each time a windrow is harvested. However, the solution to that is quite straightforward, in that compost made from green waste will need to be screened before sale, and the screening process will recover more wood chip than is lost from the base. It is a simple matter to spread the screen oversize on the site before establishing the next windrow.

As a design, albeit a primitive one, the shovel loading system has much to commend it. It is inherently flexible in that the site can be extended with no more than a few h work with existing plant and the import of some aggregate. It is certainly about as cheap as possible, and the materials for maintenance are free. The disadvantages are that almost all of the conventional windrow turners are unsuitable, and that in spite of careful attention to the standard of the working base, the whole area will become a sea of mud with deep wheel ruts at some times of the year. It will be difficult to exclude soil and rubble contamination from the finished product, and so a high standard of screening will be necessary if a retail market for the product is planned.

5.5.2 Sites for Tractor-Mounted Implements

The next level up relies upon geotextiles (see Figure 5.8). Here the design requires that the site be first graded and leveled, with a fall of about 1 in 50 to one end. Ideally, the ground should be rolled down with vibrating rollers. Next, 2 in. (50 mm) of sand is spread on the prepared ground and covered with a woven geotextile designed to resist stone penetration. Then a membrane of 500-gauge polyethylene is laid, with overlaps of 3 ft (1 m) between sheets. Finally, on top of that a load-bearing geotextile is spread, and covered with 8 in. (200 mm) of wood chip or straw.

This is again a system that does not create a really hard standing, and therefore it is unsuitable for many of the conventional windrow turners. The design also requires care when harvesting; otherwise, it is easy to catch the geotextile with the teeth of a shovel bucket. There are, however, several

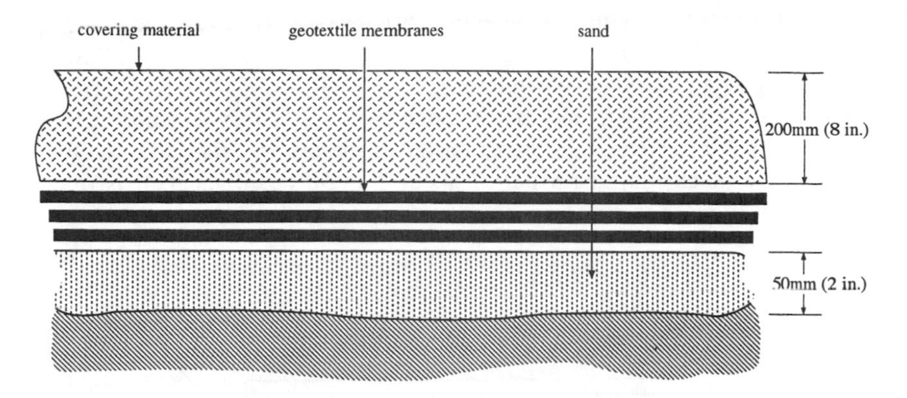

Figure 5.8 Geotextile-based turning apron.

advantages. Leachate control is quite straightforward, since the liquors seep through the working base and the load-bearing geotextile to the membrane. There they will move down the site fall to the collector channels; they will not stay in the bottom of the windrow to cause problems. Also, the principle means that the site is never permanently affected. The ground underneath is not contaminated by rubble or any other damaging material, and so temporary sites can be constructed without later cost implications. Site restoration is simply achieved by removing the geotextile layers and ploughing the soil. Expansion of the site only requires more geotextile with the appropriate overlaps, and in the event of major damage or deterioration, maintenance is only a matter of replacing a few sheets.

Because the design is essentially free-draining, it is not usually necessary to "top up" the working base of wood chip once the process is under way, since the compost itself will eventually roll down quite firmly to replace it without effecting permeability. It can be used in areas of some groundwater sensitivity, because the impervious membrane protects the sub-base.

Improving sophistication leads to the construction of a concrete slab. The design is exactly as defined for aerated static piles, and so we will not dwell upon it other than to observe that almost any type of windrow turner can be used with confidence, and more material can be processed than is permitted with static piles, simply because windrow turners take up less room. The disadvantage is of course that it is costly to construct, and the design does not lend itself to temporary sites, since the restoration costs can be significant.

5.5.3 Sites for Large Straddle and Site-Cutting Turners

The final level of cost and complexity is one used on occasion in Germany particularly, and here the design involves the use of concrete drainage blocks upon a prepared base (see Figure 5.9). There are many ways of preparing the base, but the most obvious is to excavate down to about 16 in. (400 mm), and to lay aggregate to a depth of about 6 in. (150 mm) on a rolled sub-base. Next, 2 in. (50 mm) of lean mix blinding concrete is laid on the aggregate, followed by a 500-gauge polyethylene membrane. On top of that, about 2 in. (50 mm) of sand is tamped down, and the drainage blocks are then set upon it in a staggered pattern. Water collection trenches or drains are laid out at the two ends of the block run, and a fall of about 1 in 50 is sufficient. It is often convenient to make this all to one end.

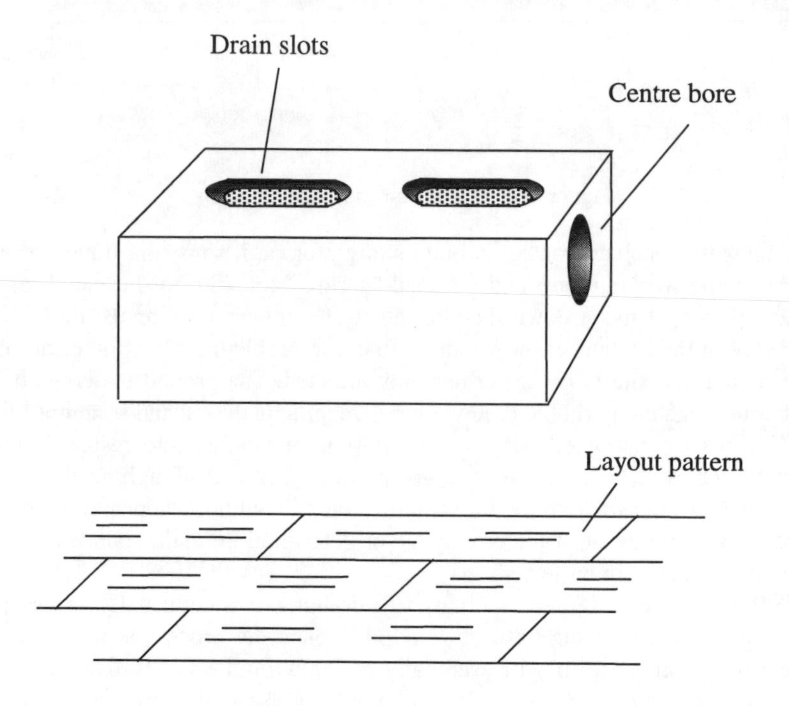

Figure 5.9 Concrete drainage block.

As a design for the idealist, it has much to commend it. Drainage is very much improved, and there is little possibility of water lying about over the site. Leachates can be delivered to one place, and so the drainage arrangements can be quite simple, and maintenance in the event of damage may be no more involved than lifting and replacing a few blocks. Against that is the matter of cost. Laying several thousand blocks is expensive, and not suited to a facility expected to produce only low-grade, low-value composts.

5.6 ESTIMATING SITE CAPACITY

Whatever the technology chosen, estimating site capacity is far from simple. It is an equation with rather too many variables. Some of the factors involved are

- duration of processing
- volume loss due to increased granularity

As a material decomposes, carbon is converted ultimately to carbon dioxide and water vapor. And more than 50% of the organic mass may be lost that way. Yet at the same time, the bulk density of the material may at least double, and it follows that for both parameters to apply there must also be a significant volume loss. Unfortunately, there is no precise way of calculating how the effects will take place, since the process is not linear in any sense. There will be an acceleration of volume reduction during the first few days, especially where the windrows are turned mechanically, and then a more gradual and progressive loss as the material moves toward maturity. Overall, a mechanically turned process could be expected to experience a volume reduction of anywhere between 5:1 and 12:1 over a period of weeks, depending upon the turning regime. An aerated static pile may lose less, although even then local materials and conditions will make a good deal of difference.

Since aerated static piles remain in position for the full term, and are not added to, the estimation is fairly simple. The site must be large enough to accommodate the input volume of material over the chosen composting period plus the roadworks and machinery spaces, and that is all. Mechanically turned facilities are, however, another matter. Generally, they operate by the construction of the first windrow, and then by progressively moving it across or along the site as it is turned. Thus, every 130 yd^3 (100 m^3) may become less than 13 yd^3 (10 m^3) eventually, and the first windrow becomes smaller as it is moved.

There is, however, a calculation that gives a reasonable answer for most practical purposes:

$$\text{Site area} = A + (fA \times (1 - f^{n-1})/(1-f)) \quad (1)$$

where A is the ground area of the first (fresh waste) windrow, n is the planned number of weeks of composting, and f is the volume loss factor, representing the pile volume at the end of a week as a percentage of that at the beginning.

Example (1)

A site is to take in 1300 yd^3 of waste a week with a volume loss factor of 0.8, and compost it for 8 weeks, starting with a windrow 123 ft long, 30 ft wide, and 9 ft 6 in. high. Establish what the site area occupied by windrows will be.

Answer

$$\text{Area} = 3695 + (0.8 \times 3695 \times (1 - 0.8^7)/0.2)$$

$$= 15.371 \times 2 \text{ ft}^2$$

where a total contained volume is required instead, replace the expression for initial pile area (A) with the input volume per week.

The calculation is based upon a simple geometric progression, which assumes in this case that the composting mass will lose 20% of its volume, or in other words 80% of the volume will remain after each week. It mirrors fairly closely the fact that the greater volume loss occurs in the early stages of the process. If it were taken to extremes, using a very large number of turning operations and a long duration, then it suggests that the material would reduce eventually to very small quantities indeed (without ever actually reducing to zero). In theory, of course, that is exactly what would happen, certainly with soft organics, but at some point the calculation would start to break down. Sooner or later there would be insufficient nutrients for the bacteria, and any further material loss would depend upon much slower fungal activity. So the equation should not be used beyond about 15 weeks, which is much longer than the average process would ever require. Method of evaluating volume loss factors for different materials is provided in Figure 5.10. Its derivation was the result of observation of the behavior of selected materials, and the establishment of indices which, when used in a mathematical model, produced a result closely comparable with actual events. It must, however, be stressed that a windrow composting process cannot ever be modeled with precision. The figure, and the mathematics described, should only be used as a guide.

Figure 5.10 may be used in either of two ways. In the first, estimate the likely mix of "hard" and "soft" organics in the windrow, then look up the volume loss factor in the left-hand column most nearly equivalent to the estimate. Thus, for example, the volume loss factor for a pile containing about 40% hard and 60% soft organics is 0.76. Alternatively, using the table as a means of evaluating a single species of organic material, consider which fraction of the material is likely to be cellulose and which is mainly water. Again, look up the left hand column figure most nearly equivalent to the proportions.

Thus apples, for example, could be considered to be 90% water, and the table suggests that a mixture of 10% hard and 90% soft fractions results in a volume loss factor of about 0.715. Hedge prunings, meanwhile might be 70% hard and 30% soft, suggesting a volume loss factor of 0.805. This implies that composting apples would result in a much greater volume loss than would composting hedge prunings, as is the case.

Figure 5.10 introduces a further consideration; that of the volume loss ratio. This is a comparison between the original volume and the final volume, where a loss ratio of 7.0 means that the original volume is 7.0 times the final. The figure is valuable in that it provides an intuitive check upon the validity of calculations, in that the greater the fraction of soft organics in the mix, the greater the volume loss. But more importantly, it permits a direct estimation of the amount of product that can be manufactured from the waste received over a period, in that if the loss ratio is 7.0 and the waste received per year is 30,000 yd^3 (23,000 m^3), then the compost products yield will be one seventh of that.

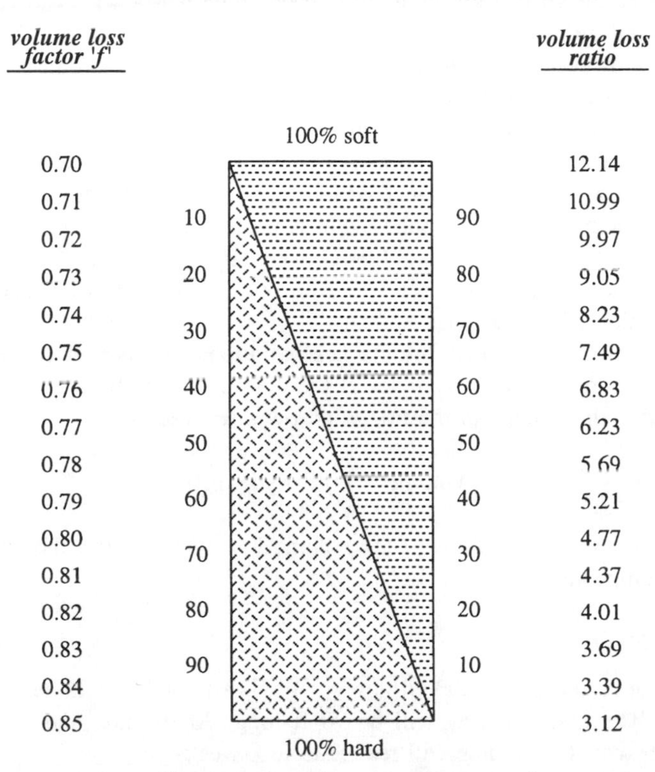

Figure 5.10 Volume loss factors.

In applying the data to the calculation, it is important to distinguish between the effects of turning and of biological decomposition; in other words, when volume loss is most immediately apparent. When a windrow is turned, there will be an initial but small increase in volume, due to the material having been fluffed up, but it will begin to collapse again almost immediately as a result of compaction, and continue to do so as a result of biological activity. The biological activity will be accelerated and maintained by turning, and so the mass loss rate will increase as the population of bacteria develops. Then over a period of days or weeks, without further turning, the pile will begin to stabilize. The biological activity will reduce as the available oxygen is used up, and the mass and volume loss rates will decrease again.

In the calculation, the variable n is taken to mean the composting period in weeks, although perhaps intuitively it could be taken as the number of turning operations. The reasoning behind the choice is that the turning supports the biological activity, but the latter consumes the material. Provided that the windrows are turned sufficiently to maintain the biological activity at its peak, as indicated by the core temperature, then the volume losses will be as projected. This relationship is confirmed by experience of actual operations, where it has been shown that an unturned pile of green waste will lose very little volume and suffer very little decomposition even when stored for several months. However, once regular turning is commenced, decomposition and volume losses occur rapidly.

Once a figure for site area is arrived at, some allowance must be made for the aisles between the windrows, and for space for product refining and storage. The amount of allowance will depend upon many factors, such as the nature of the market to be supplied, whose machinery is to be used, the shape of the site, etc. The aisles should be between 10 ft and 13 ft (3 m and 4 m) wide and the full length of each windrow, but all other spaces are entirely at the discretion of the designer.

As a progression, the formula has variations that permit several other calculations to be made. Examples of the possibilities follow.

To find the volume of the first windrow after week *n*,

$$\text{Volume} = V(f^{n-1}) \text{ yd}^3 \text{ (m}^3\text{)} \quad (2)$$

where V is the initial volume, f the volume loss factor, and n the number of weeks composting.

Example (2)

A site is to take in 1000 yd³ (765 m³) of waste with a volume loss factor of 0.80. The first windrow will be 90 ft long, 30 ft wide, and 10 ft high. Establish what the volume will reduce to in 8 weeks.

Answer

$$\text{Volume} = 1000 \times 0.80^7$$

$$= 209.7 \text{ yd}^3$$

$$= 765 \times 0.80^7$$

$$= 160.4 \text{ m}^3$$

To find the waste composting capacity for a site of given area,

$$\text{Capacity} = [(1 - f)A_t / (1 - f^n)]H \text{ yd}^3 \text{ (m}^3\text{) per week} \quad (3)$$

where f is the volume loss factor, A_t the total site area available for composting, n the number of weeks composting, and H the height of windrow.

Example (3)

A site has an area of 15,000 ft² (1380 m²) available for composting, and will receive waste with a volume loss factor of 0.80. What volume of waste can it accommodate each week using windrows 10 ft (3 m) high and composting for 8 weeks?

Answer

$$\text{Capacity} = [(1 - 0.8) \times 15{,}000 / (1 - 0.80^8)] \times 10$$

$$= (3000/0.8322) \times 3$$

$$= 10.814 \times 7 \text{ ft}^3 \text{ per week}$$

$$= 1201.63 \times 63 \text{ yd}^3 \text{ per week}$$

An obvious criticism of a mathematical formula used in an attempt to mirror a complex process in which a number of factors are involved is that it starts with a fairly bold assumption. In this case, the assumption is that a volume loss of 20% per turn is both realistic and consistent throughout the whole period, whatever that is. Inevitably, there must be some variations caused by changes in materials, or moisture contents, or almost anything else. However, the volume loss has to follow a geometric progression of some sort, otherwise the material would either disappear altogether in a few weeks or not achieve the loss rates actually identified in every process.

Generally, windrow turning operations involving the more usual mixes of green wastes experience a volume loss of between 7:1 and 10:1, depending upon the process duration. This suggests that the proportions of hard to soft organics range from about 35:65% to 10:90%.

5.7 LEACHATE/SURFACE WATER COLLECTION AND TREATMENT

All of the designs for either static piles or turned windrows assume that the surface water runoff and leachate will be delivered to the sides or ends of the prepared area, and the first consideration is of how it should be collected. Where the designs use an impervious membrane, as most of them described here do, then the easiest way is to construct French drains (Figure 5.11).

At the end of the apron toward which the fall lies (or even all around if desired), a trench 20 in. (500 mm) wide and 3 ft (1 m) deep is excavated and lined to a depth of 2 in. (50 mm) with sand. The antipenetration and polyethylene membranes are extended down into the trench and up the far side, to at least a 3 ft (1 m) overlap on the ground beyond. A 4 in. (100 mm) perforated land drain pipe is then laid in the bottom and the trench is infilled with 2 in. (50 mm) pebbles. Each trench is arranged with a fall to one end, and all of them should be led into a collecting chamber with a capacity of about 140 ft^3 (4 m^3) for every 1000 ft^2 (100 m^2) of working apron. The overlap of membrane on the ground beyond the French drain should be held

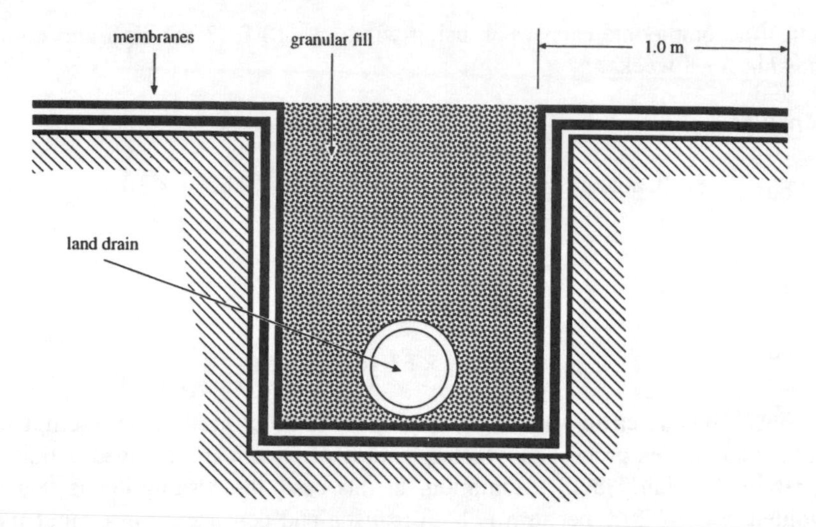

Figure 5.11 French land drains.

down with either a thick layer of soil or, for a presentable-looking site, paving slabs. The collecting chamber, meanwhile, can simply be a pond, constructed from a 2 in. (50 mm) layer of sand and a butyl liner, or a buried purpose-built tank.

Again, there are variations upon the theme, and one of the most sophisticated uses a somewhat enlarged version of standard road drains. Here, conventional drain grilles and chambers are arranged at close intervals along the sides of the apron, and are run down to very large bore concrete drain pipes. There is no collecting chamber or pond as such, because the pipes themselves provide the storage facility.

The storage suggested here permits containment of 1.5 in. (40 mm) of rainwater over the whole site, and at first glance that seems high. However, as an example, reference to standard meteorological data suggests that on average, rainfall in the U.S. ranges from 16 in. p.a. (4.1 mm) in the mid-continental states to between 32 in. (81 mm) and 48 in. (122 mm) in eastern states. Western states receive between 32 in. (82 mm) and 64 in. (163 mm), while Miami records the highest rainfall with an average of 9 in. (230 mm) in the rainy period over September and October.

The data suggests that the containment described would be able to hold at least 1 month of rainfall in mid-continent states, and up to 10 days of rainfall in Miami even if all of the precipitation occurred evenly every day. The windrows, meanwhile, will need to be turned every 7 to 10 days, at which times water will have to be added, probably even when there is a lot of rain. So in normal circumstances, the chamber will never be more than half full at best, and there will be plenty of buffer capacity to allow for treatment and cleaning.

Not all installations will be in an average region of the U.S., or even in the U.S. at all, and so the suggested capacity should not be taken as absolute. The figure of 140 ft^3 (4 m^3) per 1000 ft^2 (100 m^2) of apron is simply a starting point. What should follow, then, is reference to local meteorological data to see if that estimation gives reasonable 10-day storage, and to add to it or subtract from it accordingly.

Once the water and leachate have been collected, the initial treatment is to reduce the BOD, best accomplished by aerating it in one of two ways: blowing air through it, or spraying it into the air.

On face value, blowing air through the liquor seems easiest, since all it needs is a compressor and a length of perforated hose. That will work, although not very efficiently, and unless the compressor has an oil separator, it will create a colorful but potentially harmful film on the surface. Ideally it should be achieved by microdiffusers on the floor of the chamber, but even then there may be a problem with settlement of suspended particulates, which will need to be cleaned out every now and again. A simpler alternative is to use a pump to draw water from the chamber and return it back through coarse spray nozzles. It will become adequately aerated simply in the process of falling back again into the collector, and the system has the added advantage that it tends to keep particulates from settling.

The pump and sprayer principle also reduces the cost: All aerobic composting operations lose moisture rapidly to the atmosphere, and that loss must be made up or the material will become too dry for biological processes to continue. In fact, the problem is more often one of where to obtain sufficient water than of how to dispose of water one already has. So the water from the collector should be stored in preparation for pumping back onto the windrows, and that is where pump and spray aeration is the obvious choice. It uses the same equipment for both purposes and returns most of the particulates to their place of origin. In fact, the only purpose of aerating the water in the first place is to eliminate odors caused by anaerobicity and to clean it up in the rare event that it must be discharged to main drains.

The type of pump employed is not particularly important, but some general principles merit consideration. First, the pump must be designed to work with contaminated water because of the level of particulates that may be encountered, and it should either be portable or enclosed in a weatherproof housing. On an above-ground pump, the casing should have drain cocks so that it can be emptied between periods of use. Otherwise, since the composting operations are outdoors, water may freeze in and crack it.

Submersible pumps are free of risk from freezing, since they are below the water level all of the time. But they do need attention every now and again, and so they should be sunk in an accessible position, preferably with a lifting frame over them so that they can be removed. They should be installed so as to eliminate the possibility of damage to the floor or liner of the chamber, caused by the twisting of the pump under starting torque, and it is advisable to enclose them in a wire mesh frame to keep larger particles out of the pump filter.

While either type of pump is suitable for the job, above-ground, self-priming machines have advantages. They can be directly driven by internal combustion engines, making them independent of external power supplies, and they can be removed to secure and weatherproof storage when not in service. Submersibles can be manufactured with the facility for engine drive, but the power transmission to them is hydraulic and somewhat cumbersome. Also, with an above-ground pump, the only thing in the tank or collector is the suction pipe, foot valve, and strainer. The potential for damage to the base is reduced, and provided that it is done with care, the suction can be moved about regularly to keep the tank clear of sediment. And the final advantage, and one that soon becomes obvious to the site operators, is that in the event of a suction strainer blockage there is nothing heavy to lift out.

Where centrifugal pumps are selected, they should be of an open impeller design specified for particulates, and should not be run at more than 1500 rpm. Higher speeds tend to cause the fibrous material in the water to "ball" in the impeller center and create blockages.

Finally, once the pump has been chosen, there is the matter of its controls. Some operators are attracted by submersible pumps and an electrical power supply, because these lend themselves to automatic operation via float switches. The pumps will start up when a predetermined water level has been reached in the collector, and will spray the water back onto the pile or pump it into the main drains. It is an attractive idea, but on balance one best avoided. The most evident disadvantage of such a system is that the times when the collector is at the chosen level do not necessarily coincide with those when the windrows need water. Unusually heavy rainfall may cause the collector to fill with what is effectively clean water, but the same rainfall may be overwetting the piles. Also, it is fairly apparent that the level controls will either have to be switched off when the site is being worked, or the pump discharge must be redirected to a main drain, since the operating staff will not take kindly to being sprayed unexpectedly. If the controls are to be out of service at least half of the time available, one must question the point of having them at all. Equally, pumping to a main drain from the sedimented base of the collector without supervision might well infringe on discharge consents.

A better principle to consider is to install an overflow channel with a weir and a strainer to catch any floating material, to deal with extremes of rainfall (as long as it goes to a main drainage system and not to an open watercourse). Site staff can then decide when to pump onto the windrows, when to aerate, and as the occasion demands, when to discharge into main drains, having carried out the necessary quality tests. That way, all of the variables in the considerations are covered in a way that a simple level control cannot do.

5.8 COMPOSTING SITE OPERATION

5.8.1 Aerated Static Piles

Generally, the fan systems in aerated static piles are designed to suck as well as to blow, since experience has indicated that a regime that alternates between the two gives the best aeration together with a minimization of fissuring and moisture loss. What the actual regime is — how long to suck and how long to blow — depends upon the material, and it is difficult to lay down precise rules about it. Some practitioners believe that the fans should be running all of the time in one mode or the other, but that tends to supply more air than the biological process needs, and may cause excessive moisture loss. A better way is to run the fans on intermittent suction for a few h one day, then a few h on intermittent blowing for the next, and so on. Thermometer probes extended into the pile at regular points will reveal the effect of the regime, but the purpose should be to try to achieve a fairly steady temperature above 130°F (55°C) for as long as possible. Whatever principle is used, too much blowing will dry the pile and tend to cause fissuring, while too much suction may chill the outer layers and prevent them from ever reaching an acceptable temperature.

Where water needs to be added, it is best to do it with the fans shut down. There are obvious dangers in too liberal watering with them running in the suction mode, since their rotors will almost certainly be damaged by a slug of liquid introduced without warning. Watering with the fans blowing may well prevent the water from penetrating the pile at all. Leaving the fans shut down until water is seen leaking from the base of the pile, and then setting them to blow for a few h is as good a rule of thumb as any.

How long the complete decomposition cycle will take is again somewhat difficult to predict. It will depend upon the nature of the wastes, the desired product quality, and to some extent upon the ambient temperatures, but at least 12 weeks should be allowed. Even then, much of the material will be insufficiently decomposed and will have to be screened out, to be returned to the new pile.

5.8.2 Mechanically Turned Windrows

One holdover from the days when all outdoor composting operations were aerated static piles is the fixed belief that temperatures must be maintained between 130°F (55°C) and 150°F (65°C), or the bacteria will die and the biological decomposition will cease. If that were so, then mechanically turned windrows would simply not work, since they all regularly achieve temperature of up to 176°F (80°C) and are difficult if not impossible to modulate.

While a certain amount of research has been conducted over the years into composting processes, they are still not fully understood. This is mainly because nature has a way of breeding an almost limitless number of strains of bacteria to occupy almost any conceivable environment. It is, however, true to say that viable bacteria have been found at considerably higher temperatures than those occurring in composting operations, and if greater than 160°F (70°C) kills them, then it is difficult to explain how a pile can maintain such a temperature for weeks on end.

It is also questionable whether composting is entirely a biological process anyway. Certainly, with some materials it is not, and paper reduction is a good example. It is perfectly possible to make compost out of, say, newsprint, and to do so without the addition of any organic nutrients. There is in theory at least nothing for the bacteria to eat, so there should be no decomposition. Of course, eventually fungi will break down the cellulose and produce suitable nutrients, but this takes quite a long time. Yet paper composting operations reach high temperatures in a day or so given the right moisture content. Oxidation is the answer to the riddle.

With sufficient carbon, some moisture, and air, any material will begin to oxidize. As a result, the temperature will rise, and the cellular carbon structure will begin to break down. The material will turn brown. From that it is not a great step of the imagination to suppose that any composting operation combining cellulosic and vegetable materials will involve both bacteria and oxidation. It may be limited biological attack that causes the temperatures and moisture contents to rise slightly in the first place, to a level at which oxidation becomes possible. A process may even go through the various stages of bacteria growth to a point at which oxidation takes over.

Behavior of this sort would not be easy to identify, since it is characterized by indicators almost identical to those of biological digestion. In an oxidation process, carbon combines with oxygen to form carbon dioxide and liberate heat. Aerobic bacteria absorb nutrients and liberate, again, carbon dioxide and heat. So, at any one time, it would be a little difficult to establish which reaction is the dominant one. But the purpose of the exercise is to produce "brown stuff" which has an acceptable level of plant nutrients while being free of pathogenic organisms and weed seeds. Biological digestion and oxidation both do that, so one must pose the question, Does it matter which?

Translating this theory to a mechanically turned windrow, one faces the fact that the temperature will rise to around 176°F (80°C) whatever one does about it. More frequent turning will simply cause faster particle breakdown, which will expose more surface area to the bacteria, and they in turn will multiply faster and generate even more heat. Note, however, that there is a difference here between heat and temperature. Greater heat generation, once 176°F (80°C) has been reached, will only cause greater moisture evaporation, not temperature rise. The pile is in that sense self-stabilizing, because the water vapor conducts the heat away as fast as it is generated. Piles never heat up indefinitely, but just dry out instead, to the point where eventually biological growth or oxidation become impossible or reduced and the temperature drops again.

It follows that to some extent too much turning can become self-defeating. The process will not be accelerated, but watering may become more critical, and may make the granularity too fine too soon, resulting in a finished product with too low an air-filled porosity (AFP). Such compost will not grow plants easily. The AFP ideally should be around 15% for compost and over 25% for mulch, and excessively fine granulation will place it well below either of these figures. It will make good house-bricks but poor growing medium.

So how to decide when turning and watering are necessary? Again, temperature probes will give some indication of the former. The temperatures will start to drop when any one of three situations have occurred: (1) the material has run short of oxygen, (2) it is becoming too dry, or (3) all of the available nutrients have been consumed. There is a fourth possibility, that the pile has become waterlogged, but that almost never happens unless there is too high a level of soft organics, and the odor soon makes that apparent. Even then, it is still a matter of insufficient air causing the problem.

Generally, a mechanically turned process will take 8 weeks to produce an acceptable landscape gardening product, and up to 12 weeks to make a high-quality retail grade. Theoretically organic materials, given an ideal environment, will decompose sufficiently in about 2 weeks, and they will certainly *appear* to have done so in 4, but they will not have become sufficiently stable. Also, there is more to compost than just the stage of decomposition, and the longer it is processed the better the appearance. The "look" of the product has as much influence upon the buyer as any other factor.

In the first weeks, a slow temperature drop will be due to either lack of oxygen or lack of moisture. If the material feels warm and damp to the hand, then it is wet enough, and it just needs turning. At that point, it is good practice to wet it thoroughly and then begin the turning operation, which seems a strange thing to do if it is already wet enough. But anyone who has seen the dense clouds of steam that rise during windrow turning will realize why it is necessary. Windrowing without watering is probably the quickest way of stopping the process.

Sooner or later, of course, the available nutrients will be used up, and the third variable comes into play. One could of course measure nutrient levels by analysis, but it is not a particularly accurate measure of compost maturity. In fact, the whole subject of maturity has been argued about by scientists for years. From the operator's point of view, however, there is a simpler way. Every time the temperature starts to drop, the operator should water and turn, until the point is reached where doing so no longer results in an immediate temperature rise.

Here it is worth introducing another variable into the equation. At what point does a mass of organic material become compost? The answer to that is when it is sufficiently decomposed for its intended purpose, since any fully decomposed material has disappeared altogether! The measure of suitability is in fact when the granularity, color, and nutrient levels are as desired, and if that point is reached while turning still causes temperature rises, stop anyway. Whatever is done, the material will never cease to be biologically active. All organic composts remain so until they are entirely absorbed by plants. But in the absence of turning, the compost will run out of air and moisture and the bacteria will become dormant. While it will stay warm in a stockpile for a very long time, it will cool rapidly in bags, and will become effectively stable until it is used in the soil.

That, of course, provides an interesting guideline to estimating how the process is going, and permits a choice of products to be made. For example, a seed compost needs to have a fairly low level of nutrients, good granularity, and good moisture retention. Too many nutrients and the seedlings will be too "leggy." So to make a suitable seed compost, it follows that the process should continue for longer, so that the bacteria consume more of the nutrients. A shrub compost, however, needs a good AFP, a fairly high level of nutrients, and should be reasonably free-draining. So in this case, the duration of turning operations should be reduced. A multipurpose compost should have good granularity, good moisture retention, and a reasonable but not excessive nutrient level, and that can be achieved by turning more frequently than the temperature drop alone indicates. Doing so will tend to reduce the particle size to the right level before the nutrients are consumed, and that is the point at which to stop.

5.8.3 Operations with Side-Cutting Turners

So far, all of the earlier discussions have been upon how to quantify site areas, and how to evaluate the process. It has also been observed that mechanically turned windrows using side-cutting turners are gradually moved across the site by the action of the machine. Since the windrow will lose volume as a result of the increased biological activity, the width of a 30 ft (10 m) windrow may reduce by 6 ft 6 in. (2 m) each time it is turned, assuming that the optimum height is maintained. Therefore, the rate of travel across the site would appear to slow down, which suggests that eventually there would be no longer sufficient space to accommodate the fresh waste. The windrows would move in one direction, but the whole operation would gradually creep in the reverse.

The optimum height for a side-cut windrow is 9 ft 6 in. (3 m), since that results in the best heat and moisture retention without creating undue compaction. If a site is to accommodate 1300 yd^3 (1000 m^3) a week for 40 weeks a year (equivalent to about 10,000 t of green waste) and compost it for 8 weeks, then a windrow created from the first week's input could for example be 120 ft (37 m) long, 30 ft (9.4 m) wide, and 9 ft 6 in. (3 m) high. If that were then turned regularly, the width would reduce to 25 ft (8 m), then to 20 ft (6.4 m), and so on until it became too narrow for practical purposes. At the end of week 8, the turner would be working eight piles of varying width with an access gap around each, and the gaps would be growing larger as each windrow lost volume.

A production system based upon a series of windrows, one for each week's input, is therefore clearly impractical. Eventually, the piles would separate sufficiently to take up the whole site. The solution lies in the fact that a side-cutting turner operates independently of the width or length. It is irrelevant whether the pile is 32 ft (10 m) or 320 ft (100 m) wide; all that matters is the capacity of the machine in terms of cubic meters per working week. Thus for example, a machine with a capacity of 500 yd^3 (380 m^3) would process some 18,000 yd^3 (13,600 m^3) a week, and this is equivalent to 14 weeks' input to the chosen site. In other words, it could turn the full 8 weeks' input every week and still be operating for no more than 3 days at a time.

Therefore, there is no point in building eight individual windrows at the rate of 1 a week. It would be more logical to place the first waste in the center of the site and keep adding more to one side of it, while working the opposite side with the turner. The result would in due course be one large windrow containing 8 weeks worth of waste, where fresh waste is continually being added to one side and finished compost is being harvested from the opposite.

As this one large windrow is being constructed, the fresh waste will gradually accommodate the available site in one direction, while the processing face will move in the other. Eventually, by week 8, the maximum area will be utilized, and there will no longer be any travel in either direction. Individual sections of waste will be moving across the site, but the windrow will not. Thus it follows that the result of the area calculation defined in Section 5.5 needs to be increased by an amount sufficient to provide a maneuvering space around the windrow, say 32 ft (10 m) each end and 65 ft (20 m) at the sides.

There may of course be reasons why one large windrow is inappropriate. Perhaps the shape of the site does not lend itself to such an operation, or perhaps the input of waste is too large. However, the target should always be to have the smallest number of the largest windrows possible, up to the full capacity of the machine. Once that machine capacity has been fully committed, then operate a second, processing a further windrow in the same manner. Note that it is impractical to operate two machines on a single windrow; they obstruct each other and in so doing reduce each other's capacity.

5.9 WASTE PREPARATION AND PRODUCT REFINING

5.9.1 Preparation

When the waste is delivered to the composting site, it is probable that much of it will not be in a condition that permits consistent biological digestion. Green or yard wastes are likely to contain quite high levels of tree branches, hedge clippings, and other woody materials in quite large lumps. A windrow made from such wastes as they stand, whatever system is used, will at least do very little and may even be beyond the capability of the machinery. Some degree of shredding may be necessary, and there are three main machinery designs in common use: "tub" grinders, horizontal rotor flail mills, and brushwood chippers.

Tub grinders (Figure 5.12) consist of a large-diameter, slowly rotating conical drum mounted upon a trailer frame, with a short flail rotor offset to one side in the base. Material is loaded into the top of the drum and is slowly carried around to the point where the flails intrude into the cone. The flails catch it and shred it, taking it down to a set of perforated plates (grids) under the rotor. There the shredded product is ejected through the holes onto a collecting conveyor underneath, and the hole size determines the degree of granulation.

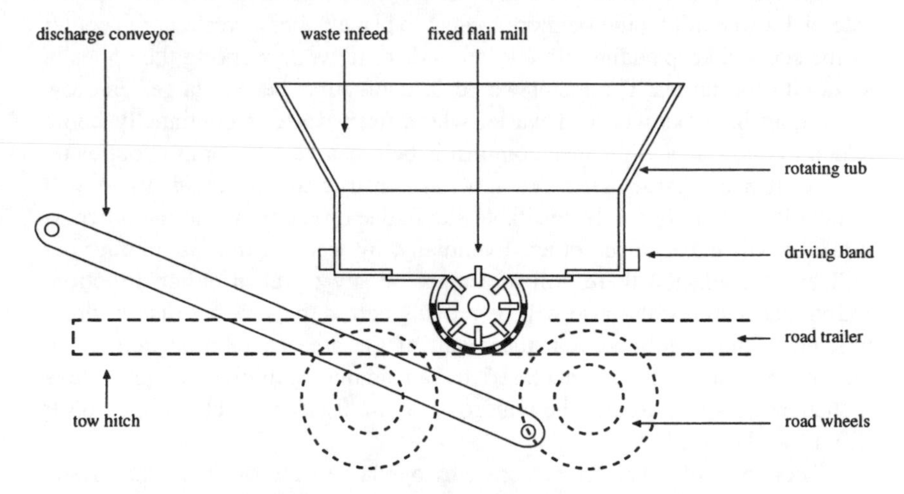

Figure 5.12 Operating principle of tub grinder.

Horizontal rotor flail mills (Figure 5.13) dispense with the conical drum as a feeding mechanism, generally replacing it with a steel belt conveyor. The rotor is invariably much larger, and they have no grid plates, so they will accommodate much larger items. In fact, machines of this type will deal quite happily if spectacularly with a complete tree trunk 20 in. (500 mm) in diameter. To deal with the dangers of overfeeding and stalling the rotor, and to prevent material being thrown back, most horizontal rotor flail mills have a load-sensing control on the feed conveyor which causes it to slow down or even reverse if the power requirement becomes excessive. In their basic form, the machines generally discharge straight onto the ground underneath the rotor, necessitating their being moved back regularly, but there is often a provision for an inclined collecting conveyor if desired.

Brushwood chippers (Figure 5.14) are at the small end of the machinery spectrum, but even so they may be sufficient where the number of branches, etc. are small and the size is limited. In these machines a fast-rotating cutting disc with short, sharp flails on its periphery runs inside a cylindrical casing.

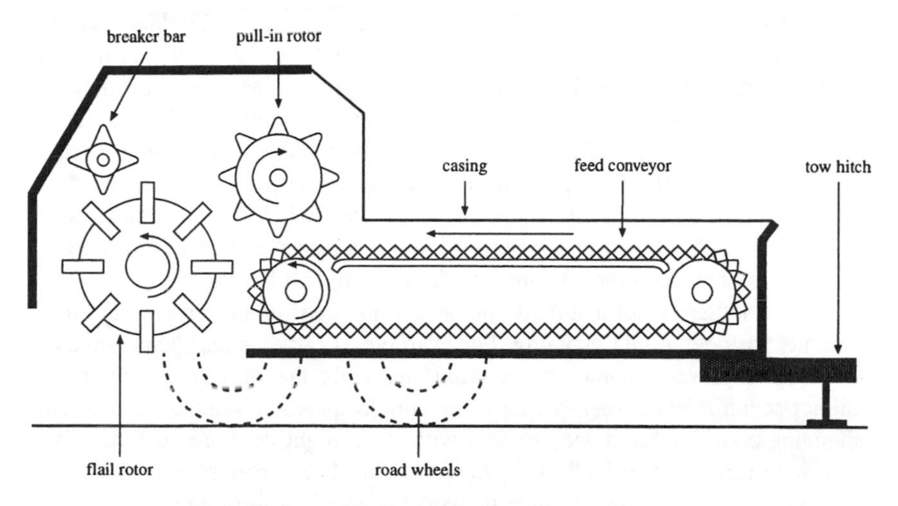

Figure 5.13 Typical horizontal granulator.

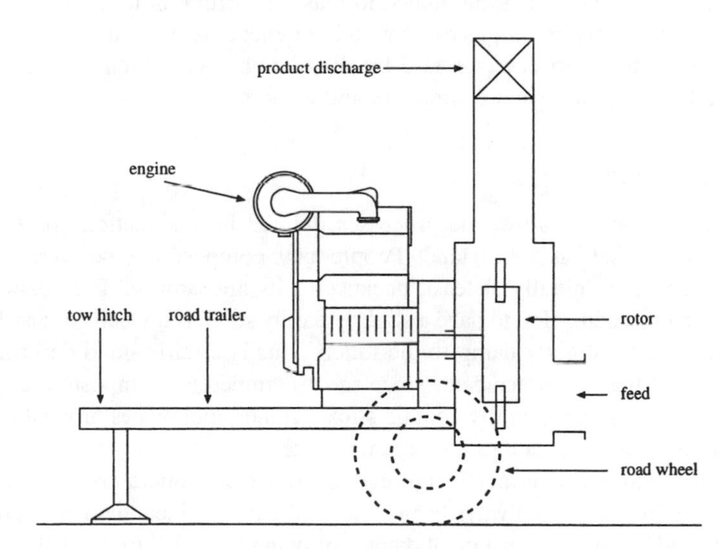

Figure 5.14 Brushwood chipper.

Items for shredding are fed into the center of the rotor and, upon being chipped, are thrown out through a square, tangential discharge tube with some force.

All of these machines are quite large power users, and by far the greater in that respect are the horizontal shaft mills with up to 350 hp (260 kW). Tub grinders, with their smaller rotors, come into the medium power range, while brushwood shredders are down in the tens of horsepower area.

While much of the waste, upon delivery, may need pretreatment, this does not follow inevitably. There is a common misconception that no composting operation can possibly work without a shredder, and that a windrow turner or fan system is an optional extra. In fact, the reverse is true. Shredders do not make compost because aeration needs to be carried out regularly, and if used for that purpose they would simply clog up on material that is part of the way down the composting process. Anyway, a large shredder will cost about \$150,000 and have a capacity of around 80 yd^3 (60 m^3)/h. A side-cutting windrow turner will shred partially composted waste at a rate of up to 520 yd^3 (400 m^3)/h and will cost between \$37,500 and \$67,500. Such a turner still needs about \$45,000 worth of tractor to power it, but the tractor, with a fore-loader, is essential on the site anyway for loading and harvesting.

So, whether or not a shredder is necessary occasionally or even at all is a matter for operational planning. For example, if only a small percentage of the incoming waste contains oversized materials the turners or windrows cannot accommodate, then perhaps one should question whether it is worth spending considerable money to deal with it. It might be more cost effective to simply remove it to landfill. If, however, say 10% or more is in the form of branches of trees, then it might be worth stockpiling those items off to one side until sufficient have accumulated to make it worth renting a shredder for a day. If only fairly small pieces of wood are encountered in any quantity, it might be sufficient to buy or rent a brushwood chipper, which is a machine stocked by most landscape contractors and dealers.

5.9.2 Refining

The products of all of the processes defined in this section are almost impossible to sell as they stand. People buy compost not because of its composition but, initially at least, because of its appearance. They expect it to be dark brown in color, to have a slightly earthy smell if it smells of anything at all, and to be slightly damp to the touch. This is a fairly good description of peat, and that is no accident, since most commercial composts are peat-based. Peat does not naturally contain sticks, stones, and pieces of plastic, nor does it have lumps of congealed organic matter.

This, regrettably, is a fact ignored by too many operators. The waste processing industry has always been bedeviled with the idea that if something is "recycled," then there is a great danger of being trampled in the rush to buy it, whatever the quality. Anyone who still believes that is possible is advised to take a black trashbag of crude compost from a windrowing facility to one or two professional distributors or retailers and invite them to purchase it. The fact that they are all becoming heartily tired of hearing about such products will be made fairly plain quite early in the conversation! People will not buy things simply because they are "recycled," whatever the price. They think *other* people should, but *they* will not.

So the product has to be refined to at least create the right cosmetic appearance, and at a minimum it requires screening. A good-grade growing medium for retail sale should have a consistent particle size not greater than $^1/_4$ in. (5 mm), and to achieve that it needs to be run through a $^1/_2$ in. (10 mm) screen. The machines used for the purpose are most usually drum screens, that is, a rotating drum consisting of a mesh-covered frame which rotates upon trunnions and is driven by an engine or an electric motor. The desired product falls through the screen onto a collecting conveyor underneath, while the oversized material reports to a second collecting conveyor at the drum end.

Drum screens, as typified in Figure 5.15, are always made with square perforations; round holes blind too easily and the percentage open surface area (the ratio of hole area to total area) is too low. Even then, screening all of the product from the windrow through a $^1/_2$ in. (10 mm) screen at once is time consuming. The smaller the holes, the lower the capacity in terms of volume per h. For that reason, a better approach is to remove the coarse rubbish first by running the material through a 1 in. (25 mm) screen, then changing to the $^1/_2$ in. (10 mm) for the final polish. The advantage of screening twice is that two products result. The first, obviously, is the compost which was the whole point of the exercise, but since the oversize of the $^1/_2$ in. screen is actually less than 1 in. and greater than $^1/_2$ in., it is potentially an excellent mulch with the right AFP. The material greater than 1 in. from the first screening is generally useless and should be dumped..

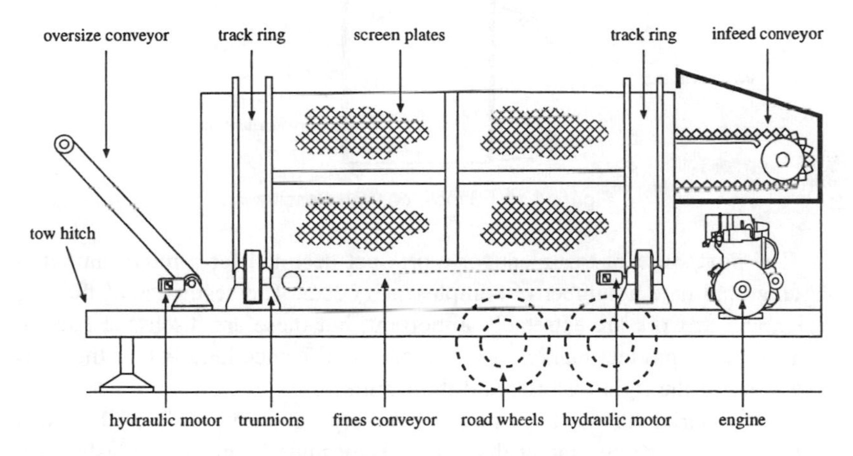

Figure 5.15 Typical drum screen.

When the product has been screened, it may still contain unwanted tramp materials, of which the most common is plastic film from the bags in which some of the waste arrived. Unfortunately, efforts to exclude them at the beginning are rarely successful, and pieces of plastic inevitably pass right through the process. If they exist in the compost, and are fairly limited in quantity and small in size, their presence may not matter greatly. Compost, after all, is generally buried, and so they will not be seen for long. Mulching products, however, are used on the top of the ground, and it is to them that much of the plastic will report and stay visible. Fortunately, taking it out is quite easy, since the film at that point is very light in weight while the mulch is not. A simple air classifier (Figure 5.16), where the material is introduced into a rising air flow in a square casing, is sufficient. The air is supplied by a centrifugal fan with a variable displacement control, and the flow rate can be adjusted so that the product falls to the bottom of the casing while the air carries the plastic film away and out of the top. Some operators achieve that effect quite naturally and at no cost by simply carrying out their screening operations on windy days. It is not a very responsible way of doing the job, but it works.

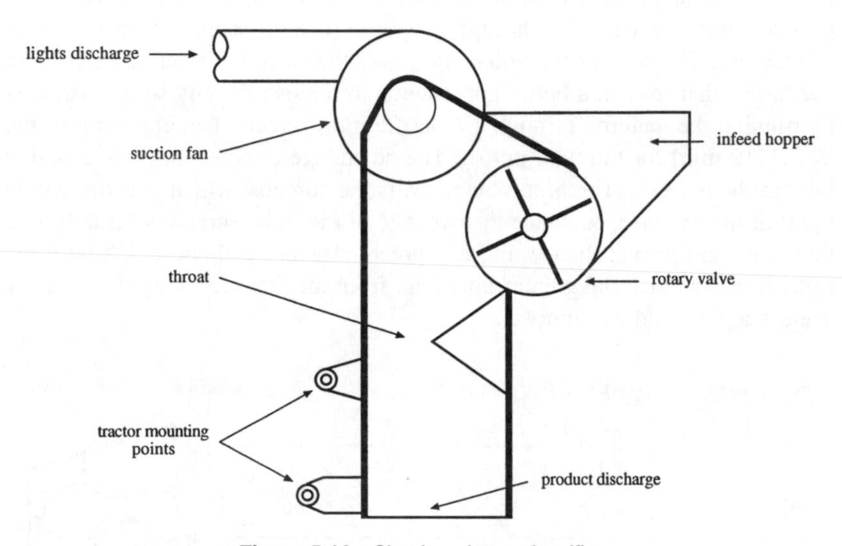

Figure 5.16 Simple column classifier.

The next stage of treatment for a product destined for a retail market is bagging, and here a number of complexities occur. The aesthetics of the bag design and artwork are extremely important, but these are discussed later in Chapter 11 on product marketing. Of more significance here is that the bags have to be of the right material and the right size.

The material for the bags must obviously be waterproof, or they will simply fall apart sooner rather than later. They must be made of plastic. But the material inside needs to breathe, since it still has some residual biological activity going on, and if it is prevented from doing so it may become anaerobic, which would be harmful to anything planted in it. So the bags must be perforated with holes large enough to allow breathing but small enough to prevent the contents from escaping. Some bag manufacturers achieve this by supplying microperforated film, where the bag is punctured with tiny holes. Others achieve it by somewhat larger holes down the side gussets, on the principle that once the bags are stacked on pallets, only the side holes will allow aeration.

Having chosen the type of perforation, the next point is the type of film. Ideally, the inside of the bag should be black, to exclude light from the product, but if the outside is to be printed and to look attractive, then a two-ply film is necessary. Some manufacturers offer three-ply film, where the center layer is tear-resistant, the inner is black and also smooth to minimize product adhesion, and the outer is matte finished for ease of printing and stability on a pallet.

Then there is the question of bag size. Some volumetric bagging plants simply fill a given bag to a predetermined level. If the bag can contain 40 l, then that is what will go into it. Unfortunately, it is not simply a matter of buying bags that have that enclosed volume, since the very act of bagging will temporarily aerate the material and reduce its density. Then, when the bag is stacked with others for storage, the product will compact again, so that when the customer finally opens it there will be somewhat less than 1.5 ft^3 (40 l) in it. A declaration of "1.5 ft^3" will be regarded somewhat skeptically if the bag is only half full. But even an oversized bag is not necessarily the answer, since once it has been filled it probably will then have to be palletized for transport, and European standard pallets for example are commonly 1200 mm by 1000 mm. The bags must permit stacking on such a pallet in a way that interlocks them for stability, without any of them overhanging the sides.

There is no arbitrary rule in choosing suitable bags that meet all of these constraints, since the nature of the product itself and the target market will decide most of them. The only way to find out what dimensions and film types are best for a given product is to try some on the bagging plant that will be used, and most bag manufacturers are well aware of it. They will generally supply sample bags for the purpose on request, and can usually give advice upon a starting point. Cross any manufacturers that will not do so off of the procurement list.

5.10 LEGISLATION AND COMPOST STANDARDS

5.10.1 Pollution Control Legislation

Composting operations will almost invariably have to comply with some environmental protection controls, which vary among countries. However, a common thread of logic applies everywhere.

An outdoor windrowing process threatens the environment in three ways: It could potentially cause damage to watercourses and groundwater by releasing leachates and particulates to them. It could cause atmospheric pollution with dust, microorganisms, and odors. It could, theoretically at least, cause soil pollution by the products of anaerobicity or by increasing nitrates beyond permitted levels.

A well-designed and professionally operated site will not threaten the environment in any of those ways, but a poorly designed one will do more than threaten. For that reason, regulatory authorities are inclined to have a jaundiced view of windrow composting in general. They do not like, and generally will not tolerate, the activities of the "get rich quick" operator whose idea of composting is to simply make a pile on the open ground and leave it there until it rots. Such people should be condemned not just by the authorities, but also by all serious processors who have a due regard for the environment and who see composting as a way of safeguarding it. Unscrupulous operators do immense damage to the industry by alienating the public, saturating the product markets with rubbish, and encouraging the enactment of legislation that is hastily conceived in response to their behavior.

While controlling legislation exists almost everywhere, it is beyond the scope of this book to record all of it. In fact, a series of volumes could be written on that one subject alone. Perhaps what is more to the point for an operator wishing to establish a facility is what steps he should take to ensure that his plan and designs meet the relevant national and local requirements. The following checklist will cover most eventualities, and although it is not exhaustive, it will expose any special regulations of which the operator may not be aware.

5.10.1.1 Legislation

1. Have you obtained permission from the water authority for drainage arrangements for leachates and surface water?
2. Have you consulted the pollution control authority?
3. Have you investigated any need for licensing under waste disposal legislation?
4. Have you consulted the trading standards officers on the packaging you intend to use?
5. Will your chosen products meet all relevant quality controls?
6. Have you decided how you will accommodate health and safety regulations?
7. Do you have planning consent for the operation?

5.10.1.2 Indemnities

1. Have you investigated product liability insurance?
2. If you are going to treat other people's materials using your machinery, have you considered contractor's liability insurance?
3. Have you arranged insurance to protect you in the event that your operation damages either the environment or other people's property through something going wrong?

5.10.1.3 The Raw Materials

1. Have you established which wastes you are going to process, and are you fully aware of the differences between them in respect to how they must be handled?
2. How are you going to collect the wastes? A separate collection from households and commercial properties or a "bring" system? How are you going to ensure that you only get the materials you want and can handle?

3. How are you going to shred the wastes? Shredders are large and noisy machines that can present very considerable dangers to the public if not handled carefully. Please do not consider towing a machine around to all of your local household waste sites to which the public have access.
4. How are you going to aerate the wastes? There are many types of machines on the market and some are more suitable. Do you have experience of aerating your particular material?
5. How are you going to refine the finished products? What grades and granularities do you expect to achieve?
6. For how long will you have to compost the material to achieve the required quality?

5.10.1.4 Markets

1. What will be your chemical and physical specification upon which you sell? Can you guarantee that it is repeatable?
2. Is the product biologically stable?
3. Can you guarantee that it does not contain toxins?
4. Can you show the results of approved growing trials using your intended product?
5. Have you designed your point-of-sale literature and can you certify that the claims it makes are true?
6. Have you established a procedure for dealing with customer complaints?
7. What is your policy with respect to guarantees?

5.10.1.5 Compost Standards

The question of what is a compost has vexed legislators for a number of years, since it is almost impossible to define in any meaningful legal way. It is an unfortunate fact of modern life that as soon as anyone attempts to harness a natural process, a legislator with a burning desire to invent a regulation will come along! All organic materials will decompose and eventually become a source of nutrients for new organic material, and that has been going on since the earth was formed, so why do we need standards at all? Environmental and consumer protection is why.

From the environmental point of view, nature generally deals with one thing at a time rather than all together. Leaves falling from a tree in a forest decompose with all the other leaves. They do not usually become mixed with sewage sludge and food wastes first. The natural process that works quite happily with individual materials may be significantly modified when faced with mixtures, and not all natural processes are benign. For example, it is quite possible to create a compost so laden with heavy metals that it would do positive harm to the soil and to anyone eating the products grown in it. Or it may be infected with pathogenic organisms or parasites — they are also quite natural phenomena. Then there is the matter of physical contamination. Glass does not decompose in a composting process, and its presence in the final product could pose a significant environmental and health risk.

The difficulty with establishing controls is that, while it is possible to set quite arbitrary limits for physical contamination from glass and other potentially harmful materials, chemical and biological controls are another matter. For example, some fairly recent European legislation established rules for levels of heavy metals in soils. It was not aimed at compost particularly, although theoretically at least it did ensnare it. Unfortunately, it was later found that soils exist that already have higher heavy metals levels than the legal limits. For that reason, there is a tendency nowadays to think in terms of the capacity of a material to *increase* soil concentrations over a period of time, rather than to concentrate upon an arbitrary base limit.

Not least of the problems is defining at what point an organic material becomes compost. The biological process is one in which nutrients are digested by microorganisms, and so the longer the process goes on the lower are the nutrient levels available to plants. Some decomposition is necessary to release the nutrients in the first place, but too much reduces them. Therefore, in theory an only slightly decomposed vegetable waste should have the highest level of nutrients and be the better growing medium. Unfortunately, it may still at that point be contaminated by plant pathogens and weed seeds, and if nothing else it would not have a very attractive appearance and would be highly unstable.

In spite of the difficulties, some European countries have taken the initiative and imposed standards of some sort, even if they are less than optimal, and there are a number of EC directives that at least impinge upon the manufacturing and use of composts. Meanwhile the European Committee for Standardization (CEN) approved resolution number BTS3.32 in 1990 establishing Technical Committee number 223. Its purpose is to consider "the standardization of two types of material used in agriculture, horticulture, gardening, and landscaping." The two materials are soil improvers and growing media — in other words, composts.

The growth in composting installations, particularly those based upon windrowing technology, has been so considerable that even where detailed standards do not exist at the moment, one can be assured that they will in due course. If nothing else, they will present a "level playing field" for all operators, excluding those who are careless or indifferent, and that can only be a good thing for the industry. In the meantime, the legislation that may cover compost manufacturing and marketing is very wide ranging, and we could not possibly define all of it here. Instead we offer the following list, with the recommendation that any potential operator who believes he may be effected should obtain copies of that which is relevant. The list only covers European legislation and even then does not contain everything which might apply in any circumstance. To conduct the exercise worldwide would be impossible.

5.10.1.5.1 Level 1 — European Community Legislation — While there is currently no Community-wide legislation on compost as such at the moment, the following enactments may well impinge upon it and may be used as a framework for more specific rules in the future.

Official Journal

Reference	Covering
O.J. L20	Groundwater protection
O.J. L181	Sewage sludge
O.J. L024	Directive 46/116/EEC relating to fertilizers
O.J. L213	Sampling and analysis of fertilizers
O.J. L222	Directive 78/659/EEC — physical and chemical characteristics of two categories of water
O.J. L20	Directive 80/68/EEC — introduction to groundwater of dangerous substances
O.J. L181	Directive 86/278/EEC — application of sewage sludge (maximum quantities of heavy metals in soils)
O.J. L175	Directive 85/337/EEC — environmental impact assessments

Level 2 — National Legislation

Country	Standard
France	AFNOR standard U44/051 — 1980, basic standard to be attained before a product can be called "compost"
Germany	Bundesgutegemeinschaft Kompost e.V, standards for the assessment of compost quality (see also "Blue Angel" and LAGA standards)
Austria	Gutekriterien fur Mullkompost ONORM S-2022 levels of contaminants in municipal waste composts (see also Austrian standard S-2100, definition of suitable refuse material)
United Kingdom	Code of Practice for the Agricultural Use of Sewage Sludge. Defines the limits of potentially toxic metals in soils (see also Environmental Protection Act 1990 — Prescribed Processes and Substances Regulations)

5.11 HEALTH AND SAFETY

Health and safety considerations are properly an indivisible part of modern human activities, and as such composting operations pose a number of threats: Any undertaking that employs people in the presence of earthmoving plant and a hot, biologically active material must present risks. Therefore, it is worth exploring the nature of the risks.

Health and safety considerations are not restricted to employees. They also cover members of the public and possibly even animals. "Health" in this context means the recognition and accommodation of risks from the material resulting from the biological, fungal, and oxidation processes. "Safety" means protection from injury caused by the equipment, the site, the material, and any ancillary activities. The scope is wide, and there is no excuse for operating a site without a properly constructed health and safety policy that sets out to identify as many risks as possible and to deal with them.

5.11.1 Health Risks

Composting operations have been proved to liberate microorganisms and fungal spores in significant quantities. So far there is no proof that they are harmful, although there is some evidence that people who already have respiratory weaknesses may find their condition aggravated by them. Research into the airborne emissions from mushroom compost manufacturing has suggested that fungi implicated in farmer's lung disease may be detectable, as may those responsible for mushroom worker's lung. Obviously, aerated static piles are less likely to liberate large quantities of fungal spores and microorganisms during normal processing than are mechanically turned windrows, but both will do so during pile construction and harvesting.

In addition to the threats from living organisms, the dusts that may be released from time to time can be harmful. There is the risk of eye damage or infection caused by windborne particles, and respiratory problems caused by breathing them in. Organic dust toxic syndrome may be associated with intense exposure to compost dusts, and fever could be caused by inhalation of endotoxins. Those who wish to explore the allergic and infectious reactions would do well to read the work by Lacey et al.[1]

In view of these risks, however slight and unsubstantiated, a careful site operator or employer would be wise to consider their possibility, and to be on the lookout for symptoms. For that purpose, the following risk summary may be of value.

1. Infection by fecal coliform bacteria, streptococci, and salmonella. Symptoms are diarrhea and abdominal pains. Such infections are unlikely, since the temperatures experienced in composting operations can generally be relied upon to destroy pathogenic organisms very quickly. In addition, a compost pile does not provide a suitable environment for pathogens, which are dependent upon the human body for survival.
2. Lung infections caused by aspergillus molds. In this case, there appears to be a need for the subject to be predisposed to lung infections.
3. Fever and influenza-like symptoms caused by inhalation of bacterial endotoxins.
4. Bronchitis and obstructive pulmonary disease exacerbated by smoking and susceptibility to allergy.
5. Allergic responses such as rhinitis and asthma in predisposed subjects.
6. Alveolitis caused by prolonged exposure to airborne spores.
7. Organic dust toxic syndrome resulting from massive exposure to dusts.
8. Inhalation mycotoxicosis due to the inhalation of poisonous fungal metabolites. This is rare in humans and in a properly controlled composting operation should present a negligible risk.

9. Weil's disease (leptospirosis ictero-haemorrhagica). This is a very serious, often fatal disease contracted from the urine of infected rats. It is not normally associated with processes involving only vegetable wastes, but it could be present where the materials contain proteinous items (kitchen waste, for example). The symptoms are initially very similar to those of influenza, pneumonia, rheumatic fever, and nephritis. Later they develop a similarity to hepatitis and gallstones. It is good practice to ensure that every worker on any waste disposal site, and particularly one involving the composting of domestic waste, is issued with a small booklet that identifies the symptoms of Weil's disease. In the event of his developing any of the symptoms listed he should be advised to show the booklet to his doctor immediately.

These risks apply to anyone entering the site or its environs, not just to the employees. The risks may be small, but there is at least theoretical evidence that they are possible. So from the point of view of precautions, only essential personnel should be allowed into the working areas during turning, loading, and harvesting. Turning operations should be discontinued when weather conditions suggest that airborne material is likely to progress beyond the site boundary. Employees involved in any of the operations listed should be encouraged to wear dust masks and eye protection, and dusts should be suppressed wherever possible by regular washing down.

5.11.2 Safety

The machinery used for composting is inherently dangerous, since it is difficult to guard it so completely that it becomes impossible to make contact with any moving part. Where mobile plant (tractors, loaders, etc.) is concerned, the risks are fairly obviously those of contact with the machines, and it would be pedantic to go into them in detail here. Other machines, however, pose less obvious risks to the uninitiated, as does the compost itself, and the following summary lists some of them.

1. The rotors or cutter chains on side-cutting windrow turners run very fast and are fitted with cutter teeth that may easily catch in clothing, in which event the machine will not stop. They can also, on occasion, find something hard in the material and eject it with some force. All personnel should be disciplined to keep well clear when the machines are operating, and the drivers should be under instructions to stop whenever anyone approaches.
2. Straddle turners, while much slower moving, are still powerful machines that will not hesitate because someone has become trapped by their rotors. The driving position may make it difficult for the operator to see anyone close by, and so the same rules as for side-cutters should be applied.
3. Some tub grinders have a habit of ejecting hard items vertically to a surprising height, and a piece of timber on its way down from 30 ft (10 m) can inflict a serious injury. Operating personnel should wear safety helmets and eye protection, and the machines should be sited well away from other activities.

4. Horizontal rotor shredders eject material horizontally at or just above ground level with, at times, considerable velocity. The author once witnessed an injury to a colleague who was standing some 250 ft (80 m) from such a shredder, when a stone was ejected. No one should be allowed to stand in front of the discharge aperture, whatever the distance, when the machine is running, and again, operators should wear head and eye protection. A further but more obvious risk from horizontal machines is that if anyone falls onto the feed conveyor, it will not stop. It is designed to handle tree trunks, and a human body will cause it no problems at all. Therefore, standing on the framework to release blockages, etc. should be absolutely forbidden.
5. Compost windrows become very hot, and the habit of pushing a hand into them to gauge the temperature should be discouraged. More than one operator over the years has suffered an unpleasant scald by ignoring the precaution. For the same reason, personnel should never be allowed to clamber over the top of a windrow. Any soft spot or void covered by a layer of material could trap them into a very unpleasant fate indeed.

5.12 CAPITAL AND OPERATING COSTS

The purpose of this section is to provide the means by which the costs of constructing and operating a facility may be established, rather than to suggest global figures. Where prices are shown, they are the average of those in effect in 1995, and they should be adjusted to take into account local circumstances and escalation.

5.12.1 Site Construction Costs

The accurate establishment of site construction costs requires the preparation of a detailed bill of quantities, from which other costs such as design, labor and supervision may be calculated, normally as a percentage addition. In the following analyses it is assumed that the outline site area has been calculated using the methods listed in section 5.5.

Only two basic design principles have been considered: a sealed concrete and a geotextile base. They are the only currently available systems which offer protection against pollution of ground and water. Each system is then further subdivided into coverage of the whole operational site and of the turning pad only. Where the whole site is to be covered, it is most convenient to apply a percentage to the actual area of compost windrows calculated as shown in Section 5.6.

In establishing the bills of quantities, the following materials are considered:

Concrete Surface

Slab concrete: 1:2:4 mix compacted with vibrating pokers and tamp-finished.
Membrane: 0.01 in. polyethylene impervious film.
Blinding: Lean-mix concrete of 1:15/24 mix, generally applied dry and rolled.

Reinforcing fabric: 0.5 lb/ft^2 welded steel mesh.
Backfill: Screened rubble, >4 in..
Stone: 2 in. pebble.
Drainage pipe: 3 in. perforated UPVC land drain.
Shuttering: $^1/_2$ in. shuttering plywood.

Geotextile Surface

Top textile: Load-bearing textile suitable for heavy vehicle traffic. Recommended specification: woven polypropylene, circa 6,000 lb/ft (90kN/m^2) tensile strength, weight 0.09 lb/ft^2 (425 g/m^2), thickness 0.06 in. (1.5 mm).
Membrane: 0.02 in. (500-gauge) polyethylene impervious film.
Bottom textile: Antipenetration textile. Recommended specification: woven polypropylene, circa 1,265/633 lb/ft (18/9 kN/m) tensile strength, weight 0.02 lb/ft^2 (90 g/m^2), thickness 0.01 in. (0.3 mm).
Sand: Plain construction sand.
Stone: 2 in. (50 mm) pebble.

Bill of Quantities — Concrete Cover

(a) *Turning Apron Only*

Material	Quantity
Concrete	Apron area (ft^2) ×0.0185 yd^3
Blinding	Apron area (ft^2) ×0.0185 yd^3
Backfill	Apron area (ft^2) ×0.022 yd^3
Membrane	(Apron length + 52 ft) ×(apron width + 52 ft) ft^2
Reinforcing fabric	Apron area ft^2
Stone	(2 × Apron length + width) ×0.33 yd^3
Pipe	(2 × Apron length + width) + 6.5 ft
Shuttering	(4 × Apron length = width) ×3.5 ft^2

(b) *Whole Site*

Where the whole site is to be covered, it is usually sufficient to apply a multiplier of 1.5 to the turning pad area. In other words, the whole site should be 1.5 times the area of the turning pad to allow for product refining and storage space, etc. Therefore, the bills of quantities may be arrived at from the figures in (a) above as follows.

Material	Quantity
Concrete	(a) ×1.5 yd^3
Blinding	(a) ×1.5 yd^3
Backfill	(a) ×1.5 yd^3
Membrane	(Site length + 50 ft) ×(Site width + 50 ft) ft^2
Reinforcing fabric	(a) ×1.5 ft^2
Stone	(2 × Site length + 6 ft) + (2 × Site width + 6 ft) × 0.33 yd^3
Pipe	(2 × Site length + site length + 24 ft) ft
Shuttering	(a) ×1.5 ft^2

Bill of Quantities — Geotextile Cover

(c) *Turning pad only*

Material	Quantity
Top textile	(Apron length + 50 ft) ×(Apron width + 50 ft) ft^2
Membrane	(Apron length + 50 ft) ×(Apron width + 50 ft) ft^2
Bottom textile	(Apron length + 50 ft) ×(Apron width + 50 ft) ft^2
Stone	((Apron length + 6 ft) ×(Apron width + 6 ft)) × 0.33 yd^3
Sand	Apron area ×0.05 ×2 t
Pipe	(2 × Apron length + Apron width + 25 ft) ft

(d) *Whole Site*

Once again, a multiplier of 1.5 is applied to the turning pad area as a convenient method of establishing whole-site quantities.

Material	Quantity
Top textile	(c) ×1.5 ft^2
Membrane	(c) ×1.5 ft^2
Bottom textile	(c) ×1.5 ft^2
Stone	(c) ×1.5 yd^3
Sand	Apron area ×1.5 ×0.05 ×2 t
Pipe	(2 × Site length + Site width + 25 ft) ft

5.12.2 Unit Capital Costs

The costs of the various materials used in site construction will inevitably vary quite considerably depending upon distance from suppliers or from their points of origin. Sand, for example, may be very inexpensive in desert states, but costly in largely agricultural ones, and it would be impossible here to cover every variation nationally or internationally. Instead, the purpose is to show how a cost budget may be developed and to indicate the levels of cost which could be anticipated. Example guide costs of materials are:

Material	Cost/U.S.	Cost/U.K.
Ready-mixed concrete	\$46/yd^3	(£40/m^3)
Blinding	\$37/yd^3	(£32/m^3)
Sand	\$8/t	(£6/te)
Stone	\$7.50/t	(£5.50/te)
Reinforcing fabric	\$275/t	(£200/te)
Backfill	\$5.75/yd^3	(£5/m^3)
Pipe	\$17/100 ft	(£36/100 m)
Water tankage	\$6/yd^3	(£5/m^3)
Pumps, etc.	\$3,000	(£2000)
Shuttering	\$1.50/ft^2	(£10/m^2)
Top geotextile	\$0.17/ft^2	(£1.20/m^2)
Center membrane	\$0.05/ft^2	(£0.30/m^2)
Bottom geotextile	\$0.05/ft^2	(£0.33/m^2)

To the total of the appropriate costs listed above it is then necessary to add 15% for design, and then 10% each for supervision and contractors' profit margin, followed by 20% for labor. A final contingency sum of 10% of the total cost is a worthwhile precaution against cost over-runs and unexpected problems.

5.12.3 Operating Costs

Possibly the most meaningful way of representing the total operating costs is in terms of a cost per cubic yard or cubic meter of finished compost. Such a method provides a good indication of the likely profit or outcome of the venture, since the market values which the product must achieve to break even can then be compared with those of competing products. It also provides a means of estimating where economizing may be necessary if the project does not initially appear viable. However, so many factors affect operating costs that it is difficult to form an accurate estimate, particularly if the operating budget is split into only a small number of headings. Generally it is best to apportion costs over as many activities as possible, since that way inaccuracies are minimized. An example of a budget for a facility capable of manufacturing 50,000 yd^3 a year of compost is as follows:

Cost Item	Cost/yd^3
Establishment	
Labor — 3 employees, 52 weeks/ann	$1.60
Supervision — 1 employee, 52 weeks/ann	$0.60
Employment overheads — pension plans etc.	$0.33
Insurances	
Product liability, employer's liability, public liability, business interruption insurance	$0.50
Consumables	
Utilities — water, drainage, electricity	$0.20
Fuel	$0.15
Tires	$0.03
Mobile plant hire	$0.10
Lubricants	$0.015
Maintenance	
Mobile plant	$0.15
Site	$0.06
Cleaning/housekeeping	$0.15
Administration	
Records	$0.04
Quality control	$0.18
Personnel	$0.02
Accounts	$0.05
Postage	$0.01
Telephones	$0.02
Marketing	
Packaging	$14.00
Literature	$0.06
Advertising	$0.08

Cost Item	Cost/yd³
Transport	$7.50
Invoicing	$0.02
Commercial	
Finance interest (6%)	$1.37
Depreciation	$0.75
Profit (30%)	$6.85

In the above analysis the total cost per cubic yard is $22.84, and the target market price becomes $31.80. The highest components by far are packaging and transport, and clearly the pressure upon market value could be reduced substantially by supplying the product in bulk rather than in packaging. However market values and presentation are interactive. Bulk materials attract a considerably lower price than packaged products, and it is necessary to examine the local market in detail while preparing the project budgets.

REFERENCE

1. Lacey, J., Williamson, P.A.M., and Crook, B., Microbial emissions from composts made from mushroom production and from domestic waste. AFRC Institute for Arable Crops Research, Rothamsted Experimental Station, Harpenden, Hertfordshire, U.K.

6 MORE SOPHISTICATED MECHANICAL COMPOSTING SYSTEMS

6.1 PRINCIPLES OF AEROBIC AND ANAEROBIC DECOMPOSITION

6.1.1 Aerobic Decomposition

Aerobic decomposition is a naturally occurring biological modification to the chemical and physical structure of an organic material in the presence of oxygen. This process encourages the development of colonies of bacteria, and is characterized by the generation of heat. At temperatures up to 95°F (35°C), psychrophiles predominate. Between 95°F and 130°F (35°C and 55°C), mesophiles become active, and beyond that range and up to about 185°F (85°C), thermophiles develop.

Emissions from the process are usually limited to carbon dioxide, water vapor, and occasional traces of ammonia. Emissions of volatile organic compounds (VOCs) are possible. An abundance of water (50 to 65% wet weigh) and aeration is essential.

Residues are dry (30 to 40% moisture content), dark brown in color, and friable. Phytotoxins (plant poisons) are possible in small concentrations, but are rarely sufficient to cause any measurable effect when the product is used as a soil ameliorant. Short periods of maturing (4 weeks or so) may be sufficient to enable use in agriculture.

6.1.2 Anaerobic Decomposition

Anaerobic decomposition is again a naturally occurring biological phenomenon that modifies the chemical and physical structure of organic materials, but in this case does so in the absence of oxygen. It is a cold process in that there is no significant generation of heat, but the addition of heat to raise the reactor temperature to about 85°F (30°C) is beneficial.

Emissions are mainly methane and carbon dioxide, with traces of hydrogen sulfide and VOCs. Any oxygen ingress will increase the carbon dioxide emissions at the expense of the methane.

Residues are usually recovered in the form of a slurry at between 40% and 60% solids in water, and are very dark brown or black in color. They are usually contaminated with plant phytotoxins occurring as unstable fatty acids, and with methane and hydrogen sulfide residues. Before use as soil ameliorants, the residues must be dried, and matured usually for several months to permit the breakdown of the phytotoxins.

Anaerobic processes are often followed by a short-duration aerobic treatment to accelerate maturing and to reduce phytotoxins and gaseous contaminants.

6.2 AEROBIC DECOMPOSITION PLANT DESIGNS

Waste composting is one of the oldest methods of recycling, and as a result innumerable large-scale process designs have been developed over several generations. Almost all of them have been superseded, have been failures, or have simply gone out of fashion. In describing those which have survived in theory if not always in practice, the term "more sophisticated" remains something of a misnomer. In engineering terms, they are largely anything but sophisticated, but they are at least systems that have been built into very large plants, which distinguishes them from static piles and windrows.

Large composting plants became very fashionable during the first half of the century, when the engineering resource and the scale of technology was sufficient to build them. They almost invariably processed crude municipal waste, sometimes with some element of hand sorting to recover recyclable or inert materials, but nothing was done to remove glass and metal fragments. The resulting product was unhygienic, visually unappealing, inconsistent in quality, and likely to contaminate the land upon which it was used with glass and metal fragments. Later, as plastics were developed and found their way into the waste stream, yet another source of contamination arose. It is almost an understatement to suggest that the products did not sell well.

Although there remained some applications for humic materials derived by those early methods, generally where the land was so poor that almost anything would have been an improvement, the technology stagnated for some years. However, with better waste separation, and particularly with separation at source, it became possible to achieve a cleaner organic fraction, at which point composting became attractive once more. Many of the early process designs which had been developed for crude waste had survived, at least on paper. Developments of these are the basis of the current generation of designs.

It is not the purpose of this book to emulate a trade journal, and to identify and evaluate each individual proprietary design. Rather, the purpose is to define design concepts in waste processing. We will therefore restrict ourselves to the general principles to which most current major plant designs conform.

In dealing with aerobic composting systems, there is some difficulty in establishing at what point a design become "more sophisticated" or suitable for a large process facility. Section 5.2 discussed several types of mechanically agitated, aerated bay systems, many of which are used in very large plants and are mechanically quite complex. Therefore, the distinction chosen for this section on composting is one where the process is both intensely mechanical and agitation of the raw material is *continuous*.

6.2.1 Static Digesters

While there have been a number of variants of the static digester, the general principle is one of a large, enclosed vessel, with a means of forced-aeration and an entirely mechanical means of unloading. Raw organic fraction, or even still in some circumstances crude waste, is introduced through the top of the vessel. It is mixed intimately with air inside, and then discharged some time later from the bottom. Decomposition times are very rapid, reducing the materials to a usable compost in 10 to 15 days.

Unfortunately, simply allowing the waste to lie in the vessel for that period of time is unworkable. Apart from the danger of self-compaction making the material impenetrable to air, the decomposition process creates quite large volumes of water. This also seals off the potential air spaces between particles very rapidly, and brings the process to a stop. The aerobic bacteria are inactivated, and the process slowly becomes anaerobic.

One early attempt to overcome that result was to divide the digester into a series of floors, each of which could be moved to allow the waste to fall through onto the next below. The design was sound in two respects: the repeated tumbling action overcame self-compaction and restored aeration, and there was never a sufficiently large column of material in any section to compact greatly in the first place. The disadvantage was that a very large digester could only be utilized to a reduced volume, due to the allowance for void space in each section and for the floor mechanisms. In spite of the rapid decomposition rate, a number of very large digesters were needed to accommodate significant quantities of waste.

A later design developed in the early 1980s involved very large digesters. These were manufactured primarily as agricultural silage silos, and the concept was to install a perforated floor in the base, fill the vessel with waste, and blow air through the floor using a large reciprocating air compressor. The first vessels used at the pilot stage of development were 80 ft (25 m) high, and when filled the waste compacted under its own weight to the extent that it became impossible to remove. The situation was made worse by the inferior quality of the waste separation and granulation. Large pieces of textile and even complete tires become entrained in the mix, rendering all of the unloading methods attempted inoperative.

A similar concept developed in more recent times in Britain, and was more successful. Here again, large agricultural silos were used, assembled from glass-lined plates, and air was forced through the mass from beneath a false floor. However, the air was provided in much less copious quantities, by means of high-pressure fans. The height was constrained to no more than twice the diameter, and unloading was by means of a horizontal cutter mounted in the floor. The system made compost, and it advanced the technology sufficiently for a digester system to be designed with some confidence.

6.2.1.1 Biological/Chemical Processes

The biological processes occurring inside a static digester are similar to those associated with open-air windrows. If organic waste is aerated adequately, and if the moisture content is suitable, then biological decomposition will begin almost immediately. During the first few h, the material will become colonized by mesophylic bacteria, and there will be a progressive increase in activity for up to 2 to 4 days. That period is evidenced by a steady temperature rise, the rate of increase of which begins to fall off toward the end of the period. Mesophyllic bacteria are not particularly tolerant of heat, and when the temperature of the material reaches about 130°F (55°C) the activity ceases. At that point, the material becomes colonized by thermophilic bacteria, and the temperature then proceeds upward again at an increasing rate, reaching about 175°F (80°C). During the period from first charging the digester, a graph drawn of temperature against time then shows a rapid increase to 120°F (50°C), sometimes followed by a slight curve toward temperature stability. But then the graph increases again, and a final stability curve occurs at 175°F (80°C). That temperature may then be maintained for up to 10 days before beginning to decline quite rapidly. Figure 6.1 shows a typical curve.

The rate of development, and the period of continuation, of biological activity is determined to a large extent by the granularity of the material. Many very small particles make more surface area available to the bacteria than do large ones. Any intermediate treatment that progressively reduces the granularity both accelerates and prolongs the decomposition, as fresh surfaces are exposed.

During the period of biological activity, there is also a parallel period of chemical oxidation of the lignin in the cellulosic fraction, particularly of any paper in the mix. As far as the end product is concerned, the result is very much the same in that a friable, dark brown material is created. The oxidized cellulosic fraction does, however, perform two very important functions. It absorbs and retains some of the moisture liberated during the biological activity, thereby maintaining porosity in the mix, and it adds a peat-like texture to the final product. Therefore, if the cellulosic fraction is too small, then the mixture is likely to become wet, to lose porosity, and to coagulate to the extent that anaerobicity may result. Even at best the process will inevitably slow down dramatically, and conditions that should occur in days may take weeks.

Recent research has suggested that there is no upper limit to the cellulosic content of the waste beyond which the process is adversely affected. In fact, experiments in composting granulated newsprint, with less than 5% by weight of organics, have produced very good quality soil conditioner indeed, even if it could not legally be called "compost" in countries where standards exist. There is, however, a lower limit if excessive moisture, coagulation, and anaer-

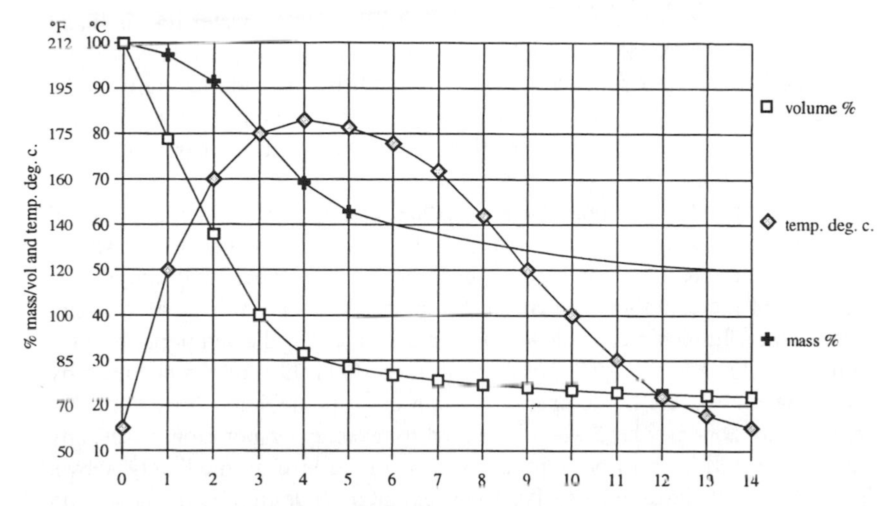

Figure 6.1 Characteristics of aerobic disgestion.

obicity are to be avoided. The transition point between an unstable system and an active aerobic one occurs when the cellulosic fraction is 20% or more by weight of the mix.

6.2.1.2 Physical Behavior

Although the ratio of soft organics to cellulosics may be satisfactory, the soft material will still tend to compact under its own weight to the point where unloading becomes difficult if not impossible. In this sense, while cellulosics are of course themselves organic, the expression is used to distinguish the fact that they are not readily attacked by bacteria. During composting, a mass loss of 50% and a volume loss of 70% may be experienced. The cellular structure of the organic material collapses as bacteria absorb the nutrients contained within it, and the cellulosic material degrades under oxidation, losing structural strength. The result may be an increase in bulk density from an original 17 lb/ft^3 (270 kg/m^3) to 44 lb/ft^3 (700 kg/m^3). At the higher density, the compacted and highly fibrous mixture becomes extremely difficult for any harvesting mechanism to penetrate.

If very high compaction is allowed to occur, then liquors begin to leach from the digester contents. The first effect of this is to soak the lower strata nearest to the air admission and to reduce the porosity in that region. This may result in insufficient oxygenation of the higher levels, creating a trend towards anaerobicity. At the same time, liquids with a high biochemical oxygen demand (BOD) are likely to begin to seep from digester seams and through unloader orifices. The liquids are corrosive and malodorous, and they may contain pathogenic bacteria. Even in small quantities, their black color and slightly oily nature makes them extremely disfiguring if allowed to spill out onto walkways, etc. They are highly polluting if released into watercourses.

In spite of the release of water from the lower strata, water loss is likely to be a problem. As already stated, the biological processes cease without sufficient moisture to support the bacteria, and the high temperature of the process liberates water vapor very rapidly indeed. A conservative estimate suggests that a digester process treating 30,000 t of waste a year may, in the same period, emit 6000 t of water as vapor.

Unfortunately, an appropriate moisture content cannot be maintained throughout the mix by the addition of water as necessary. Figure 6.2 shows the temperature regime in a hypothetical digester, and reveals that roughly the top third of the chamber is occupied by material at little more than ambient temperature. Immediately below that is a strata where the temperature rises rapidly to about 175°F (80°C), followed in the final third of the chamber by a cooling zone where the temperature again drops to ambient. Since the high-temperature zone is where water is needed to make up evaporation losses, any make-up would have to be introduced through the level immediately above. That level would have to be soaked to the extent of saturation before any could percolate further downward, yet saturation would so reduce the porosity as to prevent aeration and stop the process.

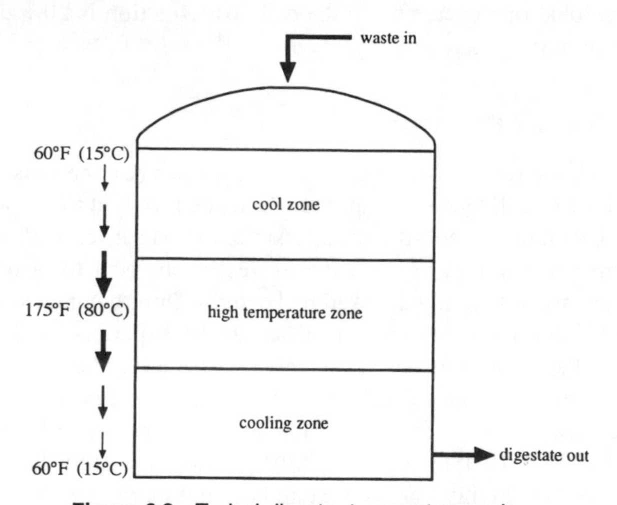

Figure 6.2 Typical digester temperature regime.

Water cannot be introduced through the sides of the digester either. To do so would require some means of delivery that could reach into the vertical center, and the material is too fibrous to be penetrated easily by temporary pipes forced in horizontally from outside. Permanently fitted pipes would be broken almost immediately as the material moves down the chamber in consequence of volume and mass losses.

The inability to modify the moisture content once the material is inside the digester consequently makes adjustment beforehand essential. If there is insufficient water in the mix when it is introduced into the digester, then there will be no opportunity to correct it later. The resulting product will be immature, and prone to a resumption of biological activity and reheating as soon as any moisture reaches it.

The need to prepare the moisture content before introduction into the digester leads to a method of removing product-damaging contaminants such as batteries and glass. The behavior of both the organic fraction and the contaminants is not sufficiently dissimilar for air or ballistic classification to work with any reliability. Screening by particle size is also impossible, again because there is insufficient difference between the fractions. However, most organic materials have a *true* density close to or less than that of water, while glass, batteries, and metals have a much greater density. The organic fraction has a measurable buoyancy, while that of the contaminants is negligible. Therefore, a flotation system can be used to separate the fractions very effectively, and at the same time it can adjust the moisture content of the waste.

Removal of contaminants is vital if a usable product is to result. The presence of batteries and metals in the moist, hot, acid environment of the digestion process leads to heavy contamination from metal salts. Glass renders the product dangerous to handle and is visually unattractive when the compost is used, since it is invariably left on the surface long after the compost has all been absorbed in the soil. Plastics are of course a fairly recent phenomenon in compost production, and they also reduce the cosmetic appearance of the product, even if they leave it unaffected chemically. A flotation system will remove some plastics, but by no means all. Film will always report to the organic stream.

Fortunately, the presence of plastic film in the crude product is not a significant problem, since it is easy to remove during the essential product-refining stage preceding marketing. While the *true* bulk density of film is close to that of the compost, its granularity is quite different, as is its behavior in an airstream. In refining, the compost needs to be screened to remove any lumps, and that process also separates most of the plastic film. Air classification of the product to remove any stones that may have escaped the screen then removes any small particles of film.

The provision of both air and water in copious quantities involves something of a compromise. If the initial water content of the material is less than about 60% by weight, biological activity will be subdued and may even cease before the waste is fully composted. If, however, the water content is too high, above about 80%, then the porosity of the mixture may be reduced to the extent that air can no longer penetrate. Once again, aerobic biological activity will be subdued and anaerobic decomposition will proceed. It is therefore important to consider the functions of the air and water in the physical as well as the biological sense.

Air provides oxygen for biological respiration, and physically the means of exhausting the products of respiration — carbon dioxide and water vapor. It has little to do with temperature control, and an excess of air will not reduce the temperature in the digester. In fact it is more likely to increase it by encouraging more rapid bacteria growth. A shortage of air will succeed in reducing temperatures, but only because it inhibits bacterial growth.

Water provides the means for the bacteria to develop and breathe, since it is as important to aerobic microorganisms as it is for higher animals. It is also the major conductor of heat from the composting waste. This latter characteristic is not always immediately apparent, since a shortage of moisture again inhibits bacterial growth, without which there can be no temperature rise. However, given a moisture content of between 60% and 80% by weight, most of the water initially present and that created by biological activity will be vaporized and emitted from the digester. Water, in changing from the liquid phase to the vapor phase, absorbs a considerable quantity of heat even though the vapor itself may not appear hot, and since its thermal conductivity is much greater than that of air, it extracts heat much more efficiently.

The amount of heat energy necessary to convert water into vapor is very much greater than that necessary to increase its temperature. While the biological activity can easily generate sufficient heat to raise the temperature of the water, it must do so upon a much greater scale to cause it to change to the vapor phase. This factor makes the temperature reached by aerobic processes self-limiting. As the temperature rises, so does the input of heat to the water. Eventually a point is reached where any further increase in heat output is used in vaporization, and the temperature stabilizes. This point is reached at between 175° and 185°F (80° and 85°C). The stabilization is therefore not, as sometimes thought, caused by the temperature becoming sufficiently high to limit biological activity, since bacteria can survive in much higher temperatures than those experienced in a digester.

The operational consequences of the interaction between water and air in the mixture can therefore be summarized as follows:

Sufficient air and water — temperature stable at 175°C (80°C).
Sufficient air, insufficient water — premature temperature drop after initial rise. Slowing of decomposition and unstable product.
Insufficient air, sufficient water — limited or no temperature increase, and development of anaerobicity.
Excess air, adequate water — possible fissuring of material and excessive dryness of product, but otherwise little effect.
Adequate air, excess water — limited or no temperature increase, and development of regions of anaerobicity.
Insufficient air and water — restricted temperature increase followed by early and rapid temperature drop. Biological activity curtailed and inclined toward anaerobicity.

6.2.1.3 Digester Design and Operation

With the understanding of the biological, chemical, and physical processes occurring in digesters, a design for a large-scale facility can be approached with some confidence. In doing so, and in recognition of the phenomena already discussed, the following design constraints must be accommodated.

- The organic fraction must be thoroughly granulated sufficient to ensure a consistent mix and to expose the optimum surface area to biological attack.
- The moisture content of the feedstock must be adjusted to the optimum level for composting before introduction into the digester. At the optimum level, there is sufficient water in the mix to permit decomposition to the stage required for the finished product.
- Heavy and metallic contaminants must be extracted before biological processing.
- A biologically inert, cellulosic fraction amounting to at least 20% by weight of the total mix must be equally well-granulated and added to the organics to act as a bulking agent.
- Once introduced to the digester, the mixture must not be allowed to settle for long enough to become excessively compacted.
- There must be a provision for aerating the material.
- Since very large quantities of water vapor are likely to be emitted, there must be a facility for emissions control.
- The digester must be fitted with an unloading mechanism capable of handling material bulk densities of up to 45 lb/ft^3 (700 kg/m^3).
- The casing of the digester must be resistant to acids and alkalis.

6.2.1.3.1 Granulation and Moisture Content of Feedstock — Ideally, the reduction of the raw feedstock to an acceptable particle size is a secondary result of processing carried out for other purposes. The most common, and most effective, method used for the initial separation of the organic fraction is trommeling. A trommel screen is a large-diameter 8 to 10 ft (2.5 to 3 m) perforated mesh drum, of a length determined by the waste input volumes, and rotating at 60% of its critical speed. Section 3.1 contains a more detailed description of machines of this type.

A trommel screen with mesh plates perforated by 2 in. (50 mm) holes and a 48% open surface area is ideal. When mixed waste is introduced into the drum, the tumbling action causes the organic fraction (and at this stage the glass) to become sufficiently broken to pass through the holes. Paper, cardboard, cans, and plastic films meanwhile remain inside the drum and are discharged through the open end as oversize. As a result, the coarsely granulated organic fraction is diverted to one stream, while the oversize fraction is diverted elsewhere.

In the next stage, the contaminants within the organic fraction are removed; a fluid system is well-suited to the purpose. Water is the obvious fluid to use, since water must be added to the fraction to make it suitable for composting, and one of several designs that use it is shown in Figure 6.3. Although there are a number of ways of using the natural buoyancy of the fraction in water, they are all characterized by three features. (1) The residence time in the water must be sufficiently long for most or all of the contaminants to be flushed out. (2) There must be sufficient turbulence in the separating chamber to liberate any heavies adhering to the organics. (3) There must be sufficient water in the system to permit settlement of very small particulates at some point.

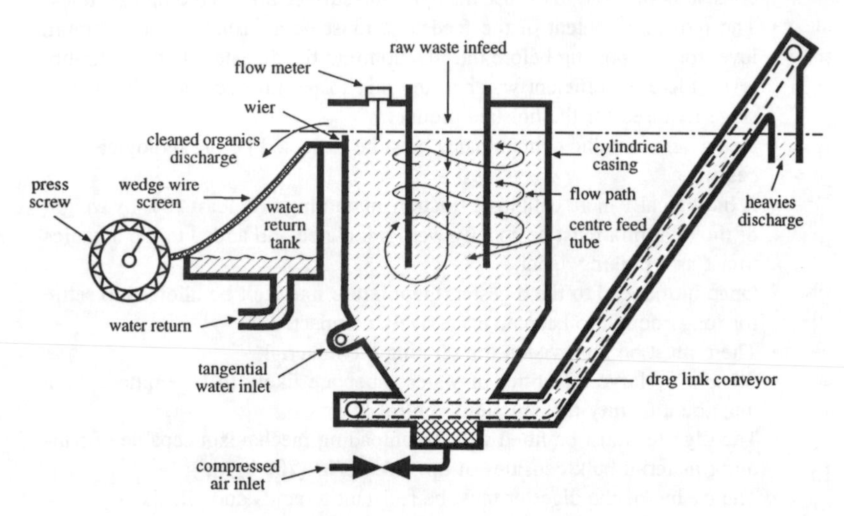

Figure 6.3 Fluid separation plant.

Obtaining an acceptable residence time can be achieved by forcing the water into a fast-rotating, circular path, as shown in Figure 6.3, or by devising a long chamber with a straight-line water flow. In either design, as the contaminants fall clear of the organic fraction they are collected from the bottom of the chamber, usually by a submerged drag-link conveyor. While the organic fraction passes on with the water to a second dewatering stage, the heavy fractions are discharged into a separate line for further processing if required.

Turbulence can be achieved by water jets, by baffle plates, or by micro-aeration of the water. Of the alternatives, micro-aeration is best because it ensures that very fine, noncontaminating fibers stay with the organic fraction rather than affecting the viscosity of the water. These fibers arise partly from household dusts, and partly from paper residues broken up in the trommel sufficiently to stay with the undersized stream. If left in suspension in the water, they will have the effect of transforming it into a thick fluid that is biologically active and increasingly begins to carry over contaminants. When the water is aerated, microbubbles of air adhere to the fibers, making them highly buoyant and ensuring their separation.

Settlement of small particulates can be encouraged either in a large, separate settlement tank, or by means of a clarifier. A settlement tank is the cheapest and least complex mechanism for encouraging settlement, but it is not efficient even when it has a very large water reserve. A settlement tank capacity of less than 10 times the circulating system volume is ineffective. In addition to capacity, the tank must have mechanical scrapers to deal with the scum that will form on the water surface. "Scum boards" for this purpose are common in the liquid waste industry.

Clarifiers are a more efficient, but much more expensive, way of dealing with contamination in the water. They basically consist of a very fast-rotating, small-diameter drum into which the water is pumped through one end. Centrifugal forces drive the water and contaminants to the drum sides, and force the rapid settling out of heavy particulates. The particulates are generally removed by a slow-speed screw, driven by a separate gearbox, while the water flows from the end of the machine under pressure from the flow behind.

When the organic fraction leaves a wet scrubbing system it is too wet for introduction into digesters. This occurs as a result of the volume of free water that attaches to the organic particles, and it varies inversely in relation to the granularity achieved. Some of the water must be removed to reduce the overall moisture content to between 60% and 80% by weight, and a screw press is a satisfactory method for achieving that. In such a machine, a helical auger similar to that of a screw-feeder conveyor is enclosed in a circular cage of wedge-wire bars. The bars are aligned along the longitudinal axis of the auger, and they terminate at an open end that is restricted by a spring-loaded plate. In operation, wet materials are introduced into the auger and forced down the circular cage, gradually being compressed by frictional resistance against the bars. At the discharge end, the materials are further compressed against the spring-loaded plate, and the degree of spring loading determines the final moisture content of the product. It is the pressing action against the end plate, combined with the rotation of the auger, which finally granulates the material to a suitable particle size.

The extent of dewatering necessary depends largely upon whether or not a bulking agent is to be added. If, for example, paper from a shredder is to be included in the process, then the fraction leaving the wet scrubber must be wetter than if it were to be used alone. Some excess moisture must be retained for introduction to the bulking agent, otherwise its dryness will reduce the overall water content of the mixture below that required for effective composting.

6.2.1.3.2 Heavy and Metallic Contamination — A fluid system as described above is probably the most efficient way of removing most of the heavy contaminants and at least some of the metallic ones. It cannot be relied upon to eliminate heavy metals salts, but it does reduce them to manageable levels provided that the circulating water is treated. However, levels of contamination at the beginning of the process are only a precursor of the final problem.

All aerobic composting processes reduce both the volume and the mass of the treated waste very substantially indeed. Mass losses of over 50% and volume losses of over 70% are possible, and all of those are constrained within the organic fraction, as are most of the contaminants. As an example, suppose the municipal waste received at a facility contains 6% by weight of glass, and 35% by weight of organics. When the organic fraction is separated, the glass will report to the same stream. Taken from 100 t of waste, 35 t of organics will now contain 6 t of glass, which implies a contamination level of 17%. If the glass is not removed at that stage, then the further mass loss during the composting process will reduce the 35 t of organics to 17.5 t, in which there will still be 6 t of glass. The percentage of contamination therefore rises to 34.2%, which is beyond the separating capacity of most conventional compost refining plant: it simply becomes overwhelmed by the quantities involved.

This example illustrates equally well what happens in the case of chemical contamination, particularly in the case of heavy metals, most of which come from batteries and vacuum cleaner dusts. A level of 0.00009 oz in every pound (50 mg per kilogram) of waste can rise to nearly 0.003 oz/lb (200 mg/kg) in the final product, which is why a fluid-based separation system is attractive. It removes batteries without breaking them and releasing their contents into the waste.

6.2.1.3.3 Bulking Agents — The most effective and readily available bulking agent is paper extracted from the waste stream, after the oversize discharge from the trommel screen. However, at that point the paper will be heavily contaminated with heavy plastics and metal cans. Granulation of the total oversize in a flail mill will begin the process of overcoming the problem.

A flail mill is effectively a hammer mill, either with a horizontal or vertical rotating shaft, upon which thin flail arms rather than hammers are suspended. The arms are located in large-diameter flanges, which are in turn keyed to the shaft by means of long rods. Normally there are four flail arm mounting positions equally spaced around the periphery of each pair of flanges. Milling the oversized fraction while it is still mixed with cans has the advantage of improving granulation while cleaning the cans of any organic residue ready for recycling.

When the fraction has been milled, ideally down to a granularity of between $^1/_2$ in. and $^3/_4$ in. (10 and 20 mm), the metal fragments and heavy plastics must be removed. Fortunately, at that stage air classification works extremely well. Introducing the material directly into a rising air column as it discharges from the mill causes the paper and plastic film to become entrained in the airstream, while the metals and heavy plastics fall clear for recovery elsewhere. The paper may now be added to the cleaned organic fraction ready for composting, and the fact that it is contaminated with plastic film is immaterial. That can be removed at the end of the process.

Ideally, the paper should be added well before introduction into the digester, since efficient mixing later becomes progressively more difficult. Usually the easiest location of paper addition is upon the conveyor that receives the discharge of organics from the dewatering screw press. Once added there, the materials will become intimately mixed at the first transfer chute they encounter on their way to the digester.

Further mixing can be achieved at the point of introduction to the digester, since there some means of distributing the material evenly across the full diameter has to be provided. A convenient way of achieving even distribution is by means of a spinner onto which the material falls from the infeed conveyor. The device is a circular steel plate, the diameter of which is equal to the width of the infeed conveyor. The plate is mounted upon a vertical, gear-motor-driven shaft suspended below the infeed conveyor discharge, and the upper surface of the plate is spoked with 2 in. (50 mm) high flat steel strips. Six spokes should be equally spaced around the plate, each running from the center shaft to the outside edge. The spinner should be driven by a variable controller such that its rotational speed cycles continuously between a slow speed and a fast speed range.

The fast speed applied depends upon the diameter of the digester, and should be capable of adjustment independently of the slow range. Adjustment is then a matter of setting the speed until material is thrown just to the sides of the digester. The slow speed can remain fixed irrespective of the diameter of the chamber, since its purpose is to allow material to fall around the vertical center.

A variable-speed spinner of the design described offers a final and efficient mixing stage, but only if the bulking agent has been introduced earlier. If the agent is introduced at the spinner instead, then the result is more likely to be a separation. Dry, granulated paper added at the top of the digester will be blown around by the spinner and is more likely to report to the digester ventilation system than to the compostable material.

6.2.1.3.4 Settlement and Capacity — As suggested earlier, organic waste mixed with a bulking agent will begin to compact immediately under its own weight. This characteristic is most significant before decomposition has become established, becoming less so as the process continues. It occurs partly as a result of a reduction of the air space between particles under pressure, and partly as a result of collapse of the cellular structure of the organics. There is nothing that can be done to prevent it occurring, but the worst effects can be reduced.

It is at this point in the design logic that a decision has to be made, since there are only two ways of dealing with compaction. Either the process is to be carried out in a single digester, or in a group of digesters operating in sequence. If the decision favors the former, then the multilevel system with intermediate floors may be necessary. If the latter is chosen, then the whole plant must be designed so that the waste moves through the digesters one after another until decomposition is complete. Either way, the materials must be moved frequently enough that compaction has not had time to reach an unacceptable density.

The multilevel digester system is no longer in widespread use, possibly because a number of very large, mechanically complex units are necessary to achieve significant throughput. A multi-digester system is in many ways more straightforward in mechanical design, although there are some operational consequences.

A multi-digester installation, based upon the design shown in Figure 6.4, is one where the waste is introduced to a series of digesters, moving from the first to the second after a period of initial decomposition, then from the second to the third after a further period, and so on. In such a design, the capacity of each digester, and the residence time within it, depends upon the degree of mass and volume loss achieved and upon the extent of the tendency to compaction.

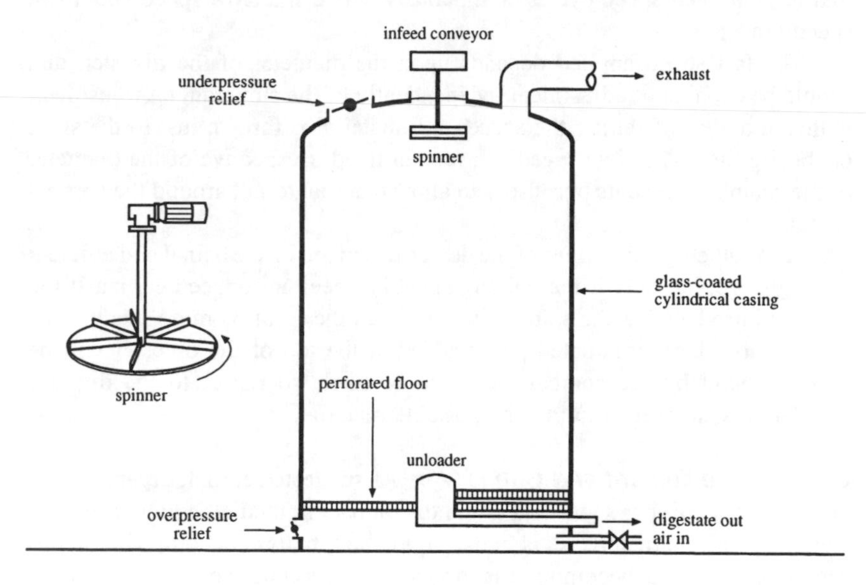

Figure 6.4 Typical digester design.

Aerobic composting plants may be designed either as batch processes, where a volume of material is fully treated and removed before further volumes are dealt with, or as continuous processes, where compost is extracted at a rate consistent with a continuous infeed of raw material. Batch processes are inadequate for dealing with major waste arisings, and they impose a requirement for raw waste storage that is difficult to meet. Therefore, batch processes are no longer widely used. Multi-digester systems lend themselves to continuous processing, and are more attractive, but the design logic path is somewhat more tortuous.

There are several manufacturers of digester-based continuous systems, and they have each derived operating principles best suited to their designs. It would be impossible in a single book to explore each of these in detail, but an examination of a process that covers the general concepts of digester composting is worthwhile.

Waste is introduced into the first digester of the series, and then after a predetermined period of time, it begins to be transferred into the second. After a further period, it is again transferred into the third, and then into the fourth digester. At the first stage, fresh waste is added as partially composted material is transferred, and that principle is extended to each of the stages in turn. As a result, once the system has reached continuous operation, raw waste is loaded continuously into one end, while finished compost is removed from the other. In theory, all that remains is to calculate the initial volume per h of the raw waste, and to multiply that figure by the residence time in h in each unit to establish the volumetric capacity to be constructed. Unfortunately, the variables affecting the calculation are as follows.

1. The loss of mass is not linear. It develops slowly at first, more or less in line with the increase in temperature. When the temperature reaches its peak, the rate of mass loss is also at maximum. Then, as the available nutrients are consumed, the temperature begins to drop and so does the rate of mass loss.
2. Volume loss is directly related to mass loss in the sense that all of the mass loss occurs in emissions, but it is also directly related to the degree of compaction.
3. Compaction depends upon both the period of time during which the material remains undisturbed and the degree of decomposition achieved. The potential for compaction is highest when the waste is first admitted.
4. The aeration of the waste is adversely affected by the moisture content.

It follows that from consideration of the variables, the material is more prone to compaction during the earlier stages of the process. Mass and volume losses are likely to be more in evidence during the middle stages, and stability occurs toward the end. Therefore, avoidance of compaction could be best achieved by gradually increasing the residence time in each digester in succession, starting with the minimum time in the first. In applying that principle, some practical engineering constraints would need to be taken into account, otherwise the digesters would all be of different sizes. Once again, in common with all waste processing systems, the design becomes something of a compromise.

Although decomposition to a satisfactory stage can be achieved in 10 days, it is sensible to design for longer. In that way there is a surplus of capacity to absorb fluctuations in waste arisings and the inevitable circumstances when something goes wrong. The compromise then arises from the need to accommodate the mass and volume losses, and consistent digester sizing, in a calculation that stores the material for progressively longer periods.

Figure 6.1 shows a typical aerobic digestion process in graphical form, suggesting a mass loss which develops slowly with the temperature rise. The mass then declines quite rapidly while the temperatures are at their maximum, consistent with the vigorous biological activity. Finally, as the temperature declines again toward ambient, the rate of mass loss also declines until it becomes undetectable.

The volume, meanwhile, declines very rapidly initially, as a result of compaction and to some extent, of mass loss. However, as the biological activity passes its peak, the volume loss rate declines rapidly until near stability is achieved. It should however be noted that volumetric stability is never completely reached. There always remains some element of compaction, depending upon how the material is handled and stored. It is, for example, possible to unload a greater volume than is actually contained in the last digester, simply because the unloading mechanism reduced the *apparent* bulk density by its cutting action.

Using the data contained in Figure 6.1 and relating them to the constraints imposed by mechanical engineering, it becomes possible to construct a mathematical model, as shown in Table 6.1. Returning to the hypothetical process plant we have used throughout, there are total municipal waste arisings of 150,000 t per year, of which 35% by weight is organic. If the plant is to operate 16 h a day, 5 days a week, for 50 weeks a year, then the total organics output will be 13.13 t/h. An addition of 20% by weight of bulking agent will then increase that output to 15.76 t/h.

Table 6.1 Digester Characteristics

	Silo 1	Silo 2	Silo 3	Silo 4
Residence time, days	2	3	4	5
Density in	17 (272)	31 (496)	31 (496)	31 (496)
Waste in per day	252 (229)	229 (208)	158 (144)	137 (125)
Volume in per day	1098 (839)	548 (418)	378 (288)	328 (251)
Mass loss %	10	45	15	10
Density out	37 (593)	37 (593)	37 (593)	37 (593)
Waste out per day	229 (208)	158 (144)	137 (125)	125 (114)
Volume out per day	459 (350)	316 (242)	275 (210)	250 (191)
Average volume	1557 (1189)	1296 (989)	1306 (997)	1446 (1104)
Cum. mass loss (%)	9	37	45	50
Cum. volume loss (%)	58	71	75	77

Note: Mass in tons (metric tons), volume in yd^3 (m^3), density in lb/ft^3 (kg/m^3).
Waste in per h = 13.13 t; bulking agent per h = 20%; working h per day = 16; working days per week = 5; working weeks per year = 50; initial density = 17 lb/ft^3 (272 kg/m^3); total mix in per h = 15.756 t.

At this point some assumptions must be made respecting of the number of digesters to be used and the residence time in each. Four digesters would be a reasonable choice, and residence times of 2 days in the first, 3 days in the second, 4 days in the third, and 5 days in the fourth would be consistent with the compaction characteristics of the material at each stage.

Upon average the organic fraction of municipal waste has an *apparent* bulk density of around 17 lb/ft^3 (272 kg/m^3). The addition of the dry bulking agent has the effect of reducing that, but then the addition of water increases it. Most of the water will finally report to the bulking agent, and so it is convenient to assume that the overall bulk density of the mixture remains consistent. Therefore waste will be delivered to digester number 1 at a rate of 252 t/day, at a density of 17 lb/ft^3 (272 kg/m^3) and a volume of 1098 yd^3 (839 m^3) per day.

Table 6.1 which is derived from experimental results in a real plant, then suggests that at the end of 2 days residence time, the waste will have lost 10% of its mass and 58% of its volume. To maintain the reception capacity within the digester, it will be necessary to begin unloading material at a rate of 229 t/day by weight, or 459 yd^3 (350 m^3)/day by volume. Thus the bulk density of the material has risen to (229 × 2000)/(459 × 27) lb/ft^3, or 36.96 lb/ft^3 (593 kg/m^3).

That volume of material is now delivered to number 2 digester, and some reduction in apparent density should be expected as a result of the intermediate handling. However, the reduction in density will only be temporary, since it results from aerating the material. It will rapidly be restored by the weight of successive layers. Experience indicates that a reduction of about 6.25 lb/ft^3 (100 kg/m^3) can be expected, and so 229 t per day at a density of 31 lb/ft^3 (496 kg/m^3) have to be added to digester number 2. In volumetric terms, that input equates to 548 yd^3 (418 m^3) per day.

After 3 days, unloading of digester number 2 begins, by which time a further mass loss of 45% has occurred, and the density has again reached 37 lb/ft^3 (593 kg/m^3). Accordingly, the daily output mass to maintain stability has become ((229/1.45), or 158 t, and the corresponding volume has reduced to (158 × 2000)/(37 × 27), or 316 yd^3 (242 m^3). That material is introduced into digester number 3, and again the loss of density due to handling is assumed to be 6.25 lb/ft^3 (100 kg/m^3).

Carrying out the same calculations for digester number 3 results in a digestate infeed of 158 t per day, at a density of 31 lb/ft^3 (496 kg/m^3) and a volume of 378 yd^3 (288 m^3). A further 15% mass loss occurs during the 4 days storage, according to Table 6.2, and the original density is restored to 37 lb/ft^3. As a result, digestate is then unloaded at 137 t a day, at a volume of 275 yd^3. Digester number 4 then receives the material at a rate of 137 t per day, at 37 lb/ft^3, and of 328 yd^3 volume. This, after 5 days, is further reduced to 125 t a day, at a density of 37 lb/ft^3 and a volume of 250 yd^3.

When the results of the calculations are tabulated as in Table 6.1, it can be seen that over the full period of decomposition, the waste has lost 50% of its original mass and 77% of its volume. The lost mass will have been liberated as carbon dioxide gas and moisture vapor, while the volume loss results from compaction, increased granularity, and the collapse of the cellular structures.

Now the model must convert the results of the calculations into data appropriate for the establishment of digester dimensions. If digester number 1 is to receive 1098 yd^3 (839 m^3) per day, and store it for 2 days, then the total input volume over the period will be 2196 yd^3 (1678 m^3). However, compaction and volume losses will increase the density of the material, and together with mass losses will create a final density of 37 lb/ft^3 (593 kg/m^3). Unfortunately, all this reveals is that the material density in the top of the digester will be the lower figure, and the higher figure will be at the bottom. The increase in density with depth is not linear, but since it is not realistic to attempt to calculate the results of biological activity to the nearest cubic metre, an assumption of linearity is appropriate. Therefore an acceptable estimate of number 1 digester capacity could be made by simply taking the *average of the input and output volumes over the period of residence*, which in the example produces a figure of 1557 yd^3 (1189 m^3).

Again, applying the same calculation to each of the digesters in turn results in an estimated capacity of 1,296 yd^3 (989 m^3) for digester 2, 1306 yd^3 (997 m^3) for digester 3, and 1446 yd^3 (1104 m^3) for digester 4. Therefore a digester erected using industry standard plate sizes, and having as a result a volume of 1765 yd^3 (1,350 m^3), would be adequate for each of the stages. The design would allow for up to four days longer digestion than strictly necessary, or for an increased input of approximately 40%.

6.2.1.3.5 Aeration — During the time in which the waste remains in the digesters, aeration is extremely important if anaerobicity and inadequate decomposition are to be avoided. That, however, creates something of a complex problem, at least at first sight.

A circular digester, possibly 32 ft (10 m) in diameter and 55 ft (17 m) high, is not a pressure vessel. Even a moderate positive or negative air pressure of less than 14.7 lb/in.^{2} (1 atm) inside it may be sufficient to cause catastrophic failure. It may either explode or implode. The air introduction mechanism must deliver sufficient pressure to penetrate the waste, but never sufficient to overpressurize the vessel. Equally, the aspiration system that deals with the emissions from the sealed top of the vessel must never be able to create a significant vacuum. And if that were not enough, the structural resistance of a cylinder is greatest for an internal pressure and least for an external one.

The possibility of creating an overpressure or a partial vacuum does not exist only in relation to the air supply or exhaust. It is possible to reach a situation where both are experienced at once in the same digester. This can happen if the material contained within it has been allowed to settle and compact for too long. An attempt at that stage to unload the digester can cause a void to form beneath the waste as the unloader removes material but that above does not immediately fall to replace it. In some circumstances, the "plug" of material may eventually slip, falling down the digester rather like a piston inside a cylinder. The result may then be a significant pressurization of the vessel beneath the material, and a corresponding negative pressure above. Either or both may easily be sufficient to rupture the casing.

Clearly, such an eventuality can never be allowed to happen, since the rupturing of the vessel under even a slight pressure would be devastating. The piston effect caused by compaction of the digester contents can never be entirely excluded, and so some means must be provided to ensure that if it occurs, venting to or from atmosphere is possible. Therefore, on the air pressure side, beneath or level with the perforated floor, a safety vent must be provided. The vent should be designed such that it stays closed under normal aeration, but can open fully as soon as overpressure is detected. In such an event, it must be capable of fully exhausting the overpressure within a maximum pressure limit.

The conditions that might occur during overpressure can be calculated. If a digester contents of 504 t (two days' storage in digester 1 in the example used so far) were to fall as a coherent mass in a vessel with no air escape device, then it would create a pressure equivalent to the weight divided by the floor area. If the vessel diameter was 32.8 ft (10 m), then the floor area would be 845 ft^2 (78.55 m^2). The pressure would therefore be 504/78.55 t/m^2, or 6.42 t/m^2, more normally expressed as 9.12 $lb/in.^2$ (0.642 kg/cm^2). While that may not appear to be a particularly high pressure, it would certainly be sufficient to disrupt the casing of an agricultural design of vessel, and worse, it may not be applied evenly over the full surface area.

The situation in the top of a sealed chamber under such circumstances is much more dangerous. The creation of an overpressure of 9.12 $lb/in.^2$ under the plug implies a negative pressure of atmospheric *minus* 9.12 $lb/in.^2$ above it. A pressure differential of that order between the outside and the inside of the vessel would again be sufficient to disrupt it, or to collapse the roof.

In view of those possibilities, there are several design essentials to consider. Normally a cylindrical container of the scale used for digesters is built with heavier plate at the bottom than at the top, simply to support the weight of the higher levels upon the lower. However, overpressure in the base may result from the piston effect of the contents, and will result from the pressure applied by the aeration system. It is therefore essential that the suppliers of the digester are aware of the requirements, and that they are provided with data upon both the piston effect and the pressure required to aerate to load. The result should be a specification calling for much heavier platework in the lower levels.

That of course poses the question of what pressure should be provided to ensure that the material is effectively aerated. It would be possible to design a system that created an aeration pressure of 9.12 $lb/in.^2$ (0.642 kg/cm^2), upon the principle that anything more would simply push the contents to the top of the digester. In other words, 9.12 $lb/in.^2$ spread over 845 ft^2 (78.55 m^2) would generate a vertical force of 504 t. That is in fact a somewhat unlikely event. The greater probability is that such an air admission would simply cause excessive fissuring in the contents of the digester. The purpose of the aeration system is to apply sufficient air for biological reduction reasonably gently, not as a sudden blast or as a sustained high pressure. If a situation occurs where the air pressure could ever rise sufficiently to lift the contained mass, then compaction and coagulation have progressed beyond redemption. The only option then is to completely unload the digester.

Research has upon occasion provided data for the air flow necessary to support biological activity in a given mass of material, but the results are somewhat academic. There is no reason to suppose that at greater than laboratory scales, the aeration of the mass would be consistent throughout. In fact, the more established manufacturers of aerobic digesters assume that it is not. Some provide for "zone aeration" — a means of selectively applying air to sections of the base of the digester in response to temperature variations across the material. In addition, the cellulosic bulking agent will require a certain amount of air for oxidation to take place, and more will be required to carry away the carbon dioxide and water vapor resulting from respiration and heating.

Air requirements, expressed in terms of the volumetric ratio between air flow in units per min and the organic mass, and ranging from 0.04 to 0.25 have been suggested. In a digester containing 1700 yd^3 (1300 m^3) of material, the air demand based upon those figures would range from 1836 ft^3/min (52 m^3/min) ($0.04 \times 1700 \times 27$ m^3) to 11,476 ft^3/min (325 m^3/min) ($0.25 \times 1700 \times 27$). Some designers of aerobic systems avoid becoming too specific about aeration requirements by providing an abundance of air capacity, and then monitoring the oxygen in the exhaust gases. In those cases the principle followed is that aeration must be sufficient as long as there is a surplus of free oxygen.

It does not follow that all of the air must be supplied by mechanical means. Once the material contained in the digester has reached the higher temperatures, and provided that it has not been allowed to compact excessively, then the thermal effect causes an air flow through the mass to take place naturally. Where that effect is occurring, the material is too friable and porous for any significant air pressure to be developed in the base of the digester.

Here the opinions of the best way to provide aeration vary somewhat. Some designs of aerobic systems have used banks of air compressors, which can provide sufficient pressure to overcome impermeability in the mass but are somewhat limited in terms of air flow. Others have resorted to centrifugal fans, which can provide ample air, but at much lower pressures. Silo ventilation fans are one option worth considering, since they combine a reasonable pressure of up to 12 in water gauge (about 300 mm, or 0.03 kg/cm^2) with a good displacement (up to 52,000 ft^3 (1500 m^3) per minute). On balance, silo ventilators are intuitively attractive, since the purpose for which they are designed is not entirely dissimilar to digester applications.

Returning to the plant used as an example, at this point in the design logic it has been established that aeration requires an air flow of between 1836 and 11,500 ft^3/min (52 and 325 m^3/min), at a pressure of up to 9.12 lb/in.^{2} (0.642 kg/cm^2). In order to avoid unnecessary fissuring of the digester contents, high pressures are best avoided, and while there is no basis of research upon which to judge when fissuring may occur, it is wise to avoid any pressure greater than 25% of that necessary to displace the material. The realistic pressure limit, therefore, becomes 25% of 9.12 lb/in.^{2}, or 2.28 lb/in.^{2} (0.16 kg/cm^2).

This is considerably above the capability of a conventional fan, but within the range of a bank of air compressors.

A pressure relief arrangement could now be designed to accommodate any pressures rising above that level, and the digester casing could be specified accordingly. The question then becomes one of the dimensions to which the relief mechanism should be designed. The greatest danger of overpressure inside the casing results from the piston effect of load movement. It is therefore necessary to establish the volume of compressed air to be relieved should sufficient of the load move to reach the relief pressure setting, and again this can be calculated.

In the example used, the floor area of the digester is 845.6 ft^2 (78.5 m^2), and a pressure of 2.28 lb/in.^{2} would result in a vertical force of 139 t (2.28 × 846 × 144/2000). At maximum bulk density, that equates to 278 yd^3 (212.6 m^3) of material and suggests a plug 32 ft (10 m) in diameter and 9.33 ft (2.84 m) thick. If that were to fall through the available height inside the digester, it could compress 38,397 ft^3 (1087 m^3) (45,903 – 7506 ft^3) of air to a pressure of 2.3 lb/in.^{2} (0.16 kg/cm^2). While such an event is extremely unlikely — such a plug would have too little thickness to remain coherent — it is a worst case to which the relief mechanism must be designed.

The application of simple gas laws, assuming no change in temperature, provides that the product of the initial pressure and volume is equal to the product of the final pressure and volume. In algebraic terms, it is the fundamental equation

$$P_1V_1 - P_2V_2. \quad (1)$$

Therefore, if a volume of 38,397 ft^3 of air at atmospheric pressure is exposed to a pressure of 2.3 lb/in.^{2} (0.16 kg/cm^2), the resulting volume will become

$$V_2 = P_1V_1/P_2$$

$$V_2 = 14 \times 38{,}397 \ / \ 14 + 2.3 \text{ ft}^3 \text{ or } = 1.02 \times 1{,}091 \ / \ (1.02 + 0.16) \text{ m}^3 \quad (2)$$

$$V_2 = 32{,}979 \text{ ft}^3 \ (943 \text{ m}^3)$$

When volume V_2 is reached, the vertical reaction of the compressed air is equal to the downward force of the mass, and further drop becomes impossible. And volume V_2, in a chamber 32 ft (10 m) in diameter, is equivalent to a column height of 41 ft (32,979/804.35). Therefore, the plug could only fall 4.7 ft (Digester height minus plug thickness minus 41 ft).

Under the influence of gravity, the plug of material could fall, in theory at least, with an acceleration of 32 ft/s^2 (9.8 m/s^2). Applying the standard formula,

$$S = {}^1\!/_2 gT^2 \tag{3}$$

where S is vertical distance, g is acceleration due to gravity, and T is time leads to the conclusion that the falling mass could compress the air volume in 0.54 seconds. Therefore, the relief mechanism must be able to exhaust the volume of air displaced, 3780 ft^3 (4.7 × 804.35), at a pressure of 2.3 lb/in.^{2}, in 0.54 seconds to avoid overpressure developing. In other words, it must be capable of exhausting 7000 ft^3/s (198.24 m^3/s) of air, and the calculation necessary to establish the dimensions of the orifice is

$$\text{Vol} = \text{Ca}.108.58\sqrt{\text{Ta}\left(1-\left(\frac{14.7}{\text{P}_2}\right)^{0.29}\right)}\,\text{ft}^3/\text{s} \tag{4}$$

where C is the orifice coefficient of discharge (0.55), a the orifice area, P_2 gauge pressure before the orifice, and Ta is the absolute temperature (°F).

Figure 6.5 shows the products of the calculation for a range of orifice diameters related to a series of pressure differential curves. The data relates to flat plate orifices, of the type most convenient for use in a cylindrical digester, and suggests that to relieve 7000 ft^3/s, at a differential of 2.3 lb/in.^{3}, an aperture of 68 in. diameter is necessary. The aperture needs to be kept closed during normal operation of course, but that is a simple matter of installing a cover plate and gasket, held closed by a counterbalance weight. If the weight is mounted so that it is adjustable, then the closing plate can be finally set after erection of the digester.

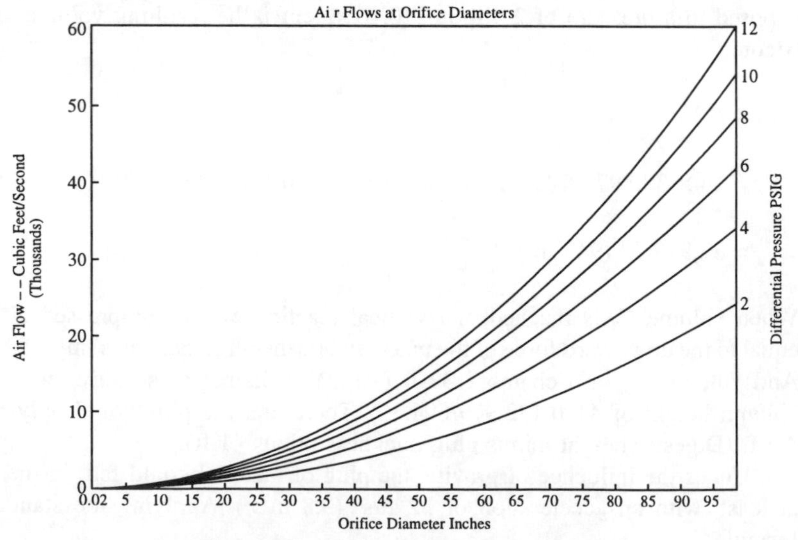

Figure 6.5 Air flows at orifice diameters.

Having protected the digester lower casing, the protection of the roof structure needs attention. As observed earlier in this section, the roof is more likely to experience a negative pressure inside, although a breakdown of the exhaust system could reverse that. Caution therefore dictates that the relief mechanism in this case should react to either a positive or a negative pressure beyond those of normal operation. The closing flap must be capable of swinging in either direction, and should be counterbalanced in the mid-position. Again, a vent of 68 in. diameter would serve the purpose, although rectangular cross-section vents are more common. In that case, and bearing in mind that the infeed conveyor will be located on and will discharge through the center of the roof, it is useful to place two 3 ft square vents (3 ft per side) of the transfer chute. With such a design, the vents also provide useful access to the spinner and to the conveyor discharge.

6.2.1.3.6 Emissions Control — When in operation each digester will produce considerable quantities of carbon dioxide and water vapor. During the first charge of the first digester, there are also likely to be detectable quantities of ammonia emissions. In addition, since 6% of the bacterial loading of municipal waste has been shown to be pathogenic, there is a strong potential for some of those organisms to be present in the high bacteria count that will inevitably exist in the exhaust.

Each of these factors provides sufficient justification for an exhaust handling and emissions abatement system. Without one the digesters, particularly that receiving crude organic waste, would emit a large steam plume, considerable volumes of carbon dioxide, and water droplets containing potentially harmful microorganisms. A plant designed to process the organic fraction of municipal waste could liberate up to 12,000 t of water as vapor in a year.

Since the pretreatment process described requires considerable volumes of water for the removal of heavy and metallic contaminants, it would be illogical to throw away such large quantities. An emissions abatement system that will recover the water, remove at least some of the carbon dioxide and ammonia, and capture the bacteria, is needed. Probably the lowest cost, and technically least complex, way of achieving that is by using a wet scrubber of the type described in Section 3.2. While machines of that type are not highly efficient at capturing submicron particulates, they are efficient at condensing saturated water vapor. They will remove any larger particulates such as the inevitable pieces of plastic film that become entrained in the airstream, and they provide the means of absorbing carbon dioxide in some quantity. If the water bath of the scrubber is in the flow circuit of the pretreatment plant, then any bacteria captured are likely to be returned to the waste, to be dealt with in the digester. Most of the water recirculates around the digestion process, and so the need for make-up is limited.

It is normal, where wet scrubbers are used in aerobic digester applications, to install one scrubber per digester. Since they are proprietary machines, they are not normally available in exactly the capacity required, and so it is customary to install the nearest size *above* the air throughput of the digester. In the event that the digester operates at something less than full capacity at any time, then the scrubber will simply pull free air down through the roof relief vent, and so no air flow controls are necessary.

An alternative often proposed is the use of biological filters. These were again described in some detail in Section 3.2, where they were explained to be large containers of biologically active material through which exhaust gases may be passed. They are efficient at capturing bacteria and hydrocarbons, but may become saturated by an excess of water vapor. They do not lend themselves to water recovery and recirculation, and since by definition the digesters themselves are actually large biological filters, their value is somewhat uncertain.

Electrostatic precipitators are viable for digester emissions abatement, but while their ability to recover submicron particulates and even bacteria is unsurpassed, they are extremely expensive, energy-hungry devices. Unless they are manufactured from corrosion-resistant materials, which will add significantly to their already high cost, the large volumes of water vapor acidified with carbon dioxide are likely to cause severe corrosion to the internal structures.

Conventional filters should be avoided for digester applications. While they may be able to deal with the larger particulates, and even some of the submicron particulates, they would rapidly become overwhelmed by the quantity of water vapor and would cease to function.

6.2.1.3.7 Unloading — There are very few ways of unloading an aerobic digester effectively, or even at all. The material is dense, fibrous, moist, and anything but free-flowing. It is likely to be compacted to a bulk density of 37 lb/ft^3 (600 kg/m^3), or even more if process difficulties have occurred. In view of those constraints, while there are a number of attributes the unloader must possess, there are some things that it must not do.

The unloader must not push into or through the material in any way. If it attempts to do so, then it will simply create increased compaction to the point where and further movement becomes impossible. It must not provide a line of least resistance for ventilating air to gain access to the outside; otherwise, air will simply blow out of it rather than ascending through the digester contents. It must not be designed so that the only access to it is through an empty digester. Unloaders do break down or jam, and attempting to empty the complete contents of a large digester to reach one is an almost impossible task. It must not impose any forces upon the walls of the digester, since if it does so, not only will it threaten the structural integrity of the vessel, but it will also create void spaces around the material through which the ventilating air may escape.

The unloader must be capable of cutting complete layers of material over the full diameter of the digester, and removing them to the outside. Here the significant word is "cutting." The action must be as if a series of slices were being taken from the bottom of a vertical, cylindrical cake. In this case the cake is dense, tough, and initially at least, fibrous. It does not cut particularly easily, and the presence of sheets of plastic film worsen that characteristic. For those reasons there are few unloaders capable of the task.

Conventional grain silo unloaders based upon an auger, sweeping around the silo and mounted upon a central casing, do not work at all in aerobic digesters. They do not normally have sufficient power to both cut and sweep at the same time, since they are manufactured for granular materials that flow relatively easily. Their augers are prone to picking up film and binding it around their shafts, blinding the screw. In general they are only accessible for maintenance from inside the silo, posing the risk of a major digging-out exercise in the event of a breakdown.

Reciprocating unloaders are similar in operation to the moving floor conveyors described in section 3.3, and are generally based upon crescent-shaped wedge sections that sweep back and forth across the diameter of the vessel, dragging material into a central discharge aperture. They work extremely well with grain, but they are essentially pushing devices, and their usual effect in digesters is to compact the material beyond any possibility of discharge.

Unloaders based upon toothed cutting chains are successful. They are rather like large chain saws, lying upon their sides and mounted on a central pivot over a discharge chain conveyor (see Figure 6.6). In operation the cutter arm, driven by a ratchet device from an external geared motor, pivots upon a bearing housing in the center of the digester and slowly sweeps around the complete floor area. The cutter chain, mounted upon the arm and also driven by the external gearbox via a long shaft, cuts material from the base of the stack and conveys it to the central housing. There the material falls through a transfer chamber into the extraction chain conveyor, which carries it to the outside in an enclosed trough beneath the floor. The whole assembly is so designed that, using the conveyor trough as a guide, it may be removed and inserted into the digester irrespective of the quantity of material inside.

Atlas® and Goliath® unloaders manufactured by the A.O. Smith Harvestore Products Inc. of the U.S. are typical of the type, and are specifically designed for fibrous or sticky materials. Their ratcheting action is reversible so that they can cut in either rotation, and they can be supplied with a range of cutter-tooth profiles to suit the material to be unloaded.

Unloaders of this type are capable of imposing very heavy loadings upon the structure of the digester. While they are installed in the perforated floor, they must not rely only upon that for support. Rather, it is normal for the trough and mountings which locate them to be fixed to mass concrete foundations that are independent of the digester. Without such a rigid structure, there is a possibility of the cutter arm climbing at an angle and distorting the floor.

Some aerobic digester designers use instead a development of grain silo auger unloaders, but while the operating principles are similar, the construction

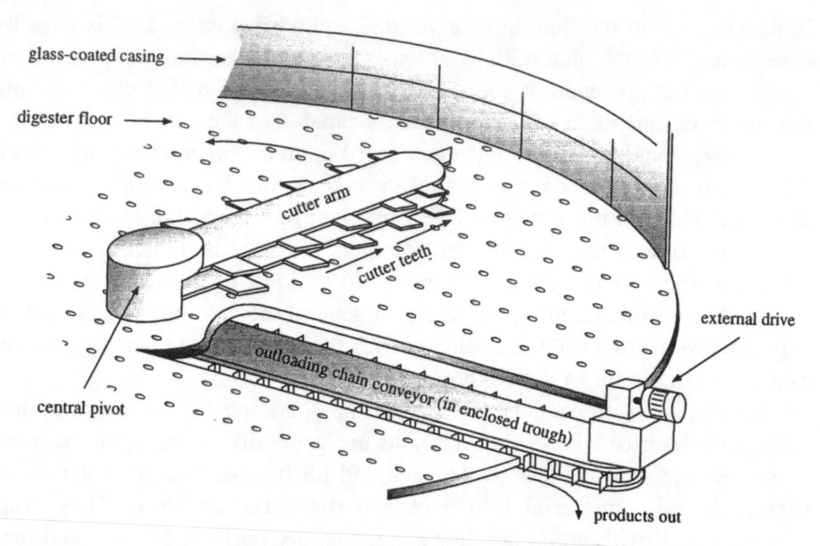

Figure 6.6 Cutter chain unloader.

of the machines is quite different. In the Weiss* design a double-rotor auger is mounted upon a central pillar that allows the assembly to sweep the floor of the digester, but instead of being entirely enclosed within the chamber, the end of the augers protrudes through the sides at the base. The digester design is such that the whole vessel is mounted upon a ground-level machinery installation, effectively with an open base. The unloader augers are driven by a shaft to the central pillar, but carried round the floor by a hydraulic ratcheting device on a circumferential rail. The auger is of extremely heavy-duty fabrication, and may be fitted with serrated flight edges to improve the cutting action.

During the unloading process it is normal for a chain cutter to create a void space above itself, supported by material remaining at the sides outside the sweep of the cutting arm. As the product slips down in the digester, more is removed from around the void, but the free space tends to remain. This action is both normal and beneficial, in that it reduces the load upon the arm and makes cutting easier. It also reduces the likelihood of jamming due to overloading. The characteristic does not mean that the machine only removes material from the sides. In fact, the roof of the void is almost continuously collapsing, but with the result that the material so released loses some compaction and apparent density in the process.

Until continuous operation has been established, it is a worthwhile practice to remove the unloader from the digester, if the design of the machine permits it. During the period of first charge, particularly in the digester receiving crude organics, compaction will be at its greatest. It is then, more than at any other time, that settlement on top of the unloader may cause such a loading that the

* Nordfab-Weiss Bio-Anlagen GmbH, Dillenburg-Frohnhausen.

machine cannot sweep properly, or even at all. The cutter chain machines particularly lend themselves to such an operation, since when first introduced into the digester, on a rail system provided for the purpose, they cut their way into the center mounting position, and the possibility of jamming is eliminated.

6.2.1.3.8 Digester Casing — The material inside the digester is hot and moist, acidic at some stages of the process, and quite highly alkaline later. It adheres strongly to bare steel, and as a result is quite corrosive in spite of the oils and fats contained within it. It is also fairly abrasive, to the extent that a painted protective coating would be removed quite rapidly as the material moves down the digester.

Some manufacturers produce glass-coated vessels, which are ideal for the purpose. They are constructed from curved steel plates, bolted together in rings which are usually about 5 ft (1500 mm) wide, and which accordingly develop digester heights in 5 ft (1500 mm) increments. The glass coatings are applied in paste form to prepared plates, which have already been rolled to the correct curvature, drilled for bolts, and degreased. The plates are then subjected to high temperature in large ovens, with the result that the glass is fused into the surface of the steel and down into the bolt holes. The process is effectively one of stove-enameling, and it produces a very hard weather, corrosion, and abrasion-resistant coating which has an almost unlimited life. The process does, however, produce plates that are almost impossible to cut by conventional methods, and so any apertures for pressure relief, pipework, temperature sensors, unloaders, etc. must be prepared in the plates before enameling. There is almost no opportunity to change the design after the digester is erected, and any attempt to do so will at least reduce the corrosion resistance of the platework where cutting has been tried.

Digesters are normally made to stand outdoors, rather than inside a building, and so the roofs need special attention. It is normal for roofs to be slightly domed for structural strength, but even then they are the weakest point of the whole assembly. While they have sufficient strength to withstand a limited amount of pedestrian traffic, and will of course tolerate wind, rain, and snow loadings, they are not usually designed to support the weight of the infeed conveyors and other equipment that may cross them. Were they to be so, then their support structure would be large enough to make the positioning of spinners and internal fittings much more difficult.

For those reasons it is more normal, at least where conventional agricultural-type vessels are used, to mount the conveyors on the closing rings around the top of the cylindrical casing. The closing ring is a rolled steel angle flange bolted to the edges of the last ring of casing plates, and because a load upon it places the casing plates in compression, it can withstand significant weights. The inevitable result of locating conveyors in that way is that they must then be capable of quite large, unsupported spans themselves. However, where drag link conveyors are used, that is not a problem, since the nature of their rectangular casing provides very considerable structural strength. For other types of conveyors, or in cases where the conveyor terminates over the center of the digester, a steel A-frame can provide adequate support. Designed with forethought and fitted with an attachment for lifting gear, the frame can prove very useful for maintenance, especially where a new conveyor belt has to be fitted. Figure 6.7 shows a typical A-frame span.

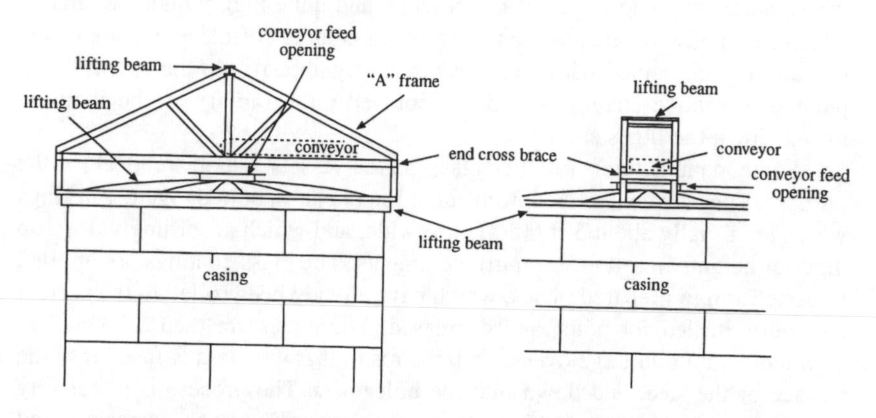

Figure 6.7 Shows a typical A-frame span.

Some aerobic digester designs provide for insulation on the outside of the casing in order to minimize heat loss through the plating. To be effective, a considerable thickness of insulation material is required, and again a compromise becomes necessary. If lightweight insulation (glass fiber, for example) is applied, then at least a 6 in. (150 mm) layer is desirable. An outer skin of thin-gauge, plastic-faced aluminum sheeting or similar product is then necessary to keep the weather and any possibility of mechanical damage away from the insulation. The outer skin itself then presents a maintenance problem for the future, since it is much less durable than the casing of the digester itself.

Alternatively, a thinner layer of much denser insulation — strawboard or asbestos cement sheeting perhaps — could be used. Both are structurally quite strong and they are less prone to mechanical damage. They would still need protection from weather by some form of cladding, but would normally be applied in thinner sheets. Both types of insulation are however much heavier than glass fiber, and their resistance to heat transfer is lower by a factor of 7:1.

Glass fiber insulation therefore provides better heat retention and imposes less weight upon the structure, but is more prone to mechanical damage, even when encased. Dense insulation offers better mechanical strength, at the expense of considerably more weight and insulation resistance, in an installation where each digester already may weigh 800 t when full.

Since insulating the digesters adds nothing to their appearance or durability, and at least may create a maintenance obligation where none would otherwise exist, there are reasons for wondering if it is worth fitting at all.

Much depends upon the final use of the product. The only reason for attempting to retain heat in the digester is to prevent a cool zone from becoming established close to the casing, where pathogenic bacteria, parasites, and seeds could survive. That is however not as likely as one might suppose, since the heat generation rate in the composting mass is sufficient to make up losses and still keep the casing surprisingly hot. If the product is to be used for bulk applications such as land restoration, then it is probable that the risks from pathogen survival are too small to justify the expense of full insulation. If, however, the intended market is a retail one, then the risks may be unacceptable.

In general, without effective separation at source and a considerable amount of refining and postcomposting treatment, composts manufactured from municipal wastes have heavy metals concentrations that place some limitations on their use for food production. They are not, therefore, ideally suited to the retail market, even in the unlikely event that one exists. The most attractive applications are in land reclamation and restoration, in landscaping schemes, and large-scale nonfood agriculture. Therefore, the value of digester insulation, while never being discounted out of hand, should be the subject of critical evaluation. It may simply create more problems than it solves.

6.2.2 Tunnel Composters

Many of the more modern continuous aerobic decomposition systems are an attempt to overcome the compaction experienced in large vessels, and tunnel composters are an example. Such a device is effectively a long, perforated heavy-duty conveyor enclosed inside a sealed casing of approximately square cross section. Air is blown through the conveyor pan, and is exhausted from the casing top, while waste is loaded onto one end of the conveyor and compost is discharged from the other. In common with cylindrical static digesters, a high-temperature zone through which the material must pass exists, but in this technology its position depends upon the speed of the floor conveyor.

A tunnel composter is a fully continuous process, in that the waste is being moved through the machine at a consistent rate. A static digester, even with continuous recirculation through a series of units, can never quite match that — there is always some point at which the material is moving more slowly than elsewhere. However, a tunnel composter is limited by the maximum practical width to which it is possible to build a conveyor, and by the power requirements to drive it. In addition, power must be provided to overcome friction between the material and the side casing, while in a static digester that power is supplied by gravity.

Frictional resistance against the casing does not necessarily apply throughout the system. The material will lose volume in any design of system, and in this case, as it does so, it will tend to shrink away from the walls as well as from the roof. Although it may not actually lose contact with the walls, at least the pressure against them will be reduced.

While tunnel composters have much to recommend them, they do have something of a capacity limitation. The running speed of the floor conveyor is determined by the infeed rate of raw material, not by the volume loss during the process as it is in a static digester. It is not possible to slow the material down progressively as its volume reduces. Therefore, the total enclosed volume must be equivalent to the volume of waste input multiplied by the number of days of treatment. In the example used throughout this text, the process is to receive 1200 yd^3 (917 m^3) per day and to treat it for up to 10 days. The appropriate tunnel composter volume would therefore become 42,120 ft^3 (12,838 m^3). If its cross section dimensions were 9 ft 6 in. × 6 ft 6 in. (3 m × 2 m), the composter would have to be 1.3 miles (2140 m) long.

A number of devices operating in series could overcome the problem of excessive length, and running each successive machine at a slower conveyor speed could take advantage of material volume losses. Even then, if four machines were used, each permitting a material residence time similar to that of the static digesters, overall lengths ranging from 675 ft to 980 ft (211 to 305 m) would be necessary. Once again, because of the loadings upon the conveyor chain and the motive power demand of such a system, the lengths are unrealistic. Splitting the process into two groups of four tunnels becomes somewhat more manageable, but the mechanical complexity is now beginning to develop beyond reasonable operating reliability. There are becoming simply too many moving parts.

As a general criticism, it would therefore appear that tunnel composters, while they can overcome the traditional compaction problems experienced with static digesters, do so at the expense of capacity and mechanical simplicity. They also occupy much more floor surface, since they are long rather than high, and may therefore bear an additional cost penalty.

6.2.3 Container Composters

A container composting system could be likened to a static digester, one where there is a large number of individual digesters and no recirculation between vessels. The typical design has one or more long product conveyors out-loading from a pretreatment facility, and supplying a number of short distribution conveyors. A large, open-topped container of up to 80 yd^3 (60 m^3) capacity is placed under each distributor conveyor discharge, and an underfloor space in each container is connected to a forced air supply. The containers are filled in turn with raw organics, and the air supply is turned on. The material then resides in the containers until sufficient composting has been achieved.

Such a facility is continuous in the sense that the containers are filled with waste and emptied of compost in a sequence that is being constantly repeated. It overcomes compaction problems to the extent that each individual container possibly only stores waste to a height of 6 ft 6 in. (2 m), but it imposes a capacity limitation. Also, although there is no mechanism associated with each container, mechanical complexity still exists. In this case, the whole container is moved when the process is complete, not just the load. Large and powerful container handling plant and vehicles become necessary, and logistics become all-important given a large number of containers with contents each at a different stage of treatment.

Once more applying a comparison with the model used for static digesters, if 917 m^3 of waste input per day are to be stored for 14 days, then the total waste storage capacity must be 917 × 14, or 12,838 m^3 again. That suggests a total of 214 containers, which would occupy a very large site by the time provision was made for vehicle operation. It also suggests that reducing the container numbers by "topping up" as volume loss occurs would become logistically unmanageable. Each container could, in that event, have layers of material all at different stages of decomposition.

Again then, while container composting plants have some attractions in their versatility and a limited number of fixed structures, they do impose quite severe capacity restrictions. While they are associated with more sophisticated pretreatment and sorting of waste, their actual composting process is somewhat primitive, yet can be logistically quite complex.

6.2.4 Drum Composters

The drum which gives the name to processes of this type is more properly a pretreatment device, rather than being the heart of the technology. In a such a composting facility, a large-diameter [9 ft 6 in. (3 m) or larger] slowly rotating drum is fed at one end with crude organic waste, water, and in some facilities, sewage sludge. As the waste mixture is tumbled in its passage down the length of the drum, a process of attrition gradually breaks down the material and mixes it intimately with oxygen and water. The progressive increase in granularity causes a rapid development of biological activity such that, although the waste may spend no longer than a few h in the drum, decomposition becomes well established. When the period of reduction in the machine is completed, the waste is generally windrowed until a suitable stage of composting has been reached.

Drum-based systems are among the oldest of the mechanical processing technologies, and after many years of being underused they are again beginning to be viewed with favor by the waste processing industry. Possibly their major attraction lies in their ability to co-compost a mixture of sewage sludge and municipal waste, with an ease almost impossible to achieve with other systems.

In the U.S. the Bedminster system offered by the Bedminster Bioconversion Corporation (see Chapter 5, page 160) appears to have gained most acceptance. The system employs drums 12 ft 6 in. in diameter and 180 ft long to precondition municipal wastes and sewage sludges for a period of 3 days before windrow composting in large processing halls.

When sewage sludge is mixed with municipal waste, or with the organic fraction from it, there are considerable advantages to be gained. There are also a number of significant disadvantages. The sludges contain high levels of nitrogen, which greatly assists the biological activity, and the excess of water contained within the sludges is absorbed by the solid wastes, thereby reducing the need for extensive sludge dewatering. The sludges are treated aerobically, eliminating the possibility of methane emissions, when in any other circumstances they would always become colonized by anaerobic bacteria.

While the addition of sludges may initially make the solids too wet for treatment by any other process, since excess water effectively blocks the air spaces between particles, the continuous turning in a drum system overcomes any such tendency. In a static digester, for example, particularly where the solid wastes have already been subjected to a fluid separation process, the addition of sludges would make the mixture so wet that aeration would become difficult if not impossible. Compaction would become so severe under the extra weight of water that unloading would be severely restricted at best, and the increased nitrogen content may accelerate anaerobicity.

Among the disadvantages of sewage sludge addition, possibly the greatest is the potential for a substantial increase in heavy metals concentrations. Composts made from municipal wastes, and even from the cleaned organic fractions, already contain high levels of zinc, lead, and copper. With careful processing they may be held below most standards for maximum concentrations in soils, but sludges may increase those levels above permissible limits. They may also introduce contaminants that would otherwise not be present in the raw material at all. Composting processes are always characterized by large mass and volume losses, both of which occur exclusively in the organic fraction. Metals and toxic chemicals do not decompose biologically, and so the effect of their presence in the raw material is one of progressive concentration. A zinc presence of 0.4% by weight in the feedstock may become 0.8% in the finished product by the time 50% of the mass has been liberated as carbon dioxide gas.

The risk of excessive contamination is further increased by the variability of sewage sludges. In rural areas, where there is little or no industry, the sludges may be fairly consistent. In cities and industrial areas it is rarely possible to be sure what contaminants may be present. It depends upon what is flushed into main drains, and even upon the weather to some extent in that periods of heavy rain are likely to increase lead concentrations, in runoff from roads.

Problems can also be created when the finished compost products are to be marketed. It is rare for both the production and end-use of the materials to be under the control of the treatment facility. More normally, the compost is sold to independent users, who will have some expectations with respect to quality. Agricultural and land restoration undertakings are more likely to be concerned with the possibility of a buildup of heavy metals concentrations in soils. Domestic and landscaping users are likely to be repelled by the origins of the original raw materials. In many Western societies, marketing recycled municipal waste is difficult, although most people support the concept in theory if not in practice. They tend not to show the same enthusiasm for what is perceived as recycled human wastes.

Heavy metal and other chemical contamination is not the limit of the problem. Sewage sludges frequently also contain significant quantities of vegetable seeds. Tomato seeds, for example, remain viable after passing through the human digestive system, and so the addition of sludges to municipal waste may substantially increase the numbers of viable seeds that have to be destroyed. Although the high temperatures experienced in aerobic systems should achieve that reliably, the higher the seed concentration in the feedstock, the greater the risk that some will escape destruction.

While drum composters lend themselves to sewage sludge and municipal waste codisposal, they are by no means restricted to that application. In fact worldwide probably as many installations handle municipal waste alone as treat mixes. However, as processes for treating solid wastes they leave something to be desired. In pretreating the waste rather than completing the whole of the decomposition process, they combine the cost of a fairly sophisticated mechanical installation with the labor-intensity and land demand of windrowing technologies.

6.2.5 Capital Costs

Composting processes for the organic fraction of municipal waste are developing almost continuously, as the interest in the technology becomes reestablished. The basic concepts of the most common systems have been discussed here, rather than detailed evaluation of proprietary designs. It is therefore somewhat difficult to provide specific capital costs.

A single static digester developed around a glass-coated, agricultural-type vessel of the type described, complete with floor, unloader, temperature monitoring, and roof, would range in price between $220,000 and $260,000 U.S. at 1995 prices. An installation of four digesters, complete with infrastructure, conveyors, pre- and posttreatment plant of the capacity considered in the example, could be expected to cost between $4.8 and $7 million U.S. The complete plant installation cost would depend upon a large number of variables, among which are the extent to which the waste is sorted at source and the ground conditions where the facility is to be erected. In the latter case, each digester could impose a ground loading of 800 t or more over a circle 32 ft (10 m) in diameter. In most ground conditions, such a weight would demand piling. In soft ground it would require very heavy piling indeed, and piling costs can add many thousands of dollars to the total.

At least at the initial stages of any major engineering project, it is customary to consider only the machinery installation cost. The design of buildings, the nature of the groundworks, and the construction of access roads and drainage are all local considerations that can only be evaluated when a specific design and site have been chosen. In waste disposal engineering it is usual to speak of the cost per ton of crude waste to be processed. Thus a plant costing $4.9 million and capable of treating 50,000 t of organic waste a year has a unit installation cost of $98 per ton.

Upon that basis, a realistic cost range for the processes discussed here is between $80 and $200 per ton. Within that range, container composters would appear at the lower price range, with agricultural-type static digesters and tunnel composters appearing at about the midpoint. Large, auger-unloaded static digesters built over complex sorting and refining installations appear with drum composters at the higher end of the price range. However, price observations of this type can be misleading. For example, it is possible that a container composting facility could transfer from the lower price range to the highest as a result of its demand for land, if it happens to be built in an area of land shortage.

6.2.6 Operational Complexity of Aerobic Composting

Since the purpose of all of the systems described is solely to provide an ideal environment for the microorganisms that do the actual work, there is an implication of operational simplicity. Unfortunately that is not entirely so, and the more complex the mechanical process becomes the more the variables which affect the outcome. However, apart from the more obvious possibilities of mechanical failures, there are a number of operational parameters that must be both met and monitored continuously. Doing so is not always easy.

Aeration is necessary for the survival and development of the bacteria, and it must be provided in abundance throughout the decomposing mass. Any failure to do so will inevitably result in anaerobicity either in localized areas or throughout. The consequences of anaerobicity are the production of methane and hydrogen sulfide gas, which may form and remain in the material or be emitted with the digester exhaust. When remaining in the material, both gases are poisonous to plants. In the digester exhaust they may easily form an explosive mixture.

The risks of methane and hydrogen sulfide generation are significant only in static digesters. Tunnel composters generally do not have sufficient mass in any one place to create gas in quantity. Container composters again have insufficient mass in one place, but rather have a large number of small masses, each of which decomposes in isolation. Drum composters do not retain material for long enough, and their continuous tumbling action makes inadequate aeration almost impossible.

The presence of localized pockets of anaerobicity in static digesters does not actually create a major problem, since they are most likely to occur during the first few days of decomposition when compaction forces are at their greatest. Later, as the material is disturbed, harvested, and stored for distribution to markets, it is usual for sufficient air to be admitted to reverse any anaerobic tendencies. However, anaerobicity upon a larger scale carries with it the danger of fires or explosions, and at least may sufficiently contaminate a whole production run to the point where the product becomes useless.

Clearly, both aeration and methane gas emissions must be monitored. The former should ideally be measured continuously, and the output from the sensing devices should be linked to the aeration plant control system, so as to vary air pressures and flows according to demand. Methane detectors need to be linked to an alarm system and be capable of responding to levels of the gas well below the explosive mixture range, which is generally between 5% and 14% at atmospheric pressure. Unfortunately, the detection of anything in the exhaust from aerobic digesters is made difficult by the very high levels of water vapor, and to some extent by the elevated temperature. Some designs of static digester do boast oxygen monitoring, for which the designers claim some success. Methane detection, however, is not usually very successful at all, in the experience of the writers. Fortunately, there is an alternative, if more primitive method for ensuring aerobicity.

Aerobic processes are characterized by the creation of high temperatures and the liberation of considerable amounts of water vapor. Anaerobic processes are cold, and do not emit appreciable amounts of water. Temperature monitoring in static digesters is relatively simple, and can be achieved by installing long pyrometers in stainless-steel tubes, suspended vertically from the digester roof. Any number can be used, and at very little risk of physical damage, since the material slides down their outer casings. Provided that there are no external supports other than at the top, each pyrometer should operate reliably for the life of the plant. Temperature measurement can be achieved at all levels of the digester by the expedient of inserting a number of pyrometers of different lengths into each tube. Monitoring then becomes a matter of linking the pyrometer outputs to a controller that examines each output in turn and compares it to a predetermined minimum for each level.

Monitoring temperature in this way reveals a considerable amount of information, and much more than just whether or not the process as a whole is remaining aerobic. It also indicates the absence of anaerobicity in regions which would always be cool or cold in any case. Informed examination of the temperature regime provides a relatively complete picture of what is going on at all levels within the digester. The picture emerges as follows.

Reference to the temperature graph in Figure 6.1 indicates the conditions that should be expected throughout an aerobic static digester process, and they apply irrespective of the number of digesters in series. The creation of heat depends upon the growth of bacteria and the availability of nutrients upon which they feed, given of course sufficient air and water. Therefore, in a four-unit-series plant, the normal temperature regime would be as follows.

Digester No. 1 — Temperature increasing progressively from ambient at the top surface of the material, to about 140°F (60°C) at a point approximately one third down the mass. This temperature should then be maintained fairly consistently down to about 1 m from the base, where a reduction to about 130°F (55°C) is usual.

Digester No. 2 — Temperature 150°F (65°C) approximately 3 ft (1 m) down from the top surface of the material, increasing rapidly to between 175°F and 185°F (80 and 85°C) with depth, reducing again to about 158°F (70°C), 3 ft (1 m) from the base.

Digester No. 3 — Temperature rising again to 175°F (80°C), 3 ft (1 m) down from the top of the material, then declining steadily with depth to approximately 120°F (50°C), 3 ft (1 m) up from the base.
Digester No. 4 — Temperature about 104°F (40°C), 3 ft (1 m) down from the top of the material, possibly followed by a slight increase to between 113°F and 120°F (45 and 50°C) in central regions, but falling rapidly to ambient toward the base.

From this information, it is possible to evaluate the quality of the process. In digester no. 1, the newly admitted waste should be cool because there has not yet been sufficient time for thermophilic bacteria to become established. If there is sufficient aeration, the center one third should be hot. If it is, then there should by inference be sufficient air in the higher regions, otherwise the temperature gradient would either not exist or would rise suddenly. If the moisture content of the waste is adequate, emissions of steam should be such that a flashlight beam through an inspection port in the roof should not penetrate more than 1 ft (300 mm). Slight cooling should exist close to the base of the digester, but only as a result of some chilling caused by the air flow. When the process becomes continuous, and material is being transferred to digester no. 2, the material should be too hot to handle comfortably, and it should be steaming vigorously. Again, if the center region is at the expected temperature, then there must be adequate air for the lower region, since the air must pass through the center.

A high temperature zone that is predisposed toward the base of the digester suggests an inadequate air flow, with all of the available oxygen being used before it reaches the middle levels. Insufficient moisture is likely to result in the high temperature zone not being maintained over more than 6 ft or 9 ft 6 in. (2 or 3 m) of depth.

When it is admitted to digester no. 2, the waste should have lost some temperature as a result of the temporary change in volume. However, restoration should be rapid, indicating no loss of bacteria population, and temperature should then be maintained fairly consistently down to the base. Provided that the temperature is restored rapidly at the top of the vessel, then there must be adequate air at all levels. Water vapor emissions should not be detectably different from those in digester no. 1 if the moisture content of the material remains adequate.

Again a high temperature zone predisposed toward the base suggests inadequate aeration, while a steady decline in temperature down the mass indicates insufficient moisture.

In digester no. 3, the higher temperature should be regained rapidly after transfer, but then an almost linear decline should occur down to the air-chilled region immediately above the floor. If the overall temperature drop is less than about 68°F (20°C), the implication is that digester no. 2 is inadequately aerated, while a decline greater that about 95°F (35°C) suggests either insufficient aeration or moisture in digester no. 3. Since moisture vapor emissions should still be hardly less than those from the earlier digesters in the series, one can visually assess which of the two parameters is out of limits.

Provided that optimum conditions have been achieved in all of the digesters, the temperature regime in digester no. 4 should remain low, and should decline steadily to the bottom one third of the vessel as available nutrients are absorbed. The bottom level should be at only little more than ambient temperature, and the reduction in moisture vapor emissions should be very apparent. A very rapid decline in temperature and almost no vapor emissions again suggests inadequate moisture content in the earlier stages of decomposition. An increase in temperature at or above the central region suggests inadequate aeration in digester no. 3.

In the earlier stages of decomposition, in digesters 1 and 2, the correct temperature regime indicates that anaerobicity is not occurring to any great extent. That does not mean that there is none at all. In reality it would be somewhat surprising if air could reach every particle of material all of the time. However, small areas that may have temporarily become weakly anaerobic are unlikely to have any effect upon the process operation or the quality of the final product. The mixing that will later take place as a result of transferring to the next vessels in the series will both dilute any anaerobic materials beyond identification, and will at the same time admit oxygen to any congealed lumps. At the later stages of the process the material becomes increasingly friable, and anaerobicity becomes progressively less likely.

Finally, at the completion of the decomposition process, some slight anaerobicity may develop in the finished compost, particularly if it is stored without through-ventilation. At such times it is unlikely to develop as virulent, localized activity. It is more probable that a general tendency toward anaerobicity may develop throughout the whole mass. Provided that the stockpile is examined regularly to ensure that any such tendency is not accelerating, that is not necessarily harmful. Slight activity will create acid by-products which to some extent counteract the high alkalinity of the product, and is beneficial. Very active anaerobicity during later stages of the process, and during storage, is in fact somewhat unlikely if adequate aeration and moisture contents have been ensured throughout. As aerobic decomposition progresses, nutrients are absorbed. Eventually there is insufficient left for any heightened biological activity.

Although the risk of anaerobicity is slight in stored, aerobically composted material, long-term storage in an excessively humid environment can cause sufficient to reduce the quality of the product over a long period. Increased acidity, and slight methane and hydrogen sulfide contamination create the conditions which, to the layman, suggest that the compost has "gone sour."

Single digester designs, where all of the process takes place in one vessel, follow similar rules to those described for series systems. The most fundamental difference is that the whole of the temperature range occurs in a single mass, varying from ambient at the point of infeed to near-ambient at the unloader. The temperature will again rise to about 175°F (80°C) in the center, and follow approximately the curve shown in Figure 6.1.

The overall process image which the temperature data reveals, when observed together with the level of moisture evaporation, is also similar. The high temperature zone cannot be created or maintained without adequate oxygen and moisture, and so if it exists, then the conditions for the lower regions must also be adequate. If the temperature gradient from the top of the digester to the high temperature region follows the correct pattern, it again follows that there must be a surplus of oxygen escaping the central levels. Otherwise there would be insufficient to establish the biological activity that creates the increasing heat.

Equally, if the temperature gradient for the lower levels is progressive toward ambient, then again there must be sufficient oxygen and water there. Otherwise, the gradient would show a much more rapid fall than the temperature curve shown in Figure 6.1 permits.

Finally, any of the processes described may vary from the prescribed conditions as a result of lengthening or shortening the process durations, and the static digester model provides a convenient example. Shortening the process duration will have the effect of moving the temperature profile toward the last digester in the series, or toward the bottom of a single digester. The high temperature zone will occur later in the series or nearer the floor, and finished compost is likely to be unloaded hot. It will remain active during storage, where anaerobicity will become much more likely. The risk of product loss through souring will increase. Conversely, lengthening the process will result in the profile moving toward the first digester, or closer to the top of a single unit. Compost manufactured in a shortened process will have a shorter shelf life and will be less stable. It may lack a satisfactory granularity and appearance, but it will contain higher levels of nutrients. Compost from a lengthened process will be more friable and much more stable. It will have a satisfactory appearance, but may be quite alkaline (pH 8.5), and will have a very low level of nutrients.

6.3 ANAEROBIC DECOMPOSITION

Anaerobic decomposition is a process of biological degradation that takes place in the absence of oxygen. Although the bacteria involved are similar to those occurring in aerobic processes, anaerobic decomposition does not generate heat to any great extent. The by-products of the bacteria activity are methane, carbon dioxide, and traces of hydrogen sulfide gases. Solid residues are generally much darker in color than those from aerobic processes, and may be almost black.

The technology is well understood and widely used in the waste water industry, where it has been used to treat sewage effluent for over 100 years. Digestion takes place in sealed vessels, and in many of the plants used for this purpose, the methane gas is recovered, compressed, and used to generate electrical power on site. Normally, even from such a high-nitrogen material as sewage effluent, the quantities of gas recovered are sufficient only to make a contribution to the overall electrical load of the facility. They do not replace mains power, and there is not normally a surplus for export.

Disposal of the solid residues usually presents a problem rather than satisfying a market. While the residues are often referred to somewhat optimistically as "compost," they are rarely perceived to be such by potential consumers. Disposal upon agricultural land is the most common outlet, but even in those cases it is normal for the utility to pay the land owner for the privilege.

There has been a considerable amount of research internationally in the last 10 years, to explore the potential for anaerobic decomposition in the treatment of municipal wastes, but the process is still not widely used for that purpose. Attention has generally been focused upon the creation of a fuel gas as a means of energy recovery from a waste fraction that would otherwise have a negative value. In theory, even if the solid residues are landfilled after decomposition, significant volume reduction can be achieved and the wastes become stabilized to some extent. Methane pollution of soils can be considerably reduced if not entirely eliminated, and leachates may lose at least some of their toxicity.

In spite of the perceived attractions of the process, and although a small number of designers offer anaerobic systems, early expectations of large-scale developments have yet to be realized. The calorific value of the gas mixture is low at around 550 Btu/ft^3 (20.5 MJ/m^3), compared with natural gas at between 990 to 1154 Btu/ft^3 (37 and 43 MJ/m^3), and it has a high moisture content that must be reduced before use. Volumes produced are reported as being between 10 ft^3/lb and 14 ft^3/lb of solids (0.6 m^3/kg to 0.9 m^3/kg).

The gas generation rates appear attractive at first sight, since apparently significant volumes can be produced from modest quantities of organic wastes. However, the value of the waste as an energy resource could be compared with alternative fuels. In other words, what is the potential calorific value of a quantity of waste used to produce gas, compared with an equal quantity of, say, coal? Since the gas mixture has an average calorific value of 550 Btu/ft^3, and the highest reported generation rate is 14 ft^3/lb, then the resulting potential calorific value of the waste is 7900 Btu/lb (18.5 MJ/kg). The calorific value of coal ranges from about 11,000 Btu/lb to 12,900 Btu/lb (26 MJ/kg to 30 MJ/kg), which suggests that anaerobic digestion cannot be justified purely upon the basis of its value as an energy source.

If the process is not a viable energy resource in its own right, then its value must be judged in terms of waste disposal. In those circumstances the greater the quantities of gas available, the more income (or savings on energy) could be used to offset disposal costs. However, such an evaluation must consider the nature of the waste, how much gas is likely to be available from it, and what the necessary process plant design will be to obtain it.

Chandler et al.[1] propose the use of the "biodegradable fraction" as a measure of the suitability of the waste for high-solids anaerobic digestion, and have derived the following formula:

$$\text{Biodegradable fraction (BF)} = 0.83 - (0.028 \times \text{LC}),$$

where LC is the lignin content of the waste expressed as a percentage of dry weight, and the biodegradable fraction is expressed on a volatile solids basis. Following are the lignin contents of the materials commonly found in the organic fraction of municipal waste: food wastes, 0.35; newsprint, 21.9; and garden waste, 4.07.

The formula indicates clearly that the best material for methane production is food waste, followed at some distance by garden waste. Newsprint is not very biodegradable at all, and so the higher the percentage of newsprint in the mix, the lower will be the volumes of gas available. Therefore, in a mix containing 10% newsprint, 60% food waste, and 30% garden waste, the BF as a percentage of the total volatile solids is 74%. If the mixture contains 60% total solids, of which the volatile solids are 80%, then the BF expressed as a percentage of the total solids becomes 74% of 80% of 60%, or 35.52%.

Gas generation rates have been revealed as between 9.6 and 14.5 ft^3/lb (0.6 and 0.9 m^3/kg) of biodegradable solids, and so in the example mixture, between 7.4 and 11.3 ft^3 (0.21 and 0.32 m^3) of gas is potentially available from every pound of the dry waste. That, however, is something of an approximation, since it supposes that all of the available biodegradable fraction is capable of being converted into gas. In reality that is somewhat unlikely, since the process will never be 100% efficient when scaled up to process plant levels. There will always be some biodegradable matter left in the residues, and so it is sensible to reduce the expected gas generation by about 5% to accommodate that.

In terms of plant size, for design purposes, the weight of total solids is somewhat misleading. No decomposition process can work upon dry material, and anaerobic systems need more water than aerobic ones do. How much they actually need depends upon the type of process, and currently three are considered worthy of research. These are low-solids, high-solids, and multistage systems.

6.3.1 Low-Solids Systems

All sewage effluent anaerobic treatment installations are low-solids systems, operating at a total solids percentage of less than 10%. In them the organic mass is suspended as a water slurry and is digested in enclosed vessels fitted with some means of continuous stirring. It is common to provide some heating to maintain the slurry either in the mesophylic temperature range 95°F (35°C) or even in the thermophilic 130°F (55°C).

In the sewage industry the purpose of the treatment is sanitization and odor elimination rather than gas production as such. The gas is simply a useful by-product of an essential process, and it is often used to provide supplementary heating for the digester vessels, either directly through combustors and exhaust gas heaters, or indirectly via electricity generation and electrical heat exchangers. Heating increases the speed of digestion substantially, and as a result maximizes the capacity of the treatment plant.

At the end of the process the slurry may have been reduced to about 4% dry solids content, which means that the residues, although they may be biologically nearly stable, are most certainly not compost. Drying is a massive problem, and the energy costs of attempting to remove sufficient water to achieve a more normal compost moisture content of 40% by weight would be impractical. For that reason, sewage plant sludges are normally either removed in road tankers for injection into agricultural land, or are left to dry in large lagoons over many months.

Considered as a principle for dealing with municipal wastes, low-solids systems become somewhat cumbersome. Comparatively dry solid waste has to be turned into a low-solids slurry, with a corresponding substantial increase in volume in order to produce a limited amount of gas, most or all of which will be used to maintain the temperature of the process itself. Upon the completion of the treatment, disposal facilities will be needed for very considerable quantities of sludges which are difficult to handle and unattractive to agriculture or landfill site operators. If the waste has not been well separated before treatment so that it contains significant levels of contamination by glass and metals, then any disposal option other than landfill may be excluded completely.

6.3.2 High-Solids Systems

High-solids systems are a recent innovation resulting from a considerable amount of research activity in the last 20 years, most of which is still continuing. The principle was examined in an attempt to discover the maximum levels of solids contents at which gas could still be produced reliably in an anaerobic process. Work carried out by the University of California, Davis suggests that the maximum practical solids concentration is about 40%,[2] although the rate of digestion become affected at concentrations above 32.5%. In the system described by Professor Jewell, increases in solids from 25% to 65% were reported as a result of the decomposition process.

The process researched by U.C. Davis involved both anaerobic and aerobic decomposition in order to achieve the maximum possible utilization of the waste (see Figure 6.8). The organic fraction was introduced to an anaerobic reactor at a solids concentration in excess of 25%, and was heated to maintain a temperature of 130°F (55°C). The reactor contents were mixed on average for $2^1/_2$ min every 20 min, and were retained for a nominal 30 days. At the end of that period the waste was introduced to a forced-ventilation aerobic reactor and was again heated to 130°F (55°C). Agitation was reduced to 30 seconds every 15 min, and the contents were retained for 3 days. The bulk density of the resulting biomass was found to be 35 lb/ft³ (560 kg/m³).

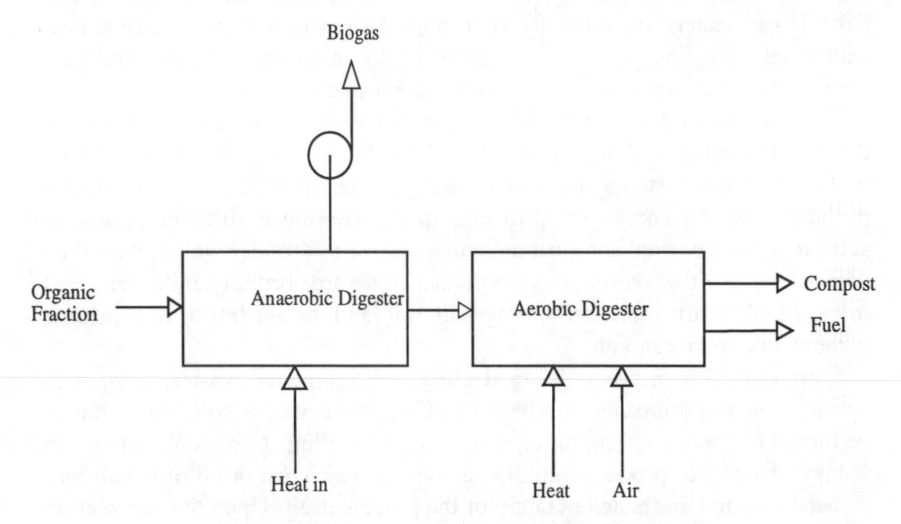

Figure 6.8 U.C. Davis anaerobic/aerobic reactor.

An interesting and potentially commercially valuable discovery resulting from the high-solids digester research was that the final humus product had a calorific value of 6360 Btu/lb (14.8 MJ/kg), and that it appeared to be stable enough to be used as a solid or granulated fuel in its own right. From that observation it could be concluded that a large-scale, high-solids digester plant could, by creating both methane gas and combustible humus, achieve an attractive energy balance. If, for example, such a plant could be optimized to achieve 12.8 ft³/lb (0.8 m³/kg) of gas, and could leave a residue of dry solids reduced to 64% of the original with a calorific value of 6360 Btu/lb (14.8 MJ/kg), then the overall energy content available would be 13,416 Btu/lb (31.2 MJ/kg) on a dry solids basis.

Unfortunately, no figures were produced to suggest how much of that available energy would have to be used in applying sufficient heat to the reactors to stimulate the process. Even so, one would expect some energy surplus in such a case. At least the final aerobic stage would overcome many of the problems associated with the digestate from anaerobic systems and make it a much more attractive option as a growing medium.

On the debit side, the process takes rather a long time to achieve volume reduction compared to aerobic systems — twice as long in the case of the U.C. Davis research. That study also reported that the system appears to suffer from accumulation of the volatile fatty acids common to all anaerobic processes. Some research in the early 1990s suggested that the latter problem could be overcome to some extent by chemical treatment later, if the humus product were to be used for soil amelioration. The process duration and two-stage philosophy, however, is another matter.

Since the solids concentration is considerably lower than that appropriate for an aerobic digester, the high-solids anaerobic vessel must already be larger for a given input of organic waste. If that waste then has to reside under treatment for 30 days instead of 15, two vessels will be needed instead of one. The aerobic stage then needs at least another vessel, and so in direct comparison with aerobic systems, the high-solids anaerobic design appears to need three vessels instead of one. That disadvantage leads to further complications, in that the greater the number of reactors, the larger and more complex the conveyor and handling infrastructure must be to serve them and the more land is occupied with plant.

In the event that the humus product is not considered suitable for a combustion process, and in Europe at least it would be difficult to establish it as such under the emissions legislation, then it may have to be landfilled. The marketing of high-quality organic compost is already difficult and competitive, and there are almost no sales outlets for heavily contaminated material. If that situation occurs, then the only remaining benefit is that derived from the production of gas, at which point a simpler process presents itself.

Landfill involves a process of high-solids anaerobic decomposition without the need for any application of heat. It is capable of creating very large volumes of gas for long periods from equally large volumes of mixed waste without the problem of where to dispose of contaminated residues. Where a large-scale anaerobic digester plant may be able to create energy in the order of kilowatts, a well-engineered landfill with gas extraction can create megawatts. Such a technology is well understood and, nowadays, well controlled. Sophisticated anaerobic digestion processes would be attractive if they avoided the need for landfill, but in general they do not. They would be even more attractive if they could offer a better environmental option more cheaply, but again the evidence is to the contrary.

6.3.3 Multistage Systems

Multistage systems have currently progressed little further than the research and development stage, and so there is little that can be said about them. Their general principle is one of physical or chemical treatment of the waste, followed by hydrolysis and then by anaerobic decomposition, and so they appear to be derivatives of low- and high-solids processes with some raw waste refining added. Any treatment that increases the volatile solids content of the organic waste is likely to correspondingly increase the quantities of gas produced, and so in that sense the process becomes more efficient. The end product also is likely to become more attractive in the soil ameliorant market place if it is free of glass, metals, and plastics.

Disadvantages however remain, and are inescapable. Pretreatment of the waste, either chemical or physical, adds cost and energy consumption to the process. It can only be justified if it also adds considerable value. Since most of the contaminants in the average organic fraction of municipal waste are extremely dense, their volumetric impact upon a digestion process is not great. Removing them does not necessarily mean a major improvement in efficiency.

Equally, the absence of serious contamination in the final humus product may make it more attractive, but in many countries there is almost no market for such a product in any condition. If the product cannot be sold for a price which more than recovers the cost of the pretreatment, then there is no point in pretreating the waste at all.

6.3.4 Capital Costs

Because of the few systems actually in use for the treatment of municipal waste at the moment, and because these all tend to reflect individual manufacturers' ideas rather than a standard design, it is difficult to discover true costs. Some reporters have suggested operating cost figures of $16 to $25 per ton of waste processed, but those prices are generally derived from low solids processes with no pretreatment and little or no market for the humus product. As more sophisticated technology is introduced, costs must increase rapidly. It would be perfectly reasonable to double the cost estimates in those cases, without necessarily obtaining a significant extra income from products.

Installation costs are also difficult to establish, since there is little history upon which to base an observation. However, an anaerobic process must employ larger and structurally stronger digesters than those of an aerobic system. While the most primitive designs may avoid the need for aeration plant, they need gas cleaning and compressing equipment instead. They require sludge dewatering plant to restore the residues to a manageable condition, and large maturing areas if the humus is to be used as a soil ameliorant. Therefore, using a "benchmark" principle of estimating, if an aerobic system costs between $90 and $150 per ton of waste to build, then a low-solids anaerobic plant of the same capacity could cost $120 to $180 per ton.

A high-solids plant is of course likely to be even more expensive, since it needs much more digester capacity. It also involves aeration later in the process, and requires some or all of the gaseous product for process heating. Upon that basis it would be expected that the installation costs would be very high indeed, and may exceed $200 per ton.

6.4 COMMON PROBLEMS WITH PLANT OPERATION

As with any complex mechanical process plant, there are many things in a composting facility that can develop faults. Conveyor chains may break, electric motors may burn out, drive belts may wear, and transfer chutes may block. It is beyond the scope of this book to consider every possible eventuality, and so this section will concentrate upon problems characteristic of composting processes in particular, rather than of mechanical engineering in general. For convenience, the observations are listed in tabular form.

6.4.1 Aerobic Composters

Problem: *No increase in temperature immediately after first charging the digester.*

Observations: Check aeration system back pressure. Check running h of digester base sump pumps. Check water vapor emissions in digester exhaust. Check position of digester roof vacuum relief valve.

Conclusions: If the aeration system back pressure is at the maximum of fan or blower capablities, the sump pump is running frequently, there is no water vapor in the exhaust, and the vacuum valve is admitting air, then the charge is too wet. Unless corrected, anaerobicity is almost inevitable.

If the back pressure is low, the sump pump is not running at all, there is a little water vapor in the exhaust, but the vacuum valve is closed, then the waste is too dry. Aerobic decomposition will become established, but will proceed very slowly.

Solutions: If the charge is excessively wet, recirculation should commence immediately. Begin unloading the digester and reintroducing the material back to the same digester. The agitation caused will allow the material to drain, while at the same time admitting sufficient air to cause aerobic conditions to become established.

It is possible for the same conditions to occur if there is insufficient inert bulking agent in the charge mixture. Lack of bulking agent is the same thing as excessive moisture. Therefore, during recirculation more bulking agent should be added until a temperature rise of at least 50°F (10°C) is established.

If the waste is too dry, then again the only solution is immediate recirculation. Water should be added through sprays directed to the distributor in the digester roof to ensure adequate mixing, and should be discontinued when a temperature rise of at least 50°F (10°C) has been established.

Again, since the moisture content is directly related to the amount of bulking agent in the mixture, excessive dryness implies too much bulking agent. The solution is still irrigation as described, but since the bulking agent cannot be removed, irrigation will be required at each stage of a series system. In a single stage process recirculation, irrigation, and addition of neat organics should be considered.

Problem: *After an initial increase from ambient, temperatures stabilize at around 50°C and then begin to fall slowly.*

Observations: Check for abnormally low level of material in the digester. Check aeration system back pressure. Check if the digester roof vacuum relief valve is open.

Conclusions: An abnormally low level of material suggests that excessive compaction has taken place. This may be the result of retaining the material in the unit for too long, or of insufficient bulking agent in the mix. Excessive compaction is confirmed by an open vacuum relief valve and a high aeration system back pressure. As a result, there has been insufficient aeration to permit the process to develop into the thermophilic stage.

Solution: Immediate recirculation must be commenced, otherwise the material may continue to compact to the extent that unloading may become difficult or impossible. Visual examination of the material as it transfers will suggest if there is insufficient bulking agent, but if there is any doubt in that respect then more should be added. All aerobic processes will continue to operate with too much bulking agent. None will operate with too little.

Problem: *Sudden temperature "crash" occurring over 1 day or less.*

Observations: Again, the aeration back pressure and vacuum relief valve position should be checked. If sufficient sensing points exist, an attempt should be made to identify the temperature gradient over the full depth of the material. Check the depth of material in the digester.

Conclusions: There are three basic possibilities: (1) chemical contamination sufficient to inhibit bacteria growth has occurred, (2) the waste was sufficiently moist to establish biological activity but insufficiently so to sustain it, or (3) excessive compaction has restricted air flow. The first is extremely unlikely but theoretically possible. Insufficient moisture to maintain biological activity is most likely. Excessive compaction is possible, but unusual if the initial temperature rise was as expected.

In a case of chemical contamination or drying out, aeration pressures and vacuum relief valve will remain normal. The former is likely to create more rapid cooling in one zone rather than throughout. The latter is likely to reveal a temperature profile biased abnormally toward the top of the vessel. Excessive compaction could occur as a result of overloading the vessel, and will be revealed by a high aeration pressure and a partially open vacuum relief valve.

Solution: Immediate recirculation if the evidence discounts chemical contamination. Observation of the material as it is unloaded will suggest if dryness is the problem, since the first out will be dry and dusty. Excessive compaction should be treated as described above, while suspected chemical contamination cannot be safely treated. In that case, the contents should be disposed of.

Problem: *Inconsistent temperature profile across the diameter of the digester, but relatively normal profile in each vertical zone.*

Observations: Aeration pressure and vacuum valve setting should be checked. Water vapor should be observed through roof inspection aperture or through relief valve.

Conclusions: A normal aeration pressure and vacuum relief valve position suggests that fissuring is occurring, allowing the air to escape through vertical paths in the waste. Insufficient aeration in some regions then causes biological activity there to be inhibited. Fissuring is most often the result of the material coagulating and shrinking, without subsiding down the digester. It is more likely to occur close to the digester casing. Observation of the upper surface of material inside the vessel is likely to reveal narrow regions of considerable water vapor emissions with very little elsewhere. The phenomenon is most often caused by failure to operate the process on a continuous basis.

Solution: Operation of the unloader will often dislodge the material sufficiently to collapse any fissures. Alternatively, the application of copious quantities of water via the infeed distributor is often sufficient to obstruct fissures and to force the air to create other paths. Provided that the overall temperature profile is not too far from normal, and that in general the high-temperature zone is at about 160 to 175°F (70 to 80°C), then the resulting excess of water will be evaporated in due course. It will not penetrate deeply enough to have any adverse effect.

Problem: *Finished compost is dry, dusty, and extremely hot when unloaded from the digester.*

Observations: Observation of the temperature profile throughout the system will reveal if it has become predisposed toward the discharge end or if it is simply remaining in the thermophilic range for longer than it should.

Conclusions: Predisposition toward the discharge end suggests that residence time in the digester or digesters is insufficient for completion of the decomposition cycle. If the temperatures are remaining in the thermophilic range for an unexpectedly long period, the percentage of biodegradable fraction (BF) volatile solids may be higher than was assumed. Changes in BF are not unusual, since municipal waste is not a consistent material. The organic characteristic and content can change seasonally or even daily. A high temperature in the final product is not unusual and is not necessarily harmful. It simply suggests that the product has retained more nutrients at the end of the process.

Solution: If the purpose of the process is to manufacture a product that has low levels of nutrients and is required to remain relatively stable in long-term bulk storage, then some action must be taken. All that can be done is to extend the residence time at each stage of the process and to increase the aeration. Otherwise, no action is needed, since when the material has been unloaded into storage, the lack of aeration will cause it to lose temperature quite rapidly, usually over two or three days.

Problem: *Although the temperature profile throughout the system is normal and the product is cool and relatively dry when unloaded, it reheats rapidly as soon as it is placed into storage.*

Observations: The temperature profile in store should be measured as soon as the problem is identified.

Conclusions: The phenomenon is not unusual or particularly harmful provided that the reheating is not excessive. It is generally caused by the degradation of granularity and the extra aeration caused by the unloading, transfer, and storage handling systems. Although the material may have cooled in the digester, it is still potentially biologically active, and the exposure of any untapped nutrients by increased granularity will start off the activity again.

Solution: If the temperature profile suggests that the product is reheating well into the thermophilic range again at depths below 1 m (3.2 ft), then the stockpile should be turned over occasionally using a shovel loader. This will ensure sufficient aeration to prevent coagulation, or condensation of moisture in the upper regions.

Problem: *Visible evaporation of water vapor in isolated regions of stored finished product, leading to suspicion of high temperature zones developing.*

Observations: Exposure of those regions is necessary to observe the physical condition of the material within them. There are two possibilities: It appears moist, light brown in color but remains friable, and has an earthy smell. Or it is becoming very dark brown or nearly black, has an odor reminiscent of charred paper, and is coagulating in layers.

Conclusions: The first possibility is not unusual. It results from heat retained in the lower levels liberating moisture, which condenses nearer the surface where the temperature is lower. Since the levels nearer the surface have more access to air, an increase in moisture there is likely to reestablish some biological activity. The phenomenon is not harmful to the product or dangerous to the plant. There is no significant risk of spontaneous combustion.

If, however, the second possibility appears to be developing, then oxidation of the lignocellulosic fraction is occurring. Again, the cause is initially heat and moisture from the lower levels, but in this case the phenomenon can severely reduce the quality of the product. Spontaneous combustion is possible, although somewhat unlikely. At temperatures up to about 175°F (80°C) the carbon in the lignocellulosic fraction increasingly forms relatively stable carbon–oxygen compounds with some release of heat. Above 175°F (80°C), however, the stable compounds begin to break down, releasing carbon monoxide, carbon dioxide, water vapor, and considerable quantities of heat. In effect, a low level of combustion begins to develop, and given sufficient oxygen, it may escalate.

Solution: In the event that some condensation is occurring and stimulating minor biological activity, then unless the problem is severe, no action is necessary. The only option is to mix the material using earthmoving plant, but that is likely to stimulate renewed activity throughout the stockpile. It is often better to simply keep the pile under observation to make sure that the second possibility is not developing, and otherwise leave it undisturbed.

If oxidation is occurring, then action must be taken. It may develop in small, localized areas, usually only 1 ft (300 mm) or so down from the surface, where sufficient oxygen may be drawn in to sustain the process. Left alone, the localized areas can develop into a clear strata throughout the length and width of the stockpile. The most suitable action is again immediate mixing, since that eliminates the self-heating effect and breaks the strata down into particles too small for it to be reestablished. It also redistributes the residual moisture throughout the stockpile, overcoming the condensation which was probably the original cause.

Problem: *Localized areas on or immediately below the surface of the stockpile of finished product change color to gray–white or even pink.*

Observations: Investigate ambient humidity within the storage building.

Conclusions: The stockpile is becoming infected with fungal growth in the lignocellulosic fraction. This is a natural phenomenon, since the lignocellulosic fraction is decomposed by fungi rather than by bacteria. In the presence of sufficient humidity, airborne fungal spores may be able to colonize the material. When they do so, they begin to break down the cellulose into nutrients that become accessible to bacteria. Biological activity then recommences, creating more moisture to encourage continued fungal growth. The most usual cause of the problem is inadequate ventilation of the storage area, possibly together with unreasonably long storage in bulk in such an environment.

Solution: Improve the ventilation and mix the material.

6.4.2 Anaerobic Digesters

Problem: *Early cessation of biological activity, before the available carbon in the biodegradable organic fraction has been converted to methane.*

Observations: Ensure that the mixing/stirring equipment is functioning correctly, and that the predetermined temperatures are being maintained.

Conclusions: Anaerobic processes are sensitive to some contaminants, and particularly to chlorinated hydrocarbons. Mixing the organic solids in a slurry increases the possibility of exposure of the whole batch to chemical contamination. The processes are also sensitive to temperature, in that they are at their most vigorous at the lower end of the thermophilic range 130°F (55°C). Low temperatures can subdue the reactions very considerably, while excessively high temperature may effectively sterilize the digestate. Lack of effective mixing in a low-solids system may result in settlement of the organic fraction, thereby limiting access to the nutrients.

Solution: If the mixing/stirring equipment is functioning normally and the temperatures are being maintained, then chemical contamination is probable. The process will have to be discontinued and the batch removed from the digester. It is possible that if the event has occurred in a high-solids system, aerobic treatment may be effective. In a low-solids system, increased dilution may have some effect, but the most probable final resort will be disposal of the batch.

If either the mixing or the heating equipment has ceased to function, then it must be restored as soon as possible.

Problem: *Increased generation of carbon dioxide, and reduced production of methane.*

Observations: This is unlikely to occur in a low-solids system, but is possible in a high-solids one. The solids concentration should be investigated, and the heating system should be checked to discover if the heat demand has reduced.

Conclusions: An increase in carbon dioxide production with a corresponding reduction in release of methane suggests that the process is becoming somewhat aerobic. At solids contents above about 40%, such an event becomes increasingly likely, particularly as the heat supplied may set up sufficient convection currents within the mass to introduce oxygen. The development of aerobicity may reduce the demand upon the heating system.

Solution: Increase the water content of the mix and thereby reduce solids concentration to 20 to 25%.

Problem: *Increased acidity of the dry residues.*

Observations: In a high-solids system, there will be reduced generation of methane in the anaerobic stage, and of carbon dioxide in the aerobic stage. Residues smell "sour."

Conclusion: High-solids systems are prone to the accumulation of volatile fatty acids at higher solids concentrations.

Solution: Reduce the solids concentration in the digestate to about 20 to 25%.

6.4.3 Pretreatment Plant

Problem: *Circulating fluid in wet separation system becomes increasingly viscous. Heavy, inert contaminants begin to be carried over into the organic fraction.*

Observations: Visually inspect samples of the fluid under a microscope at $\times 100$ magnification. Visually inspect fluid surface in settlement tanks. Examine air injection into the separator to ensure that microatomization is being maintained. Ensure that clarifiers if installed are running and producing "cake."

Conclusions: Wet separators are very prone to biological colonization of the circulating fluid. The effect becomes more prominent in the presence of cellulose fibers in suspension. Under a microscope, very thin fibers up to 1 mm long appear to be contained in a clear, thin gel, from which water slowly separates. The presence of a thick, light-brown scum on the surface of settlement tanks supports the conjecture, and inadequate microatomization of the separator air system is a likely cause. Clarifiers that are running apparently normally but are producing less solid residues may also indicate increased biological colonization of the fluid.

Micro-aeration of the separator is the best defense against increasing viscosity, since it limits or eliminates the build up of fibers in the fluid. In normal operation, micro air bubbles attach themselves to cellulosic fibers, causing them to become highly buoyant and to rise out of suspension.

Solution: Reduce the throughput of mixed waste temporarily, while increasing the aeration of the separator. Increase the fluid flow rate and, if necessary to restore operating conditions, consider pumping a proportion of the fluid into the top of the aerobic digester, making up the loss with clean fluid. In the case of the latter action being necessary, the effect upon the digester will be minimal provided that it is done with a certain amount of circumspection. The fluid must never be dumped to main drains, since its suspended solids levels will be almost certain to breach discharge consents.

Problem: *Reduced quantities of cleaned organics reach the digester, although the waste input level remains unchanged. Fluid in wet separators rapidly turns to slurry that is difficult to clarify. Increasing difficulty in dewatering the organics output from wet separators.*

Observations: Examine the waste before introduction to fluid separators. Examine the output from the first-stage screening process if installed.

Conclusions: The problem is usually caused by "old" waste that has reached an advanced stage of decomposition before reaching the plant, as evidenced by an immediately apparent increase in the amount of paper in the organic fraction. Where that is observed at the output from a first-stage screen, the volumetric impact may be sufficient to overfill conveyors.

Solution: Either discard the waste or mix it with a substantial quantity of fresh waste.

6.5 PRODUCT REFINING

The products derived from the biological treatment of municipal wastes are unsuitable for any practical application as they are. The modern fashion of avoiding the title "compost" and using instead such terms as "humus," "soil ameliorant," and "soil conditioner" do not change that fact. Even with extensive pretreatment and separation of the organic fraction, they are generally too heavily contaminated with inert materials to be at all attractive to potential users. This reality is demonstrated throughout Europe at least by the general lack of markets, and by either the enactment of legislation forcing the use of the products, or limited sales to low-grade outlets.

That a waste processing facility is a manufacturing *business* cannot be too strongly emphasized. A manufacturer of consumer products normally expends a great deal of energy in ensuring that his raw materials meet rigorous standards, that his process is tightly controlled, and that his end-products meet the highest standards of consistency and quality. A manufacturer of waste-derived products must do the same if he wishes to survive in the real world of commerce and industry. There is no point in making an inconsistent, shoddy product, and expecting the market place to respond to it simply because it has a "recycled" label.

This comment is particularly true in respect to digested organic wastes. The material that comes out of the aerobic or anaerobic digester is still waste in the view of the market, even if it has changed color somewhat during processing. The marketplace is dominated by very large companies that are extremely competitive and professional, and their activities have generally created a surplus of good quality compost, at least in the developed countries.

The introduction of waste-derived growing media into the existing market is made more difficult still by the large quantities involved, although that seems difficult to believe immediately. One of the major established manufacturers of conventional products may, for example, sell a million tons a year, and the output from a waste processing plant appears rather small in comparison. But the established manufacturer does not attempt to sell all of his production in one place. His is a national operation at least. A waste processing plant meanwhile, with a waste input of 150,000 t a year, may produce 26,000 t of digestate a year, and that is a substantial quantity in one area even by an existing major compost manufacturer's standards.

The operator of a single waste plant is unlikely to have the facility to establish a national marketing network, and so he can only compete on a local level. There he is faced with the possibility that his production may equate to a quarter of all of the composts sold in his region. If it is to have any chance of market penetration, his product must either be better for the same price or cheaper for the same quality. Either way, extensive product refining is almost inescapable.

The purpose of a refining process is to improve the chemical and physical nature and the visual appearance of the material, until it is of a quality that meets the requirements of the market to which the final product is targeted. Low-grade markets demand less product refining than high-grade ones, but all potential markets require better quality than the core processes can deliver. Therefore, before enlarging upon the mechanics of refining systems, it is worth examining those characteristics which need attention.

6.5.1 Chemical Considerations

Crude digestate from an aerobic process is likely to be highly alkaline (about 8.5 pH), and to have a correspondingly high conductivity. It is generally supposed that plants do not respond well to soils where the pH is greater than about 8.0, even when they are alkali-tolerant. While that is not entirely borne out by experience of using waste-derived products, it is the market view and therefore a matter for attention. High conductivity does have a measurable negative effect. It causes moisture to migrate from seeds into the soil rather than the reverse, thereby preventing germination.

Fresh organic composts also contain levels of phytotoxins, irrespective of the process that manufactured them. While that fact has been recognized for generations, there was in the past very little understanding of what phytotoxins actually were, or by what mechanism they affected plants. It was known that a period of maturing, where the materials were stored unused for some months, seemed to eliminate phytotoxins, but there was little research to discover why. However, in 1990, Keeling et al.[3] carried out research into phytotoxins in aerobic digestates at Birmingham, England, and identified short-chain fatty acids among others as likely culprits. It is known that anaerobic digestates are more initially phytotoxic than aerobic ones, and earlier work at U.C. Davis[2] suggested that in high-solids anaerobic systems, fatty acids could accumulate. This is true even for aerobic systems.

Even with extensive pretreatment, both anaerobic and aerobic digestates are likely to be contaminated to some extent by metals. Zinc, copper, lead, and iron are likely to be the most concentrated. Many countries impose limits upon metallic pollution of soils, without direct reference to soil ameliorants. Some also impose limits upon concentrations in materials intended to be added to soils.

Trace metallic elements are necessary for the growth and normal development of plants. Because of their possible concentration, particularly in waste compost, the following foodstuffs are of importance: nitrogen, phosphorus, potassium, calcium, magnesium, iron, manganese, copper, zinc, and boron.

Nitrogen: Nitrogen is present predominantly in an organically linked form, which produces a slow-flowing source of nitrogen. The amount of NH_4-N and the ratio of ammonium nitrogen to nitrate nitrogen are an indicator of the state of decomposition. The target condition is as low an NH_4-N and as high an NO_3-N content as possible.

Phosphorus: Depending on the process feedstock, phosphorus is present both in organically and also inorganically linked form.

Potassium, Calcium, Magnesium: Again, depending on the feedstock, these elements are present both in slightly soluble (e.g., silicates, carbonates) and also in available form. The origins of the raw material frequently cause wide variations in calcium and magnesium in waste compost.

Iron, Manganese, Copper, Zinc: These elements are present in the feedstock predominantly in inorganic form. In the course of decomposition, forms which link with organic substances are also possible, and this can bring about a change in the availability of these metals. The concentrations of copper and zinc can also reach the phytotoxic range if the basic material is unsuitable for the production of waste compost. A common precursor of high concentrations is inadequate pretreatment.

Boron: Boron is a trace element essential for plant growth and, insofar as the given range is not exceeded, is a factor in assessing waste compost.

6.5.2 Physical Considerations

The physical analysis and the appearance of the finished product are closely interlinked. Excessive contamination with metals, glass, and plastics is immediately apparent and very unappealing to a potential user. Glass particularly detracts from the quality on two counts: it is potentially dangerous when the product is handled, and is left on the surface of the soil by the worms which take the humus below. Users run the risk of physical injury from shards, while the use of excessively contaminated composts eventually leads to a surface layer of glass that is both obvious and disfiguring. Attempts to somehow grind the glass in such a way as to conceal its presence are rarely successful, since the particles of compost surround and protect the shards.

The presence of plastic film is equally obvious as soon as the product is used, often because the wide range of colors are so distinctly different from the basic color of the compost. White and red are particularly noticeable as flecks that remain entrapped, while larger pieces are usually liberated by weathering. Some plastic materials make the product very unattractive because their origins are immediately apparent. Users react with distaste to identifiable tags from used diapers, tapes from sanitary products, and the end reinforcing rings of condoms.

Metals are not always so immediately obvious, but the presence of nails, wire, and beverage can ring pulls at least presents a safety hazard. They offer considerable potential for cuts and skin penetrations, and may have unwanted chemical effects where the material is used for planting.

Wood chip is likely to be present in some quantity where the pretreatment process involves shredding. It is not chemically harmful in any sense, but it is disfiguring. Its greatest effects are in causing an irregular granularity, giving an impression of an inconsistent product, and making the compost somewhat more difficult to handle in small quantities.

6.5.3 Visual Appearance

As with many others, the compost market is subject to preconceived ideas. If an average user is asked to describe the appearance of compost, it is likely that he will actually give a fairly accurate description of peat. Compost is supposed to be dark brown in color, to have a flaky texture with a granularity similar to that of soil, and to posses an earthy odor. A material in which the color ranges from black to gray, and where the particle sizes range from 1 in. down to dust, is simply not considered to be compost. Neither is a material that exhibits the sour smell of ammonia or releases wisps of steam when the user first exposes it to air.

Even when it is used for low-grade landscaping undertakings, the general principles of user perception still apply. A contractor may not be unduly concerned about steam emissions from large piles when they are first tipped, but he is unlikely to be encouraged by windblown plastic littering his site. He will not be impressed by a granularity so inconsistent as to make spreading difficult. He will be equally unimpressed by a product that causes complaints from his employees, when they find some of the more unappealing contaminants adhering to their earthmoving plant.

6.5.4 Refining Processes

All refining processes, whatever the product, are by nature something of a compromise and subject to the law of diminishing returns. This is particularly true with waste-derived compost, where there is even some limitation upon what refining can achieve. Excessive contamination cannot generally be corrected later, particularly if it is of a chemical nature. Even too much glass and tramp metal in the product may be difficult to deal with.

Initially, refining involves the separation of contaminants from the bulk material. However, no separation system can ever be so efficient that it only removes the unwanted fraction without any loss of product. The better the system is at recovering contaminants, the more product is likely to be removed as well. It would, for example, be possible to guarantee that the product contains no plastic film, but that can only be achieved by a series system of screening and air classification. Some product would be carried over with the reject from the screen, while a classification air flow sufficient to remove even wet and dirty film would also extract the finer compost particles. The compromise then is one of choosing to lose all of the contaminants and some of the product, or to retain all of the product and some of the contaminants.

The concept of diminishing returns then applies. The more polished the product, the higher the cost of obtaining it. Removing the last 5% of contamination is likely to be more expensive than removing the previous 95%, and it is the target market that determines how far along the refinement route the process goes. There is little point in sophisticated separation and treatment if the sole market is landfill restoration, where some glass and plastic contamination will make very little difference to the quality of the finished site.

Identification of the product market has to be considered in the light of the pretreatment as well as the refining processes. Biological decomposition results in considerable mass and volume reduction, and thereby concentrates contaminants. A feedstock containing 16% glass may, after decomposition, contain over 30% by weight. Attempts to remove such large amounts from the product would result in a correspondingly large product loss, even if the sheer quantity did not simply overwhelm the separation equipment. It is therefore not possible to create a high-quality compost by means of extensive refining if the pretreatment design is crude.

When all of the necessary compromises have been considered, and the desired product quality has been established, a refining process can be designed. Figure 6.9 outlines a complete process line which begins with maturing and ends with chemical treatment. Increasing quality is achieved after each stage until a product, similar at least in appearance to high-value commercial ones, is created. The final design needs only to include those stages leading to the quality required.

Each stage may then be considered in detail as follows.

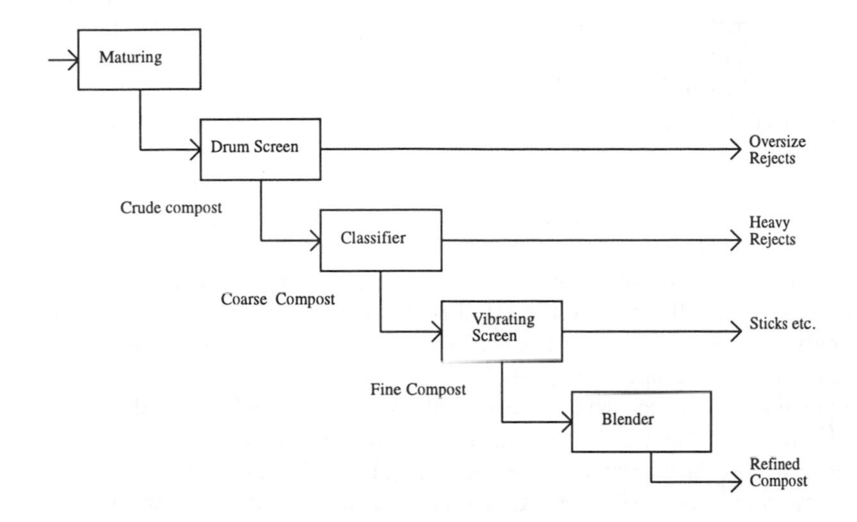

Figure 6.9 Compost refining flowline.

6.5.4.1 Maturing

Storage of the crude digestate leads to the breakdown of unstable fatty acids and to the reduction of phytotoxins. In the case of aerobic products, which are not generally highly phytotoxic, maturing does not need to be undertaken for more than 1 month at the longest. Low-grade applications for aerobic products are unlikely to need any maturing at all. Anaerobic digestates, however, need extensive maturing to permit the release of entrapped methane, eliminate phytotoxins, and reduce the moisture content to an acceptable level.

In general, the products need treatment that is the reverse of the biological process which created them. Those from aerobic digesters benefit from mildly anaerobic maturing, while those from an anaerobic process need aeration. In the former, anaerobic maturing leads to some reduction of alkalinity. In the latter, aerobicity assists in eliminating methane and hydrogen sulfide contamination, improves the odor of the product, and at least begins the process of moisture-content reduction.

In both cases, the process reduces phytotoxins, although the chemical and biological pathways are not entirely clear. However, because of the factors needing correction in anaerobic materials, they need much longer maturing times than do the aerobic ones. Simply weathering them in an environment where they are exposed to the atmosphere but protected from rain will have the desired effect eventually. It is a lengthy process, with a timescale of at least 6 months, but it can be reduced to a period almost of days by effective aeration.

Unfortunately, it is impossible to determine at what stage a compost becomes mature. There are many opinions about what constitutes maturity, and years of research have not yet completely resolved the question. Many of the ways in which plants absorb nutrients are not clearly understood. While the effects of some trace elements are known as a result of observation, others are recognized as important without any clear understanding of what they actually do. Some elements seem to be essential in low concentrations, but become positively harmful at higher levels. Zinc, for example, appears to be beneficial in the organic form, but toxic in the metallic.

The maturing of aerobic digestates requires the least expenditure in energy. Generally, it is sufficient that the materials are stored in a well-ventilated area where they are protected from rain. Regular examination is necessary, particularly to depths of 3 ft (1 m) into the stockpile, to ensure that oxidation is not occurring. Occasional turning is beneficial if the storage quantities are large and maintained for periods of longer than 3 months, in order to prevent anaerobicity from becoming too well established. Temperature monitoring is a sensible precaution, since excessive core temperatures may cause moisture condensation nearer the pile surface, leading to coagulation and fungal growth. Substantial increases in temperature 68°F (20°C) plus can usually be corrected by mixing the pile with shovel-loaders. Slower than normal temperature loss is generally best left alone, since any mixing then is likely to create an immediate increase.

Anaerobic digestates need initial, vigorous aeration to create conditions leading to maturity. In many designs this is achieved by placing the crude materials into a "store," which is in practice an aerobic digester in its own right. Large maturing sheds with slatted wooden walls and drained concrete floors are one commonly used approach. There the material is laid out in windrows and each is turned regularly by means of dedicated machines. Anaerobic plants have been constructed where the compost store is almost identical to a grain drying store, with a perforated metal floor grating through which high-pressure fans deliver ventilating air.

6.5.4.2 Primary Screening

Even when matured, waste-derived composts are still unattractive as they are. The concentrations of tramp materials are generally much too high, and inevitably there will be congealed lumps of insufficiently digested products. Primary screening is necessary to establish an initial, coarse granularity. Drum screens, where a large-diameter, perforated-mesh drum rotates upon trunnion wheels or upon a longitudinal shaft, are the most common devices for the first stage of mechanical treatment. While there are a number of proprietary designs available, there are also several basic essentials. Drum screen design is a more complex business than simply constructing a rotating cylindrical vessel and filling it with material.

The first consideration is that of how a drum screen actually works, and how it must treat the product for the best effect. It is important that it lifts the material to some height before allowing it to fall, otherwise the effect will be more akin to one of rolling a sausage. In that case, particularly if the compost is somewhat moist, the screen will *create* lumps rather than reduce them. Considering the end of the drum as a clock face, it should ideally elevate the material to approximately two o'clock before releasing it. By that action, the trajectory will then be such that the material is always falling upon unobstructed screen plates (see Figure 6.10). Most drum screens are fitted with lifters, which are usually no more than horizontal bars arranged along the longitudinal axis of the machine. Their purpose is to prevent the material from slipping back as the drum rotates, and not, as commonly supposed, to elevate it. Elevation is created by the speed at which the drum rotates, and is in turn a function of the critical speed of rotation.

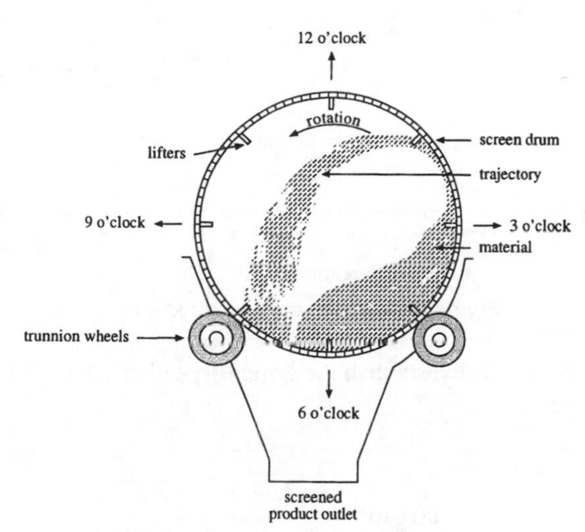

Figure 6.10 Action of drum screen.

The critical speed of a rotating drum is that speed at which the centrifugal and centripetal forces exerted upon the material are exactly in balance. When the material is being turned in a cylinder, centrifugal forces tend to hold it against the inside wall. As it progresses in the circular path, a point is reached where the gravitational attraction becomes greater than the centrifugal force, and at that point the material falls back. At the critical speed of the drum, because the forces are in balance, the material completes the circular path and never falls back at all. It is necessary to know what the critical speed is for the drum diameter to be used, since without that knowledge the performance of the screen cannot be optimized.

The formula for calculating the critical speed is

$$v(\text{rpm}) = \frac{\left(\sqrt{gr} \times 30\right)}{\pi r} \tag{5}$$

where v is the critical speed in revolutions per min, g is the acceleration due to gravity, and r is the radius of the drum. Figure 6.11 shows the range of critical and operating speeds for a series of drum diameters.

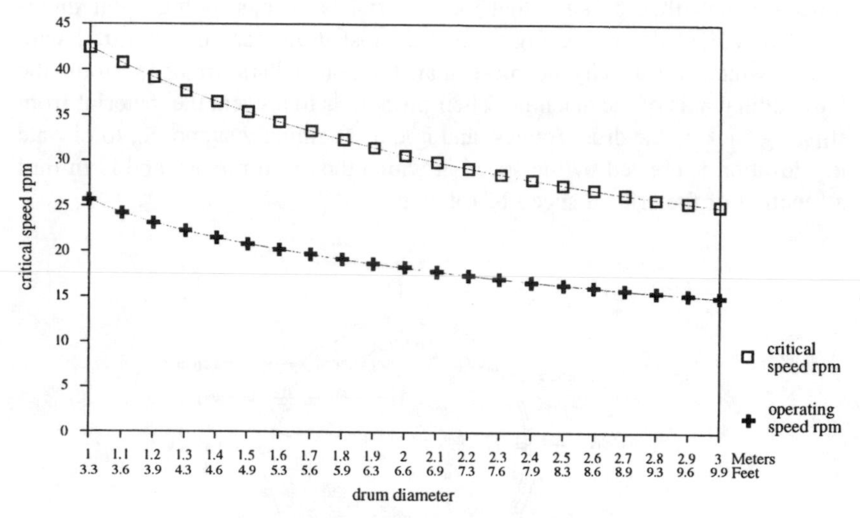

Figure 6.11 Drum screen speed ranges.

Using formula (4) to establish the critical speed of a 8 ft (2.4 m) diameter drum results in

$$v(\text{rpm}) = \frac{\left(\sqrt{32.2 \times 4}\right) \times 30}{4 \times \pi}$$

$$= \frac{11.34 \times 30}{12.568}$$

$$= 27.06 \text{ rpm}$$

It should be noted that the result is independent of the mass of the material. It is entirely a question of circular velocity and the radius of the circular path. Therefore, it follows that the larger the drum, the slower its critical speed. An everyday analogy supporting that conjecture is whirling a weight on the end of a string — the shorter the string, the faster the weight must be whirled to keep it from falling.

Once the critical speed is established, it becomes necessary to decide at what fraction of it the actual running speed should be to obtain the desired effect. The result of the calculation has shown at what speed the material just stays against the drum at its highest point, without quite falling off. If the release position is chosen to be at 70% of the arc between the lowest and highest points, then the speed must be 70% of the critical. In the case of the example above, it must be 18.9 rpm.

For all drum screens, a rotational speed of 70% of critical gives the best results. Faster, and the material impacts upon the opposite side of the screen at an increasingly oblique angle. Slower, and it increasingly falls upon material lying in the bottom.

The necessary diameter of the drum, meanwhile, is determined by the screening capacity which is required, and again a compromise is called for. Effective screening takes place when the screen is 10% full of material. At higher levels of fill, the material does not lift clear of that which falls. It blinds the mesh and restricts effective separation. At lower rates of fill, the machine is simply underutilized. If the fill rate is to be used as the measurement of capacity, then it becomes necessary to consider the relationship between three parameters: the diameter and length of the drum and the length of time in which the material is to remain inside (the "dwell" or "residence" time).

The appropriate residence time depends upon the material being screened, in that the longer the residence time, the more often the material is turned, and the more often it impacts against the side of the drum. Therefore, a product likely to be coagulated or to need a considerable amount of attrition, will require a longer residence time in the machine to achieve the best result. A fine, granular material with loosely mixed, oversized particles will require a very short residence time. Upon that basis, an estimate has to be made of a suitable duration, and from it the drum enclosed volume can then be calculated as follows

$$V = \frac{C_p \times T_r \times 100}{F}, \tag{6}$$

where C_p is the required processing capacity in volumetric units per min, T_r is the residence time in min, and F is the fill percentage.

When a prospective screen volume has been obtained from this exercise, it must be viewed objectively to see how realistic it is. It is necessary to consider what ratio of length to diameter would produce that volume, and then to decide whether the resulting diameter equates to an operating speed within practical limits. If it does, then it becomes necessary to establish how many revolutions the drum will make during the chosen residence time, since that relates to how many times the material will be thrown against the screen mesh. The more often it is thrown, the greater will be the degree of attrition and the higher the screened product yield compared with the level of rejects.

To illustrate this concept with an example, supposing the product is to be processed at a rate of 70.6 ft^3/min (2 m^3/min), and that a residence time of 1 min has been selected in a drum 16 ft 6 in. (5 m) long. Then from equation (6),

$$V = \frac{70.6 \times 1 \times 100}{10}$$

$$V = 706 \text{ ft}^3$$

It then follows that a drum 16 ft 6 in. (5 m) long, and with a volume of 706 ft^3 (20 m^3) must have a diameter of

$$D = \sqrt{\frac{4V}{\pi 1}} \tag{7}$$

where D is the diameter of the drum, V is the volume, and l is the chosen length. Then

$$D = \sqrt{\frac{4 \times 706}{16.5\pi}}$$

$$D = 7.38 \text{ ft (2.25 m)}$$

And

$$v = \frac{\left(\sqrt{32.2 \times \frac{7.38}{2}}\right) \times 30}{\left(\frac{7.38}{2}\right)\pi}$$

$$v = 28.2 \text{ rpm}$$

And

$$\text{Operating speed} = v \times 0.7$$

$$= 19.74 \text{ rpm.}$$

In this case, the calculations and assessments suggest a drum screen 7.38 ft (2.25 m) in diameter and 16 ft 6 in. (5 m) long, running at an speed of 19.74 rpm and providing a residence time of 1 min, during which period the material would be turned 20 times.

In considering this result, the drum is rather more than twice its diameter long. It is therefore well balanced. A drum where the length is less than the diameter cannot be mounted upon trunnions, since it is longitudinally unstable. A ratio of 2:1 in length to diameter is acceptable.

The material will fall through a distance of approximately 70% of the resulting diameter, or about 5.2 ft. That is not strictly accurate, since the trajectory is a curved path, but it is sufficient to provide intuitive guidance.

An operating speed of 17 rpm is realistic. It is not so fast that windage would become a problem, or so slow that drive reduction gearing would become complex. Turning the material 17 times is also realistic, in that if a particle has not found one of the many thousands of openings in the screen mesh in 17 attempts, then it is likely to be too large.

There is, however, a seeming anomaly in the calculations. None of them make any allowance for the amount of product that will fall through the mesh almost as soon as it enters. The mathematics assume that everything stays in the drum until the end of the residence time. The anomaly could be overcome by using a detailed knowledge of the particle size distribution in the material and calculating — using probability equations — when various percentages of the total are likely to pass through the screen. If one knew the answers to those questions, then probably the material would not need screening in the first place! It is for those reasons that a rotating drum screen almost always has a greater actual than calculated capacity. That, given the inconsistencies encountered in dealing with waste-derived composts, is no bad thing.

Having concluded the mathematical model for the screen, there are some entirely mechanical points which should be addressed. Not least among them is the profile of the screen plate perforations.

While steel perforators can supply sheet in a very wide range of profiles, the most common are square or round. The plates are specified in terms of hole diameter or square side, and open surface area (OSA). The latter term is the percentage of the plate area which is holes, and so a 3 ft (1 m) square plate with a 48% OSA has a total hole area of 4.3 ft^2 (0.48 m^2). Waste-derived compost is a very fibrous material, and for some inexplicable reason it blocks round holes very rapidly, while not blocking square ones to anything like the same extent. Therefore, it is common practice for compost drum screens to be fitted with square-perforated mesh plates.

Even then the mesh can become blinded if the material is unusually fibrous or damp, and some manufacturers seek to overcome that by installing long rotary brushes at a position 45° round in the direction of rotation from the material release point. The brushes are sometimes driven by transmission belts from the drum itself, and occasionally are simply free to rotate as the drum passes beneath them. Either way, the intention is that the individual bristles project through the mesh plate perforations and clear them of any material that may be causing bridging.

An alternative that has been used with some success occasionally, if with rather more noise, has been to install "rapping" hammers. These are simply rows of short steel arms, pivoted on a shaft which runs parallel with the longitudinal axis of the drum. As the drum rotates beneath them, raised projections upon it lift the hammers and then allow them to fall to strike the mesh plates, dislodging any material which may be adhering to them.

Sometimes the composts may be sufficiently moist that even a screen with cleaning brushes or rapping hammers may block. Or the material may coagulate so well that it will not break up sufficiently to be screened. In circumstances where that is expected to be an ongoing problem, drum screens are occasionally fitted with flail chains. These are simply lengths of steel chain welded to the lifters and allowed to hang freely inside the drum. During operation they are continually in motion, falling with the material but restrained inside the drum. They increase the rate of attrition significantly, and tend to prevent lumps forming. With dry materials they may also cause an excessive breaking and carryover of contaminants into the product.

The size of the perforations does not directly establish the granularity of the product. For example, a $^1/_2$ in. (10 mm) screen will produce a material granularity in which 90% is less than $^1/_4$ in. (5 mm) and the remaining 10% is less than $^1/_2$ in. This occurs because most of the particles are smaller than $^1/_4$ in. in the first place, so a $^3/_4$ in. (20 mm) screen will still produce 90% less than $^1/_2$ in., but now 10% less than $^3/_4$ in. For primary screening, $^3/_4$ in. square perforations are satisfactory for a coarse soil conditioning product to be sold in bulk loads. Perforations of $^1/_2$ in. will provide a finer material, but one still suitable only for bulk soil conditioning.

A further consideration in the selection of the screen perforation is that smaller meshes create larger reject levels. A $^1/_2$ in. screen rejects everything greater than 10 mm, even though it may be perfectly good compost for some applications. Processing rates, however, are not significantly affected, since a small-mesh screen may have the same OSA as a large-mesh one. It is the OSA that determines the rate at which screening takes place, not the size of the individual perforations.

In factories that manufacture peat-based composts, it is normal to mill the peat before any screening processes. Doing so leads to more efficient screening and a higher product recovery rate. However, it is not so advantageous when dealing with organic digestates. Milling is likely to reduce the particle sizes of plastics and other contaminants, making them more difficult to screen out or to classify. It will also reduce the granularity of the organic fraction of the compost, creating a material that very quickly turns to mud in the presence of water. Most importantly, it is likely to reinvigorate the microorganisms by making fresh surfaces available to them and by providing surplus aeration. Aerobic decomposition will recommence, the product will reheat, and its initial stability will have been lost.

6.5.4.3 Ballistic/Pneumatic Classification

Primary screening can only separate particles of greater dimensions than those of the screen mesh. It makes no distinction between densities or chemical nature, and so the product will still contain plastic, metal, and mineral particles that happen to be smaller than the chosen mesh. Ballistic classification may be able to remove some of the particles that are much more dense than the compost. Air classification can remove very light materials. Neither can remove both heavies and lights at the same time, because the particle densities in the compost are likely to be quite close to many of those of the contaminants.

Grain stoners (see Figure 6.12) combine both ballistic and air classification into one machine, and they work quite well with compost. Their design purpose is, as the name implies, the removal of stones from grain. In operation, a perforated deck vibrates rapidly in such a way that a point upon its surface describes a small-diameter circle — the "throw" of the machine. Pressurized air is blown through the deck perforations into the casing above, and the deck is sloped at an angle of up to about 10° to the horizontal. Rotary valves are fitted at higher and lower ends of the deck, and at the point of admission of product for treatment.

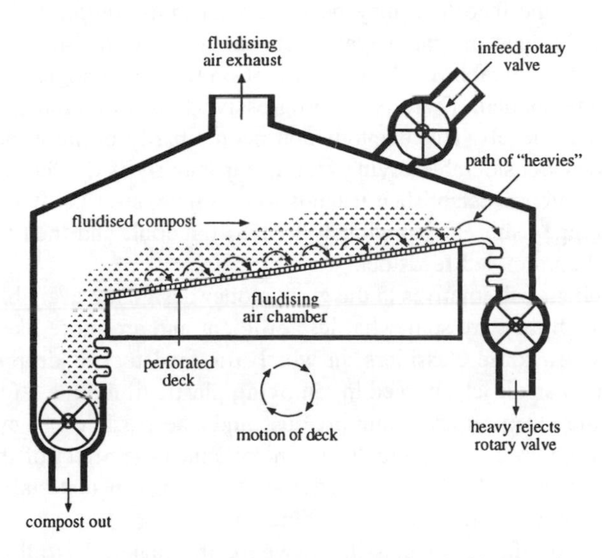

Figure 6.12 Pneumatic/ballistic separator (grain storer).

When a contaminated, granular material is introduced onto the top of the deck, the air pressure may be adjusted until the material fluidizes. Each particle becomes suspended in the air, and the material begins to exhibit some of the characteristics of a fluid — it will flow down a slope, and spread to exactly fill the surface area available to it.

If the material contains heavy contaminants, it can no longer support them in its fluidized condition; they will fall through to the deck of the machine. There, because of the action of the "throw," they will be conveyed toward the rotary valve at the raised end. The deck acts like a vibratory feeder. The fluidized material, meanwhile, flows down the slope to the rotary valve at the low end. While the air pressure is sufficient to fluidize the granular material, it is too high to maintain the lighter particulates in the bulk. They are forced to the top of the bed, and then carried clear by the exhausting air. Therefore, in treating a compost containing metallic and mineral particles together with plastic film, the compost will be discharged through the lower valve, heavies through the higher, and plastic film with the airstream through the casing hood.

While machines of the type described are efficient at refining composts, they do have the disadvantage that a certain amount of the finest particles of compost are lost with the plastic film. The more dry the compost is, the worse that loss becomes. If the purpose is to manufacture a very high-quality product, that may not be a bad thing, since too much dust is unattractive to users. A moist compost releases less dust, although the fluidization pressure needs to be higher, but there are limits to how high it may be permitted to go. A material with a moisture content of greater than about 40% by weight becomes difficult to fluidize at all. It is more likely to homogenize upon or just above the deck, and to become pierced with fissures that allow the air to escape.

Although the feedstock may be somewhat moist initially, that does not mean that fluidization can never occur at all. As the moisture content approaches the critical level, the transformation from a homogeneous to a fluid mass is quite sudden. There is no progressive transformation as there is at lower moisture levels. This phenomenon occurs partly because the fluidizing air has a very considerable drying effect, particularly if the feedstock is hot. Once fluidization is established it tends to continue, so it is often possible to start a machine using dry, previously treated material, and then to gradually introduce the unrefined feedstock.

A number of alternatives to the grain stoner design are available and often used, although they are somewhat less efficient and are more likely to reject excessively. Air knife classifiers, in which the feedstock is dropped down a chute through an air jet, succeed in removing plastic film quite effectively, but as before they also remove compost dust and fine particles. They are crude devices and so are rather difficult to control, since the effect of the air jet is dependent upon the feed rate of material. Variations in material feed create immediate variations in separation efficiency.

Column or zigzag air classifiers, where the material is allowed to fall through a rising column of air, are somewhat more efficient than air knives. Again, they are sensitive to variations in feed rate, and changes in product density caused by moisture content result in rapid deterioration of performance. How effective (or ineffective) they are depends largely upon what is to be separated. If the intention is to remove plastic film in the air flow, with the compost falling through, then quite high efficiencies are possible at air speeds of 16.5 ft/s (5 m/s). If, however, heavy particles, glass for example, are to be removed, then the compost will be the lighter fraction and that which must be entrained in the air flow. In that mode, air classifiers are inefficient, since it is then that they are most sensitive to product variations. No air classifier system can separate into three categories, and so the separation of film, heavies, and finished product is impossible in a single machine.

6.5.4.4 Secondary Screening

The quality of the product from the second, ballistic classification, stage is still of low quality. It is still of an inconsistent particle size, but is substantially free of contaminants. As such, it is suitable for use as a coarse landscaping material, but its appearance limits its attractions for higher-grade markets. A further improvement can be achieved by another screening stage.

The design of the secondary screen is somewhat dependent upon which characteristic of the product needs attention. If the purpose is solely to improve the consistency of the granularity, then a drum screen similar to that used in the first stage of refining is sufficient. Again, square perforations are necessary, and in this case an aperture size of $^1/_2$ in. (10 mm) will provide a satisfactory result. The calculations necessary to determine screen dimensions and capacities are as described for primary screening.

A flat-decked, vibrating screen is an alternative often used in the manufacturing of peat-based composts, particularly where the purpose is to eliminate sticks and twigs. Rotating screens will not provide separation of those materials, since they are as likely to be presented to the mesh plates end-on as side-on. Thus any twig of less than $^1/_2$ in. will pass through with the product. A flat-deck screen, however, can be designed and operated so that twigs are always presented side-on and are always rejected.

A typical flat-decked screen is shown in Figure 6.13, and is a variation upon conventional designs in that it includes a spreader deck at the loading end. The purpose of that section is to allow the material to spread out evenly across the width of the machine before it is presented to the screening mesh, and to do so to the minimum thickness possible. Provided that the "throw" of the machine — the circle described by a point upon the surface — is smaller in diameter than the length of the sides of the mesh perforations, then twigs will always lie flat upon the deck. Any that are longer than $^1/_2$ in. cannot therefore pass through.

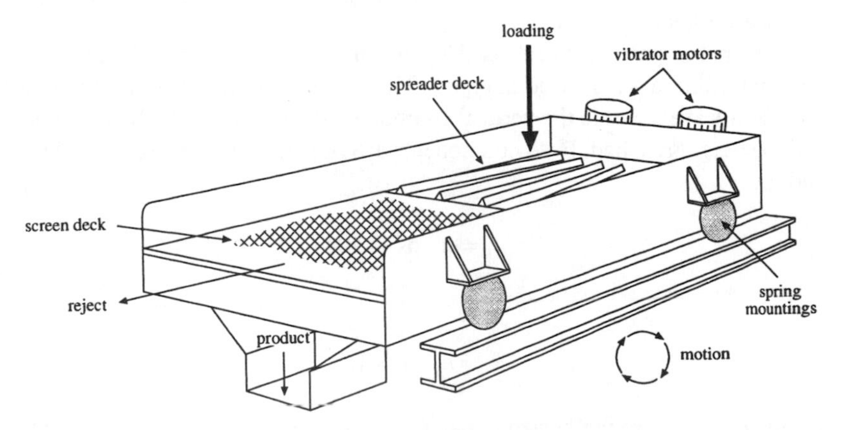

Figure 6.13 Vibrating screen.

The calculations necessary to establish the productivity of the machine are more complex than those for a drum screen, and they have to be based upon an assumption of the feedstock particle size. Therefore, there is some considerable margin of error. The influence of the particle size is exhibited in the depth of material on the deck, since that establishes what the minimum depth can be. The first calculation then necessary establishes the feed rate of the machine, or in other words, how rapidly the material will travel down the deck.

In order to explain how a vibrating screen works, it is necessary to imagine that the deck of the screen is actually a series of circles, the diameter of each of which is that of the throw. Each circle touches that immediately before it and following it. An imaginary line then drawn between the center of the first and the last circles represents the deck. If a particle of material is introduced to the horizontal line at the point where it intersects the first circle, then it will be accelerated around the circumference of the circle to some angle, before being thrown forward toward the next circle.

The linear velocity u at which the particle will be carried is calculated from

$$u = \frac{S \times \pi d}{60 \times 12 \text{ ft/s}} \tag{8}$$

where S is the speed of the vibrator motors in revolutions per min, and d is the throw circle diameter in inches.

The point at which the particle is released and allowed to follow a trajectory is a matter of conjecture. The deck is constrained by drive motors and spring mounts, while the particle is not. As a point on the throw circle approaches the vertical, the material will tend to continue upward at some angle as the deck drops from under it. If the material is fine and moist, then a negative pressure will be created between it and the deck, keeping it from flying free. A coarse-grained material through which air can easily penetrate is not subject to such a negative pressure. Therefore, the former is likely to be released later than the latter.

When the particle is released by the throw, its velocity will possess a horizontal and a vertical component. The horizontal component determines how far it travels, while the vertical component determines both the time taken and the height reached. Both components are functions of the angle of release, and are

$$u_h = u \cos \Phi \text{ ft/s} \tag{9}$$

and

$$u_v = u \sin \Phi \text{ ft/s}, \tag{10}$$

where u_h and u_v are the horizontal and vertical components of velocity, and Φ is the angle of release.

Using the vertical component, the time T during which the particle is in free flight is

$$T = \frac{u_v}{0.5\ g}\ \text{s} \tag{11}$$

The distance R over which the particle will travel in free flight is given by

$$R = u_h T\ \text{ft} \tag{12}$$

If this series of calculations is carried out for a range of release angles between 0° and 90° at a fixed velocity, it can be seen that the distance R is the same for a release angle of 30° and 60°, while the greatest distance is achieved at 45°. The equations suggest that the screen may be able to deliver a fine, moist material at the same rate as a coarse, dry one, if the former is released at 60° while the latter is released at 30° around the circular path.

The frequency of the machine — how many throws per second it delivers — has some influence upon the release angle. The higher the velocity of the particle, the more inclined it will be to travel in a straight line rather than to follow a curved path. Therefore, the higher the velocity, the smaller the release angle, at least to some extent. There is no practical way of calculating what the angle will be, however, since the physical characteristics of the particle also have an influence but cannot ever be established accurately. Therefore, to be able to proceed it is necessary to make an assumption about the release angle.

As in all process machinery design, it is safer to underestimate the capacity rather than to overestimate it. A machine designed against an underestimate will always be capable of the desired throughput. Therefore, bearing in mind that waste-derived compost is moist and somewhat sticky, it is reasonable to assume that the release point occurs at 60° around the throw circle.

Having established how far a particle will travel under those circumstances, one could then arrive at an approximate feed rate by simply multiplying the distance R by the number of throws per min, S. In most circumstances the answer will be correct, while in others it will be wrong. Returning to the analogy of the screen deck being a series of touching circles, if the distance traveled by the particle is not a multiple of the circle diameters, then sometimes the particle will fall on the deck when it is describing the lower arc of the throw circle. There will then be a delay of a fraction of a second before the particle is accelerated again. The speed of the particle along the deck becomes erratic. Where the distance *is* a multiple of the circle diameter, then the particle is in motion continuously. Figure 6.14 demonstrates the point by showing graphically how the feed rate increases with frequency and amplitude, and the calculations from which the curves were derived were adjusted to account for the phenomenon.

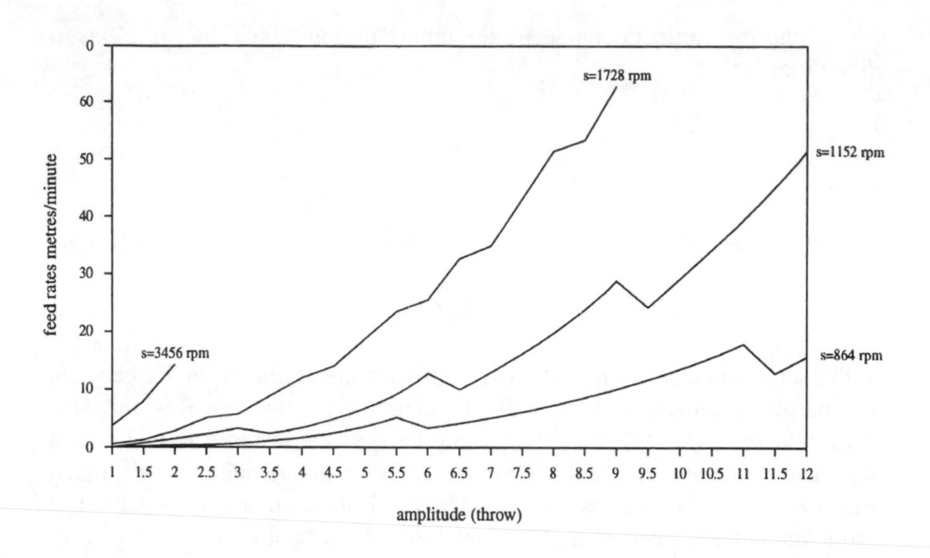

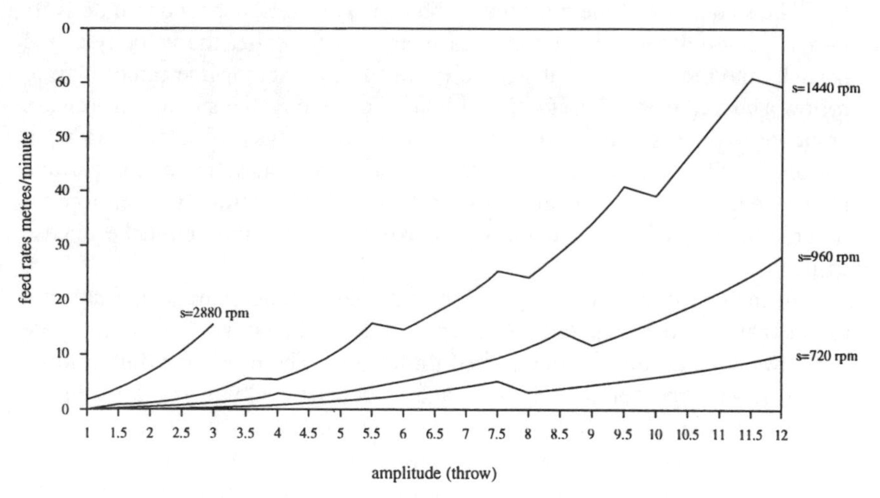

Figure 6.14 Vibrating screen feed rates. (a) 60 Hz. (b) 50 Hz.

In many ways the point is academic, since the actual release point on the arc of the throw circle can never be known with confidence, and because there are a large number of particles that may all react against each other. It can however be demonstrated upon a vibratory feeder where either the amplitude or the frequency of the vibrating motors is variable. There the feed rate of a granular product can be seen to increase and then stabilize cyclically as the amplitude or frequency is gradually increased.

The most readily available vibrating motors for driving screens are of a fixed frequency, but their amplitude can be varied by adjusting the position of pairs of out-of-balance weights. They are not continuously variable, but can be readjusted with the machine out of service. In that case, the machine being designed will be of fixed amplitude and frequency for all practical purposes. Ideally, therefore, the machine should be constructed so that the theoretical distance it projects particles is a multiple of its amplitude, and the time of particle free flight is a multiple of the frequency. Provided that the chosen parameters occur at about the mid-range of the vibrator motor amplitude adjustment, then fine tuning will be possible when the machine is first put into service. It is possible to obtain further control by the use of an inverter controller, changing the power supply frequency to the motors and thereby changing the speed. It is however a more expensive option and hardly necessary if the screen is correctly designed.

The number and positioning of the vibrator motors is a matter for careful attention, since the way in which they are mounted effects the path of the throw. Two motors placed side by side and wired to contrarotate produce a straight line motion at right angles to their longitudinal axis, rather than a circular one. One or more motors rotating in the same direction only produce a true circular throw if they are mounted at the center of gravity of the structure; elsewhere they produce an increasingly elliptical motion. These effects are however modified by the mounting springs and are therefore not generally apparent unless the structure is significantly off-centered.

Vibrator motors are manufactured in a wide range of centrifugal force capacities, depending upon the balance weights with which they are equipped. They are also offered in a range of running speeds dependent upon the number of poles in the motor field windings. Two-pole motors provide 3456 rpm at 60 Hz (cycles per second) power supply frequency and 2880 rpm at 50 Hz. Four-pole provide 1728 rpm at 60 Hz and 1440 rpm at 50 Hz. Six-pole deliver 1152 and 960 rpm, respectively, and eight-pole 864 and 720 rpm.

As rotational speeds and vibration amplitudes are increased, the dynamic loadings upon both the structure and the motor itself increase in proportion. Very high speeds and amplitudes may appear mathematically to be attractive, since they offer rapid feed rates. They do so however at the expense of machine reliability, since structural stresses may become so high that fatigue cracking becomes the norm. For that reason motor manufacturers advise limits upon the acceptable amplitudes, based upon the formula

$$d = 1.059 \times (1000/S)^2 \text{ in., or}$$

$$d = 26.9 \times (1000/S)^2 \text{ mm,} \qquad (13)$$

where d is the maximum permissible amplitude (throw), and S is the vibrator motor speed in rpm.

The feed rate curves shown in Figure 6.14 terminate at the maximum permissible amplitudes for given speed ranges. Even where the amplitude is within limits, however, it should be noted that the greater the speed of feeding down the deck of the machine the greater the trajectory of the particles of material. Amplitudes and velocities creating a trajectory in excess of the screen mesh dimensions may be counterproductive in that material is likely to pass over the mesh too fast to be screened properly.

When the linear velocity has been calculated within these guidelines, and the maximum bed depth has been chosen, then the screen width can be calculated for a required capacity from the formula

$$W = C / \mathrm{RSD}, \quad (14)$$

where R is the travel of the particle in free flight, S is the number of throws/min, D is the depth of material on the deck and C is the required capacity. Note that at this stage the calculation is independent of the length of the screen — it will deliver material at the same rate whatever its length. The only relevance of that dimension is one of ensuring that the mesh area is sufficient to achieve complete screening at the travel speed designed.

A final check upon the likely height of projection of particles using the mathematical model is worthwhile. After all, the purpose of the machine is to eliminate twigs and sticks. If the height of projection is greater than the side dimensions of the screen mesh, then twigs could turn over and become presented end-on, in which case they could again pass through with the refined product. The projection height is a function of the vertical component of velocity:

$$h = u_v^2 / 2g \quad (15)$$

In the somewhat unlikely event that the model suggests that the twigs will pass through, then the solution is to reduce the screen amplitude. However, most conventional vibrating screen motors provide amplitudes of up to $^1/_2$ in. (10 mm) maximum, and more commonly operate at about $^1/_4$ to $^3/_8$ in. (5 mm to 8 mm), so the risk of excessive vertical projection is not great.

Having calculated the screen dimensions from the preceding formulas, the deadweight of the machine and its load can be assessed. That is simply a matter of adding together all of the weights of the individual structural components, etc. Then using that weight and the selected frequency and amplitude of oscillation necessary to achieve the feed rate, the vibrator motors can be selected. The motor manufacturers publish centrifugal force data for their products, and it is by using their tables that the motor specification is established. The calculation in this case is

$$\mathrm{Cf} = \frac{A \times S^2}{0.07 \times 10^6} \times L \text{ lbsf}$$

or (16)

$$\text{Cf} = \frac{A \times S^2}{1.786 \times 10^6} \times L \text{ kgf}$$

where A = Amplitude (in./mm)
S = Motor rpm
L = Load in lb. or kg

In order that the screen may oscillate, it must be installed upon elastic mountings, and so the next stage in the modeling procedure is to establish the characteristics of those mountings. Two types are common — springs or rubber bushes. The considerations leading to the design are common to either type, and they concern themselves with the spring "rating," that is, the amount by which the spring will compress for increments of load. A too-stiff spring will not allow the screen to oscillate at all, while a too-soft spring will permit uncontrolled oscillations during start-up and shut-down.

The necessary spring rating is a function of the static weight of the screen and the load upon it, and the dynamic load imposed by the vibrating motors. As the motors rotate, they impose a load in terms of centrifugal force that is continually changing direction. At one point it will be directly opposed to the weight of the screen, while at the opposite point of rotation it will be in the same direction as the weight. From the mounting spring perspective, the screen will gain and lose weight once per revolution.

The mountings must both allow the screen to oscillate and must isolate the oscillations from surrounding structures. Conventionally, 95% isolation is considered satisfactory for most purposes. Using that figure, the allowable deflection of the mountings under the load of the vibrators, the structure, and the material is derived from

$$\text{Deflection} = \frac{\left(\dfrac{100}{100 - I_s\%}\right) + 1}{\left(\dfrac{c_f}{415.45}\right)^2} \text{ in.}$$

or (17)

$$\text{Deflection} = \frac{\left(\dfrac{100}{100 - I_s\%}\right) + 1}{\left(\dfrac{c_f}{950}\right)^2} \text{ mm}$$

where $I_s\%$ is the degree of isolation required.

Therefore, a set of mountings which together deflect by that amount under the total imposed load are required. A screen would normally have four. Spring deflection under load is calculated from

$$\text{Deflection} = \frac{8D^3N_aP}{Gd^4}\text{ in. (mm)} \tag{18}$$

where D is the mean diameter (outside + inside diameters/2), N_a is the number of active coils, P is the applied load in pounds (newtons) (0.25 × total load in a four-spring machine), G is the modulus of rigidity for the spring material (11.5×10^6 lb/in.^{2} or 79.3×10^3 MPa) for the most common, and d is the wire diameter in inches (millimeters).

If the acceptable deflection obtained from (17) is substituted in (18), then the formula can be transposed to calculate either the wire diameter given a mean diameter or vice versa. It is usually the case that the calculation needs to be carried out for several assumed dimensions until a practical spring has been identified. Such a spring is one where the outside and wire diameters diameter is between 7 to 14 times the wire diameter, and where the number of coils is such that the load causes less than 80% of the available deflection to closed length.

The screen designed by the methods described is assumed to be horizontal, but there are some advantages in its being slightly inclined upward toward the discharge end. Such an inclination has the effect of creating a more rapid leveling of the bed of material on the deck, and improves the screening quality. It reduces the feeding capacity, but since an inclination of a maximum of 5° is sufficient, the reduction is only marginal.

6.5.4.5 Blending

Blending installations are almost always the final process stages of peat-based composts, in order to introduce the trace elements and nutrients that do not naturally occur in peat. Waste-derived composts are less likely to need blending because, being entirely organic, they do contain available nutrients. In addition, markets for them are likely to be less sophisticated than are those for peat-based products, and so exact repeatability is less important. However, the installation of a simple blending line is a wise precaution, since it provides the facility for correcting any chemical imbalances that might otherwise prejudice a newly discovered market outlet.

The most convenient blending system is based upon a belt conveyor, which must exist at the end of the flowline if for no other reason than to remove the products to storage. Since the finished product is dense 30 lb/ft^3 to 37 lb/ft^3 (500 to 600 kg/m^3), continuous belt weighers can record its passing accurately, particularly if the conveyor loading rate is made consistent by the use of a transfer hopper with a sliding gate. The output signal from the belt weigher can then be used to control the speed of simple dosing units operating through metering auger-feeders. These machines can be mounted in any number along the belt, using one for every chemical or trace element to be added, and allowing their output to trickle onto the compost as it passes beneath. Final

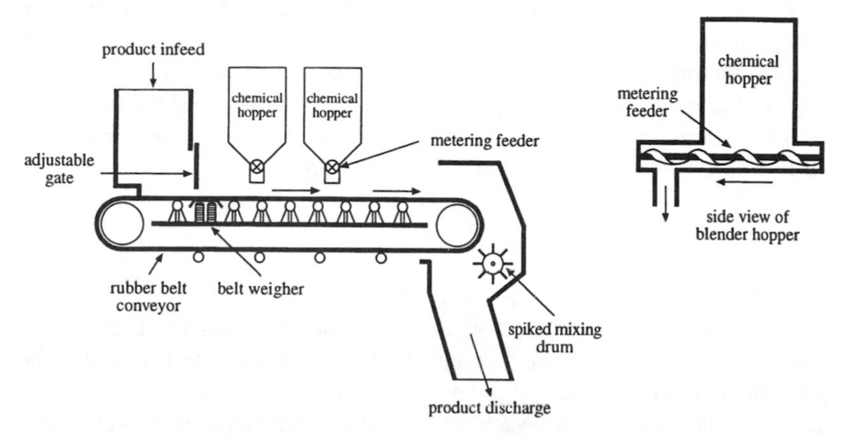

Figure 6.15 Typical blending conveyor plant.

mixing of the compost and the additives takes place in the discharge chute from the conveyor, where a rotating, spiked roller agitates the fractions. Figure 6.15 shows a simple blending line of the type described.

An alternative blending system sometimes used for proprietary composts involves paddle mixers, which are in essence large-diameter auger feeders with ribbon scrolls, or with paddles arranged helically around the center shaft. Again, trace elements can be added with the machine running, and can be linked to an infeed belt conveyor through a belt weigher. They are not, however, ideal for waste-derived composts. They have a tendency to cause the material to form lumps, thereby negating the screening work that has been carried out to remove lumps.

A final point that cannot be stressed too highly is that compost blending is a highly complex business, particularly where organic products are concerned. It is never simply a matter of tipping in some chemical nitrates if the nitrate level in the product is a little low for the intended market. Using the wrong chemical, or even the right one in the wrong form, can create completely unpredictable effects, and solidification of the whole mass of compost is one of the more common. The ways in which plants use trace elements and nutrients are the subject of almost continual research, and still to some extent a matter of conjecture. Certainly the field is too great to be covered even briefly in a book such as this. Therefore, the best that can be suggested, where a blending line is being considered, is to think carefully about whether the organization will have the skilled and detailed scientific knowledge sufficient to operate it.

6.6 HEALTH AND SAFETY

Compost is regarded as a benign product, and therefore it is easy for personnel involved in its manufacturing to underestimate the dangers. In fact, a composting process can be threatening to human health and safety for a number of reasons. The material itself can be hazardous, almost all through the complete manufacturing system, and much of the machinery used is more dangerous than it at first sight appears. Irrespective of whether the process is aerobic or anaerobic, the risks can be categorized as biological or physical, and further subcategorized as material or hardware.

6.6.1 Biological Risks

The raw organic waste from which compost is to be made is inevitably highly contaminated with microorganisms, some of which are likely to be pathogenic. Some pathogens, although by no means all, are believed to be incapable of survival outside of the human body, and to therefore pose no risk when they have entered the waste stream. However, some researchers[4] have suggested that 6% of the total bacteria in municipal waste are pathogenic, and among these are *Pseudomonas, Klebsiella*/enterobacter, *Serratia* spp., and *Escherichia coli.* Many are fecal bacteria, and if they can be identified, then they are both viable and potentially hazardous. At various stages throughout the process, any or all of the species of bacteria may be released into the surrounding atmosphere. In addition, fungal spores of the species *Penicillium* spp., and *Aspergillus fumigatus* have been identified in emissions from municipal waste. As composting progresses, other species of bacteria and fungi are released, although in aerobic processes pathogens are destroyed in a matter of h or days, depending upon the temperature regime.

Many of the liberated bacteria and fungi are not considered essentially harmful, although they may cause allergic reactions in predisposed people. There is also some evidence that allergies can be developed as a result of exposure to large concentrations over a period of time. Gram-negative bacteria and streptococci have been recorded as exceeding the concentrations likely to give rise to symptoms through the inhalation of endotoxins — cell wall lipopolysaccharides.

In addition, there is a perceived risk of Weil's disease (leptospirosis) in handling crude municipal waste. Weil's disease is a viral infection, mainly contracted from the urine of infected rats. It is similar in many ways to influenza, and develops symptoms of jaundice. It is a potentially fatal disease easily mistaken for a less threatening one, and although once fairly rare, it is now becoming more common in many developed countries. The greatest risk of contagion occurs from contact with liquid effluents from waste, and particularly from contact with fluid separator circulating water. It is a sensible precaution to issue all employees with a Weil's disease card, requiring them to keep it at their homes and to make sure that it is given to their medical practitioner if they ever become ill, particularly with symptoms that appear influenza-like. An example of a suitable card is shown in Appendix A.

Personnel protection against the risks described depends upon good engineering design, competent process operation, and well-planned operating procedures. Worker training is vital, and all employees should be instructed in personal hygiene. Their medical practitioners should be made aware of the work they do, and should be informed of the health risks.

Engineering design can minimize some of the risks, but it can never eliminate them. Adequate ventilation, air cleaning and filtration in areas where bacteria may be liberated is vital, but any of those measures can only attempt to capture organisms that have already entered the atmosphere. There is no obvious design route by which emissions can be prevented completely. Neither is there any design route through which any possibility of contact with the waste, or with liquors arising from it, can be eliminated. All machinery can be, and should be, totally enclosed, and leakages of liquids should be dealt with immediately. But plant failures do occur, and blockages in systems handling organic wastes are not uncommon. Maintenance work involving contact with waste residues has to be carried out, and there is always a risk of contagion. Protective clothing, a secure operating procedure, and an effective design can minimize the risks. Nothing can eliminate them.

As the process proceeds and the wastes begin to decompose, the nature of biological threats changes. It does not, however, cease altogether, however innocuous the material begins to appear. Pathogens may have been destroyed by high temperatures in an aerobic process, or by the actions of competing bacteria in anaerobic processes, but the emissions of bacteria into the atmosphere may actually increase. Some researchers[5] have shown that when aerobic waste-derived composts are being harvested, thermophilic actinomycetes including *Faenia rectivirgula* are likely to enter the atmosphere in concentrations that may exceed recognized safety levels. *Faenia rectivirgula* is a source of farmer's lung antigens, as is *Thermoactinomyces* spp. which is also usually present.

In the absence of adequate air movement and air cleaning, atmospheric concentrations of bacteria and fungi may be expected to exceed levels at which allergic alveolitis occurs in humans. Sensitization is not necessarily an immediate result of inhalation of bacteria and fungi, but may gradually develop if any worker is in prolonged contact with them. Once developed, allergic reactions may become increasingly severe, and so a cautious employer is well advised to keep detailed records of employee health, noting even apparently minor ailments that do not cause absences from work.

6.6.2 Physical Risks

The possibility of entrapment or injury involving heavy rotating plant is a matter that must always be at the center of good engineering design practice, and most countries have established legislation to deal with it. The hazards associated with a drum screen are the same in the waste disposal industry as in any other, and so it is not proposed to deal with such matters here. The purpose of this section is to explore risks specific to waste composting that may not be immediately obvious.

The first of those risks has to do with the material temperatures experienced in aerobic processes. The contents of an aerobic digester may, and indeed will, reach a temperature of up to 185°F (85°C). The digestate in such circumstances is moist, and so the temperature is quite sufficient to cause painful scalds. In addition, the physical nature of the material is such that it tends to adhere to skin or machinery equally well, increasing the area of contact and worsening the injury.

The risks are greatest at the digestion and unloading stages of the process. In the digester, particularly where a continuous process involves unloading and feeding at the same time, voids may form temporarily but not be visible from an examination of the surface. Such voids collapse readily if disturbed, and so anyone unwise enough to step upon the surface could easily become submerged in very hot digestate. In addition, the emissions from aerobic processes are carbon dioxide and water vapor. Even with an excess of ventilating air there can be no guarantee that there is sufficient oxygen above the material to sustain respiration. Access into digesters is essential for maintenance purposes, but the design should locate them is such a way that they cannot be used when the digester is in service. This constraint is normally achieved by putting them at the level of the perforated floor, so that they are covered when the digester is working. Access from the top of the digester is never necessary, and so while observation ports are useful, manholes are simply an invitation to take risks.

At the unloader discharge, the material may again be very hot, depending upon what stage of the process is involved. Unfortunately, at that point it does not always appear to be so, and it is very tempting to insert a hand to take samples. There have been many cases of minor scalds resulting from this practice, and again, the design should minimize the opportunity. Where sampling is necessary, the provision of a circular port too small to admit a human hand is a useful precaution. Samples may then be taken using a long, semicircular scoop without risk of direct skin contact. Otherwise, casing doors over unloader discharges should be bolted on, not simply hinged and locked.

Chain unloaders pose significant risks of more than just scalds. They can be removed and extracted with the digester in operation, but that exercise requires the machine to be mounted upon guide rails temporarily bolted to the digester casing door. The unloader then has to be under power, with its cutting and unloading chains running, so that it may cut its way into or out of the digestate. While that is happening, the cutter chain, with its prominent and sharp cutting blades, cannot be guarded and is extremely dangerous. At the same time, hot composting material will be ejected in a somewhat uncontrolled manner and in larger than normal quantities. There is little that the engineering design can do to minimize the risks, since the operation is essentially a maintenance one. The machinery has to be exposed to carry out the task. However, there should be a provision for the temporary installation of guard handrails or barriers to keep sightseers at bay, and one responsible member of staff assigned to the task of observing the whole operation. The sole duty of that staff member should be to position himself where he can watch every employee involved, and can easily reach power shut-off switches in the event of an emergency.

Anaerobic digesters also pose a number of threats that are not always obvious. Among those which perhaps are obvious is the atmosphere of methane and carbon dioxide inside the digester. If there is likely to be inadequate oxygen in an aerobic vessel, it is almost certain that there will be none inside an anaerobic one. If oxygen were present inside, then there would be a very grave risk of an explosive-range mixture developing. For both reasons, any essential access ports should be so designed that their use while the digester is in service is impossible, not simply difficult.

When anaerobic digestate is unloaded from the vessel, it does not suddenly cease to be biologically active. As long as it still contains some nutrients, and has sufficient water content to exclude oxygen, it will continue to emit methane gas. Smoking anywhere in the vicinity of the unloaders should be absolutely prohibited, and storage and handling should take place in the open wherever possible. If any part of the handling must take place under cover, then a very high order of ventilation must be provided. Every fixed handling machine should be both totally enclosed and aspirated, and the building itself should be exhausted through absorbing filters.

Methane is considerably lighter than air, while carbon dioxide and hydrogen sulfide are both heavier. Methane will therefore rise to the roof space of the building, while the other gases will stay at or around floor level. Therefore, all light fittings and electrical services in the roof must be flameproof and certified for use in gaseous atmospheres, and the extraction ducts should be run so as to provide the maximum coverage at the highest points. Careful attention must be paid to air movement to ensure that there are no dead spaces where gases could collect.

Anaerobic reactors have been shown under laboratory circumstances to produce methane at a rate equivalent to 5% of the biomass volume per day. While it is not suggested that the same rate of generation should be expected to occur in a digestate handling room, it would be wise to use the figure as a maximum in assessing risks. Therefore, a mass of 35,300 ft^3 (1000 m^3) could potentially release 1760 ft^3 (50 m^3) of gas, which would then occupy any free space. If that space was 35,300 ft^3 (1000 m^3), then the gas/air mixture could reach the lower explosive limit (5%) in 1 day. At first sight, the risk does not therefore appear great, and it would seem that even a small-capacity ventilation system would be adequate do deal with it. However, if the process is continuous, then there is always likely to be some gas present. Four air changes per h would ensure that the gas concentration never rose above 0.05%, and would provide a realistic safety factor.

An alternative to positive ventilation is to provide roof vents that simply allow any methane generated to escape to the outside atmosphere. In safety terms, that would be the most secure option, since the vents could be placed to eliminate dead spaces in which the gas could collect. However, methane is a significant atmospheric pollutant. It would be difficult to justify releasing such a pollutant from a process that should be justified as a best environmental option.

While the products of either treatment technologies carry with them certain physical risks, so too do the intermediate processes. Fluid-based separation equipment can be particularly hazardous. In a fluid separator, heavy materials fall through heavily contaminated circulating water for removal by drag link conveyors. Those materials will be a mixture of broken glass, ceramics, stones, metals, and batteries. There is also a strong possibility of the presence of syringes and the residues of drugs. Much of the glass will be in the form of shards ranging from particles as large as 2 in. (50 mm) down to pieces almost too small to be seen with the naked eye. Any contact with them presents a severe risk of skin penetration, possibly leading to infection.

The metals encountered are most likely to be in the form of small, heavy items such as nails, screws, and cutlery. Almost all will be sharp and penetrating, and again may not always be clearly visible. In addition, many of the metals will be contained in batteries, in which other chemicals may be caustic. Alkaline batteries containing cadmium and mercury will be present, and while they are not easily subject to breakage, the action of the extraction conveyor may be sufficient to do so. Both metals are extremely toxic, and easily introduced into the body accidentally if contact with the waste is permitted.

If the residues are to be further refined to create end-products, for example by milling the glass to sand and screening it from the contaminants, then it would be normal to install some form of washing system. Washing with fluids containing disinfectants would eliminate risks of contagion, but would do nothing to minimize the physical hazards. Therefore, the process design should again ensure that all handling plant is totally enclosed and that operational contact with its contents is avoided.

Key-operated interlocks are advisable for security of enclosures. Here the principle is that an enclosure can only be opened by a special key, which is trapped in the electrical control system. The key cannot be removed to open an enclosure without disconnecting the power supply to the machine. Many designs then take that concept further by establishing a time limit before the machinery stops and can be opened. Thus, once the key release mechanism has been activated, infeed to the process line is stopped and the plant runs on for sufficient time to become emptied down. The key is not released until that time period has elapsed.

REFERENCES

1. Chandler, J. A., Jewell, W. J., Gossett, J. M., Vansoset, P. J., and Robertson, J. B., Predicting methane fermentation biodegradability, *Biotechnology and Bioengineering Symposium,* 10, 93, 1980.

2. Jewell, W. J., Future trend in digester design, *Proceeding of the First International Symposium on Anaerobic Digestion,* Cardiff, Wales, 1979, Applied Science Publishers, Ltd., London, 17, 1979.
3. Keeling, A., Mullett, J. A. J., and Paton, I., GC-mass spectrometry of refuse-derived composts, *Soil Biology and Biochemistry,* vol. 26, 1994 Elsevier Science Ltd., London, England.
4. Lacey, J., Williamson P. A. M., and Crook B., Microbial emissions from composts made for mushroom production and from municipal waste, AFRC Institute for Arable Crops Research, Harpenden, Hertfordshire, England, 1990.
5. Lacey, J., Williamson, P. A. M., King, P., and Bardos, R. P., Airborne microorganisms associated with domestic waste composting, Warren Spring Laboratory, Hertfordshire, England, 1990.

7 BIOLOGICAL ASPECTS OF COMPOST PRODUCTION AND UTILIZATION

7.1 INTRODUCTION

Composting can be defined as the aerobic (oxygen requiring), biological decomposition of organic waste to yield a stable, hygienic material that is beneficial to soil and plant growth. The products of the decomposition are carbon dioxide, water, and solid matter undergoing the process of humification, i.e., being converted into soil organic matter whose major components are humic and fulvic acids. The process has been widely studied,[1] but is still not fully understood. The organisms involved in the composting process are abundant both in air and soil, so decomposition of dead organic matter will occur without human intervention, as is clear from the generation of odors during storage of putrescible matter for a short period of time. Placement of mixed kitchen waste into small compost bins will yield material decomposed sufficiently to lose the identity of individual waste components over a few months, but little or no control is exerted over the kind of compost ultimately produced, and its chemical structure is very variable. The key to successful large-scale compost production is careful control of the process from beginning to end, not simply relying on nature to take its course. Failure to control the process adequately will result in a number of problems which, if not addressed, rapidly could result in closure of the production plant, or reduce the marketability of the product.

It is important to ensure that incoming organic matter for composting is of an adequate quality. Specifically, it needs to be low in contaminating inclusions, particularly glass, plastic, and sources of heavy metals. A major source of heavy metals is batteries. Ideally, therefore, the components of a compost feedstock would be raw green waste from parks, gardens and agricultural processes or well-separated kitchen waste. Organic wastes high in lignin and cellulose (e.g., wood and paper) may be included in mixtures to balance wastes containing high quantities of nitrogen, for example, grass cuttings. Although precise definition of the quantities of each type of incoming waste may not always be possible, sufficient analysis should be carried out to enable a fairly accurate definition of the overall composition of the feedstock. This will allow an estimation of some of the likely properties of the final product.

7.2 BIOCHEMISTRY OF COMPOST PROCESSES

Compostable matter arises from once-living tissue of either plant or animal origin. The composting process essentially involves the use of this matter as nutrients for microorganisms, i.e., the organic matter is transformed by microbes into a readily usable food source. The major components of plant and animal organic matter are listed in Table 7.1.

Table 7.1 Plant and Organic Matter

Plant Matter	Animal Matter
Cellulose	Lipids
Hemicellulose	Nucleic acids
Lignin	Proteins
Lipids	Starch
Proteins	
Nucleic acids	
Starch	

In the composting process, microorganisms degrade the macromolecules into monomers which can then be built into microbial biomass. Plant materials are largely composed of the polysaccharides cellulose (typically 30% w/w), hemicellulose (typically 40% w/w), and lignin (typically 20% w/w). In the majority of compost processes, these raw starting materials predominate. They are broken down during the composting process as shown below.

Cellulose hemicellulose → Monomeric and dimeric sugars → Microbial polysaccharides and energy production

Polyphenols → Quinones → Humic acid-like substances
↑ ↑
Lignin (polyphenylpropane) → Phenol derivatives → Polyphenols → Quinones

The following substances are metabolized similarly in organic matter from plant and animal tissue.

Starch → Simple sugars → Microbial polysaccharides and energy production

Nucleic acids → Nucleotides and bases → Microbial nucleic acids

Ammonia → Nitrate
↑
Proteins → Amino acids → Microbial proteins

Lipids → Fatty acids → Microbial lipids and energy production

The breakdown of lignocellulose yields a number of intermediary metabolites of significance in the plant-growing properties of the product. Cellulose and hemicellulose can yield acetic acid, acetone, ethanol, and citric acid, while lignin can yield acetic acid, cresol, dimethylsulfide and vanillin derivatives.[2] Water and dichloromethane (DCM) extractable substances in refuse-derived compost have been characterized. By far the most abundant water-extractable substance appears to be acetic acid, derived from a range of polysaccharides and lipids. Unstable (biologically active) composts contain an abundance of substances extractable into DCM.[3] These fall into four main categories: volatile fatty acids (propionic, butyric, pentanoic, valeric, isovaleric, etc.), long-chain fatty acid esters, phthalate esters, and alkanes (waxes). The source of some of these materials is unclear, especially the phthalates, which are widely used as plasticizers and could conceivably be contaminants in the compost. However, in stable compost the only extractable substances are waxes, suggesting that the other materials are metabolized or incorporated into evolving humic-acid-like substances.

According to some authorities, there is a substantial (sixfold) increase in the humic acid content of organic matter after composting.[4] However, humic acids are diverse in nature, depending on origin, and it may be more appropriate to define such materials in compost as humic-acid-like substances. The formation of humic material in soils takes place over extended periods under ambient conditions, while organic matter in composts may typically experience temperatures of between 60 and 80°C for a number of weeks. While composts do contain dark matter, it is probable that a proportion of this results from the Maillard (browning) reaction involving a random and general condensation of sugars and amino acids at high temperatures. It follows that humic-acid-like substances in compost contain a wider variety of chemical structures and functionalities than typical humic substances of soil. As such, they may act as a slow release sources of nutrients, especially nitrogen.

7.3 BIOCHEMICAL ASPECTS OF COMPOST PRODUCTION

The aim when undergoing large-scale compost production is to control the process in order to minimize emissions to the environment, most particularly to air and water, and allow at the same time the rapid production of a useful compost. The emissions to air that may arise are volatile organic compounds and ammonia, some of which may be malodorous and result in a public nuisance. Water and land pollution could arise from uncontrolled leachates from compost piles.

The most important considerations in satisfactory compost production are:

1. *Availability of oxygen.* In order for decomposition of organic matter to occur at the maximum rate, the microorganisms must have a consistent and high oxygen supply. Absence of oxygen leads to a preponderance of anaerobic organisms, which produce phytotoxins (plant poisons) and malodor, and slow the composting process considerably. Oxygen may be supplied by various means, but the most effective involve agitation of the material. Alternatively, oxygen can be provided by blending with calcium granules.

2. *Availability of moisture.* Water content in a compost pile is important. Above 65% water, the material becomes waterlogged and is difficult to break up and aerate. Below 25% moisture, microbiological activity virtually ceases, and the material enters an inert state until rewet. Active compost piles will lose water at a high rate, since temperatures can come close to boiling point. In these cases, it is necessary to ensure that piles can be sprayed to maintain the required moisture level.
3. *Attainment of high temperatures.* Temperatures in large compost piles (with a volume of greater than 1 m^3) can naturally rise to temperatures in excess of 167°F (75°C). This heat is generated through the metabolic activity of microorganisms. Different microorganisms live and proliferate at different temperatures: psychrophiles are active up to 95°F (35°C), mesophiles are active between 95 and 130°F (35 and 55°C), and thermophiles predominate at higher temperatures. The biological decomposition is believed to proceed most rapidly around 130°F (55°C). However, the generation of high temperatures is important to achieve product *sanitization*, i.e., the killing of potential plant and human pathogens. Most effective killing of pathogens occurs at the highest temperatures, but these may have an untoward effect on the progress of composting. In compost production, however, it is important to ensure that all material has been through a thermophilic phase so that pathogen kill is achieved; it has been suggested[5] that the process should give complete inactivation of salmonella species, a 4-log reduction of parvovirus and streptococci (indicator microorganisms), and 5-log reduction of enterobacteria. It has also been proposed that in order to yield a hygienically safe product, the composting process must be operated for at least 4 weeks with at least 1 week at 150°F (65°C) or above.[6]

7.4 COMPOSTING SYSTEMS

These can be broadly divided into open and closed systems.[6] Open systems involve the formation of piles in the open which are exposed to the elements. By definition, they are the most difficult to control since components are more easily lost to or gained by the composting material. However, they may be more easily accessible and simpler to provide adequate aeration.

1. *Turned piles.* These involve the mechanical turning of piles and may be carried out in windrows using either front-loading diggers or specially designed windrow turners. While front-loaders may be adequate for small quantities of compost, they cannot effectively turn the pile or break up large chunks of material to produce a homogeneous product. Specially designed turners are available that may be able to process up to 4000 yd^3 (3040 m^3) of compost per h. However frequently a compost pile is turned, it rapidly becomes oxygen-depleted. It is proposed that piles are turned at least once every 2 days, though this may not always be practicable. For shredded green (yard) waste, turning once weekly is perfectly adequate. It is important, however, that the turning process regularly exposes material from the outer

layers of the piles to the centers so that all material passes through the high temperature phase and receives sufficient oxygen. The optimum height for piles balancing temperature requirements with aeration appears to be about 3 m. Since piles may reach high temperatures, much water can be lost through evaporation, so it is necessary to ensure that there are facilities for rewetting the material. Since leachate from the piles may occur, it is advisable to collect this via runoff channels and store it for respraying.

2. *Static piles.* Static piles are not turned but are dependent upon blown or sucked air introduced via pipes at the base of the pile. In bottom suction systems, air is sucked through the pile via perforated pipes, and in such cases the limiting factor for successful aeration is the size of the pile, which should ideally not be greater than 8 ft (2.5 m) high. The piles may be stacked on porous materials such as woodchip to improve airflow and increase pile height. In order to ensure exposure of all the material to temperatures high enough to enable full sanitization, it is necessary to cover the fresh composting material with stabilized compost, which insulates the piles. The difficulty with this procedure is ensuring a uniform product, since inevitably channeling of air will occur and some zones of the pile will receive more air than others. Also, such systems are technically difficult to operate with large quantities of organic material and it is difficult to move piles without damaging the aeration system. The same is true with bottom blowing systems in which air is forced from the bases of the piles to the exterior.

In-vessel (closed) systems may be subdivided into two types: continuous vertical reactors and horizontal reactors.

1. *Vertical reactors.* These are useful in that large quantities of organic matter (up to 2600 yd^3 [1976 m^3]) can be processed in a relatively small surface area. The reactors can be as tall as 32 ft (10 m). Material can be loaded into the vessels from the top using a spinning device for even distribution. Oxygen can be provided by forced aeration of the material from the base of the reactor. Unfortunately, when large quantities of material are present in the reactors, aeration of the material is difficult to achieve: the introduced air tends to find the least resistant routes of flow around the edges of the pile. The only really effective way of aerating large volumes of material within a vertical reactor is, as in windrows, to regularly move the material to introduce new surfaces to air. This may be achieved by recirculation of the material within the reactors, provided mechanical cutters and conveyors are of the correct specification. In enclosed systems, control of odors may be easier to achieve, since air exit points may be fitted with biological filters of peat, activated carbon, or mature compost.
2. *Horizontal reactors.* These are cylindrical reactors placed on their sides with the organic matter distributed along the length of the vessel. In these cases, the depth of material is between 6 ft 6 in. and 9 ft 6 in. (2 and 3 m). The advantage of horizontal reactors is the ease with which the composting process can be controlled. The vessels can be rotated to maintain even aeration and temperature control.

7.5 COMPOST QUALITY

Depending upon the end uses of the product, a number of criteria have been suggested as guidelines for compost quality control, and for any compost, a specification should be available. Relevant criteria have been proposed[5] and are outlined below.

1. *Organic matter content.* A compost specification should include the organic matter content of the material.
2. *Moisture content.* Moisture content should be stated. A lower moisture content helps in spreading operations. Also, dry material (25% moisture or less) is preferable for bagging, since it is less likely to become moldy or malodorous.
3. *Inert material content.* Inerts include glass and plastic, and particulate metal. These should be stated in any specification but should always be minimized to facilitate marketing. The physical nature of the inert material is important. In general, the smaller the particle size, the smaller the effect on the appearance of the product. Glass, if present, can be reduced in particle size to an innocuous physical form through ball-milling.
4. *Mineral content.* This should be stated and ideally be at the following minimum levels.

Nutrient	Minimum content
N (nitrogen)	0.6%
P_2O_5 (phosphorus pentoxide)	0.5%
K_2O (potassium oxide)	0.3%
CaO (calcium oxide)	2.0%
$CaCO_3$ (calcium carbonate)	3.0%
MgO (magnesium oxide)	0.3%

5. *Salinity.* If this is very high, crops may be damaged, particularly in calcareous soils. A salinity level of not more than 0.2 oz/ft^3 (2 g/l) salt (expressed as NaCl) should be present, and the individual quantities of Na and Cl should be specified. The harm that may be caused by high salinity is dependent on the quantity of compost to be used in a given application, and the maximum application rates can be determined from growth trials.
6. *pH.* For plant growth, pH levels between 5 and 8 are acceptable.
7. *Heavy metals.* While many of these are vital trace elements for growth, they are often present in unacceptably high levels in composts, especially if the compost has been derived from refuse or sewage sludge. The following table gives proposed limits of metal contents in soil and compost.

The figures given above are the recommended limits for compost to be used for growing ornamental plants (column 2), and the maximum concentrations in sewage-sludge-treated agricultural soil. A compost with high levels of metals may be used in a range of applications provided the total final metal concentrations recommended are not exceeded.

Metal	Max. in compost (lb/ton)	(mg/kg)	Max. in soil (lb/ton)	(mg/kg)
Zn (zinc)	3	1500	0.6	300
Pb (lead)	2	1000	0.2	100
Cu (copper)	1	500	0.2	100
Cr (chromium)	0.4	200	—	—
Ni (nickel)	0.2	100	0.1	50
Hg (mercury)	0.01	5	—	—
Cd (cadmium)	0.01	5	0.006	3

7.6 STABILIZATION OF COMPOST

Composting is characterized by a high degree of microbiological activity initially, which gradually ceases as all available nutrients are utilized. A compost with a low biological activity is described as stable. It is generally believed that compost should be well stabilized prior to applications involving plant growth, since an unstable compost may contain substances poisonous to plants (phytotoxins). Assessment of stabilization is not easy, since there is currently a lack of understanding of the processes involved. However, a number of methods have been used[10] and are summarized below.

1. *Carbon-based analyses.* These involve measurement of organic carbon levels and C:N ratios. These methods suffer from a lack of sensitivity, and will vary depending on the C and N content of the starting material.
2. *Humification indicators.* Measurements of humic substances in a compost may provide an effective measurement of stabilization, especially humic acid (HA) carbon to fulvic acid (FA) carbon ratios. HA content rises as composting proceeds. Complex equipment is required for these tests.
3. *Molecular size determinations.* Molecular sizes may rise as humic substances are formed, but will vary depending on the nature of the incoming organic material. Again, specialized equipment is required for such tests, with highly technically competent staff.
4. *Enzyme assays.* Different enzymes alter in concentration during the composting process, e.g., alkaline phosphatase, invertase, hydrolase, and protease. More research is required to show how these could be used in stabilization determinations.
5. *Respiration measurements.* As composting proceeds, there is a fall in respiratory activity. It has been proposed that a compost may be considered mature when its oxygen consumption is less than 0.0006 oz/lb (40 mg/kg) dry matter per h at 20°C.
6. *Phytotoxicity assays.* During the composting process, a number of substances are generated that may be harmful to plant growth. Such substances are generated in higher quantities if conditions are allowed to become anaerobic. The most abundant phytotoxics tend to be volatile organic acids, especially acetic, proprionic, butyric, and valeric acids, and aromatic compounds such as phenolic acids. The presence of phytotoxicity may be assessed using cress seed emergence tests,[4] and of individual phytotoxins

using gas chromatography/mass spectrometry. Initial studies of unstable and stable composts suggest that stable composts contain very few low molecular weight components and this itself may provide a useful indicator of stabilization.[3]

7. *Cross-polarization magic angle spinning ^{13}C-nuclear magnetic resonance spectroscopy (CPMAS ^{13}C-NMR).* This has been described by Inbar et al.,[10] and involves the analysis of bulk organic matter. Spectra provide carbon fingerprints of organic matter from a range of sources and information is provided on a number of carbon-linked functional groups. Analysis of compost compared with fresh organic matter in manure showed the following changes in functional group content.

Functional group	% Change from fresh organic matter to compost
Carbonyl	+68
Carboxyl	+90
Phenolic	+64
Aromatic	+38
Alkyl	+42
Anomeric	–25
Alcohol	–28
C–0, C–N	–37

With all these methods, access to well-equipped laboratory facilities with trained staff is necessary. Accurate interpretation of data is also vital to ensure that compost processes are being correctly managed. It is likely that no single procedure for determining stabilization can be applied to all composts, so the most simple tests may be more appropriate for routine local testing. Perhaps the most amenable for local analytical laboratories is phytotoxicity testing using cress germination assays, or assessment of the change of temperature profiles in compost piles.

7.6.1 Routine Phytotoxicity Testing

This is easily carried out as described by Keeling et al.[8] Samples of moist compost (20 g) are treated as follows:

1. Shake with 50 ml distilled water for 1 h.
2. Centrifuge liquid at 3000 rpm on bench centrifuge.
3. Filter through Whatman No. 6 filter paper.
4. Place a Whatman No.1 filter paper in a 10 ml petri dish.
5. Add 3 ml of compost extract to each petri dish. For controls, add 3 ml distilled water.
6. Place 20 cress seeds (*Lepidum sativum*) in each petri dish.
7. Place lids on petri dishes and leave in laboratory on bench. Do not place dishes on top of each other.

8. After 48 h, measure the total length of each cress seeding in each dish. If a seed has not germinated, its length is taken to be O.
9. Add together the total length of shoots in controls and test plates. The germination index is given as

$$\frac{\text{Total length of shoots in test plates}}{\text{Total length of shoots in control plates}} \times 100\ (\%)$$

Ideally, a compost extract should give a germination index of 100% or higher. It must be remembered, however, that a good germination index of a compost does not necessarily mean it will have good plant-growing properties. The ultimate value of compost as a growing medium component is dependent on several factors, i.e., the blend that is to be made with other materials, and the proportion that compost will contribute. Virtually any compost can have good plant germination and growing properties if it is mixed with sufficient sand or lime.[8] A better test of phytotoxicity would involve tests of the compost in its final blended form on a set species (possibly *Helianthus annuus*). Percent germination and seedling fresh weights could then be used as a measure of phytotoxicity, comparison being made with, for example, a fertilized peat compost. A weakness of this method is that the long-term growing potential of the compost is not measured, and this may not be what might be predicted from early growth results. This was shown in experiments by Keeling et al.,[8,9] which showed that unstable compost, while initially toxic, enhanced plant growth in the long term.

7.7 COMPOST AS A SOURCE OF ORGANIC MATTER IN SOIL

The beneficial effects of organic matter in soil have been reviewed by Miele.[11] Generally speaking, if manure or some other form of organic matter is not administered to cropped soils in the quantities to counterbalance humus loss, the soil becomes impoverished and its fertility falls markedly. Compost as a source of humic material effectively encourages soil structure, aeration, permeability, etc., which are vital for crop production. Application of more easily decomposable residues also leads to the ultimate formation of humic colloids and good, crumbly soil structure. In poorly structured soil, the ability of plants to take up nutrients can fall considerably. Humus, by improving soil structure, increases the ability of soil to resist erosion from wind and water. Soil organic matter has a direct effect on soil fertility through provision of nutrients such as nitrogen, phosphorus, and sulfur, and the availability of phosphorus is improved. It also encourages the presence of beneficial soil bacteria, fungi, protozoa, and earthworms.

Humic substances have a direct effect on the growth of plants through increasing the permeability of cell membranes and promoting nutrient absorption. It has been shown that they stimulate root development.[12] Humic substances appear to possess growth-stimulating (auxin-like) properties.

An excess of unstable compost or undegraded organic matter can have harmful effects, such as reduction of nutrient availability due to immobilization (especially nitrogen), formation of phytotoxins, sustained activity of pests and diseases, and deactivation of agrochemicals such as herbicides. Farmers who wish to add organic matter to their soil prefer to compost residues, since this leads to a reduction in volume without reduction in quality and enables the destruction of pathogens and weed seeds. The desirable levels of stable humic substance in soils are >2% for sandy soils, 2.2 to 3.5% for loamy soils, and 3 to 4.5% for clayey soils, and these levels can be maintained through the regular addition of less stable organic matter. The amount of stable humic substance ultimately formed appears to be related to the lignin content of the organic matter added.

The effect of municipal solid waste composition on the fertility of clay soils has been investigated by Avnimelech et al.[13] It was shown that yields of wheat were related to compost application rates up to the highest range of application 212 yd^3/acre (400 m^3/ha), which showed a slight reduction in yield. A similar pattern of yield increase was observed with corn, but there was no yield decrease at the highest level of application. An increase in salts, organic carbon, and ammonium in the soil was observed, but a decrease in nitrate concentration. The loss of nitrate was believed to be due to anaerobic conditions and denitrification induced by the organic matter addition. It was also noted that decomposition of organic matter at low temperatures (in winter) led to the production of organic acids, while at high temperatures the products were humic substances. The positive effects of the compost application were only significant over a single growing season. It was concluded that a superficial application of compost maintained a better aerobic environment and healthier conditions for plant growth and development.

As well as replenishing the organic matter content of soils, composts provide a wide variety of soil microflora and -fauna that will aid in increasing soil fertility and increase the rate of nutrient cycling. The microflora and -fauna content of a fresh yard waste compost is outlined in Table 7.2.

Table 7.2 Microflora and Microfauna Content of Fresh Yard Waste Compost

Paramenter	Number per g (fresh compost)	Mass per g (fresh compost)	Number per g (12 week mix)	Mass per g (12 week mix)
Total biomass		495 mg/C/g		910 mg/C/g
Protozoa:				
Flagellates	2.7×10^6	31.36 mg/g	2.5×10^4	0.29 mg/g
Amoebae	6.0×10^4	5.08 mg/g	4.48×10^4	3.76 mg/g
Ciliates	0	0	1303	4.88 mg/g
Nematodes	8.8	—	54.7	—

It can be seen that a fresh compost possessed a large range of organisms that proliferate when part of a growing medium. In the example shown, the fresh compost has undergone exposure to thermophilic digestion but had not stabilized. The 12 week sample had compost incorporated into a grit/sand base and was sustaining the growth of perennial ryegrass. Initially high concentrations of amoeba and flagellates fell rapidly, while numbers of both ciliates and nematodes increased. These organisms are involved in nitrogen cycling and may improve the availability of nitrogen to plants.[14]

In addition to inoculating soil with beneficial microorganisms, compost has the potential to stimulate nitrogen fixing activity. Recent research in Australia has shown that the incorporation of high-carbon residues into top spoil significantly enhances the activity of free-living nitrogen-fixing organisms such as *Azotobacter* and *Azospirillum* and *Klebsiella*. Experiments were carried out using straw, which consists of cellulose (32 to 45%), hemicellulose (32 to 34%), and lignin (14 to 17%). The cellulose and hemicellulose are easily decomposable to simple sugars such as glucose and mannose, which nitrogen-fixing bacteria can use to provide metabolic energy for nitrogen fixation. Peak nitrogen-fixing activity was noted at 30 to 45 days following incorporation of straw.[15] If soil was fed with glucose directly, nitrogenase activity peaked at 2 days following glucose treatment.[17]

Fertile soil is a complex ecosystem with the interactions of many types of soil organisms. Of particular interest are those implicated in nitrogen cycling. In recent years, interest had centered on the interactions between bacteria and protozoa. Cutler and Bal[7] showed that in the presence of ciliate protozoa, nitrogen fixation by free living *Azotobacter* was increased significantly. Griffiths[14] showed that the presence of bacteriophagous protoza increased nitrification (by stimulating the release of NH_4+ and NO_3– from the bodies of bacteria).

From Table 7.2 it is reasonable to expect that some composts at least will be able to stimulate nitrogen fixing through the supply of energy-rich compounds and action of bacteriophagous protozoa (ciliates).

7.8 USE OF COMPOSTS IN DISEASE CONTROL

Plant pathogens are effectively destroyed by the thermophilic phase of composting.[16] In addition, when composts are added to other substrates such as soil, spores of pathogens (e.g., *Pithium* and *Phytophthora*) are prevented from developing (i.e., dehumaindomant) by general suppression. This is due to the high populations and biological activity of the native microorganisms. Biological control of "damping off" pathogens (e.g., *Rhizoctonia*) may also be achieved by adding compost, but it is important to ensure that the compost possesses a high microbial species diversity. The pathogen-suppressing action of compost can be modified by its chemical properties; for example, composted municipal sludge with a salinity of higher that 10 mS (milli Siemens) can increase root rot in beans. In general, the pathogen-suppressing qualities of composts are hard to predict. Maximal disease suppression occurs in partially but not fully stabilized compost products.

7.9 COMPOST MARKETS AND END USES

As with any product, it is important that end users understand the strengths, weaknesses, and limitations of the compost. Broadly speaking, there are three main areas of application: landscaping, agriculture, and horticulture.

7.9.1 Landscaping

Using compost as a soil conditioner is a high-quantity, low-value application of compost. Applications of organic matter will improve soil structure and improve conditions for plant growth. "Hard" landscaping schemes include road construction and motorway verges and surfacing of landfill sites. For this application, the quality of the compost may be quite poor, with the presence of some glass being permissible. Also, the degree of stabilization of the compost may be perceived as of little importance. "Soft" landscaping involves the utilization of compost in, for example, parks and gardens, playing fields, and golf courses. Here, it is important that glass content is kept to a minimum and that the product is well sanitized. Again, the level of stabilization of the material is less important, since biological degradation of phytotoxins in the soil will occur rapidly. Work performed in controlled trials on amenity ryegrass has shown that unstable compost is a valuable source of slow-release nutrients, giving consistently high growth rates for at least 6 months after the initial compost application. Where such compost is applied, the nutrient content is clearly able to reduce the need for applications of conventional fertilizers, leading to financial savings. The total amount of compost that can be applied is virtually unlimited, since ornamental rather than crop plants are to be cultivated on treated land.

7.9.2 Agriculture

Compost can be used as a source of organic matter for increase of humus content and also as a nutrient source. In this application, higher quality compost is required, ideally with minimal heavy metal contamination. Also, if poorly stabilized compost is applied, crop seeds should not be applied for several months to allow degradation of phytotoxins that might adversely affect germination. A potential problem with the agricultural market is that farmers are used to being given waste for application to land, as in the case of sewage sludge. It is important that these materials are seen as valuable resources in terms of nutrient content that can reduce the need for synthetic fertilizers. The use of compost as a nitrogen source may be valuable in reducing the nitrate leaching problem associated with fertilizer use in some areas.

Raviv et al.[18] have reviewed the value of composted organic matter in container growing media. Many of the biological properties of organic matter can be attributed to the presence of humic and fulvic acids (the final breakdown products of lignin and hemicellulose). In addition, composts appear to supply growth hormones to plants. Levels of auxin controlling cell growth and root initiation correspond to soil fertility.

7.9.3 Horticulture

Here, compost may be valuable as an alternative to peat as a growing media component. For this application, the most important criteria for the compost are hygiene and absence of glass. The physical form of glass is of greater importance than the absolute content; if there is any risk of wounding and infection, then a media containing compost will not be marketable. Numerous growing trials have been carried out in media containing refuse-derived compost, and these have frequently evidenced better long-term results than more conventional or peat-based media. Trials have also demonstrated that a low percentage of unstable compost (20% by volume) can result in better long-term growth than using conventional inorganic fertilizers [see Primula (primrose) growth trial result table, below].

Growing medium	With fertilizer	With 20% refuse-derived compost
Peat	3.67*	3.65
Peat/bark	3.56	4.18
Coir	3.03	3.22
Bark	3.13	3.59

* Growth score (maximum score = 5).

Unstable compost can achieve better growth than stable compost over the long term. A recurrent problem with using unstable compost with ornamental plants is the development of fungal growth and infestation with fruit flies. Use as a fertilizer is the highest value, but probably the lowest volume use of compost.

The following table shows the effects of applying municipal solid waste compost to crops.

Crop	Maximum compost application (kg/ha)	Maximum yield increase
Corn	200	135%
Sorghum	150	25%
Cereals*	200	173%

* With added fertilizer.

In all cases, compost application ultimately has a beneficial effect wherever it is applied.

Markets for waste-derived composts in many countries in northwestern Europe are poorly developed, while in Mediterranean countries such as Spain and Italy, soils are short of organic matter and large quantities of compost are produced and utilized. In the U.K., there is still considerable resistance to compost production, since markets are not developed and there have been a number of marketing failures, usually related to glass contamination. Separation of organic matter from other wastes at source or at civic amenity sites would enable domestic organic waste to be a better compost feedstock. The anti-peat lobby is providing valuable momentum in the development of alternative products, and a well-produced composted waste would be eminently suitable.

7.10 CONCLUSIONS

As a waste management procedure for organic matter, composting is reasonably well understood, relatively easy to achieve, and can yield valuable products in a range of potential markets. However, compost producers must be able to define specific systems for individual requirement and guarantee markets. Composting must be economically viable to justify its expansion. This requires extensive market development and further investigation of product applications. The effects of compost on soil biology and biochemistry also need to be studied in greater detail in order to determine long-term effects on soil properties.

As a waste management procedure, composting has been successfully conducted for many centuries, but the true value of the products are only recently becoming clear. Many soils are now short of organic matter, and fertility is largely dependent on a soil's humus content. High crop yields and maintenance of soil fertility can be obtained by the administration of compost supplemented with lower quantities of more conventional fertilizers. The application to land of compost, rather than raw waste, is preferable, because raw wastes can be contaminated with insects, disease, and weed seeds. High temperatures associated with controlled composting sanitize the material naturally.

Finally, the momentum for composting is being driven, in many countries, by new waste management legislation, particularly with regard to landfill. It is now recognized that problems associated with landfill management can be largely attributed to the presence of organic matter which ultimately causes noxious leachates and landfill gas. A number of counties and municipal authorities in the U.S. and elsewhere, are moving toward policies to exclude organic matter from landfill. The only waste management procedure which could cope with the organic matter other than composting is incineration. There is reason to believe that incineration will continue to be unpopular, despite guarantees by operators of its safety. Should this be the case, composting may be viewed as the best available technique for reprocessing organic wastes.

REFERENCES

1. Golueke, C. G., *Composting: A Study of the Process and Its Principles,* Rodale Press Inc., Emmaus, Penn., 1972.
2. Lynch, J. M., Utilisation of lignocellulosic wastes, *Journal of Applied Bacteriology Symposium Supplement,* 71S–83S, 1987.
3. Keeling, A. A., Mullett, J. A. J., and Paton, I. K., GC-Mars spectrometry of refuse derived composts, *Soil Biology and Biotechnology,* 26, 773, 1994.
4. Giusquiani, P. L., Patumi, M., and Businelli, M., Chemical composition of fresh and composted urban waste, *Plant and Soil,* 116, 278, 1988.
5. De Bertholdi, M., Civilini, M., and Comi, G., MSW standards in the European Community, *Biocycle,* Aug., 60, 1990.
6. De Bertholdi, M., Zucconi, F., and Civilini, M., Temperature, pathogen control and product quality, *Biocycle,* Feb., 43, 1988.
7. Cutler, D. W. and Bal, D. V., Influence of protozoa on the process of nitrogen fixation by Azotobacter chroococcum, *Annals of Applied Biology,* 13, 516, 1926.
8. Keeling, A. A., Paton, I. K., and Mullett, J. A. J., Germination and growth of plants in media containing unstable refuse-derived compost, *Soil Biology and Biochemistry,* 26, 767, 1994.
9. Keeling, A. A., Mullett, J. A. J., Paton, I. K., Bragg, N., Chambers, B. J., Harvey, P. J., and Manasse, R. S., Refuse-derived humus: a plant growth medium, in *Advances in Soil Organic Matter Research and the Impact on Agriculture and the Environment,* Wilson, W., Ed., Royal Society of Chemistry, 365, 1991.
10. Inbar, Y., Chen, Y., Hadar, Y., and Hoitink, H. A. J., New approaches to compost maturity, *Biocycle,* Dec., 64, 1990.
11. Miele, S., The role of organic matter in agronomy practice and proposals for improving the humic balance of the soil, in *Humic Substances, Effect on Soil and Plants,* Burns, T. G., et al., Eds., Reda Edizioni per l'agricoltura, 1986.
12. Schnitzer, M. and Poabst, P. A., Effects of a soil humic compound on root initiation, *Nature,* 202, 598, 1967.
13. Avnimelech, T., Cohen, A., and Shkedi, D., The effect of municipal solid waste compost on the fertility of clay soils, *Soil Technology,* 3, 275, 1990.
14. Griffiths, B. S., Enhanced nitrification in the presence of bacteriophagous protozoa, *Soil Biology and Biochemistry,* 21, 1045, 1989.
15. Roper, M. M. and Halsall, D. M., Use of products of straw decomposition by N_2-fixing (C_2H_2-reducing) populations in bacteria in three soils from wheat growing areas, *Australian Journal of Agricultural Research,* 37, 1, 1986.
16. Hoitink, H. A. J. and Grebus, M. E., Plant disease control, in *Composting Source Separated Organics,* J. G. Press, Emmaus, Penn., 1994.
17. Roper, M. M., Straw decomposition and nitrogenase activity (C_2H_2-reduction): effects of soil moisture and temperature, *Soil Biology and Biochemistry,* 17, 65, 1985.

18. Raviv, M., Chen, Y., and Inbar, Y., The use of peat and composts as container media, in *Developments in Plant and Soil Sciences. The Role of Organic Matter in Modern Agriculture,* Chen, Y. and Avnimelech, Y., Eds., Martinus Nijhoff, The Hague, Netherlands, 1986.
19. Morel, J. L., Colin, F., German, J. C., Godin, P., and Juste, C., Methods for the evaluation of the maturity of municipal refuse compost, in *Composting Agricultural and Other Wastes,* Gasser, J. K. C., Ed., Elsevier, 56, 1985.

8 MATERIALS RECOVERY FACILITIES

8.1 MATERIALS THAT MAY BE RECOVERED

Perhaps the question is not what can be recovered from waste, but rather what is worth recovering. It is technically feasible to select any of the components, either manually on picking lines, or for some fractions, mechanically. Materials recovery facilities (MRFs) are quite cheap in comparison with the more sophisticated waste processes, and regrettably they have encouraged a mood of recycling for its own sake.

In reality, municipal waste recycling is very often both a waste of time and of resources, and it is especially so if fractions are recovered in pursuit of some environmental ideal rather than for their value. In some cases the result of recovery may actually be counterproductive in an environmental sense. For example, the separation of paper-based packaging is argued by many to fit into such a category. Much of the paper recovered from that fraction is incapable of being used to manufacture new paper, and even if it were, there is no shortage of suitable wood pulp. It is suggested by the paper industry that many more trees are planted to generate wood pulp than ever would exist if all paper was (or could be) endlessly recycled.

The German "Green Dot" scheme is an excellent example of environmentalism overwhelming commercialism. The principle arose from threats by the German government to impose controls requiring retailers and manufacturers to accept back packaging materials after sale. The ambition was to recycle almost everything and make Germany the recycling champion of the world. The consequence was the establishment of a recycling company, Duales System Deutchland (DSD), and a license fee of between 1 and 20 pfennigs depending upon quantity and type of material, paid to the company by the manufacturers.

In June 1993, the magazine *Der Spiegel* reported that DSD was on the verge of bankruptcy, and needed an injection of DM 300 million to survive. It suggested that "thousands of tons" of plastic wastes that could not be processed were piling up, and that the higher product prices to support recycling were costing each consumer between DM 200 and DM 500 per annum. Considerable quantities of reclaimed plastics were being exported to the Far East, mainly Indonesia, where elderly plants of a questionable environmental standard were reprocessing them into low-value consumer items such as sandals for manual workers. DSD was accused of "dumping" the problems caused by Germany's recycling initiatives upon other less fortunate peoples.

Plastics manufacturers became less than enthusiastic as the project progressed. The managing director of the plastics company Zipperling was reported as observing that "all recycling processes (for plastics) known today gobble up more energy than the production of new plastics." Taking that point, *Der Spiegel* went on to note that

- Rinsing of German waste products was using 60 million cubic meters of drinking water a year.
- Every ton of reprocessed waste paper recovered from municipal waste was yielding half a ton of metals-contaminated sludges.
- Glass recycling was consuming large quantities of energy, and there was experience of furnace damage caused by metallic inclusions.
- Metal recycling was releasing poisonous gas emissions containing lead, cadmium, copper, sulfur, and zinc.
- Waste from the German automotive industry alone was releasing 50,000 t a year into the atmosphere.
- One ton of recovered aluminum was yielding half a ton of saline slag containing dioxins.

Not, perhaps, a good track record for an initiative that began with the highest ideals.

Almost invariably, the recycling of municipal waste actually means "downcycling." The fractions cannot be used again for their original purpose, either because it is technically impossible to do so, or because parallel legislation forbids it. Food packaging cannot again be used for food because doing so would contravene food hygiene regulations. The materials must therefore all be used for some lower-grade (and lower value) purpose, and that places an automatic limit upon the number of times anything can be recycled. It cannot be down-graded for ever, either because the characteristics which make it of value are lost, or because the value has become too low to support the recovery cost.

In Chapter 1 it was suggested that a can is in the waste stream in the first place because nobody wants it, and that simply recovering it will not necessarily confer any value upon it. That must surely be the overriding consideration governing the selection, design, and operation of an MRF. It forces a hard-headed look at all of the available fractions to discover if sufficient value can be added to at least cover the running costs, and more importantly the energy expenditure, of a facility. In the following we will examine each fraction of municipal waste from that viewpoint.

8.1.1 Paper

In most developed countries, the paper in municipal waste is a mixture of newsprint and packaging, in which newsprint predominates. Packaging materials have very little value to the paper pulping industry. They are likely to be contaminated with organic wastes, may be plasticized or contain aluminum foil, and frequently are too heavily printed for de-inking to be worthwhile or possible. Newsprint and clean white paper have a value if they are recovered before it has been in contact with organic contaminants.

Even then, the paper recycling industry in most countries has long been subject to fluctuations in demand. When a market for reclaimed paper develops, recycling organizations and particularly municipalities allow their enthusiasm to overrule their common sense, and soon everyone is busy collecting it. Oversupply then deflates the price paid, and soon the recycling facilities are operating at a loss. Worse, they are again consuming energy mostly derived from nonrenewable fuels to recover that which no one wants.

In recent times, Germany has also become a good example of the consequences of excessive zeal in paper recycling. There, government has encouraged the recovery of paper and board packaging material irrespective of the market demand. Germany has become a net exporter of reclaimed paper, and has been accused of dumping in other states at prices below its costs simply to be rid of the problem. It has succeeded in deflating prices in those countries, often to the extent that local recycling schemes have been unable to compete.

In many Western countries, paper recycling appears to be financially viable, but often it is so only because the collection is carried out by voluntary organizations that bear no labor charges. Equally, often private contractors make a profit from the establishment of transfer stations, collecting points, and transport to pulp mills, but only because they receive a subsidy from the municipality. In Great Britain, there have even been alarming reports of paper recovery schemes where most of the materials have been landfilled immediately after collection, simply because a viable market within reasonable transport distances did not exist.

It appears that often the decision to recover paper is based upon statistical evidence of market prices, without the recognition that those prices relate to high-quality scrap, not to the quality that can be recovered from municipal waste. The municipal sector can produce acceptable quality, but generally only at the expense of separate collections or paper banks, and even then only with the full cooperation of the public. Paper suitable for repulping is not usually available from comingled waste.

During the recovery process, considerable care has to be exercised over what is sorted for recycling, since paper can only be reprocessed into new paper two or three times. After that, the fiber length becomes too short. Even before that point is reached, the value declines at each stage. It is possible for clean but short-fiber paper to be reused, in the manufacturing of low-value products such as egg cartons, for example. However, it should be recalled that those products are simply packaging. They are an overhead cost to a manufacturer, not a value-added item in their own right. It would therefore be unrealistic to expect the raw materials from which they are made to have a significant value.

8.1.2 Cardboard

Clean cardboard is usually worth recovering, although often the difference between the value it realizes and the cost of recovery is marginal. Not all forms are of use, and aluminized, elasticized, or bitumen-coated boards are not generally worth considering. Most reclaimed cardboard is reprocessed to make lower-grade packaging, and as with paper, that which has been used for food products is generally excluded from the same use again. Both paper and cardboard are usually recovered in MRFs by hand sorting.

8.1.3 Ferrous Metals

The most readily identifiable ferrous metal fraction in municipal waste occurs in the form of food or beverage cans, although its use in the latter had reduced dramatically over the last 10 years. Provided that it is reasonably clean, the metal has a significant value. It has some advantage over other recycled products in that the essential characteristics of the steel are not changed by use or by recycling. However, contamination with other metals creates problems if ferrous scrap is returned for remelting without intermediate treatment.

Intermediate treatment usually involves the removal and recovery of the tin coating, which then leaves clean metal for recycling. It also presents a number of challenges to the sorting facility: first of all, it is extremely unlikely that the detinning plant will be anywhere near the steel-making plant. Loose cans have a low bulk density, and so transport costs are already high without the added cost of two-stage movement. Payloads can be increased by baling the material, but detinning works do not generally prefer to receive it in that form, since before they can process it they need to fragment the bales. Shredding for transportation is a possibility, but that involves a significant extra cost to the sorter.

Whatever means are used to prepare the materials for shipment, as recycling becomes more widespread, so the quality demanded increases. Steel manufacturers can become more selective in what they will accept and at what price. Therefore, it is now common for intermediate recyclers such as detinners to demand a recovery rate of 95% metal from every load. Lower yields attract considerably lower prices, to the extent that only a few percent reduction in metal content can reduce the value of a load to below production cost.

It is therefore essential that before a ferrous recovery line is installed, note is taken not just of the national average value of ferrous scrap, but where the outlets are. In many cases it will, regrettably, be found that even at the highest prices in the range, can scrap is simply not worth recovering.

Municipal waste does of course contain a significant amount of ferrous metal in forms other than cans. Cutlery, fasteners, white goods (kitchen appliances, etc.), redundant consumer items (bicycles, baby buggies, etc.) and toys all feature consistently. None are suitable for detinning, and they can generally go direct to steel works or to other secondary recyclers. In either case, there is again a payload penalty, and in addition there is considerable difficulty in identifying the smaller items. Even so, they do present an opportunity to avoid one expensive transport stage at least.

Baling is possible, except in the case of white goods, since provided that the ferrous scrap is reasonably free of other metals, steelworks are not overly concerned about the form in which the scrap arrives. White goods, however, generally have a better value if sold on to secondary recyclers. Many small, specialized companies profit by breaking the units down to recover the copper from electrical cabling and from electric motors, before selling the residues on to steel makers.

Considerable use is made of steel scrap by the suppliers to the construction industry, where it is used to manufacture reinforcing bar and fabric for concrete.

Magnetic separators are the most widely used form of separation, although hand picking is still sometimes used in very small or old facilities.

8.1.4 Aluminum

The almost explosive growth in the use of aluminum for beverage containers has created a market where almost none existed originally, and the material is generally in demand. It attracts high prices, but also attracts high transport costs as a result of very low bulk densities. Baling or crushing is almost always essential, otherwise a sorting facility is likely to become overwhelmed by sheer volume.

However, the value of aluminum is such that in many communities, very little of it remains in the municipal waste stream as far as the recycling facility. Most is filtered out by can banks run by collection contractors, or never enters the stream in the first place as a result of separate disposal to charitable organizations by householders.

Frequently, municipal waste at a receiving plant contains between 1% and 1.5% aluminum, and since MRFs are generally quite small by waste disposal standards (up to 200 t per day on average), the amount of aluminum actually recoverable may be no more than 1 or 2 t a day at best. Such small amounts still require expensive plant and equipment to separate them and to prepare them for transport, and sufficient storage space to accumulate full loads may be too costly to support from sales income.

Hand picking is still common in MRF flowlines, although in recent years eddy current separators have become much more widely used. The latter are, however, expensive machines which are often difficult to justify for loads 1 t or so a day. Some facility designers use mechanisms that separate aluminum by air classification or by means of chain curtains, but neither system is outstandingly efficient.

8.1.5 Other Nonferrous Metals

Municipal waste also generally contains small amounts of copper, lead, and zinc, usually in the form of defective consumer items and domestic fittings, and currency. The metals invariably occur is small particles, often attached to plastics and ceramics. They are difficult to identify and even more difficult to recover at any level of purity that would attract a manufacturer.

Eddy current separators are capable of coarse sorting, but only when the contamination from other materials is minimal. Hand sorting is very rarely used, and it is customary for any scrap in this category to be supplied to a secondary recycler, who further refines it before sale.

8.1.6 Plastics

As the German experience indicates, there is more to the recycling of plastics than the current interest in it would seem to suggest. Post-consumer plastics arise in the form of many different polymers, some that have a value, and some that do not. Market values depend greatly upon the cleanliness of the materials, where "cleanliness" means freedom from solid and liquid residues *and* from contamination by metals and by other plastics.

The building blocks of all plastics are monomers, small molecules which are one "mer" in size. Many thousands of monomers joined chemically produce polymers, long chain molecules. The vast range of commercially available polymers can be broadly divided into two classes, thermoplastics and thermosetting plastics. Thermoplastics are those which soften or even become liquid under the application of heat, while thermosetting plastics form strong chemical bonds between adjacent polymers when heated, and cannot therefore be remelted. For those reasons, the former are normally capable of being recycled into new products, while the latter are not.

Typical thermoplastics are polypropylene (PPE), polyethylene (PE), polyethylene teraphthalate (PET), polycarbonate, ABS, polyacetal, polyvinyl chloride (PVC), nylon, acrylic, and polystyrene. Of those polymers, PET, PVC, and PE are the most common in municipal waste. General purpose polystyrene is widely used in toys, light fittings, and packaging. Toughened polystyrene is used for vending cups and dairy produce packaging, together with domestic appliance linings and casings. ABS occurs mainly as plated items, giving the appearance of chromium-plated copper or brass. Unplasticized PVC is used to make bottles, pipes, pipe fittings, wall claddings, and flexible household fixtures. Plasticized PVC is used for the soles of footwear and for the insulation of electrical wiring. The polyolefides, polyethylene and polypropylene, occur as pipes, automobile plastic fittings, chairs, crates, and films and foils. PET is widely used for drink containers, and is widely recycled.

Statistically, only 4% of the annual production of crude oil is used for the manufacturing of plastics, and their replacement industry would result in an increase of 300% in the weight of the packaging component of municipal waste.

In view of the limited effect the manufacturing of plastics has upon natural resources, together with the energy and resource expenditure which recovery as clean scrap requires, it is questionable whether extensive recycling is worthwhile. When the energy cost of reprocessing is taken into account, the benefit equation moves even more toward the debit side. Were plastics to be biodegradable, there would almost be an argument for not recycling them at all. However, they are not degradable in any meaningful sense, and so as unnatural substances, they are by definition polluting. Recycling at least those which can be easily reused has some merit, even if the cost of doing so is greater overall than that of making new ones. For those reasons, PET recovery is most common, followed by films, and lastly by PVC.

Plastics recovered from municipal wastes are always likely to be contaminated, and the need for extensive cleaning plant is a limitation upon their reuse in their original markets. However, if granulated they can provide a high-calorific value fuel, and it may be that in view of the costs associated with recovery and reprocessing, that route is appropriate.

8.1.7 Textiles

Much of the textile in municipal waste is synthetic fiber, which cannot be recycled to its original use. Neither can it be used for the more traditional secondary-textile applications of wiping rags for industry, since it is largely nonabsorbent. Wool can be reprocessed, but the market for it is somewhat limited in general. There is a Third World outlet for clothing in acceptable condition and washed, but the costs of recovery, examination and approval, and cleaning are likely to absorb all of any income from sales.

Mechanical recovery of mixed textiles is possible and not difficult, but that is only a first step toward recycling. The materials still have to be hand picked. No technology exists to distinguish between synthetic fiber, wool, cotton, and synthetic/natural blends. Distinguishing polyester cotton from natural cotton, for example, is difficult even for human operators.

8.1.8 Glass

Reprocessors rarely accept glass in the form of mixed-color cullet. Frequently they require it separated into green, brown, and clear colors, each fraction having a high standard of purity. The presence of any ceramics will result in the rejection of a complete load.

Color separation is reasonably straightforward where glass is collected separately at bottle banks, etc., but difficult if it is delivered mixed to a recycling plant. Optical separators have been researched, and have upon occasions proved themselves capable of a reasonable level of color identification. Large-scale processes based upon optical technology have not yet developed, mainly for reasons of cost and the difficulty of scaling up. Hand picking is frequently resorted to, but should be treated with some caution for safety reasons.

In such cases, it is usually better to pick complete bottles and containers, and to allow fragments to pass through to the reject line. Wastage rates are likely to be high, particularly if there is much mechanical handling before the picking stage, but that is better than risking injury or even serious infection from skin penetration by broken shards.

There are alternative uses for mixed cullet, mainly in the manufacturing of insulation materials. However, in many countries where there is an established glass-making industry, the needs of insulation manufacturers are satisfied by production waste rather than by salvaged glass.

8.1.9 Organics and Putrescibles

In municipal waste, the organic and putrescible fractions are usually intimately mixed, and consist of vegetable matter and food wastes. Hand separation is impractical, and is likely to constitute a significant health hazard if attempted. Mechanical separation from other materials is possible (see Chapter 6 on composting systems). Recycling is impossible, but reprocessing into soil conditioners is widespread.

8.2 MATERIALS RECOVERY TECHNOLOGY

There are only a limited number of ways in which material may be recycled from the waste stream, and the most obvious is to avoid their entering the stream in the first place. Most MRF-based schemes require some presorting at the point of arising, followed by hand or mechanical sorting at the central facility. The importance of originator sorting cannot be overstressed if clean, good quality salvage is to be recovered at a realistic cost. The ways in which that can be achieved are more varied and often more costly than the MRF designs, and so they merit a detailed examination of both the philosophy and the techniques.

8.2.1 Household Separation

Items for recycling can be separated by the householder in the kitchen, in two or more containers outside the house, at a street corner depository, or at a central household waste site or retail outlet. In deciding which principle to use, it is necessary first to consider the psychology of the public.

The majority of the residential group, if questioned, will confirm that they are anxious to support recycling initiatives, that they do so because they are concerned for the environment, and that they do themselves actively separate their waste for recycling. If what they avow is true, then it is somewhat surprising that source-separation schemes do not enjoy a much higher success rate. The reason appears to be that privately many householders find source separation inconvenient, occasionally untidy, and properly the responsibility of the municipality rather than themselves.

That reaction is less apparent in relatively affluent areas than it is in poorer ones. Poorer people may live in premises that do not lend themselves to waste segregation. Multiple-occupancy blocks of flats, for example, may only have drop chutes from each floor down to a single container. For the occupants, source separation would involve carrying several containers of waste materials down several floors, followed by a long walk to the nearest depository. Also, there is some evidence that in poorer areas, lower educational standards play a part. Environmental concern appears to be more a matter of education than of any natural human reaction.

For the residents of affluent areas, source separation is much easier. Generally they have the space in which to locate two or more containers, and they have the will to do so. Their better education makes them more aware of the pressure upon the environment, and they derive some gratification from being seen to be frugal. They are more inclined to load materials for recycling into the family car and to take them to the local depository, if the municipality does not collect them. They can easily be persuaded to wash out cans and bottles before depositing them.

From those concepts it follows that the logic train that determines the form of an MRF should begin with an analysis of the predominant residential type. There is little likelihood of success for a scheme requiring extensive source separation in an inner-city area occupied by tenement blocks. The people are unlikely to have either the facility or the will to do what is needed. In such an area, the best that can be hoped for from a voluntary scheme is some limited separation of bottles and cans at street corner depositories. Enforcement, and there is in many countries a move toward that, can increase the recovery rate, but only at the expense of some public resentment. In addition, an enforcement regime has to be policed, or it has no lasting effect. Policing adds yet more cost to a business that is likely to be only marginally profitable at best.

In an area of more affluence, much more segregation of waste is possible, and with the support of an ongoing public relations initiative, it is likely to be much more successful. The residents can usually be persuaded of the merits of recycling, and can be relied upon to suitably prepare their waste. They will happily wash out food cans and beverage bottles, will keep newspapers and magazines separate from packaging, and PET containers separate from PVC. But whatever the extent of their waste segregation, the municipality must provide them with the means of doing so.

Segregation can be achieved in the kitchen, in the yard, or at the street corner, but as always, where and how it is done affects the efficiency of the recovery system. If the best performance is to be achieved, then the system must be simple and convenient to operate. The further away the collection point is from the source, the lower will be the recovery and the degree of separation. For that reason, a provision for keeping wastes separate in the kitchen is the ideal. A group of individual containers in the yard is likely to be somewhat less effective, because it involves the householder in making several trips where in the absence of a recycling scheme, one trip would be sufficient.

The equipment that the municipality must supply differs substantially between an in-kitchen and a yard scheme. In the kitchen, the provision of two or more different colored bags may be sufficient. In the yard, two or more containers are necessary.

Container design is becoming something of a growth industry in the developed world, as more manufacturers attempt to discover one that meets the needs of the householder, the collection undertaker, and the recycling facility. There are many views upon the right approach, but among the more common are split bins, where the container is divided into two by a partition, and a two-bin system where one is reserved for recyclables and the other for general waste. The first requires the collector to use a vehicle that has two or more body compartments. The second permits separate collections using a more conventional vehicle.

No system is perfect. All have fundamental disadvantages. Waste segregation in the kitchen is convenient for the householder, but it does result in some clutter in an area where that is frowned upon. The use of different colored bags allows recyclables to be readily identified at the recycling facility, and permits the use of a single bin per household and a standard collection vehicle. It prevents the use of a compaction vehicle if cross-contamination is to be avoided, and so it limits the payload that the collection vehicle can achieve.

Systems based upon yard segregation involve a greater capital cost, since the provision of suitable bins is expensive. Split containers are likely to be less effective than separate ones, mainly because of the possibility of the householders accidentally placing the recyclables in the wrong compartment. In such an event it is probable that they will simply leave them there. Reaching down into the bottom of a dirty refuse container to recover the error is distasteful.

Whichever principle of source separation is applied, waste collection costs will inevitably increase and the service will become less efficient. Colored bags provided for kitchen use incur a long-term revenue cost, and possibly a capital cost in changing the collection vehicle fleet to a type less likely to damage the bags. Two-bin or split-bin recycling requires either vehicles of a quite different body design to those conventionally used for mixed waste collections, or the establishment of separate collection rounds for the fractions. Clearance of a district will take longer, simply because the crews now have to take two or more containers from every household rather than one. Extending the number of collections, or increasing the time each takes, may require more vehicles in the fleet.

At the reception facility, discharging the vehicles takes longer as well. If the discharge time for mixed waste from a conventional vehicle at a landfill site is taken as a basis, then discharging loads of recyclables will add one unit of time for every category. A two-part load will take twice as long, a three-part load three times as long, and so on.

It is not the purpose of this book to suggest which of the source separation options is best, since the choice is a complex one depending upon many local factors. Rather, the intention has been to outline some of the factors that may affect that choice. All of the options are expensive in terms of capital and revenue, and it is unfortunate that if the extra costs are properly accounted for, the income from recycled materials rarely matches them. Protection of the environment and the reuse of resources is costly rather than profitable.

8.2.2 Sorting Designs

When the recyclables arrive at the MRF, the design has to accommodate their unloading and further separation. Ideally, the materials would be received in a degree of separation such that the only processing required would be baling for sale. That, however, would require a level of source separation that is little more than an environmentalists dream, and a degree of public dedication that is rarely encountered.

In discharging the collection vehicle, the design must accommodate two requirements as far as possible. It should provide for the fastest possible turnaround, so that the vehicle can be returned to productive use quickly. It should prevent cross-contamination by one category spilling over into the hopper or bay containing the next. Tipping on an unobstructed floor, of the design recommended for large-scale reprocessing plants, is unlikely to satisfy the second requirement, although it would certainly accommodate the first. Tipping into individual bays or hoppers would prevent cross-contamination, but would extend the discharge time quite considerably. Which principle is used depends very much upon the design of the vehicle — another reason for integrating the source-separation scheme with the choice of vehicle and the nature of the recovery facility. In that context, there are basically four choices.

If separate collections are established, then the vehicle will only ever visit the plant when it is carrying recyclables, and therefore mixing ferrous and nonferrous cans, plastics, and paper is worth considering. The ferrous and nonferrous materials can then be recovered by magnetic and eddy-current separation, and the paper and card can be hand-picked afterward. Tipping onto an open floor followed by shovel loading onto a feeding conveyor is perfectly appropriate.

If the vehicles are to collect both waste for landfilling and for recycling, from households issued with split or twin-bin facilities for example, then they will have bodies divided into at least two compartments. In source separation schemes of that type it is customary to collect "dry recyclables" (cans, plastics, papers, and magazines) in one compartment and "wet wastes" (organics and putrescibles) in the other. Tipping onto a level floor is again possible, provided that the vehicle can do so without shedding any of the contents destined for landfill. Alternatively, discharging into a hopper is a practical approach, particularly since it overcomes the need for a shovel loader.

Where multiple source segregation is applied, then floor tipping is impractical. There is no point in requiring the householder to keep each fraction separate if they are going to be mixed together again at the recovery facility. Discharging into some form of hopper arrangement is unavoidable, as is the extension of the time the vehicle is out of service.

The final possibility, and certainly one of the most efficient, is supported by the colored bag, in-kitchen principle. In that system householders are issued with packs of kitchen trash bags in two distinctly different colors. Green bags for dry recyclables and white for organics and putrescibles is an appropriate choice. Both colors of bags are deposited in a single bin outdoors, and so the householder does not need to consider if the right compartment has been filled. The municipality only needs to issue one bin, or even none at all if the provision was previously left to the house owner, and the collection vehicles can remain of the original design and can continue on their customary collection rounds. At the recycling facility, the mixed bag load is discharged onto a reception conveyor, and elevated to a presorting station. There the bags are reduced to a single stream where their colors can be identified either manually or by optical devices. Selected bags can then be transferred to the recycling lines again either manually or by mechanisms triggered by the optical detectors.

8.3 MRF MACHINERY DESIGNS

Materials recovery facilities are now constructed to such a wide range of designs that it would be impractical to examine all of them here. Rather, the purpose of this section can only be to consider the way in which each fraction can be sorted, and then to describe flowlines which achieve that. Section 8.1 reviewed materials that can be recovered conveniently, while Section 8.2 discussed how they may arrive at the MRF. It is now necessary to combine the principles established into a complete process, beginning with the reception facility.

8.3.1 Reception Designs

8.3.1.1 Open Floor Reception

Where collection vehicles are to discharge onto a level floor, the first problem to be addressed is that of how to keep the materials in one place long enough to be moved into the sorting lines by shovel loader. Almost all of the items likely to be found in the dry recyclables mix are of very low bulk density, and mainly cylindrical in shape. It is almost impossible to pile them to any useful height, and so large areas of floor space could be needed to accommodate them. Rather than one single apron, it is usually better to divide the floor into a number of bays separated by low walls. That way, cans etc. are restrained from rolling about uncontrollably. Figure 8.1 shows a floor layout that is appropriate for the purpose, and which, by the installation of a plate-feeder conveyor across the ends of the bays, allows a shovel loader to push material rather than having to carry it.

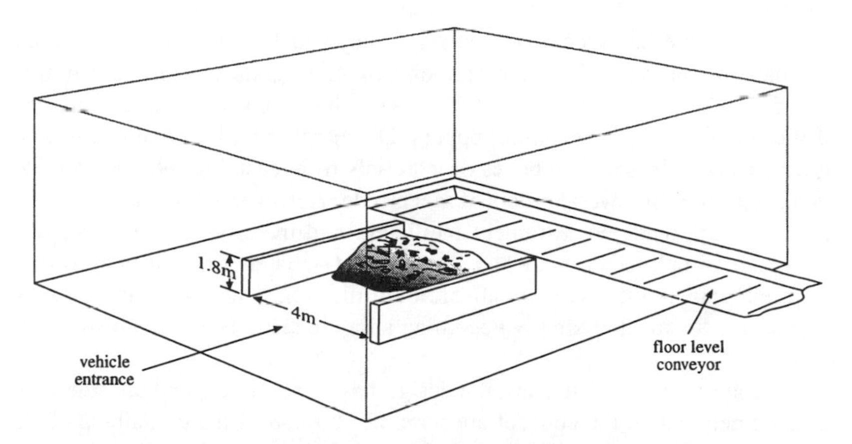

Figure 8.1 MRF reception floor.

The design has further points in its favor. The presence of the low dividing walls allows each of the bays so formed to serve a separate purpose. Cross-contamination is substantially avoided, and the conveyor can be used to feed a succession of different materials provided that deliveries are infrequent enough to permit it.

The low bulk densities of the materials received is an insurmountable problem in any MRF, and is worst in one which uses the tipping floor principle. Loose steel cans are likely to have an average bulk density of about 2 lb/ft^3 (45 kg/m^3), aluminum cans about 1.6 lb/ft^3 (25 kg/m^3), and glass and plastic bottles 18.7 and 1.2 lb/ft^3 (300 kg/m^3 and 20 kg/m^3), respectively. All of those figures are the subject of very considerable variation, since they depend upon how many of the individual items have been crushed or broken. However, they do suggest that 1 t of steel cans for example will occupy a volume of 777 ft^3 (22 m^3). A ton of PET containers represents the worst possible scenario from a space point of view by occupying 1760 ft^3 (50 m^3).

All of these materials possess a low angle of repose when tipped. That is, they do not form high piles readily. Figure 8.1 shows partition wall 6 ft (1.8 m) high, although even with a shovel loader working it is unlikely that such a height of pile could be reached consistently.

8.3.1.2 Multiple Reception Hoppers

Some vehicle designs lend themselves to discharging into hoppers rather than onto the floor. Their body designs are usually a complete departure from those of the traditional refuse truck, and are built with multiple-product recycling in mind. In one, a long wheelbase chassis is fitted with an open-sided body, in which a number of containers are sited. The vehicle has the capacity to collect and carry separately ferrous metals, nonferrous metals, mixed plastics, textiles, paper, and cardboard. It does not collect organic wastes; those are left for a conventional collection truck.

The vehicle has a comparatively low payload because of the low bulk densities of the materials it collects, and in its original design form it was intended to be unloaded by fork lift trucks, which removed and tipped each of the containers into reception hoppers. Discharging took a total of 30 min, and recovering the separate boxes of materials from each household extended collection times by several orders of magnitude. However, while the collection logistics were rendered extremely inefficient and therefore costly, the MRF design was simplified dramatically. The amount of hand or mechanical sorting required was reduced to an absolute minimum, and the plant closely approached the ideal of simply becoming a baling station for reclaimed materials.

In spite of its inefficiencies, the design has some merits, and considerable improvements upon the concept are possible. Instead of the containers being demountable, they could be fitted to the chassis in such a way that they can be tipped to the sides by a hydraulic mechanism powered by the power take-off from the truck engine. A series of hoppers, served by underfloor conveyors, and positioned such that they are in alignment with the vehicle containers, could then be installed in two rows with access for the vehicle between them. Wheel guides and an end position indicator similar to those used in automobile wash plants could be used to locate the vehicle in the correct position relative to the hoppers, allowing the operator to unload without ever leaving the truck cab. A similar design proposed for a plant in Great Britain in 1992 reduced the unloading time from 30 min to 3. Figure 8.2 shows the general design concept.

The drawing shows underfloor conveyors for convenience of illustration but, as discussed in Chapter 3 of this book, there are a number of good reasons for avoiding them in waste handling plants. It would, however, be perfectly practical to construct the tipping room with a raised floor level in relationship to the recycling plant, so that the conveyors are simply on a lower level rather than inside trenches.

The design also shows the conveyors as being offset from the centers of the hoppers, so that each hopper can be served by its own machine. The facility then exists for the hoppers to be designated for specific materials, in this case six, and for that designation to be changed whenever circumstances demand it. Offsetting the conveyors does not affect the emptying of the hoppers pro-

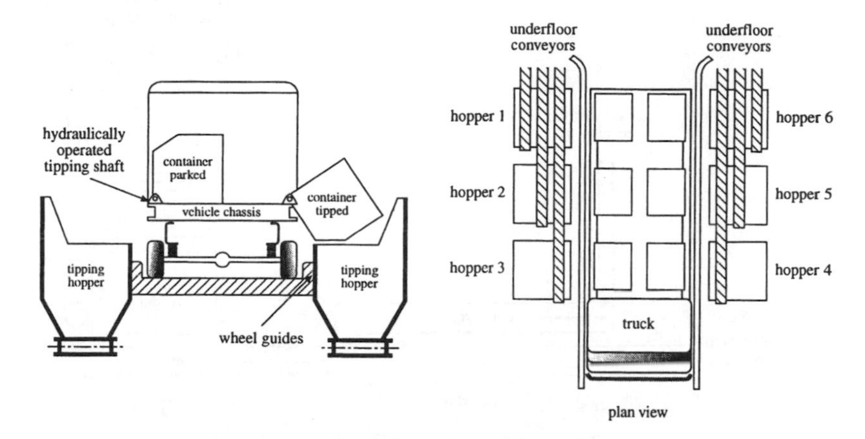

Figure 8.2 Multiple hopper discharging.

vided that they are designed carefully, with bottom angles that discourage blockages and bridging. The very low angles of repose of the dry recyclables, in this case, works to the advantage of the designer. Provided that the hopper bottoms are not at less than 60° to the horizontal, bridging is unlikely.

8.3.1.4 Single Reception Hoppers

Where a colored bag collection system is applied, a single reception hopper is sufficient. Here the hopper is simply the beginning of a reception system that sorts the bags into their respective colors and delivers them to the appropriate recycling line. The bags containing dry recyclables are directed to sorting, while those containing wet wastes are directed to composting or to landfill.

Figure 8.3 shows the general layout of the system, in which the mixed-bag load is discharged upon a plate-feeder conveyor installed in the base of an open-ended hopper. The plate-feeder elevates the bags to an inclined vibrating feeder, the purpose of which is to reduce the feed to one bag in depth. The vibrator discharges onto a belt conveyor (conveyor 1 in the illustration) which passes its end at right angles, and which has side plating such that it can only accept one bag across its width.

The bags having been reduced to a single line, they are introduced to conveyor 2, which is designed to run at $1^1/_2$ to 2 times the speed of conveyor 1. This increases the space between individual bags, and eliminates any possibility of one bag becoming lodged on top of another. Conveyor 2 passes by a presorting station, where conveyor 3 is installed in parallel to it but at a lower level. At that point, bags of the selected color are simply pushed from conveyor 2 onto conveyor 3 either by hand or by actuators controlled by optical sensors.

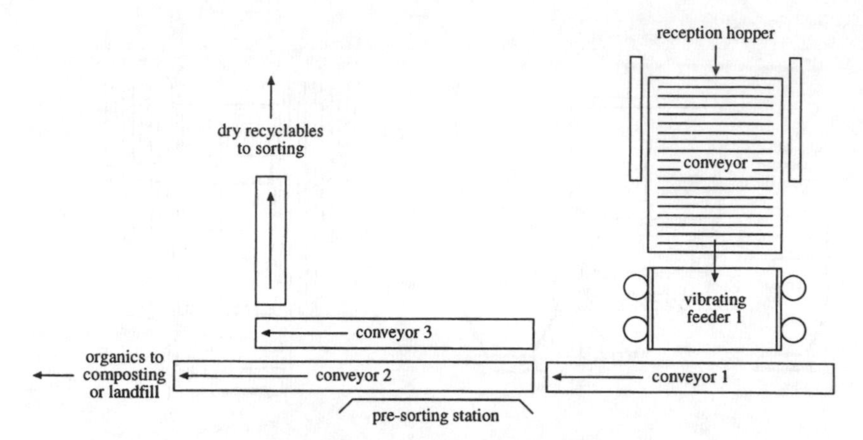

Figure 8.3 Colored bag reception.

Where optical sensors are used, the most readily available type are closed circuit, color television cameras linked to a computer. The role of the computer is to detect the desired color frequency in the signal from the camera, and to then initiate the mechanical operator.

In this design, note that the bags containing organic wastes stay upon conveyor 2. It is those containing the dry recyclables which are diverted onto conveyor 3. This ensures the freedom from contamination of the feedstock to the recycling plant. The dry recyclables bags have to be deliberately selected, and any failure to do so simply passes them through to reject. The system can even cope with bags of the wrong colors, by only selecting those of the one specific color.

8.3.2 Materials Sorting

Virtually all MRFs are deigned for a mixture of hand and mechanical sorting. Human ingenuity has yet to discover reliable methods of identifying the various types of paper mechanically, for example, and so the eyes and brain of a human operator are still essential. In the following the intention is to examine the techniques for hand and mechanical sorting, and to consider the design parameters for each.

8.3.2.1 Hand Sorting

Since the human operator is a fundamental part of the machine in this case, it is first necessary to establish his particular requirements and capabilities.

The limitations of the operator lie in his reaction time to an event, the period of time for which he can work without becoming inefficient through tiredness, and the physical constraints of the distance over which he can reach. Together with those reach restrictions, there are limitations upon the ability of the human hand to grasp items, and upon the types of materials that are safe to grasp.

Reaction time is not simply a case of how long it takes an operator to become aware of an event. In waste sorting, it is more a matter of how quickly he can identify an object or a material that he should recover. That time period may be very short if his task is to recover PET containers from mixed plastic waste, for example. PET containers are readily identified by their shape and by their very high transparency. That is not, however, the case where he is perhaps recovering cardboard from mixed paper and board. Cardboard is not necessarily identified by either color or shape, and if the mixture has originated from commercial or office premises, much of it may not be immediately recognizable.

The less easily distinguished a material is, the slower the picking feed rate must be. The operator needs time to examine items, and to discard those that do not conform to the characteristics of the ones he should recover. Table 8.1 suggests a range of reaction times for various materials mixes. It suggests that, for example, if a mix of cans, plastics, paper, and card, from which the operator is to select paper, passes his range of vision and reach in less than 3 seconds, then he will miss a substantial proportion of it.

Table 8.1 Suggested Design Reaction Times

Material mix[2]	To recover	Reaction time (s)
Fe/NFe/Pl/PACD	PET	1
Fe/NFe/Pl/PACD	Paper	3
Fe/NFe/Pl/PACD	Card	3
Fe/NFe/Pl/PACD	PVC/UPVC	2
PACD	Paper	2
PACD	Card	2
Mixed glass	Clear	2
Mixed glass	Brown/green	3
Mixed plastics	PET	1
Mixed plastics	PVC/UPVC	3
Mixed plastics	Film	2
Pl/PACD/Tx	Tx	1
Mixed Tx	Specific type	20

[a] Materials types: ferrous metals (Fe), nonferrous metals (NFe), plastics (Pl), paper and card (PACD), textiles (Tx).

The operators range of vision is another important aspect of his reaction time, and it is considerably less that many designs appear to assume. Most of the time his eyes will be focused upon that which he is recovering, not that which is coming toward him. His visual perception is likely to be limited to about 1 ft 6 in. (500 mm) each side of his body, and so it is that distance which limits the speed of the line. The piece of paper should be traveling at less than 3 ft (1 m) in 3 seconds for a reliable recovery rate, which suggests a feed rate of 65 ft/min (20 m/min).

In the event that an acceptable level of productivity cannot be maintained at slower feed rate, as is often the case, then it is necessary to increase the number of operators. In theory, what one misses will be recognized by another. That, however, leads to increased labor costs and larger machinery, so while reaction time data is not arbitrary, the design should at least take it into account wherever possible.

Operator tiredness is also a difficult parameter to design for. Everyone is different, and what makes one employee tire rapidly may have considerably less effect upon another. The point at which he becomes inefficient is determined largely by whether he is standing or sitting, how far he has to reach, how much he must move to deposit recovered items, and how rapidly materials are being delivered to him for sorting. Stress increases tiredness, and at sorting stations it is most often the result of a too-high delivery rate. The operator is placed under some psychological pressure to keep up with the machine.

Because everyone is different, it is impossible to establish working periods that are appropriate in every case. The best that can be achieved is to minimize the need to reach, to limit the movement necessary to deposit items, and to give the operator a chance to move around occasionally. A standing operator should not be required to reach further than about 24 in. (600 mm), and a seated one a maximum of 20 in. (500 mm).

Once the operator has grasped an item, he has to place it in the appropriate stream for recovery. Conventionally that is achieved by either dropping it into a bin or chute beside him, or tossing it onto a separate section of the picking line. The former requires him to be continually reaching out and withdrawing, and in that event a further 1 second should be added to the reaction times to indicate a total recovery time per item. Personnel who are required to operate in that way are likely to become inefficient after 1 h of working. Where the items are to be placed onto a separate section of the line, the amount of arm movement is considerably reduced, but that is more likely to minimize the overall reaction time rather than to extend the period before tiredness takes effect. It will achieve greater productivity, but only over the same length of time.

Even where an operator can conveniently reach an item, and can see it coming clearly, its size, shape, weight, or physical nature may make it difficult to pick up. In some cases, as with sharp-edged cans or broken glass bottles, it may even present some dangers. Paper and cardboard sheets are among the more difficult materials, since they can only be picked up by their edges. Plastic films can be gripped relatively easily, but may trap or be trapped by other items. Textiles become entangled easily, and the recovery of one type may involve pulling it loose from others. Again, there can be no accurate measure of these factors, since the ways in which any of the materials could be presented to the operator are almost endless. All that the designer can do is to take them into account and to make some allowance for them when determining delivery speeds.

In Table 8.2 an attempt has been made to define sorting rates for different materials selected by a single operator. The sheer number of variables encountered in trying to establish such data makes it inevitable that some plants will point to substantially higher sorting rates, while others may achieve lower. The purpose of the table, however, is to show that hand sorting is not always a matter of picking the material one may wish to recover. In fact, it is often more practical to pick the *reject* material until only that required for salvage is left. For example, the table suggests that picking sheets of clean white paper from a mixture of paper and card may yield a recovery rate as low as 25 lb/h (12 kg/h). If, however, the cardboard were to be selected instead, then it could be removed from the paper stream at a rate of 220 lb/h (100 kg/h). Furthermore, if the cardboard represented 10% by weight of a mix of office waste, then by removing it at 220 lb/h (100 kg/h) would yield paper at over 1 t/h.

Table 8.2 Suggested Sorting Rates Per Operator

Material	Apparent density lb/ft³ (kg/m³)	Sorting rate lb/h (kg/h) per person
PET containers	1.4 (23)	365 (160)
Paper	5.0 (80)	27 (12)
Card	5.6 (90)	230 (100)
PVC/UPVC	1.6 (25)	548 (240)
Glass	21.8 (350)	1140 (500)
Film	1.2 (20)	46 (20)
Textile	3.7 (60)	411 (180)
Ferrous metals	2.8 (45)	
Nonferrous metals	1.6 (25)	

By far the most common hand sorting designs make use of belt conveyors, either with bins or drop chutes beside them, or with a longitudinal division plate. In the former, the operators pick items and place them in the bins or chutes, while in the latter the belt carries both mixed waste and sorted items, discharging each into two separate chutes at the end of its travel. Of the two designs, that with the longitudinal division is the more efficient, because it requires fewer operator movements to recover materials.

Irrespective of which design is selected, and however imprecise the derivation of the recovery rates may be, some mathematical model of the system has to be created before any hardware can be designed. The following example illustrates how that may be achieved.

An MRF recovery line is to sort mixed dry recyclables at a rate of 10 t/h. The mix consists, in proportions by weight, of 30% ferrous cans, 10% nonferrous cans, 20% dense plastic, 5% PET, 25% paper, and 10% cardboard.

The first task is to consider in what order the materials might be picked, and that is more a matter of common sense than established rules. There is little point in creating a design in which ferrous and nonferrous cans are hand sorted. Magnetic and eddy-current separators will perform that function with much greater efficiency than any human operator. However, either form of separation is unlikely to work well as long as there is paper and card in the mixture. At least some of the metals will be buried under sheets of both, and thereby be prevented from leaving the belt.

PET is readily identifiable, and its cleanliness would be improved by removing it as early as possible. Dense plastic containers would then be easier to identify, and so they could be removed next, followed by card to expose the paper. The residues of ferrous and nonferrous metals could then be subjected to magnetic separation, which would isolate the nonferrous metal. Following that logic, a suitable order of separation then becomes PET, dense plastic, card, paper, ferrous metal, nonferrous metal.

The conveyor that will receive the mixed load is a volumetric machine, and while a mass balance already exists, it is now necessary to construct a volume balance. That will necessarily be inaccurate because the shapes and sizes of the individual items in each category are not entirely consistent. There are fewer large cans in a cubic meter than there are small ones. The air space contained in the volume is therefore greater and the bulk density is lower. The best that can be achieved is an estimate of the *average* bulk density for each fraction.

Table 8.2 suggests a range of average bulk densities that is sufficiently approximate for the immediate purpose. Using the data it contains in the formula

$$\text{Volume} = \frac{\text{Weight}}{\text{Bulk density}} \tag{1}$$

results in throughput volumes per h of 26.4 yd^3 of PET, 92.6 yd^3 of dense plastic, 13.2 yd^3 of card, 37 yd^3 of paper, 79.3 yd^3 of ferrous metals, and 46.3 yd^3 of nonferrous, all in order of sorting. The total volume to be handled is therefore 294.8 yd^3 (251.3 m^3), and so the average bulk density of the load is

$$\begin{aligned} \text{Bulk density} &= \frac{\text{Weight}}{\text{Volume}} \\ &= \frac{20{,}000 \text{ lb}}{294.8 \text{ yd}^3} \qquad (2) \\ &= 67.84 \text{ lb/yd}^3 \text{ or } 2.51 \text{ lb/ft}^3 \text{ } (40.25 \text{ kg/m}^3) \end{aligned}$$

The first conveyor in the system handles all of the load, and so it must be capable of dealing with 295 yd^3 (251 m^3/h) at a speed that permits operators to recover the PET containers first. The resulting belt speed is calculated from

$$\text{Speed} = \frac{\text{Volume per hour}}{\text{Belt width} \times \text{bed depth} \times 3600} \text{yd/sec (m/sec)} \qquad (3)$$

A realistic assumption of the depth of material on the belt (the bed depth) is 4 in. (100 mm), since that is the average diameter of a PET container, and only slightly more than that of a can. The formula therefore reveals that a 20 in. (500 mm) wide belt would have to travel at approximately 1.48 yd/s (1.4 m/s) to handle the required capacity. Every meter of belt would be passing the operator at a speed greater than his reaction time, and while that does not mean that he would never recover anything, it does suggest that his efficiency would be extremely low. A 3 ft (1 m) wide belt, however, would travel at only 0.77 yd/s (0.7 m/s), which is far more acceptable since it means that every yard would take 1.4 seconds to pass the operator.

The average weight of a PET bottle is about 1.7 oz (46 g), and so a feed rate of 1100 lb/h (500 kg/h) suggests that containers would be passing down the conveyor at a rate of 10,870 per h, or approximately 3 per second.

If the belt is traveling at 0.77 yd/s, then on every 0.77 yd (0.7 m) of belt there will be on average three PET containers, passing in 1 second. Since the individual's reaction time is assessed as 1 second, then the task of recovering the PET will need three operators. That is *nearly* consistent with the theoretical recovery rate of 365 lb/m (160 kg/h) per operator shown in Table 8.2, but the belt width has to be 3 ft (1 m), and the operator's reach is only 2 ft (600 mm). The task will need personnel standing at both sides of the belt, and so to ensure full coverage, four will be needed. If operators are to work from both sides of the conveyor, then a longitudinal division plate design cannot be used. Individual bins or drop chutes will be necessary instead. Allowing for a space of 3 ft (1 m) per operator plus 3 ft for each chute, the overall length of the first picking station will need to be 12 ft (4 m) (two operators times 2 m on each side).

Once the PET has been recovered, the volumetric duty of the conveyor reduces to 295 – 26 yd^3/h, or 269 yd^3/h (204 m^3/h). The weight carried reduces to 10 t – 0.55 t, or 9.45 t (8590 kg), and so the average bulk density increases to 2.6 lb/ft^3 (41.7 kg/m^3). A 3 ft wide conveyor to handle that capacity would have to travel at 0.65 yd/s (0.6 m/s).

That speed is not significantly different from that of the first conveyor, and so perhaps the same conveyor could be used throughout. At its design speed of 0.77 yd/s (0.7 m/s), and assuming the average weight of dense plastic containers to be about 1.8 oz (50 g), every 0.75 yd (0.7 m) of belt would contain some 11 containers, passing each operator in 1 second. If his reaction time is 2 seconds, as shown in Table 8.1, then in that time 22 containers would pass him, of which he would only recover 1. The recovery of PVC/UPVC would require 22 operators, all of whom would be picking considerably less than they should be able to, simply because the belt speed is far too high.

The difficulty cannot be overcome by simply reducing the belt speed. Without even carrying out any calculations, it is apparent that the volume reduction resulting from the removal of PET does not slow the belt sufficiently. Increasing the belt width would place some of the materials beyond the reach of the operators, and so the implication is that some mechanical form of initial separation is necessary.

A drum screen at the front end of the flowline, fitted with 6 in. (150 mm) perforation screen plates, would separate most of the paper and card from the cans, dense plastics, and PET. The resulting volume reduction upon the first picking conveyor would then be up to 50 yd^3/h (38 m^3/h) and the paper and card would be diverted onto a second line. Without the paper and card blinding them, ferrous metals could now be recovered magnetically, reducing the volume still further by 79.3 yd^3 (60 m^3). The corresponding belt speed for the residuals becomes 0.42 yd/s (0.4 m/s).

That speed would improve the recovery of PET to the extent of its needing only two operators. This is in apparent contradiction of Table 8.1, but that table assumes that material is delivered to the operator as fast as his reaction time can accommodate it. In this new situation, that is no longer so. The delivery rate is much slower.

When the PET is removed, the same belt would be carrying only dense plastics and nonferrous metals, at a volume of 139 yd^3/h (105.6 m^3/h) at a speed of 0.42 yd/s (0.4 m/s). Freed from the contamination of paper and card, the material could now be eddy-current separated to recover the nonferrous metals, leaving the dense plastics as a residue. Hand picking of the plastics would be unnecessary.

As a result of the preliminary evaluation, the design has evolved into a front-end screen and two streams. The order of separation has changed to ferrous metals, then PET, followed by nonferrous metals and finally by dense plastics in one stream. Paper and card, together with some inevitable contamination of metals and plastics, can now be sorted in parallel on the second stream.

It is impossible to judge what the metals and plastics contamination of the paper stream is likely to be, but with a 6 in. (150 mm) drum screen it should be minimal. Therefore, a preliminary hand sorting for those materials should be sufficient. The conveyor handling the paper and card will deal with approximately 53 yd^3 (40.3 m^3) of material per h (allowing for 3 yd^3 [2.3 m^3] of contaminants), and if it is also chosen to be 3 ft (1 m) wide, its speed will be 0.13 yd/s (0.13 m/s). It will take nearly 8 seconds for a 1-yd length to pass through the field of view of the operator, and so two operators could remove at least 8 items within their reaction time. That is intuitively sufficient to deal with levels of contamination that cannot be assessed, but that cannot be substantial.

Now, finally, 50.2 yd^3 (38.2 m^3) of mixed paper and card, of which 37 yd^3/h (28 m^3/h) is paper and 13.2 yd^3/h (10 m^3/h) is card, remains upon the belt. Again, 1 yd length of the belt will pass the operator in 8 seconds, in which time 0.09 yd^3 (0.07 m^3) of paper and 0.04 yd^3 (0.03 m^3) of card will pass. In this case, the designer cannot make any assessment of the shape or dimensions of any particular item, since they are likely to be too random. He can, however, form a judgment of the number of operators needed by referring to Table 8.2, which suggests that on average, one operator could recover 230 lb (100 kg) of card per h. The weight of card being handled is 1 t/h, and so the inference is that ten operators will be needed. However, the belt speed is such that, within his reaction time, each can make four selections per meter length of belt. The recovery rate is likely to be somewhat better than the average in such a case, and so a practical design would provide space for ten operators, five per belt side, with the intention that normally perhaps only four would be working. Once the card has been removed, only paper remains and no further sorting is necessary.

In this example, the word "approximately" is somewhat overused, which demonstrates the imprecise nature of hand-sorting designs. As with many of the situations occuring in waste processing, the best that can be achieved in the initial modeling is to divide the flowline up into as many small elements as possible, and to then use mathematics to *assess* what will be required of the design.

8.3.2.2 Mechanical Sorting

In the previous example, the initial model demonstrated that while hand sorting is often unavoidable, mechanical sorting is equally so if excessive labor demands are to be avoided. Within the current state of technology, mechanical sorting is available for metals, textiles, and plastics. It is also available for the removal of selected materials groups from a comingled waste stream, although such an approach is less common in MRFs. Many of the sorting devices actually do rather more than their titles suggest. For example, a magnetic separator could be used to recover ferrous cans from mixed ferrous and nonferrous. In that sense, it is therefore separating both types of material. One is recovered, one is a residue.

Occasionally some of the machines more commonly associated with major reprocessing plants are used in MRFs. Air classifiers may be installed to separate cans and heavy plastics from paper, or to eliminate labels from glass. Drum screens, as in the example given earlier, may be used for the same purpose. Various types of ballistic classification may be used to separate glass from lighter fractions. However, all of these machines are described in detail elsewhere in this book, and so in this section only those machines specific to MRFs will be considered.

Magnetic Separators — The three most common magnetic separators used in MRFs are overbands, magnetic drums, and magnetic conveyor rollers.

An overband magnetic separator consists of a high-power permanent or electromagnet, suspended above a belt conveyor, and enclosed in an endless belt (see Figure 8.4). Electromagnet machines generate an intense magnetic field using high-voltage direct current, normally from an oil-bath rectifier transformer. The magnet assembly is usually designed with a central attraction magnet, bordered at each end by a repulsion magnet. In such a machine, the only difference between the two types of magnet is the polarity of the current with which they are powered. As materials are lifted from the underfeed conveyor onto the attraction magnet, they are carried clear by the continuously running overband belt. Ferrous metals become slightly magnetized themselves in the process, and so the opposite polarity of the repulsion magnets tends to reject them.

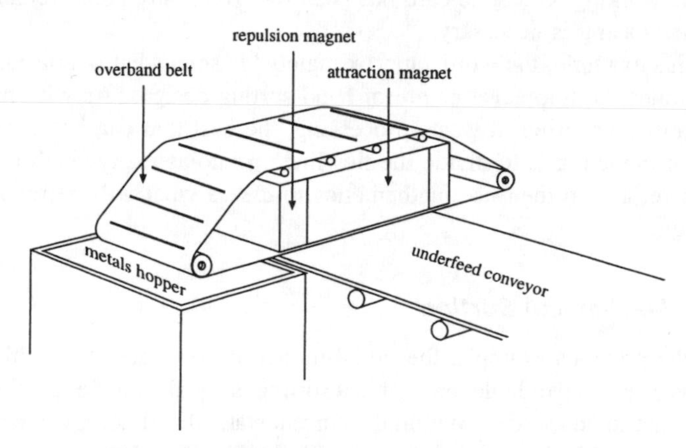

Figure 8.4 Typical overband magnetic separator.

Unfortunately, metal cans do not become sufficiently magnetized for the repulsion magnets to have any significant effect. The design arose from heavy scrap handling, and is in many ways less than ideal for light waste. Upon a conventional machine, the result of repulsion inefficiency is to prevent light metal items from falling free cleanly. There is a tendency to roll back along the belt under the continuing attraction of the main magnet. This can very rapidly damage the overband belt, and so it is customary in waste-handling machines to fit heavy rubber or aluminum flights, across the full width and to a height of 2 in. (50 mm), to the overband. Aluminum flights have the advantage that they are extremely durable, and are therefore much longer lasting than rubber. However, it is essential that where they are used, they are not continuous across the belt width. If they are, then sufficiently strong eddy currents can be set up within them to create a substantial sideways force. Permanent magnet separators are identical insofar as removal of the separated materials is concerned, but they generally do not have sections of opposite polarity, and being of somewhat lower power, they are only used for the smaller installations.

The attraction force of either type of separator is measured by the flux density generated by the magnet. This is an expression of the density of magnetic lines of force, and so the higher the flux density, the more powerful the magnet. Electromagnets are generally available at higher powers than permanent magnets.

There are two common ways in which either type are installed in a production line. They either span the underfeed belt at right angles, or they are mounted above the conveyor head drum, and in line with the belt. Where a mixture of light nonmagnetic and ferrous scrap are carried upon the same conveyor, there are some merits in the in-line mounting. In the case of such materials mixes, there is a tendency for the nonmagnetic material to hold the ferrous items down on the belt and prevent them from being removed. In addition, where the magnetic force is sufficiently strong, ferrous items may lift off taking a substantial amount of nonmagnetic with them. Where the separator is mounted over the head drum, the mix is already falling clear of the conveyor before it becomes subject to the magnetic force. It is much easier to attract the ferrous material when the mixture is suspended in air.

There are some disadvantages in such a mounting method, and one of the more significant concerns the materials of which the fabrications at the conveyor head are made. All structures and moving parts within the sphere of influence of the magnet must themselves be nonmagnetic. Otherwise they will become energized, and will attract the materials that are supposed to be recovered. There are rather more complications and costs involved in making all of the conveyor head structures nonmagnetic than there are in making intermediate structures so, and therefore spanning installations are the more common.

Magnetic Drums — Machines of this type are not as widely used in waste treatment as the convenience and efficiency of their design would suggest that they should be. Their design is one where a nonmagnetic drum encloses and rotates around a fixed-position or permanent electromagnet (see Figure 8.5). The clearance between the drum and the magnet is very small, and therefore material delivered to the machine is much closer to the magnet than it is with an overband. Separation efficiencies would be expected to be very high, and where all of the feedstock is coarsely granular, it is high. However, the very closeness of the material to the magnet tends to work against the principle where mixtures containing film are concerned. The films become trapped against the drum by the magnetic materials, and carried around to their discharge chute.

Where an overband is used, that still occurs to some extent. But much of the problem can be overcome by adjusting the clearance of the separator above the underfeed belt, thus limiting the amount of magnetic attraction to that absolutely necessary. No such provision is available in a drum, since the position of the magnet in relation to the material is fixed.

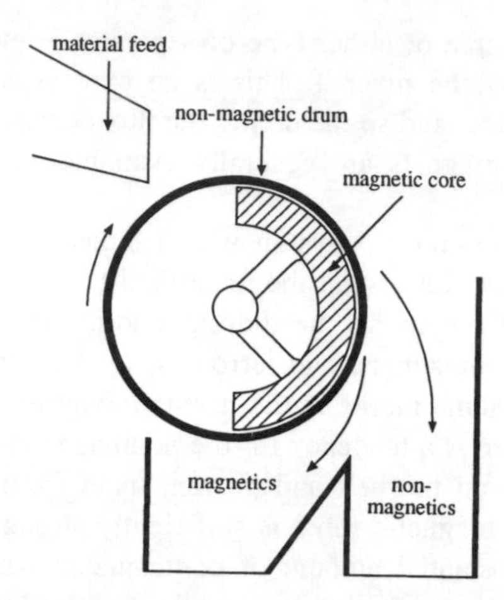

Figure 8.5 Typical magnetic drum separator.

Magnetic Conveyor Rollers — These are not machines in their own right. Rather they are a modification to an existing machine — a conveyor. In design they are virtually identical to magnetic drums, but instead of the feedstock being applied directly to them, it is carried upon the belt conveyor for which the roller is the head pulley.

Magnetic flux densities are lower for separators of this type, since there are practical limits to the diameter of a conveyor roller. There is insufficient space for a large and powerful magnet. However, equipment of this type is very common where aluminum and ferrous cans have to be separated from each other. In such cases, the magnet is built into a short elevating conveyor, which may be mobile and self-powered for touring recycling sites.

Nonferrous Metal Separators — Nonferrous metals can be separated from other materials, and to an extent from each other, by induced current technology — the eddy current separator (see Figure 8.6). The principle is one of creating eddy currents in nonmagnetic, conductive objects by either passing them rapidly through a strong magnetic field, or by passing a field rapidly over them. The electric currents that flow momentarily in the conductive objects create weak magnetic fields of identical polarity to that of the energizing magnet. The consequent repulsion ejects the materials from the region of the energizing magnet. Since different metals conduct electricity to different extents, it follows that the magnetic fields created within them differ. As a result, some metals are repelled more strongly than others.

Magnetic materials are not repelled at all. Instead they are attracted to the energizing magnet. Where its effect on nonmagnetics is created by its being rotated at high speed, magnetic objects may attempt to follow the rotating path. Quite serious damage or wear in the machine can result, and so it is important that ferrous materials are removed before and feedstock is delivered to an eddy current separator.

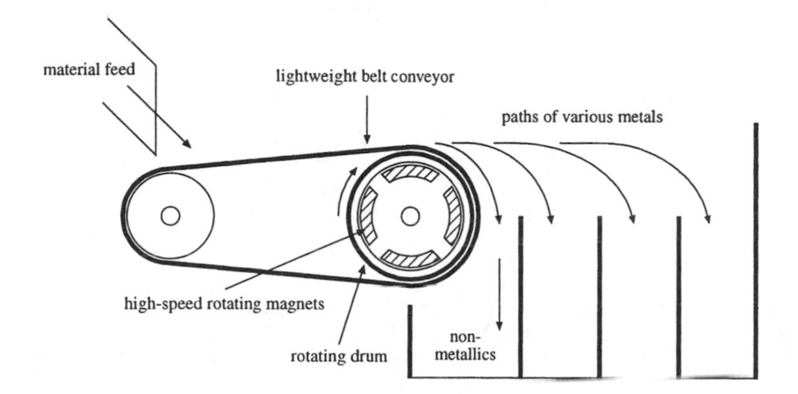

Figure 8.6 Typical eddy current separator.

Textiles Separation — Textiles can be sorted mechanically from other materials, although not from each other. Often the intention has been simply to remove troublesome materials from the waste stream before they cause blockages or interfere with process machinery. Upon occasion, putting the intention into effect has been somewhat disappointing, with some rather strange devices finding their way into sorting lines. The more traditional ways of recovering mixed textiles are successful up to a point. These are rag rollers and deragging belts.

Rag rollers are simply spiked drums that are rotated horizontally in a waste stream which is falling in a chute. The drum is generally sited so that the material falls almost at a tangent to it. In operation, textiles striking the drum are collected by the spikes and are carried round until they fall clear from the lower side. Other materials, meanwhile, either bounce off or fall through the spikes. Devices of this type work reasonably well with small pieces of textile, but they are inclined to capture plastic films and paper almost as well. Where larger textile items are encountered, binding around the drum is a common problem.

Deragging belts are a derivative of rag rollers. Here a spiked conveyor belt runs between two rollers in such a way that it is inclined at a steep angle (greater than 60°) to the horizontal. The device is normally installed in a transfer chute in such a way as to form the backplate. Once again, textiles falling onto the inclined, spiked belt are captured and carried out of the chute. Other materials meanwhile bounce off or slip through the spikes. These machines are still likely to capture film and paper, although they do so to a lesser extent than drums. They are almost totally proof against binding, since their length between centers can be made greater than any lengths likely to be encountered. Figure 8.7 depicts typical rag rollers and belts.

Plastics Separators — Plastics recovery is most often achieved mechanically as a result of the use of mechanisms of a quite different purpose. So, for

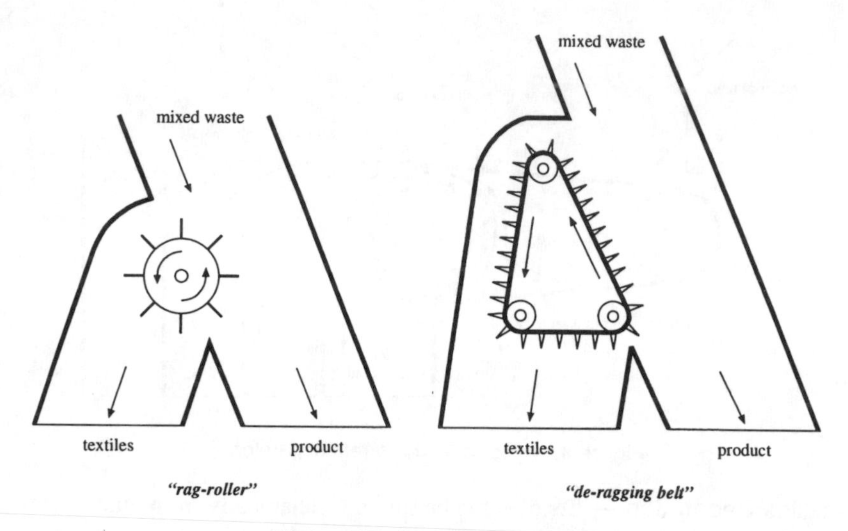

Figure 8.7 Rag rollers and belts.

example, a line with picking stations, magnetic and eddy current separators, and deragging belts may remove everything except plastics. The plastics are therefore recovered simply by being left after everything else has gone. Otherwise, where specially designed plastics removal equipment is used, the purpose is more commonly to recover PET containers.

No device for separating one fraction of waste from the remainder can ever be fully efficient. There is always a compromise between purity of the recovered material and residues of it remaining with the bulk of the waste. Plastics separators are no different. A machine intended to recover PET containers from a mixed waste stream is likely to recover also aluminum cans and other plastic containers. At best, it will eliminate the heavier items such as glass bottles and some steel cans, and for that reason devices of this type are more commonly used to prepare a feedstock for hand sorting.

In an application where chemical or smelting treatments cannot be used, there are only two ways in which one item may be separated from another. It may be identified by a fundamentally different size or shape, or by its density or weight. Plastic containers do not, with the exception of PET bottles, conform to any standard size or shape. In a mixed waste stream, they do not have a sufficiently consistent density that is different from that of other materials.

PET containers, being used almost exclusively for beverages, do tend to conform in dimensions to some standard. In Europe they are made in 1, 2, and 3 l capacities, and irrespective of the manufacturer, are all of similar shape. For that reason, there are two possibilities for separating them.

Figure 8.8 shows a device developed in the U.S. In this design, an inclined, wide-belt conveyor runs beneath rows of chains suspended above it, in parallel to its longitudinal axis. Mixed wastes are delivered to the receiving end of the conveyor and travel along it, being propelled toward the low side by gravity.

Heavy items such as glass bottles have sufficient weight to force their way through the chain curtains, but PET bottles do not. The heavy items can then fall from the side of the sorting conveyor onto a collecting conveyor underneath. Lightweight materials travel along the chain curtains until they are discharged from the end of the sorting conveyor. The quality of the sorted material is determined by the weight of the chains, and to an extent by the speed of the sorting conveyor. Lightweight chains will reject more unwanted items, but will also reject more of the wanted. Heavy chains recover all of the target product, but also an increased quantity of metallic containers.

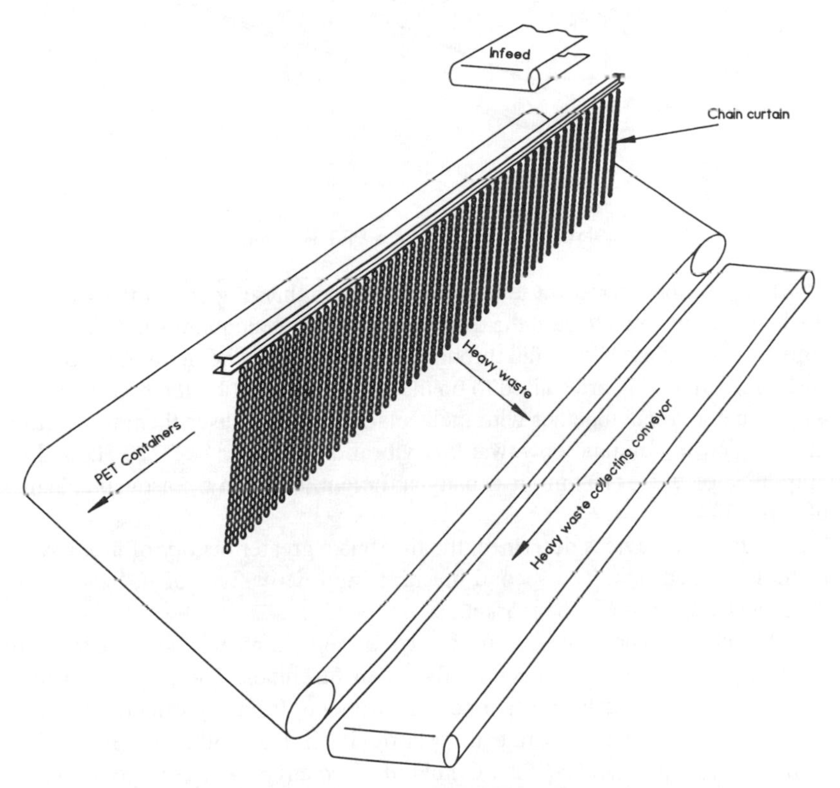

Figure 8.8 Typical belt PET separator.

Figure 8.9 offers an alternative. Here a vibrating feeder table, operating with a high vibration amplitude and low frequency, delivers mixed waste to a pair of collecting conveyors over a deck that is split across its width. The section of the deck beyond the gap is raised above the first section by some $^1/_2$ in. (10 mm). The gap in the deck is capable of adjustment by means of an extending plate, but is most effective when set to 6 in. (150 mm) throat. The deck of the machine is ribbed with inverted, 1 in. (25 mm) steel angles, the purpose of which is to align cans and containers with the longitudinal center line.

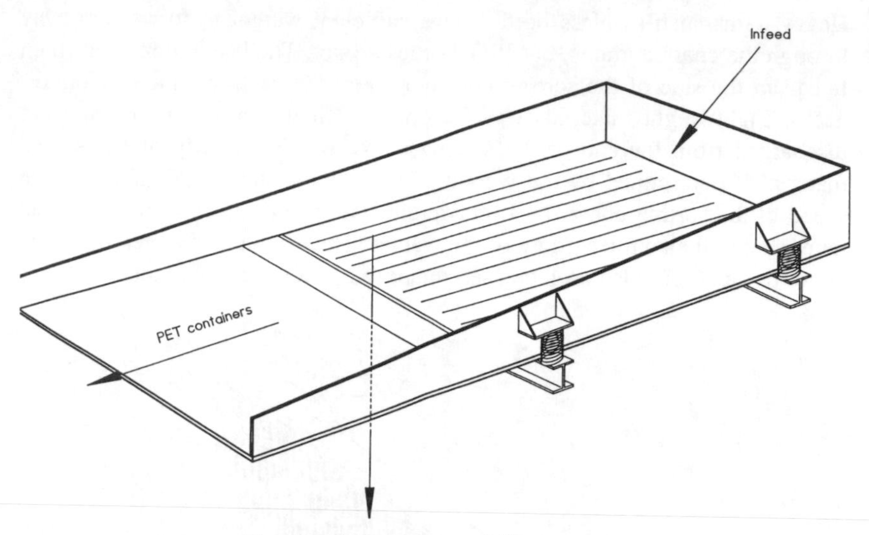

Figure 8.9 Vibrator-type PET separator.

In operation, mixed waste is delivered to the charging end of the vibrating deck, traveling down toward the discharge end. As it encounters the gap, metal cans and most glass bottles fall through. PET containers, being of much thinner and lighter material, are caused to bounce upon the deck by the high vibration amplitude, and this together with their relative length causes them to override the gap and to continue downward. A vibration frequency of 720 Hz and an amplitude of $^1/_2$ in. (10 mm) is usually sufficient to obtain a reasonable degree of separation.

Of the two devices described, the first has a greater history of use, having been developed first. The second machine, and derivatives of it, have yet to make any impact upon the market.

There are other, more complex ways of separating plastics, but they are process plants in their own right rather than machines. The purpose of their designs is more usually to separate contaminants from a particular grade of plastic, rather than to separate the plastic itself from other materials. The Refrakt® system by Anlagenbau GmbH of Germany is a case in point, where PET or other selected plastics wastes are treated in a batch process that includes washing, granulating, and fluid-based separation. The plants are claimed to yield very high-grade, clean scrap from which metallic and other contaminants have largely been eliminated. Processes of this type are somewhat more sophisticated than any normally found in MRFs, and are more suitable for secondary recyclers who deal with MRF products.

8.4 HEALTH AND SAFETY

Materials recovery facilities are potentially very hazardous places, and the increasing interest in handling comingled wastes makes them even more so. The general absence of complex, heavy machinery actually potentially makes them even more dangerous than they would otherwise be, since it implies an over-dependence on the human hand.

As has been pointed out before in this book, municipal waste is effectively a hazardous material, since the processor can never be sure exactly what is in it. Where the treatment facility involves operator contact with it, the risks are very much magnified. For example, when an operator on a sorting line picks a plastic bottle half filled with a colorless liquid, he has no means of knowing whether its contents are water or concentrated industrial acid.

Changes in modern society have altered the nature of wastes to the extent of increasing the dangers of handling it. The presence of used hypodermic needles is no longer uncommon, and the introduction of disposable diapers has brought about the inclusion of human bodily wastes, which were rarely encountered before. The range of household chemicals now readily available introduces an entirely new element of risk, since many of the chemicals are caustic, poisonous, or flammable.

Hypodermic needles are quite capable of penetrating heavy industrial gloves, and are very difficult to see when mixed with other wastes. Used diapers release contaminants onto other nearby materials, without their having done so necessarily being apparent. Not all babies are healthy, and pathogenic organisms and diseases can come into contact with employees without their realizing it. Household chemicals are sold mainly in the plastic containers most likely to be the subject of sorting, and by no means all carry adequate health warnings even if the operator had time to read them. In fact, it is a common marketing strategy to invent brand names that imply a user-friendly product when in fact the reverse is often true.

The employer has a moral obligation to all of his employees to protect them against risk to their health or safety, but increasingly he has a parallel legal responsibility. Few countries in the developed world are without some form of safety legislation, and in many infringement is a criminal act. In Great Britain, for example, the Health and Safety at Work Acts place onerous responsibilities upon employers in criminal law. Their basic tenet is that every employer is required to ensure the health and safety of any employee, or even of someone who is employed by another if on his premises. The immediate consequence of the legislation is that, in the event of injury or sickness sustained at work, the employer is by definition guilty of an offense *even if the employee disobeyed his instructions at the time.*

The Control of Substances Hazardous to Health Regulations are if anything even more onerous where MRFs are concerned. The regulations demand that every employer identifies all materials that may be hazardous in any way and used in his undertaking. He is required to advise his employees of those materials, and to provide documents that clearly define the risks, the procedures to be adopted in dealing with them, and the treatments in the case of contact with them. It is somewhat difficult to imagine how the owner of a recycling plant could possibly ever conform to such legislation, when almost any known material could appear in the waste.

In response to recognition of the moral obligation and the legal duty, MRF designers are increasingly developing employee protection measures. Unfortunately, many of these measures are so ineffectual as to border upon the ridiculous. Safety is not ensured by placing a manual picking station inside an air-conditioned booth where operators can work in shirt-sleeves! It is hand contact with the waste that is dangerous. Air conditioning will not protect the operator from penetrating wounds from glass shards or hypodermic needles.

Some measures may even be counterproductive. For example, designs have evolved where the operator stands inside an enclosed booth, with his hands and arms protruding through holes in the side. His face and body are well protected, but his hands are as much at risk as they always were. In addition, there is a very real possibility of his view being obscured to the extent that he may touch a hazardous item without even seeing it.

An additional point worth making is that in general, employees on hand-sorting lines are not of the highest caliber of labor. If they were, they would not be undertaking such an unskilled and low-paying task. They are not always the first people to be aware of risks to their health and safety. They may not be aware of the possible contents of a container simply because they are not literate enough to read the label. It is certainly unlikely that they will be giving their task their undivided attention, or that they will regard it as either challenging or interesting. Wandering minds are the single greatest cause of industrial accidents.

In view of the current worldwide interest in materials recovery facilities, it is tempting to suggest that it is only a matter of time before it becomes realized what the human cost might be. Possibly then legislation will develop in reaction to the statistics, and if it does it is unlikely to be helpful to the facility owner. Meanwhile there are some precautions he can take.

Some materials should never be subjected to hand sorting, and principal among them are steel cans and glass bottles. Steel cans almost always have sharp edges, and are contaminated with food residues that may be in the process of decomposition. If they were handled with care, using suitable hand protection, then dealing with them might be reasonably safe. Unfortunately, those circumstances are unlikely to occur on a busy sorting line.

Glass bottles are potentially even worse. If dropped inadvertently, they can shatter, spreading shards over a wide area. If they contain a residue of carbonated drinks, they can even explode with surprising violence. Shards always have some edges that are razor-sharp, and against which hand protection that permits the operator to feel anything is defenseless. Mixed cullet is never of sufficient value that hand separation from intact glass is worthwhile, and any attempt to improve its value by color sorting only increases the dangers by increasing the handling.

In dealing with the risks, it is necessary to make a clear distinction between design matters and operating procedures. Both involve careful consideration of what wastes the facility may accept, and to what degree they may contain the unexpected. The design should identify those which present dangers to operators and should seek to deal with them in enclosed, mechanical processes. Operating procedures should follow a similar logic, but should in addition clearly define how the unexpected is to be dealt with. These points are considered in the following.

8.4.1 Design for Materials

The designer should consider mechanical sorting of the types listed for the following materials.

Glass Bottles

Enclosed, hooded vibrating screens to eliminate broken glass.

Rotating drum ball mills for shard-free volume reduction for transport.

Enclosed plate feeder or drag link conveyors for transportation within the facility.

Color separation at the point of origin.

Broken Glass

Enclosed plate feeder or drag link conveyors for transportation within the facility.

Rotating drum ball mills for volume reduction.

Steel Cans

Magnetic separators where possible before any manual picking of residues.

Enclosed plate feeder conveyors for transportation within the facility.

Self-binding baling presses where volume reduction is necessary.

Aluminum Cans

While aluminum cans do not normally constitute a significant hazard, that situation can change if they have become broken open. Then very sharp edges indeed can occur. Eddy current separators should be used wherever possible in preference to hand sorting.

Transportation within the facility is acceptable upon open rubber belt conveyors.

Plastics Containers

While hand-sorting lines are acceptable, the reject discharge should be enclosed and secure, so that any item having unidentifiable contents can be left on the belt. Baling plants should not be used for rejects in view of the danger of leakage of liquids. Baling plants for selected materials should have washdown facilities adjacent to them, and drainage should be into oil interceptors.

Paper/Card

Other than the possibility of skin penetration by staples, etc., paper and card handling does not constitute a major hazard. However, where a hand-sorting station is installed, careful attention should be paid to ventilation and to dust suppression.

Textiles

Textiles have the potential to be contaminated by health-threatening residues, and so any hand-sorting should be regarded with caution. If separation of textiles is to be a part of the facility operation, and it is worth considering the economic value of such a venture, then it would be a wise precaution to install sterilizing plant.

Hazardous Materials

There is always a strong possibility that even apparently harmless materials on a sorting line may contain dangerous items. It is not at all unknown for packaging waste to have broken glass, cutlery, and even syringes in it. A fundamental rule in the design of waste processing facilities is that however good the source-separation scheme is believed to be, there will always be someone who is either careless enough or indifferent enough to dump hazardous materials in with items for recycling. Since the designer cannot know what those items are likely to be, or when they might arise, he can only attempt to produce a machinery installation which minimizes their effects. Operators on sorting lines have at least a chance of protecting themselves if they can identify hazards, and that requirement excludes raking through the material by hand as it passes. The design should be such that the operator *picks* items, not that he first uncovers them, and so the depth of material on the picking belt must be limited so that he can do so conveniently.

The bulk densities of all the materials which may be hand-sorted are very low, and so a conveyor normally capable of delivering many tons of products an h is limited by its volumetric capacity. Hand-sorting rates are also low, since they depend upon the diligence and enthusiasm of the operators. The greatest fraction of the cost of a hand-sorting line is in the head, tail, and drive. There is no direct relationship between the length of the line and its price. It is therefore tempting to install a line that employs a number of operators and which carries a substantial depth of material.

Such an installation will always involve operators in raking through the load by hand, and it will threaten an unacceptable level of injuries and minor accidents. A safe design is one where the selected depth of material is equal to the average dimensions of the items to be picked. If, for example, the line is to deal with mixed plastics containers, from which the task is to recover PET bottles, then the design bed depth should be 4 in. (100 mm), since that is the average diameter of a PET bottle. At such a depth, the PET bottles will never be buried under other materials, and there is a good chance that anything hazardous will be clearly visible. Nothing will be hidden under the mass.

8.4.2 Design of the Working Environment

All hand-sorting stations should be adequately ventilated so that the operators are not subjected to dust arising from the waste. Face masks are not an alternative, since they are uncomfortable to work in for any length of time, and sooner or later the personnel will discard them.

Adequate ventilation requires an airflow across the operator and over the sorting belt, so that all dust and odors are being carried away from him. Where operators are stationed on both sides of the belt, a central ventilating hood is essential.

Every work station should have eye-washing facilities nearby, and there should be a means of stopping the conveyor instantly from each position. Emergency stop buttons are sufficient if they are placed close by each operator in a multioperator station, but instead of simply disconnecting the power to the conveyor drive it is better if they operate a direct-current injection brake to the motor. On single-operator stations, it is safer to install trip wires along the conveyor sides, maintained under spring tension at all times. Trip wires are inconvenient as a result of the ease with which they can be operated even by an inadvertent touch, but they are preferable to an injured operator.

Since emergency stop systems are designed to deal with situations where operator safety is threatened, or has already been transgressed, it is insufficient for them to just stop the machinery. An alarm which alerts all other employees to the event is essential, and the plant control system should provide for central identification of which stop, at which station, has been activated.

The need for personal hygiene was discussed in Section 2.5, and it cannot be stressed too strongly to personnel on hand-sorting lines. However, the hygiene of the process itself is of equal importance. It is human nature that operators in a dirty plant are less careful in their own behavior, while those in a clinically clean one will respond accordingly. A picking conveyor that is difficult or impossible to clean is an invitation to health risks.

Belt cleaning involves two considerations: cleaning and disinfecting of the belt surface, and dealing with spillage from it into the working area. Figure 8.10 suggests a design that deals with both aspects and in addition has the advantage that it leaves a clear space under the carrying strand. The return strand runs beneath the picking floor, and so carry-over and its associated spillage takes place out of the way of the operators. The disinfecting sprays can be operated by remote switching at the end of a working period, leaving a clean and fresh environment for the employees of the next shift.

Figure 8.10 shows only one section of side guards on the top strand of the conveyor in order to preserve clarity. However, such protection will be along the length and both sides. It should be manufactured from easily cleaned material, and it is worthwhile to consider using high-density plastic paneling over a steel frame.

A conveyor designed in this manner does not obstruct the space beneath the sorting floor, since the return strand can be run close to the underside of the floor. It does, however, ensure that any spillage from the belt cleaner reports to the reject bay beneath, as does drainage from the disinfecting sprays.

8.4.3 Operational Procedures

However mundane the task, there is no escaping the need for training of operators. In fact, under the health and safety legislation of many countries,

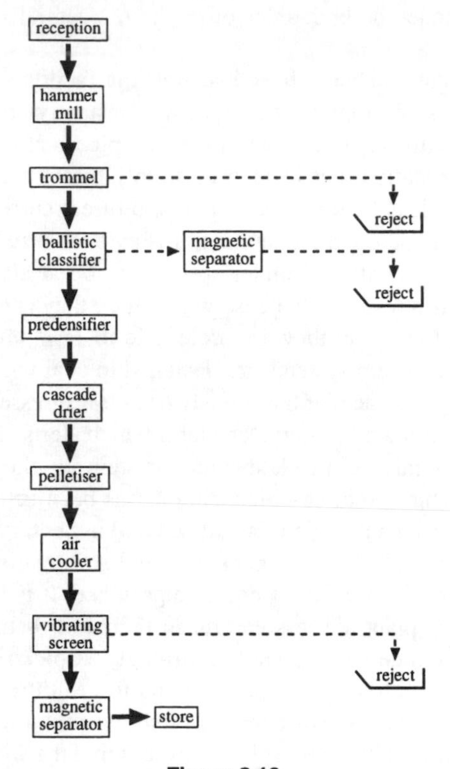

Figure 8.10

it is an indication of a failure of management duty if training is not provided. One of the most important aspects of operator training in a MRF is to teach what to leave well alone.

All personnel should be instructed in the dangers posed by waste, and keeping some samples of the more alarming items that are discovered upon occasions is always a good way of making the point. It is a wise precaution to insist that any container on a sorting line that has anything inside it should be left to go to reject without being touched.

The possible risks from Weil's disease need to be taken into account as described earlier, and a program of health monitoring should be considered. It is one of the more unfortunate records of the waste disposal industry that it spends more upon the monitoring of machinery health than on that of personnel. One of the more significant aspects of health monitoring, in view of the work engaging most of them employees of a MRF, is that of regular anti-tetanus treatment.

Surprisingly, training in how to pick items from a sorting belt is also useful. It is human nature that a conscientious employee, having been instructed upon which materials to recover, will try to recover everything, even when it is clearly out of reach. Although a properly designed picking belt is not normally a dangerous machine, it is still a piece of mechanical hardware with the power to inflict injury. Overreaching and placing a hand on the belt to regain balance may have the opposite effect. It is necessary to make personnel realize that the failure to recover one PET container in ten will make very little difference to the overall economics of the plant. The reject rates in the buyers' works are likely to be higher than that.

9 REFUSE-DERIVED FUEL (RDF) PROCESSES

9.1 THE BACKGROUND OF RDF — TECHNICAL AND LEGAL DEFINITIONS

9.1.1 Derivation of Names

Technologies intended to exploit the energy potential of various types of waste have existed for many years, and each was given a title that generally developed at the whim of the originators. Waste-derived fuel, energy from waste, fiber-fuel, bio-fuel, and refuse-derived fuel became the most common, but in the absence of any accepted international standard, they came to mean a very wide product range indeed.

"Waste-derived fuel" and "energy from waste" can suggest almost anything from burning crude municipal waste in an incinerator with some form of limited energy recovery, to milling it, separating ferrous metals, and then using it in a cement kiln. Both names have also been applied to more refined products, and as technology developed, the terminology created legislative problems. When increasing quantities of wastes were burned to produce energy, controls were obviously necessary, but were difficult to establish in view of the variation in product quality from different processes. The solution, in Europe, was to assume that the combustion of fuels made from waste was simply mass waste incineration under another name, requiring identical emission controls and handling precautions.

The situation was further confused by the predominantly Scandinavian use of the name "bio-fuel," and the american "fiber-fuel." "Bio-fuel" developed into the generic title for any fuel of nonmineral origin, and could refer to straw, bark, paper, or even crops grown for the sole purpose of fuel production. "Fiber-fuel," meanwhile, originally meant paper, but extended to cover packaging materials, and in some cases, even fibers created from the organic fraction of municipal waste. In Britain, "refuse-derived fuel" became generally accepted as a means of distinguishing quality products from those that were coarser and less refined.

Legislation was clearly important, not only from the point of view of environmental protection, but also from that of the manufacturers themselves. The more sophisticated the fuel, the higher the costs of making it, and the more sensitive it becomes to lower-cost competition. There is no point in building a multi-million dollar plant to make a high-grade product from clean paper waste, if the end-user can buy cheaper from a plant that only mills and metal-separates municipal waste. The sophisticated product manufacturers best defense is legislation that prevents the user from burning the lower-quality, higher-emission products, unless he installs suitable exhaust gas scrubbing and ash handling plant. It is the "level playing field" approach, which is intended to protect the environment, but which incidentally also protects the investment of the manufacturer.

Lawmakers, however, are not engineers, and they need specific and accurate definitions with which to work. They cannot be expected to understand the difference between the waste in a municipal collection vehicle and that which has been subjected to any one of a number of complex processes. Equally, the environmental scientists who advise them can only consider the likely effect of burning the most unwholesome wastes, unless it is proved to them that the alternative technologies can create something better.

Having proved to both the legislators and the environmental scientists that a given product is to some degree better than crude waste, one has to show that the manufacturing process is repeatable and reliable. There is no point in seeking approval for a particular grade of fuel if the quality cannot be maintained. Therefore, there is a need not just for a product specification, but also one for the waste from which the product will be manufactured. That leads to the question of what is waste, and at what point in a manufacturing process does it become a product?

9.1.2 Definitions

In general acceptance, waste is a material or mixture of materials that have been deliberately discarded by consumers who no longer have a use for them.

It follows, therefore, that for waste to become a product, it must have been substantially changed in some way such that its original characteristics no longer exist, and such that it assumes a value to an independent consumer. So, for example, simply removing the ferrous metal fraction from municipal waste does not make the waste a product. Its original characteristics remain largely unaffected, and therefore burning it will be no different from mass incineration.

At this point it is worthwhile to consider another aspect that has an impact upon a manufacturing development. Most Western countries have legislation to control the handling and transporting of waste. If the product made from it has not involved sufficient alteration from the original characteristics, then that product will still be classed as waste. The process will be considered as a waste disposal facility, not as a manufacturing one. The fuel will have to be handled and transported as waste, not as a product, leading to possible market limitation where, for example, laws exists to restrict waste movement across national or state borders.

So at what point does a material that has been discarded by a consumer who has no further use for it first cease to be waste? Perhaps, since any sophisticated fuel process must start by separating and cleaning the individual fractions, the most obvious point is that at which separation and cleaning have been achieved. There, each fraction has become a raw material from which an entirely new consumer product may be made. Each has few if any of the original characteristics of the mixture. Refuse-derived raw materials could then be defined as fractions that have been separated and treated in a way such that, if they were then remixed, the original characteristics of the waste would not be restored.

The next stage is for those raw materials to combine to some extent to become a product. That implies a further process of some sophistication, sufficient to ensure that even the original characteristics of the raw materials are substantially changed.

If, however, the waste is to first become a raw material and then a product, it is insufficient simply to require a "sophisticated process." Either the process or the raw material must be sufficiently defined such that waste regulations no longer apply, and the definitions of waste and raw materials outlined earlier are too broad for the purpose.

In Table 9.1 a simple points system is suggested, where crude municipal wastes and the main fractions within them are evaluated in terms of their potential for use as a fuel. Each of 11 main characteristics is scored out of a maximum of 10 points, leading to a possible maximum score of 110. The higher the score, the better the potential.

The purpose here is not to arrive at an arbitrary system, or to suggest that the rating given to each characteristic is anything other than flexible. Rather, it is to establish a rule of thumb for evaluating and defining waste, raw materials, and products. Each fraction is scored in the form in which it would exist if it were simply picked out of the waste.

Table 9.1 Classification of Fractions for Potential Use as Fuel

			Points[a]				
Characteristics	MSW	TXT	PLF	DP	ORG	PACD	RDF
Moisture content	2	6	8	8	0	5	10
Calorific value	2	5	10	3	0	8	9
Energy density	2	2	0	5	2	0	9
Biological stability	1	8	10	10	0	3	9
Ease of storage	1	8	1	8	0	2	10
Handling	2	2	2	5	0	2	10
Acid emissions	0	3	8	1	0	7	10
Dioxin/furan emissions	0	3	8	3	0	4	10
Ash content	0	2	8	3	0	5	8
Ash toxicity	0	2	10	2	4	7	9
Combustion efficiency	2	2	0	0	0	10	10
Totals	12	43	65	48	6	53	104

[a] Mixed solid waste (MSW); textiles (TXT); plastic film (PLF); dense plastics (DP); organics (ORG); paper and card (PACD); refuse-derived fuel (RDF).

Ideally, a clean-burning synthetic fuel that is convenient to handle and carries with it no immediate risks to human health and safety would score 110 points in the table. If a raw material from which it is to be made has been treated, such that if remixed with the other fractions, the original characteristics of the waste would not be restored, then it should score at least 70 points.

None of the fractions listed achieve that in their as-separated form, and at that point, remixing them would obviously restore the original characteristics of the waste. Therefore, further processing is necessary, leading to the conclusion that some fractions (the organics, for example) are so far below the transformation score that making them into a suitable fuel feedstock would be expensive even if it were possible. For the obvious reasons, inert materials such as glass, stones, and metals are excluded as fractions altogether, although they exist in the crude waste and do effect its score.

Some of the scores in Table 9.1 may at first sight be a little surprising. For example, plastic film appears to be close to becoming an acceptable raw material in its initially separated form. If it were to be granulated to reduce its volume and increase its energy density, many of the difficulties with storage would also be overcome, and it could easily achieve the first stage of 70 points. However, to go from there to an acceptable score for a fuel would require something to be done about the difficulty of handling it, and about its very low combustion efficiency. That latter characteristic is exhibited by the considerable difficulty of obtaining adequate oxygen mixing with the combustible gases given off by plastic film. Therefore, it will never actually make an ideal fuel with a score of at least 100 points. The table does, however, suggest that if the film were to be dispersed in the paper and card, then the combustion efficiency of the mixture might be acceptable. This is in fact a common approach in the manufacturing of refuse-derived fuels.

Paper and card, meanwhile, do not appear to reach as high a score as one would expect. The reasoning here is that in its initially separated form, it is generally in large >6 in. (150 mm), low bulk density pieces. It is moist, heavily contaminated with biologically active organic wastes, and may harbor pathogenic microorganisms. However, if it were granulated, classified, and dried, and then screened again, the moisture content could be improved considerably. The organic contamination could be removed, thereby reducing the potential dioxin and furan emissions, and its biological stability could be restored. Pathogens would be destroyed, and the energy density could be increased, in parallel with improvements in handling characteristics. The process described would, of course, change the paper and card to a form quite different from that in which it appears in crude waste, thereby permitting it to conform to the definition of a raw material.

Table 9.1 therefore serves quite well to assist in defining the meaning of the term "combustible fraction." In the general mix of municipal waste, dense plastics, textiles, and organic and vegetable materials are all combustible to some extent. However, they are mostly fairly low down in the points ratings, and each fraction has at least one major fault that would make its conversion into clean fuel almost impossible, or at least prohibitively expensive. If one were to use the gross calorific value of each fraction as a yardstick, and accepting that good-quality RDF should have a calorific value of at least 8600 Btu/lb (20,000 kJ/kg), then a suitable combustible fraction must approach that closely enough for final drying and screening to achieve it. An arbitrary lower limit of 6450 Btu/lb (15,000 kJ/kg) would eliminate the organic and vegetable fraction, and many of the dense plastics which require more energy to make them burn than they liberate upon burning. It would also eliminate most of the textiles for the same reason. All that would be left would be the paper and card, and the plastic film provided that it were minimized and well dispersed in the final RDF mix.

Having achieved the raw material stage, the fractions must then be further processed to become fuel. As stated earlier, logically the nature of the fractions must again change substantially to the extent that they can no longer be returned to their original characteristics.

Initially it may appear that the paper and card fraction, having been granulated, classified, dried, and screened, could be used directly in that form. Indeed, some processes are considered to have created refuse-derived fuel when they have not even passed through the drying and final screening process. In those cases, however, they are not strictly fuels in their own right, but are simply used as low-level mixes with conventional fuels. Thus, in many plants in the U.S., granulated, classified paper and card is blown into pulverized-fuel boilers with coal at a local power utility. Emissions for which the fraction may be responsible are highly diluted in the conventional exhaust gas.

To meet strict criteria defining a fuel, a product should be capable of being burned by itself, and not be dependent upon some other, cleaner product in a co-firing application. In Europe, and particularly in Britain, the broad definition applied to refuse-derived fuel is "fuel manufactured from the combustible fraction of municipal waste by advanced mechanical processes including the application of heat, designed to maximize the recycling potential of such wastes, and containing no more than 15% ash prior to any addition of substances to enhance fuel properties."

The reasoning behind the clause "deliberate use of heat" is that unless the material is dried, its moisture content cannot be made acceptable, and its biological stability cannot be ensured. Equally, until it is dried, organic and chemical residues adhering to the fraction cannot be screened off, and so the dioxin, furan, acid gas emissions, and ash content cannot be minimized. The definition therefore does not specifically require that the product is densified, but rather that it meets emission and residue standards. The final column of Table 9.1 (RDF) demonstrates the action necessary to achieve those standards.

The U.S., meanwhile, has adopted a somewhat different approach to the need for a classification system, and this is contained in the ASTM Standard for the Classification of Refuse-Derived Fuel (Table 9.2). The definitions applied reinforce the points made earlier in this chapter, in that the combustion of categories RDF-1, RDF-2, and c-RDF at least is simply mass-incineration

Table 9.2 ASTM Classification of Refuse-Derived Fuels

Type of RDF	Description
RDF-1	Municipal solid waste used as a fuel without oversize bulky waste.
RDF-2	Municipal solid waste processed to coarse particle size with or without ferrous metal. As a subcategory, c-RDF, which is subject to separation such that 95% by weight passes through a 6-inch square mesh screen.
RDF-3	Shredded fuel derived from municipal solid waste and processed for the removal of metal, glass, and other entrained inorganics. The particle size is such that 95% by weight passes through a 2-inch square mesh screen (also classified as "Fluff RDF").
RDF-4	The combustible waste fraction processed into powdered form, 95% by weight passing through a 10-mesh (0.035 inch square) screen (also classified as p-RDF).
RDF-5	Combustible waste fraction densified into the form of pellets, slugs, cubettes, briquettes, or some similar form (also classified as d-RDF).
RDF-6	Combustible waste fraction processed into a liquid fuel (no standards developed).
RDF-7	Combustible waste fraction processed into a gaseous fuel (no standards developed).

From Municipal Waste Combustion Study: Emission Database for Municipal Waste Combustors, NTIS PB87 206082. Ecotec Research and Consulting Ltd., September 1989.

given a more comforting name. The standard reflects the common U.S. practice of co-firing for power generation, and the general lack of interest in the products as commercial fuels in their own right. It is also a potential source of confusion, in that an American company attempting to sell RDF technology in Britain will find itself discussing a quite different product.

Over a period of time in Britain, refuse-derived fuel has gradually become accepted and understood in three categories: flock refuse-derived fuel (fRDF), crumb refuse-derived fuel (cRDF) — often taken in the U.S. to mean "coarse RDF," and densified refuse-derived fuel (dRDF). From the emissions perspective, they should all perform equally well, and the only difference between them is the degree of densification that is applied. Thus, fRDF is not densified at all, but is generally reduced to a dry particle size of 95% less than $^1/_2$ in. (10 mm). cRDF is densified to about 18.7 lb/ft^3 (300 kg/m^3), while dRDF is densified to in excess of 37.5 lb/ft^3 (600 kg/m^3). All that the densification does is to effect the way in which the fuels handle, which in turn is dependent upon the design of the combustion plant fuel handling system. Thus fRDF, used in a pneumatic system for pulverized fuel, would still score 10 for handling characteristics in Table 9.1.

In summary then, the following suggested definitions arise from the points considered in this chapter.

Waste: *A material or mixture of materials that have been deliberately discarded by consumers who no longer have a use for them. As a potential energy resource, rated at less than 20 on an ideal fuel scale of 110, and having a potential ash content of up to 40% by weight.*

Combustible Fraction: *That fraction which has a gross calorific value of at least 6,450 Btu/lb (15,000 kJ/kg) upon separation, and which can achieve at least 8,600 Btu/lb (20,000 kJ/kg) after cleaning.*

Raw Material: *A fraction that has been separated and treated in a way such that, if it were then returned to the waste, the original characteristics of the mixture would not be restored. As a potential energy resource, rated at more than 70 on an ideal fuel scale of 110.*

Flock Refuse-Derived Fuel (fRDF): *An undensified product manufactured from the combustible fraction of wastes by a sophisticated mechanical process involving the deliberate use of heat, having a granularity of at least 90% less than* $^1/_2$ in. *(10 mm), and containing no more than 15% ash prior to any addition of substances to enhance fuel properties.*

Crumb Refuse-Derived Fuel (cRDF): A product manufactured from the combustible fraction of wastes by a sophisticated mechanical process involving the deliberate use of heat, densified to a bulk density of not less than 18.7 lb/ft^3 (300 kg/m^3), and containing no more than 15% ash prior to any addition of substances to enhance fuel properties.

Densified Refuse-Derived Fuel (dRDF): A product manufactured from the combustible fraction of wastes by a sophisticated mechanical process involving the deliberate use of heat, densified to a bulk density of not less than 37.5 lb/ft^3 (600 kg/m^3), and containing no more than 15% ash prior to any addition of substances to enhance fuel properties.

By no means all of the processes described in this section achieved those specifications, but not all applications demand such a degree of sophistication. For example, a municipal incinerator, which includes power generation and/or heat recovery, may carry out a degree of raw material refinement to improve combustion efficiency and minimize combustor damage. To an extent, it is moving down the path toward coarse fRDF. It is, however, still a municipal incinerator, and must meet the appropriate emission regulations whatever else it does. A high degree of combustible fraction refining may be pointless.

9.1.3 Early Designs

Having accepted that some degree of refining was necessary to convert municipal waste into a suitable fuel, the technologies applied by Europe and the U.S. began to diverge. In the U.S., the plethora of small municipal power utilities provided ample opportunities to co-fire waste fractions that had simply been granulated, and perhaps classified. In Europe, and more particularly Britain, such small utilities did not exist, and so the approach was more toward creating a commercial-quality fuel. The more common of the various approaches is shown in Figures 9.1 through 9.4.

Even so, there were a number of common factors in the early designs, and not least of these was the almost universal conviction that before waste could be refined it must first be granulated. Upon reflection, and with the benefit of hindsight, that was very odd conclusion indeed. All waste refining processes must by nature involve fraction separation, and hammermills and granulators are extremely efficient mixers. The installation of a granulator as the first stage of the treatment process simply ensures that the waste is even more intimately mixed than it was in the first place. By doing so, it makes fraction separation later much more difficult.

In the U.S., a number of plants were constructed upon the principle of granulation before separation (Figure 9.1). There it was fairly common to first shred the waste, then to air classify it, and frequently to screen it through drum or vibrating screens. Usually a magnetic separation stage was included somewhere in the process to recover ferrous metals, and after this limited processing, the waste was considered to be a fuel. There were several flaws in the design thinking. (1) Air classifiers are not particularly efficient at separating finely granulated, mixed wastes. Air flows sufficient to recover the combustible fraction will also recover much of the organic and inert fractions as well. (2) Once the wastes have been granulated, particle sizes become more consistent. The granularity of the combustible fraction, and everything else, becomes much the same. In such a condition, screens, which differentiate between particle sizes only, cannot possibly work efficiently. (3) Granulators, in intimately mixing the fractions contained within the wastes, increase both the moisture content and the organic contamination of the combustible fraction by several orders of magnitude. The immediate result is to make the finished product much less energy-efficient, and to increase the potential for harmful emissions.

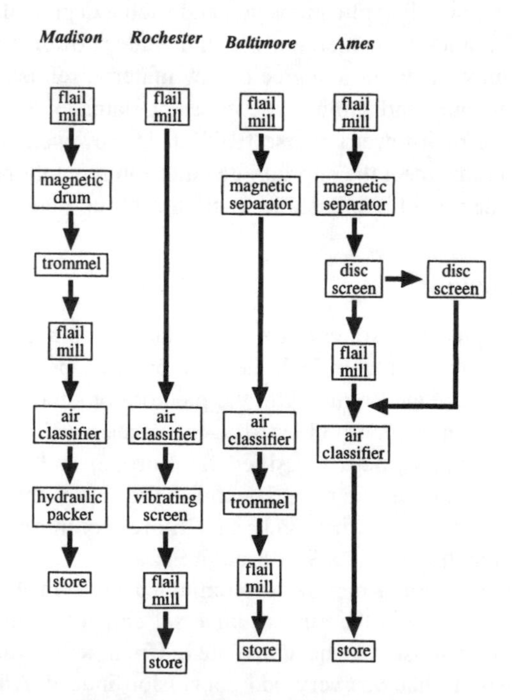

Figure 9.1 Early American systems.

Europe avoided most of the early technological development by deciding against RDF processes and concentrating upon mass incineration, often with heat recovery. Britain, meanwhile, in the 1970s and 1980s, embarked upon both incineration and RDF in the hope of discovering a solution to its growing waste disposal problems. There was a growing recognition that mass-incineration alone was unacceptably expensive for a country where landfill costs were low historically. Incineration with heat recovery was even more expensive, and although it did generate some income, it appeared to be appropriate only for large waste arisings. RDF offered lower capital costs, flexibility in capacity, and an attractive end-product that could (in theory) be sold at a profit.

Following the lead of the U.S., all of the early British plants accepted crude municipal waste and granulated it first. They then screened it, classified it, and attempted to densify it into RDF pellets using conventional animal feed pelletizers. Densification was accepted as essential in order to minimize transport costs to remote combustion facilities.

Unfortunately, one result of first-stage granulation was to make the waste too moist to pelletize properly. The loose, crumbling, and biologically unstable pellets which the processes produced were wholly unacceptable to industrial users. The solution was seen to lie in drying the combustible fraction to a moisture content that the pelletizers would tolerate, and at that point the technology diverged into two streams.

In the first, which became known as the "3-D" system (Figure 9.2), waste was granulated, screened, classified, loosely densified, dried, and finally, fully densified. Magnetic separation for the removal of ferrous metals was confined to the rejects from the classifiers, and so there was no protection for the pelletizers against metals remaining with the combustible fraction.

The initial stage of densification was necessitated by the driers that were used. Rotating cascade driers were common, wherein a large rotating drum, similar to a cement kiln, was charged with waste at one end, through which high-temperature combustion gas from a burner unit was also introduced. The waste was then tumbled in the gas stream, making its way slowly along the drier, until it reached the discharge end.

Since the combustible fraction had been granulated to an average particle size of less than 2 in. (50 mm), its bulk density was very low indeed (between 1.9 lb/ft^3 and 3.7 lb/ft^3 — 30 and 60 kg/m^3). The only way to ensure that the material stayed in the gas stream with sufficient residence time for effective drying was to increase the bulk density. Unfortunately, as observed earlier, the granulators had already significantly increased the contamination of the combustibles by organics and inerts. The predensifying stage simply locked those contaminants into the final product in such a way as to make any further refining impossible. Fuel pellets were produced with an ash content of close to 20% by weight, and equally significantly from a boiler operator's point of view, with a chlorine content of 1.5% and above.

d-RDF already has a lower energy density than conventional fuels, and so even if it had the same ash and chlorine content as a percentage by weight,

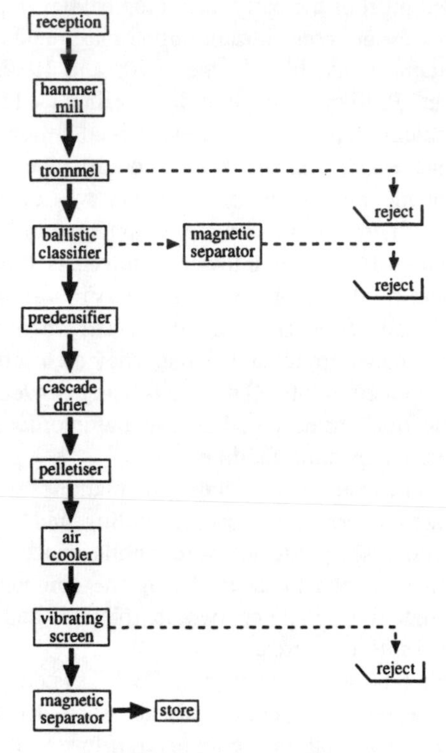

Figure 9.2 "3-D" system.

burning it would still create more of both. It is necessary to burn considerably more d-RDF to obtain the same energy output as conventional fuels, and so such a fuel, already containing three times as much ash as coal, simply overwhelmed the ash handling systems of most conventional boilers.

The other parallel development stream that took place in Britain was instigated by Buhler Miag of Switzerland, and developed in a pilot plant in Sussex. In this case, while the process initially began with first-stage milling (Figure 9.3), it later departed from this and screened the waste first (Figure 9.4). Milling was restricted to the combustible fraction alone, and thus the process became one of screening, milling, air classifying, drying, screening again, and finally densifying.

The driers used were basically pneumatic conveying systems that operated on hot combustion gas from natural gas burners, and they were called "thermopneumatic lines." They operated at temperatures 200° higher than cascade driers, and they could only work with low-density material. They were flash driers, where all of the moisture removal occurred in the first 9 ft (3 m) beyond the charging point, and from then on the product was simply being conveyed.

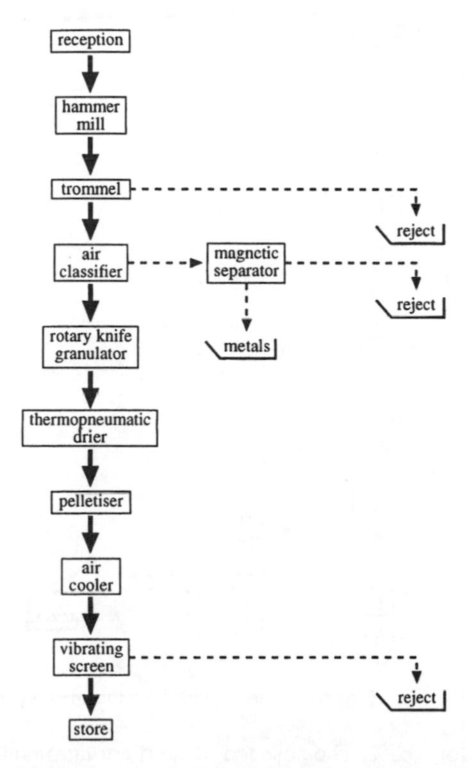

Figure 9.3 Anglo-Swiss "Eastbourne" system.

The great advantage of the thermopneumatic drier principle was that in the absence of any predensification, contaminants were not locked into the final product. Once the combustible fraction was dry, organic and inert residues could be screened out of it quite easily through fine-mesh $^1/_4$ in. (5 mm) drum screens. These worked remarkably well, since the contaminants possessed a granularity much less than the combustibles.

In the process lines that developed, the granulators did not create the mixing problems experienced earlier, since they were acting upon material from which the bulk of the contaminants had already been removed. The effect of the large-diameter front-end trommel screens was to remove almost all of the glass, ash, and organic fractions, leaving only metals, plastics, paper, and card. At that point, the mixed waste consisted of very heavy fractions and very light fractions, and was therefore ideal for air classification.

While the final screening process immediately after drying was initiated originally to eliminate inert particulates, and to reduce the product ash content, research exposed a further if unexpected benefit. It showed that most of the chlorine in the product was contained within the inert particulates, as was the heavy metals and silicates burden. With this stage in the process, d-RDF could be produced with a final ash content of less than 10% by weight, and chlorine

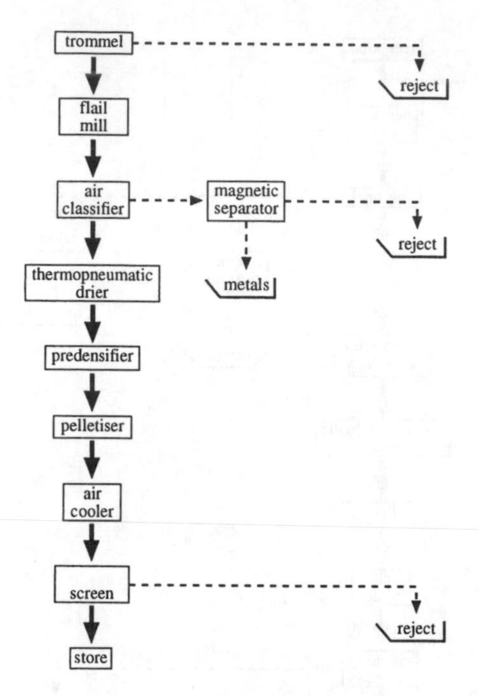

Figure 9.4 Improved "Eastbourne" system.

levels of well below 0.5%. The absence of inert contaminants such as silicates also caused the calorific values to increase substantially, so that d-RDF could be manufactured to an energy content approaching that of low-grade industrial coal. Boiler tube fouling problems, while not being entirely eliminated, were certainly reduced to manageable proportions.

The processes described here are not by any means the only ones that have been attempted. Rather, they have been chosen because they typify the general concepts. To be refined into a fuel, waste must first be separated into its constituent fractions, and there are only a limited number of ways in which that can be achieved. One fraction may be separated from another upon the basis of difference in density, difference in particle size, or sensitivity to electromagnetic forces. Techniques such as solubility, melting point, etc., which may be appropriate in other industries, are clearly impractical when making fuel from waste.

All RDF processes must therefore employ air or ballistic classification, which relies upon density differences; screening, which relies upon granularity; and metals extraction, which relies upon sensitivity to electromagnetic forces. All of the other components, such as milling and drying, are mainly there to assist those three. The order in which they are placed in a process line determines the quality of the end-product, but some combination of them is necessary to make the product at all.

9.2 RDF PROCESS MACHINERY

Before considering how the various qualities of fuel may be manufactured, it is first necessary to examine the range of machinery available for the purpose, and to establish design concepts upon which to model a flowline. In the following section, the order in which the machines are discussed has no significance for such a line.

9.2.1 Trommel Screens

In waste management terms, a trommel screen, as discussed in earlier sections, is a large-diameter rotating drum, the shell of which is formed by perforated steel plates. In the European waste processing industry there are three basic design concepts. In Britain and much of Europe two types are common, both mounted upon trunnion wheels by means of track rings at the drum ends. One process flowline favors the use of two screens in series, where the first is perforated with 6 in. (150 mm) holes, and the second with 2 in. (50 mm). Undersized material from the first is passed to the second, while oversized material is delivered directly to a granulator. The second design employs a single but larger trommel with 2 in. (50 mm) holes only. Both types are driven by a gear-motor unit mounted upon an external frame. In continental Europe, only a trommel mounted upon a longitudinal shaft has evolved. In this case there are no trunnions, and the machine is driven by a slow-speed, high-torque hydraulic motor.

The machines with 6 in. and 2 in. (150 mm and with 50 mm) holes are distinguished only by their perforations. Their general principles are largely similar, and only their application in a flowline differs. There are, however, some differences in their detailed design concepts.

One design (see Figure 9.5) available from several fabricators in Britain mounts the drum upon polyurethane-tired trunnions. The trunnions are heavy-duty cast steel wheels upon the rim of which the tire is pressed under pressure. They are usually installed in groups of two or three wheels at each of the four mounting points. The screen track rings are rolled from heavy steel plate, usually $^1/_2$ in. (10 mm) thick, and stiffened with circular flanges that rotate close to the trunnions. A smaller, vertically mounted trunnion is usually located between the track ring flanges underneath the discharge end of the machine to absorb end float due to the inclination of the drum.

Rotation of the drum is provided by a gear-motor unit, mounted upon an external frame, and driving via a shaft to one or two of the four trunnion groups. Where only one group are powered, the drive is usually to the discharge end of the drum, while two-group drive is to both ends on one side only. In both cases, the trunnions driven are those on the "seven o'clock" side of the drum when viewed on the discharge end with the drum running counterclockwise. While it may at first sight appear that there is no difference in the quality

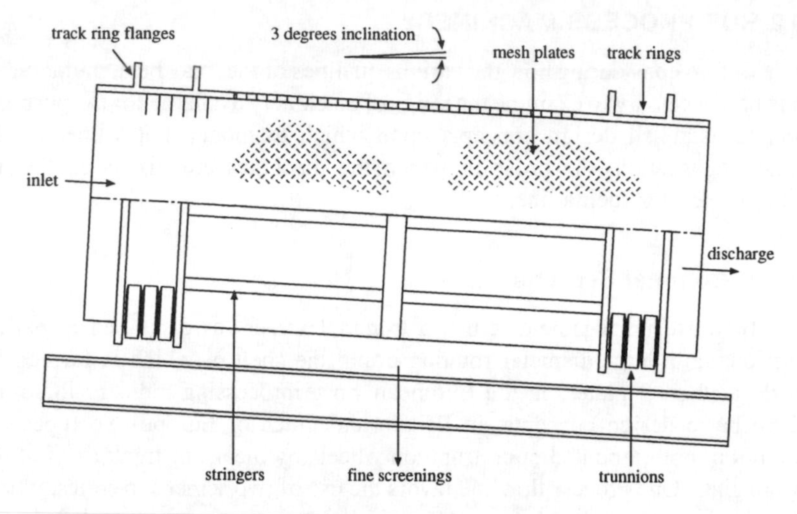

Figure 9.5 Trunnion-mounted trommel screen.

of drive whichever side of the drum it is applied, it is in fact a matter of some importance. In the location described, the driving trunnions "A" in Figure 9.6, are running clockwise against the counterclockwise drum. The torque they apply in direction "X" imposes a degree of weight transfer on the drum in the direction of arrow "Y." While the transfer is not large, it is sufficient to apply much more frictional resistance between trunnion "A" and the drum than would trunnion "B," since the latter would also generate some degree of weight transfer, but still in the same direction. This configuration is not by any means accepted by all designers, but it does improve the turning moment applied when starting the machine for the first time, particularly if it has been stopped fully loaded.

There is also some difference of opinion about whether it is better to drive on two sets of trunnions or only one. The advantage of driving on two is that the frictional resistance between the tires and the track rings is doubled, making the drive more positive and less prone to slip. The potential disadvantage is that it is actually very difficult to ensure that the turning moments at each end are identical. Torsion on the drive shaft between the trunnions may cause misalignment, as will any slight difference in the circumference or run-out of the rings. In such an event, one set of trunnions could, in theory at least, occasionally be resisting the torque applied by the other. In fact, it is possibly little more than an argument upon theory, since both types of drive have operated successfully in a number of waste processing plants.

While polyurethane-tired, cast-steel wheels are common, so too are heavy truck wheels with pneumatic tires. Each has advantages and disadvantages when compared with the other, although again both types have a long history of trouble-free operation.

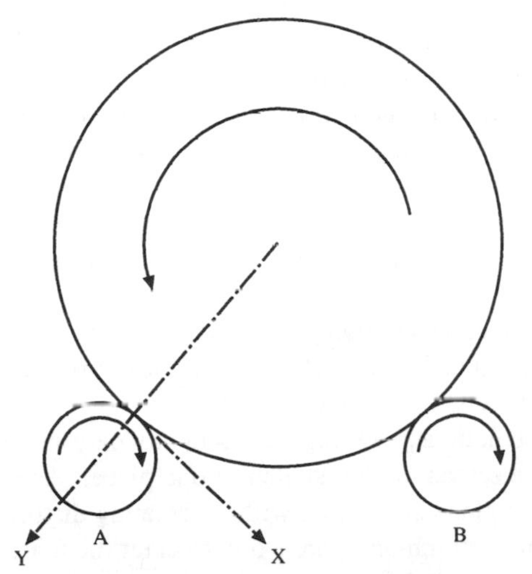

Figure 9.6 Trommel trunnion forces.

Polyurethane-tired trunnions are remarkably durable and hard wearing. Transmission of drive to them from a geared motor requires no more than a coupling, shaft, and keyways in the trunnion bores. They are reasonably easy to change in the event of repairs being necessary, and new tires can be pressed on without difficulty should the originals wear for any reason. They suffer from the disadvantage that the wheels are of a much smaller diameter than that of the screen drum — 18 in. (450 mm) is a common size — and so the contact area between the track rings and the tires is relatively small. This is not helped by the relative inflexibility of the polyurethane, which makes damage by materials becoming trapped between the rings and the tires always a possibility.

Trunnions assembled using pneumatic truck wheels and tires are of a much larger diameter in relation to the track rings, and so the contact area between the two is much greater. The flexibility of the tires increases this very considerably, providing an extremely positive drive. Tire flexibility also reduces the point loading on the circumference of the track ring, reducing the possibility of distortion and eliminating any bouncing over irregularities in the fabrication. Trapping of materials between the tires and the rings has less effect, since the flexibility of their casings allows the tires to absorb the temporary obstruction. The greatest disadvantage of the design lies in the very fact that the screen drum is held in place upon a cushion of air. Deflation of a tire causes severe and immediate misalignment, such that without safety rollers for the track ring flanges to drop upon in such an event, the whole drum could become dismounted. Even an imbalance of tire pressures at each mounting can create drum misalignment, and if the pressure differential is severe enough, it can introduce excessive torsional stresses into the drum structure.

Trommels used for the first stage of comingled waste separation are essentially very large machines. Drums of $8^1/_4$ to 10 ft (2.5 to 3 m) are necessary to achieve the correct product quality, with screening lengths of 23 to $29^1/_2$ ft (7 to 9 m) to accommodate the more common throughputs of waste. Their structural integrity depends largely upon the screen plates and the way in which the end track rings are secured. This is achieved by "stringers," which a heavy, rolled steel sections to which the curved screen plates are fixed by bolted clamps. The stringers are secured to the track rings by welding in the case of small screens, but by bolting in large ones. A screen drum of the dimensions suggested is too highly stressed for welding alone to be sufficient. Welds are likely, and in fact almost certain, to develop cracks in service, while bolts will absorb stresses without fractures.

In many installations, welding would in any case be impractical, since it would require the screen to be installed in one piece. Such a large machine is difficult to lift and place in position without creating distortions in the process of doing so. A more common approach is to erect the machine in the fabricator's works for proving trials, and then to dismantle it for transport. On site, reassembly is reasonably straightforward with the track rings mounted upon their trunnions.

The screen plates used in the applications under discussion here are usually perforated with circular holes, since that provides larger areas of metal between each hole. The purpose of a trommel screen is not simply to sieve materials, but also to break the noncombustibles to a size where screening becomes possible. For that reason, the drum critical speed is even more important than it is in the compost screening described in Section 6.2. Glass items, for example, must be able to fall onto unobstructed plates in order that they may be shattered.

To repeat the discussion in Section 6.2, the critical speed of the drum is that speed at which the forces imposed upon the contents by gravity exactly oppose the centrifugal forces created by the rotation. At above the critical speed, materials would in theory stay against the circumference of the drum throughout the whole rotation. Below it, they fall clear at some point, and where that point is depends upon the fraction of the critical speed at which the drum is rotated. The trajectory imposed by the circular motion causes the materials to fall back in a parabolic arc, striking the screen plates in a position determined by the rotational speed and the point of release.

In Section 6.2 it was said that at 70% of critical speed, material would be released from the side of the drum at about 70% of the arc between the lowest and highest point. While this is nearly true, it can be calculated surprisingly accurately and used to establish the design of a machine. Figure 9.7 shows a screen drum with the parallelograms of forces drawn for various points around the circumference. In the illustration, the vertical line "c" is the gravitational force, "b" represents the centrifugal force upon a particle caused by the speed of rotation, and "a" is the resultant force. "B" is the enclosed angle between the centrifugal and gravitational forces, and "C" is that between the resultant and centrifugal forces.

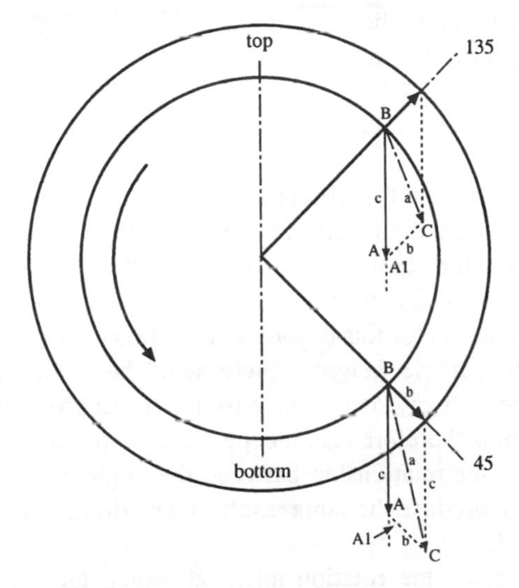

Figure 9.7 Drum screen particle behavior parallelogram of forces.

As the drum rotates, angle B increases from zero at the lowest point to 180° at the vertical, and so in the first quadrant of the rotation the direction of the resultant of the two forces is outward from the drum center. Material is therefore held against the screen plates by a combination of both forces. This situation continues into the second quadrant to an extent determined by the rotation speed, leading to the "release angle" where the angle C is 90°. At the release angle, gravity ceases to assist the centrifugal force, and begins instead to oppose it. In the drum shown in Figure 9.7, C is less than 90° at 45° rotation, while at 135° it is clearly in excess of 90°. The release point is therefore somewhere between 45° and 135°.

If the weight of a particle and centrifugal force b are known, the angle of resultant force a can be calculated for any point around the circumference of the drum. The centrifugal force upon the particle is derived from

$$C_f = WS^2/gr \tag{1}$$

where W is the weight of the particle, S is the peripheral speed, g is the gravitational constant, and r is the drum radius.

The centrifugal force is dependent only upon the speed and particle weight, and does not vary around the circle. Therefore, the resultant force can be calculated, and so too can angles A, B, and C, again for any point upon the circumference, as follows.

$$a = (b^2 + c^2 + 2bc \cos A_1) \tag{2}$$

$$\sin B = \frac{b \sin A_1}{a} \tag{3}$$

where A_1 is the outer angle between the centrifugal and gravitational forces. If A_1 is selected arbitrarily, and B is calculated, then C for any triangle is

$$C = 180 - (B + C) \tag{4}$$

It is interesting to note that if this series of calculations is carried out for a number of screen drum diameters and particle weights at a given percentage of the critical speed, the results are always the same. The angle between the centrifugal and resultant forces is only dependent upon the percentage of the critical speed. The reason for this becomes clear if one considers what happens if, for example, the particle weight is increased. The immediate result is that at any speed, the centrifugal force is increased in direct proportion. The critical speed for any drum is therefore also independent of the weight of the particle. So a calculation of the relationship between the angle of rotation and of the resultant force must produce the same result for any drum at a given percentage of its critical speed.

Additionally, since the rotation angle at which gravitational attraction begins to overcome the centrifugal force occurs when the angle between the resultant and centrifugal forces is 90°, then the release point is again the same for any drum of any diameter. It is determined only by the percentage of the critical speed at which the drum is rotated. Figure 9.8 shows the relationship between the speed and the centrifugal force/resultant force angle in graphical form, with a horizontal line indicating the critical 90° angle. It is drawn for drums operating at 70% of their critical speed, and it shows that the release point at that speed occurs at 119° of rotation.

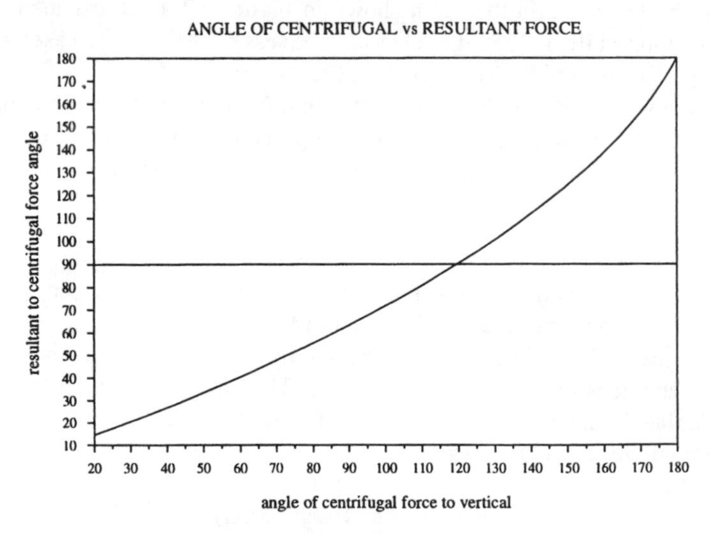

Figure 9.8 Trommel screen angle of centrifugal vs. resultant force.

Using those principles, it is possible to calculate the relationship between the percentage of critical speed and the point of rotation at which the release angle occurs (A_1), for any drum diameter. To do so it is only necessary to know two sides and one angle of the force parallelogram. Resultant force a can then always be calculated and C is 90°, and so angle A_1 can be calculated.

In these circumstances, since the triangles of forces that make up the parallelogram are right-angled, then the length of side a is shown by Pythagoras' theorem as

$$a = (c^2 - b^2) \quad (5)$$

and since angle tan A = a/b, then

$$A_1 = 180 - A \tan (a/b) \quad (6)$$

Figure 9.9 again shows such a calculation presented in graphical form. It reveals the angle of rotation at which release occurs for all possible percentages of critical speed, for any drum.

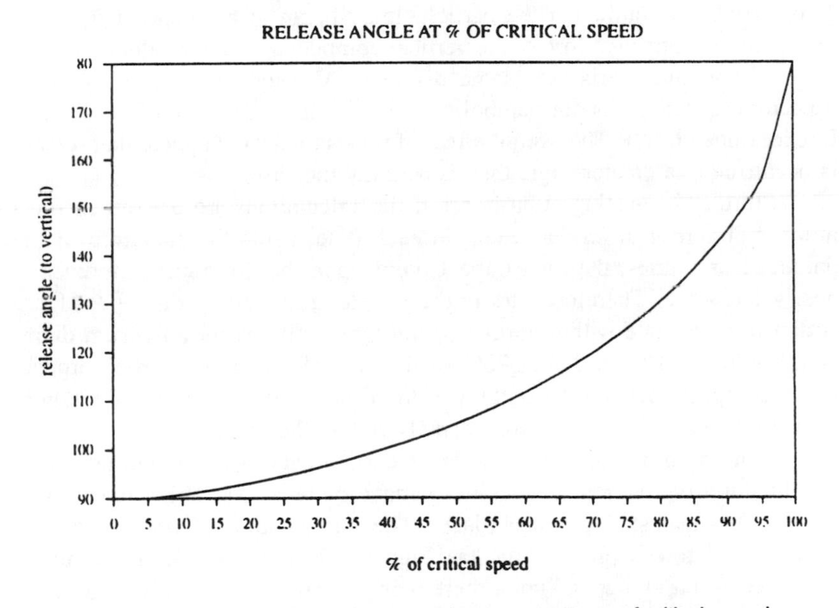

Figure 9.9 Drum screen release angle at percentages of critical speed.

These calculations lead to another mathematical opportunity. Since the trajectory of a particle is dependent upon the initial velocity and the angle of projection, it becomes possible to calculate the path of the particle as it leaves the circumference of the drum using the following formulas. That in turn

$$\text{Horizontal component of velocity} = u \cos A_1 \quad (7)$$

$$\text{Vertical component of velocity} = u \sin A_1 \quad (8)$$

$$\text{Horizontal position at time “t”} = u \cos A_1 t \quad (9)$$

$$\text{Vertical position at time “t”} = u \sin A_1 t - 0.5gt^2 \quad (10)$$

In Figure 9.10 the results of a series of calculations are plotted, using the horizontal position against the vertical position in respect to a drum screen of 10 ft (3 m) in diameter, running at 70% of its critical speed. Since the dimensions of the drum are known, it is therefore possible to establish exactly where upon the drum circumference the particle will impact. It is equally possible to design the screen to run at a rotational speed such that the particles of waste impact where the designer requires, to achieve maximum breakage — in other words, on an unobstructed area of screen plate.

Similar calculations for a range of speeds permit the preparation of a diagram as shown in Figure 9.11, where the trajectories are plotted for each. This reveals that, although the particles may begin to be released from the influence of centrifugal forces, the vertical component of their velocity causes them to continue to rise for some distance. Although, as with all particle trajectories, their paths are parabolic, they initially follow the drum circumference quite closely. The overall effect of this is to make it appear that release is occurring at a greater angle than is actually the case.

A further interesting point arises if the calculations are performed for a number of screen drum diameters. In each case, while the curvature of the parabolic arc varies, the final impact point upon the drum circumference is nearly the same. Therefore, the impact angles revealed in Figure 9.11 are sufficiently accurate within normal operating conditions for any drum diameter. There it is shown that at 70% of the critical speed, the particle impact point is slightly beyond the bottom of the drum in the direction of rotation. At 75%, however, it is quite considerably *before* the bottom.

As the drum rotates, the material is carried against its circumference in the direction of rotation. It does not behave quite as one might expect and simply slip against the screen plates. The combination of centrifugal and gravitational forces prevent that, as Figure 9.7 suggests. The calculations discussed so far all depend upon there being no slip against the screen plates at all, and in fact that is not far from the truth. Observation of a trommel screen in operation treating municipal waste demonstrates that the calculations give valid results, and so the slip must be so close to zero as to be discounted.

As a result, if the drum is filled to the correct level, the lowest point upon its circumference is largely clear of material. Since the best breakage of materials is achieved by the greatest impact momentum, the lowest is the optimum target area for the particles. That is achieved with a rotation speed about 71% of critical.

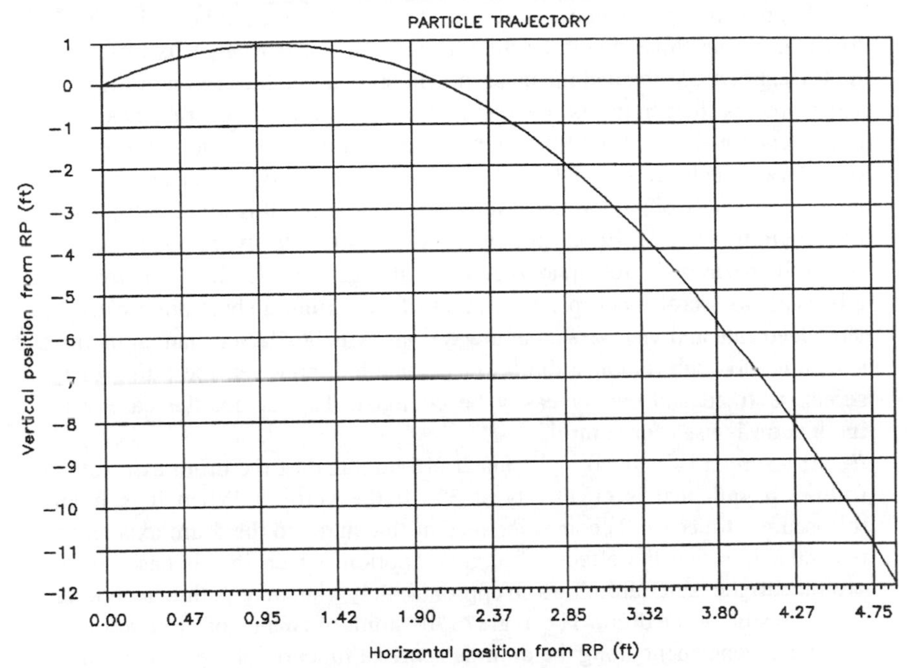

Figure 9.10 Drum screen particle trajectory.

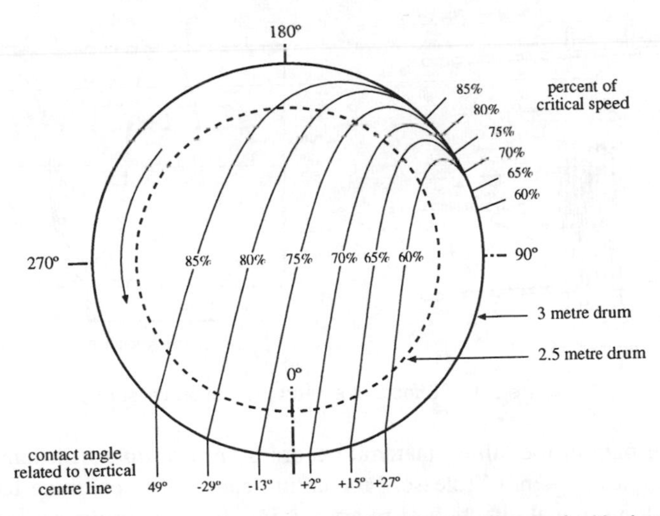

Figure 9.11 Drum screen particle trajectories at percentages of critical speed.

In order for the optimum conditions described by the mathematical model to exist, the machine should be filled to no more than 10% of its total volumetric capacity. More, and the material will spread into the region of the drum where the particle impacts are required to occur. The feed rate has to be consistent for the required fill volume to exist at all times, and is secured by one of two methods. Either the drum is inclined toward the discharge end at an angle (usually 3°) to the horizontal, or a helical scroll is fitted in the inside of the circumference. Either principle works satisfactorily, although the use of scrolls permits the feed rate to be fixed at higher levels. Since the material advances one scroll pitch per revolution of the drum, a short pitch causes a slow feed rate and vice versa, but always one which is faster than an inclined screen could achieve. Once the scroll is installed, however, the rating of the screen is fixed and cannot easily be changed. This is not the case when inclination is used for control.

If a screen is inclined to 3°, material lying against the drum as it rotates follows a path that is effectively at 3° off the vertical. When it is finally released, its trajectory becomes inclined at that angle to the drum axis almost instantly. Ignoring the effects of angular motion, which in this case are not prominent, the material follows an approximately triangular path when viewed from the side of the drum. The angle of the point of change of direction is 3°, and the advancement along the drum is nearly a function of the tangent of the angle, as shown in Figure 9.12.

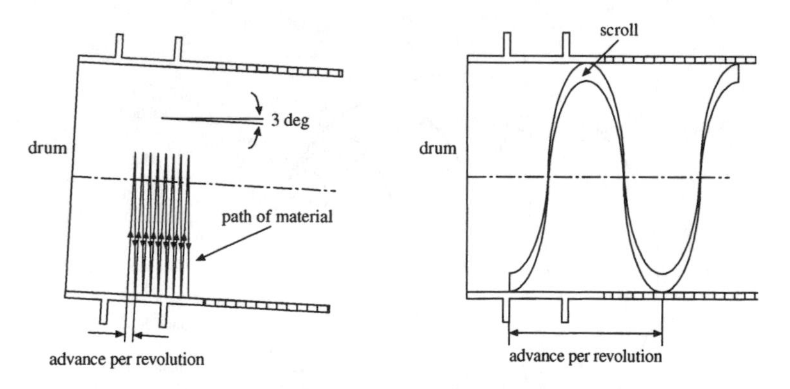

Figure 9.12 Effects of screen inclination vs. scrolls.

The path of the falling material becomes vertical *almost* instantly, but it is not so at the point of release. Momentum causes it to continue for a short time in the original direction of rotary motion. Equally, in the path described the distance of forward travel is *nearly* a right-angled triangle. It is possible to calculate both the actual path of the material and the exact angles enclosed in that plane, but somewhat more sophisticated mathematics is involved. Waste processing is never so precise as to warrant such an evaluation, and therefore this is another case where *nearly* is sufficient. It can therefore be said that the material rises to a height of 70% to 80% of the diameter of the drum, and in falling back to the circumference it advances forward by the amount revealed by the formula

$$l = h \sin \Phi S \text{ ft/min (m/min)}, \tag{11}$$

where l is the advance along the drum, h is the maximum height of the particle trajectory in meters as revealed by equations (9) and (10), Φ is the angle of inclination of the drum, and S is the rotational speed in revolutions per min.

In this case, a mathematical model is somewhat misleading, since it implies that all of the material stays inside the drum throughout its full length, advancing by distance l at each complete revolution. That of course is not actually the case, since much of it will pass through the screen plates in the first few rotations. The potential feed rate of the machine is therefore substantially greater than the mathematics would suggest.

Carrying out the calculations for a screen of 10 ft (3 m) diameter, running ATx 17 rpm and inclined at 3°, with a material trajectory rising to 8 ft 3 in. (2.51 m) above the base, implies that the material feed rate is

$$l = h \sin \Phi S$$

$$l = 8.25 \times 0.052 \times 17 \text{ m/min}$$

$$l = 7.29 \text{ ft/min (2.2 m/min)}$$

If the screen section is 29 ft 6 in. (9 m) long, then the residence time of the material within it is the length divided by the feed rate, or 4.05 min.

The total enclosed volume of the drum would be 2317.2 ft^3 (65.6 m^3), 10% of which is 231.7 ft^3 (6.56 m^3). Theoretically, the machine could therefore process material at a rate given by

$$\text{Capacity} = \frac{\text{Total volume} \times 10\%}{\text{Residence time}} \tag{12}$$

which, in the case of the above screen, is 57.2 ft^3/min, or 3432.6 ft^3/h (97.21 m^3/h). If it were to handle crude municipal waste with an apparent bulk density of 9.4 lb/ft^3 (150 kg/m^3) at the screen inlet, then its mass capacity would be 16 t/h (14.6 t/h). It should be noted that the calculated capacity remains the same whatever the length of the drum. It is determined solely by the diameter, rotational speed, and inclination, and any increase in length simply increases the residence time correspondingly. Consequently, designing an over-long screen only has the effect of limiting the screening value of the sections toward the discharge end.

It is at this point that the commercial reality of the engineering profession enters the equation. The manufacturer of the trommel screen is likely to be required to provide performance guarantees, and he will feel confident in issuing such guarantees in respect to the capacity calculated as described. In that case, he has the comfort of knowing that the screen used as an example could process 16 t/h if nothing at all passed through the mesh. It is for that reason that it is rare for a trommel screen to fail to meet its production targets.

In truth, of course, the screen described would deal with much more waste than 16 t/h. A major proportion of the waste would already be of a particle size smaller than the screen perforations, and could be expected to be discharged as an undersized fraction in the first one third of the drum length. Glass containers and larger pieces from the organic fraction would be broken by the continuous tumbling action, and they too would be released quite quickly. Unfortunately, since the initial particle sizes and the rate of breakage of other materials can never be known accurately, there is no mathematical route by which the exact capacity of the screen may be established.

It is however possible to make an informed estimate, since there is likely to be sufficient information on hand to at least provide for that. It is known that the screen will not break plastic containers or film, and will have only a minimal effect upon paper, cardboard, and textiles. Metal cans will all be retained in the oversized fraction, although some small, dense metallic items will pass through the mesh. Equation (11) reveals that the residence time in the machine in question would be 4.05 min, and so in that time the drum would rotate 68.85 times. In effect that implies that a glass container, for example, could be thrown against the steel screen plates up to 68 times from a potential height of about 8.2 ft (2.5 m). Even one impact is likely to be sufficient to cause breakage in those circumstances, and so it is reasonable to suppose that no glass would remain in the mix beyond the midpoint of the drum length at least.

Most of the organic materials will already be smaller in particle size than the screen mesh perforations, and the larger materials will rapidly become fragmented by the continuous impacts from broken glass and metal cans. Even the most robust organic items are unlikely to survive beyond two thirds of the drum length. The fine residues generally found in waste and consisting of ash, soil, etc. will be liberated very quickly, and a small fraction of the paper and card will also become fragmented sufficiently to pass through the mesh.

An increase in the infeed rate would appear feasible, but it would have the effect of overloading the first one third of the drum length and reducing the separation efficiency there. That would pass the overload on the next one third of length, and so on. Even so, it is unrealistic to conceive of a drum screen through which nothing is ever screened out.

There is of course a route via which an assessment of maximum *theoretical* capacity can be made, and that is best illustrated by continuing with the example. Assume that the screen described is to handle waste, the volumetric analysis of which is paper and card 60%, organics 15%, plastics 10%, metals

8%, textiles 1%, glass 5%, and fine residues 1%. If the machine is imagined as being in three sections, each 10 ft (3 m) long, an estimate of likely undersize in each section could be made, assuming for the moment that there was no loss of efficiency resulting from overloading. A realistic estimate for the first 10 ft (3 m) section would be paper and card 2%, organics 10%, metals 1%, glass 2%, fine residues 1%. In the second section, a further 5% paper and card, 5% organics, 1% plastics, 2% metals, and 3% glass would be expected to report to the screenings output. These figures are expressed as percentages of the total feedstock to the screen, and using them implies that 16% of the original volume is lost in each of the first two sections.

The total volume of the final one third of the drum is 772.4 ft^3 (21.87 m^3), 10% of which is 77.24 ft^3. If the initial volume reduces by 16% in the first section, and by a further 16% of the original in the second section, then the original input volume to the drum must be

$$V_i = \frac{V_3}{1 - 0.16 - 0.16} \tag{13}$$

where V_i is the input volume and V_3 is the volume in the third section. This suggests that an input volume of 113.6 ft^3/mm (3.2 m^3/min) is possible, or 6816 ft^3/h (192 m^3/h). Again, at an as-received waste density of 9.4 lb/ft^3 (150 kg/m^3), this equates to approximately 32 t/h (29 t/h), or almost double the original calculated figure. Reality will lie somewhere between the two, and will depend upon the degree of efficiency sacrificed in the first two sections by overloading them. One could therefore go on to assume that the overall loss of efficiency would be 50%, so that the input tonnage estimate becomes 16 t/h plus 50% of 16 t/h. In other words, the screen would have a minimum capacity of 16 t/h and a maximum of 24 t/h.

Instead of inclining the machine, the designer may choose to build it with scroll plates. In that event the performance becomes quite different. The scrolls will move the material forward toward the discharge quite artificially, and if pitched coarsely enough, they will influence the release point on the circumference. They may cause the material to elevate excessively, and to advance at a rate greater than the screening capacity of the screen plates.

There is also the more practical matter of the way in which the waste is collected. The use of bags is becoming common everywhere, particularly as recycled plastic becomes more available. If the scrolls are pitched closer than the average width of a bag, then there is a possibility of bags becoming jammed between them and therefore escaping, becoming split to release their contents. However, if the pitch is much larger than a bag width, the probability is that the screen will feed too rapidly. Unscreened material may be discharged with the oversize.

Obtaining the maximum capacity by overloading the first two sections of the machine is impractical where scrolls are used, since they feed the total mass through at the same rate whatever it happens to be. Running the machine in that way simply overloads all three sections, and if the final section is overloaded, then undersized materials will appear in the oversize stream.

Before leaving trommel screens, a final design requirement is often for bag-splitting in the first section of the drum. As observed earlier, waste increasingly appears at treatment facilities in bags, and they need to be burst to liberate their contents. Many plants have been built with some form of bag opening on the reception conveyors, but such devices are somewhat impractical. It is always best to keep the waste inside the bags until it can be released without spillage or emissions of dust, and the first obvious place for that is inside an enclosed trommel. Bag splitting there, by means of sharpened steel spikes welded to the longitudinal stiffeners in the first section, is extremely efficient, and any number of spikes can be fitted without affecting the screen efficiency. In 25 years of operating trommels with bag splitting spikes, the writers have never observed an intact bag in a trommel oversize discharge.

9.2.2 Granulators and Shredders

The term "granulators" used here is generic for hammer mills, flail mills, pulverizers, and the many other names that have developed with the waste disposal industry over many years. Of the many makes of granulators available, there are two basic designs for dealing with crude or screened crude waste, distinguished by either horizontal or vertical rotor shafts. Shredders have a somewhat more limited share of the waste treatment market, and in the context of this section the term is taken to mean machines that cut or shear materials rather than breaking them by impacts.

The two basic designs of granulators are shown in Figure 9.13. In the vertical shaft machine, the rotor is suspended vertically in a conical casing, mounted upon a heavy thrust bearing in the base, and directly driven by a high-power electric motor on top. The machines are always single-rotor. Horizontal shaft machines meanwhile are almost always belt-driven by separate electric motors, and twin rotors in the same casing are common.

9.2.2.1 Vertical Shaft Granulators

The vertical shaft machines often, although not always, have grinding ledges in their casings. Usually, these are projections extending into the throat of the machine from bolted lining plates, pitched in such a way that the rotating hammers pass between them. The conical reduction chamber tapers into a cylindrical grinding chamber in the base, where a final group of short hammers carries out fine granulation. The machines do not contain any provision for retaining material until a predetermined degree of granulation has been achieved, and so they are capable of passing oversized objects right through. However, where a machine is correctly specified for its task, that is a rare occurrence.

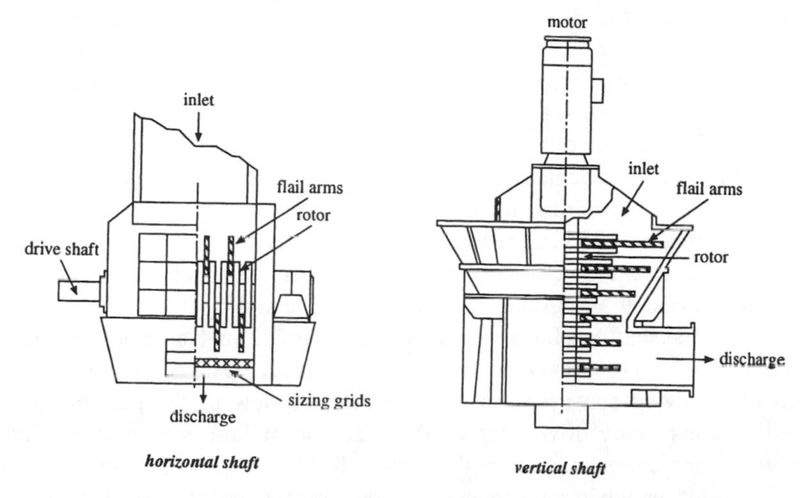

Figure 9.13 Typical vertical and horizontal shaft flail mills.

Because of the design of the grinding chamber, discharge from vertical shaft granulators is horizontal at right-angles to the axis of the rotor. Since the rotating hammers in the chamber act in a similar manner to the blades of a centrifugal fan, granulated product is effectively blown out of the machine.

All granulators rotate at comparatively high speeds, usually between 800 and 1200 rpm, and so the degree of "windage" created by their rotors is substantial. Vertical shaft machines generate windage through their casings in the direction of product flow, and so are easier to feed with very low-density materials. There is little or no backflow of air in the feed canopy. The pressurized discharge also makes them particularly suitable for use with column air classifiers, which can be bolted directly onto the discharge aperture. In such a configuration, the granulated material is blown at right angles into a vertically rising airstream, thereby offering ideal conditions for heavies/lights separation.

Machines offering waste reduction capacities of 20 to 30 t/h are common, and they are very large and heavy devices indeed. They impose heavy static loads upon their mountings, and they also impose large dynamic loadings in a horizontal plane. For those reasons, they are always installed upon deep concrete foundations, which are usually constructed independently of the foundations of the buildings in which the machines are housed.

Generally, the reduction of municipal waste in a vertical shaft granulator calls for a power ratio of about 30 hp/t, and so a 20 t/h machine would require a 600 hp drive motor.

9.2.2.2 Horizontal Shaft Granulators

Horizontal shaft machines consist of a rotor fitted with swing hammers; in this case, the rotor is supported in two end bearings in plummer blocks.

Drive to each rotor is invariably provided by a single electric motor per shaft, transmitted through flexible transmission belts and grooved pulleys. There is a much greater range of machine capacities available, offering processing rates from a few pounds per h at laboratory scale to over 25 t/h in major plants.

Twin-shaft machines are common for the higher feed rates. There, two identical rotors are mounted in parallel in a single casing, in a configuration best described as appearing like two machines welded together at their sides. Infeed is through a charging aperture above the rotors, and is delivered to the center of the machine between them. Twin rotors are usually driven to run counterclockwise to each other, and are often reversible.

Horizontal shaft machines discharge downward through a vertical aperture in the base, which prevents them from doing so directly into a classifier. Some form of receiving conveyor is necessary, and in view of the possibility of metallic items being driven out of the machine at high speed, steel plate conveyors are common. The casing immediately below the rotors houses grinding grids, which in some makes are heavy manganese steel castings rather like a sieve mesh with square or oblong holes. In other makes, a part of the casing is fitted with grinding bars, which are simply steel bars bolted to heavy lining plates around part of the periphery. In machines of those types, there is often still some form of grid, although here the purpose is simply to retain oversized items until they are reduced.

The sizing grids allow the machines to be adjusted to achieve a desired granularity by being interchangeable. Although coarse grids are used for coarse reduction and fine ones for a small particle size, it is not simply a matter of installing a grid of 2 in. (50 mm) aperture to obtain 2 in. (50 mm) particles, for example. The relationship is rather more complex, and is largely dependent upon the material being processed. Generally, 6 in. (150 mm) grids provide a granularity of >2 in. (50 mm), while 4 in. (100 mm) grids produce an average of $>{}^{3}/_{4}$ in. (20 mm).

Either type of machine is capable of very fine reduction when handling paper and card, particularly the vertical shaft versions. Granularities of 95% $>{}^{1}/_{2}$ in. (10 mm) are not uncommon. There are, however, two design factors that need to be taken into account when specifying a machine, and neither is always obvious. First, there is a direct relationship between the degree of granulation achieved and the throughput capacity. Very fine grinding requires a larger number of hammers to be installed on the rotor, and they take up much more of the available space inside the casing. As a result, there is less room for the product. Second, it is a common misconception that granulators are essentially volume reduction machines. This is probably an assumption born of memories of similar machines being used to prepare waste for landfill. In such an application, they do indeed provide some element of volume reduction, simply because they break up glass and plastic containers. However, where they are being used to process screened waste mixtures that are predominantly paper and card, they are more likely to increase the volume. The phenomenon is the result of transforming a mass largely in sheet form to one highly granular and fibrous. Depending upon the moisture content of the feedstock, it is not unusual to discover that the apparent bulk density upon discharge from a machine is reduced to nearly half that of the infeed.

While traditionally granulators used heavy cast hammers, linked to rotor discs by steel shafts and allowed to swing freely over at least 180°, modern practice in the RDF industry has moved in the direction of flail configurations. In these, the hammer heads are replaced by comparatively thin flail arms ranging in thickness from $^1/_2$ in. (10 mm) to 1 in. (25 mm), again fixed to the rotor discs by steel shafts. The flail arms may be hardened along their edges, although experience has suggested that doing so confers no real benefit in terms of extended life.

The general conversion to flail arms came about largely by the recognition that if one were attempting to shred paper and card, then doing so by hammering was somewhat illogical. The flail configuration uses less power, increases the throughput for a given size of machine, and yields a finer and more consistent product. Both the vertical and horizontal shaft machines can be converted.

Mathematical modeling of granulator characteristics is difficult if not impossible, since there are so many unknown factors influencing the way in which a machine will perform. The initial condition of the feedstock is a major contributor of variables, in that for example higher moisture contents make the material easier to fragment. Higher proportions of card in the mix tend to reduce machine capacities, because the reduction in apparent bulk density is greater. Large sheets are more difficult to deliver to the rotor through the charging aperture of the granulator.

Machine-associated variables are related to the condition of the flails and how many of them are installed, to the rotational speed, and to the spacing of any grids or grid bars. The fitting of more flails to the rotor in order to increase granulation has the effect of reducing throughput as a result of the available space becoming increasingly occupied by steel components. It also increases rotor windage in direct proportion in most machines, again making the delivery of feedstock into the granulating chamber more difficult.

Probably the factor that has most effect upon performance is the period of residence time in the machine required to achieve the necessary granularity. There is no obvious way of even assessing what that may be, since it will vary not only with the type of feedstock, but also in response to variations in feedstock quality. It is for that and the other reasons discussed above that manufacturers prefer to run trials upon a particular feedstock before pronouncing upon the capacity they can offer.

9.2.3 Shredders

9.2.3.1 Rotary Shears

Shredders fall conveniently into the two main design types shown in Figures 9.14 and 9.15 rotary shears and knife mills. Rotary shear machines generally consist of a single or pair of shafts, driven by hydraulic motors, and mounting a series of interlocking, toothed cutting discs. It is usual for forked ledges to protrude from the casing between the outer edges of the cutting discs, although they are omitted from Figure 9.14 in order to preserve clarity of the general principles. On twin-shaft machines, the shafts counter-rotate, and in common with single-shaft machines they automatically reverse in the event of overload.

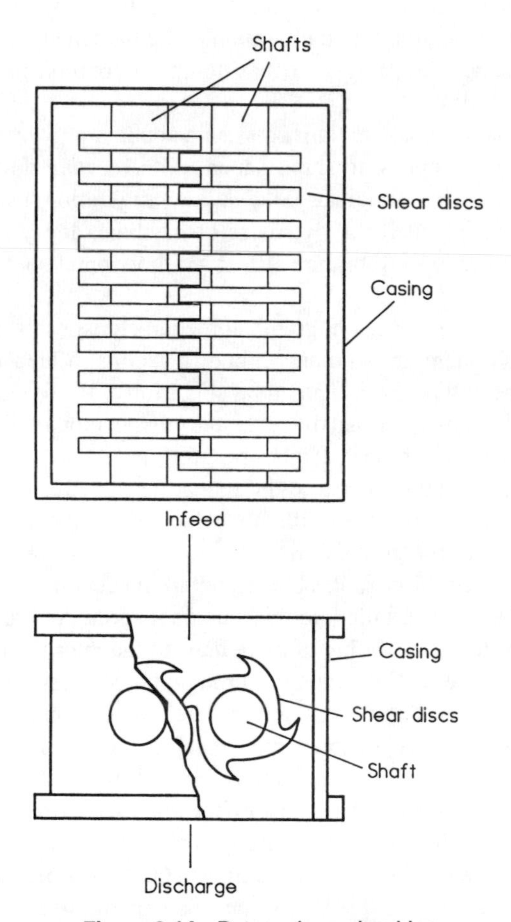

Figure 9.14 Rotary shear shredder.

Rotational speeds are very low, generally below 100 rpm, and down to 20 rpm is common. The low-speed hydraulic motors driving the shafts develop very high torques, providing considerable shearing power to the cutting discs. This power makes the machines perfectly capable of cutting through even reinforced concrete. It also offers the singular advantage, not experienced in other types of shredder, that the harder it is fed with material, the more efficient

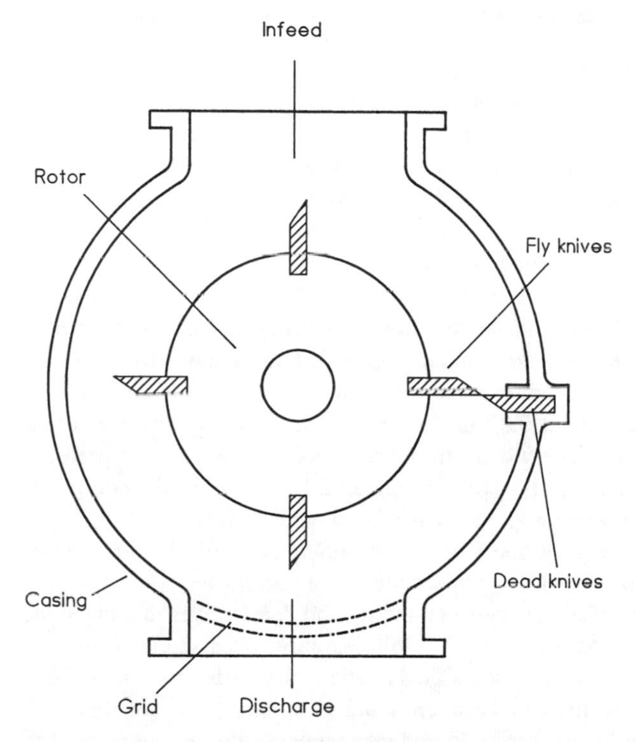

Figure 9.15 Rotary knife shredder.

the shredding action and the higher the throughput of product. For that reason, most manufacturers offer hydraulic ram force-feeders, which are installed in the charging hopper in such a way as to physically ram feedstock into the cutting chamber.

Machines of this type are by no means as common in waste processing plants as their capabilities would suggest they should be. Possibly that again is a result of tradition rather than engineering objectivity, since while offering considerably higher capacities than granulators, they invariably cost much less. Their perceived disadvantage, certainly in the formative years of the RDF industry, was that they are incapable of very fine granulation down to $>^1/_2$ in. (10 mm). Granularities of below about 4 in. (100 mm) are unusual. However, as will be discussed in Section 9.3, that is no longer a barrier to successful RDF production. Rotary shears are therefore likely to begin to make some impact in process plants.

Once again, mathematical modeling of machine performance is virtually impossible due to the large number of variables involved. Rotary shears even increase the number of variables by their use of force-feeders, which may be set to deliver material to the shearing chamber over a very wide range of bulk densities.

9.2.3.2 Knife Mills

After some early experiments, knife mills were generally discontinued in RDF flowline designs, mainly because they are too sensitive to hard or abrasive contaminants. They are also capable of a degree of granularity considerably in excess of the requirements of a densified RDF process. Granularity with these machines is achieved partly by the very high speed of the rotor, and mainly by the close proximity with which the fly knives pass the dead knives.

The design principle, as shown in Figure 9.15, mounts a series of long, sharpened knives in special clamps on a high-speed rotor. The clamps provide a means of radial adjustment in most but not all cases. A further set of knives, the dead knives, are similarly clamped in a row inside the casing of the machine, in such a way that the clearance between them and the fly knives is on average about 0.02 in. (0.5 mm). Usually, as shown in Figure 9.15, the casing immediately beneath the rotor has provision for a perforated grid, the purpose of which is to prevent material from leaving the cutting chamber until a predetermined granularity has been achieved.

It is the very fine granularity obtainable with them that originally made rotary knife machines appear attractive for waste processing. But it was the very close clearance between the fly and dead knives that made them so prone to damage. At 0.02 in. (0.5 mm) clearance, even staples from screened paper and card may do considerable damage to the knife tips, and in their as-supplied form, blade lives of between 1 and 3 weeks were not uncommon. Special blades with very hard alloy edges overcame the problem to a limited extent, but were extremely expensive. However, the end to possible applications for rotary knife machines came about when better downstream processing eliminated the need for very fine granulation.

The machines are discussed briefly here because they still appear occasionally in treatment plants, and are offered regularly for new ones. Unfortunately, they are hardly suitable for the arduous duties involved, and so should be discounted.

9.2.4 Classifiers

All classifiers, where the term by common usage has come to mean the separation of one material from others by means of a fluid flow or by ballistic characteristics, work by distinguishing the difference between densities. In this case it is the *true* density that is important (see Section 2.3), since that determines how a particle will behave in a fluid flow or ballistic situation.

In RDF process lines, fluid flow invariably means air flow, air being regarded as a fluid in the thermodynamic sense. Air classifiers are very common machines in the initial and final treatment stages, and because of their importance, a great deal of research has been conducted into their design. The result of this has been the development of four variations on the general principle of air separation: column classifiers, zigzag classifiers, drum classifiers, and air knives. Figure 9.16 describes the first two of those types.

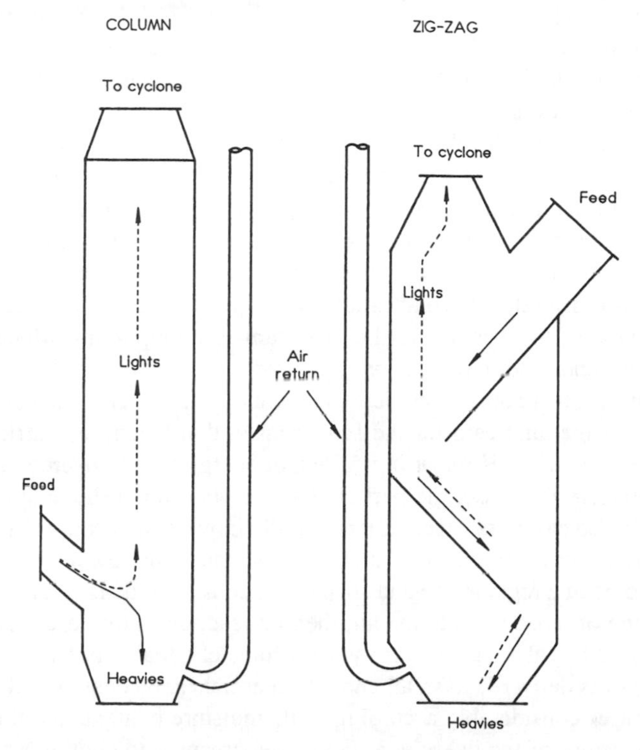

Figure 9.16 Air classifiers.

9.2.4.1 Column Classifiers

While they are invariably large structures, column classifiers are actually quite simple devices. In design they consist of a vertical duct, usually of rectangular cross-section, through which air is drawn by a high-volume, low-pressure fan. Material for separation is introduced into the side of the column a short distance above the base, entering the air flow at as near to a right angle as possible. Depending upon the air speed, which is adjustable by means of a damper on the fan suction port, some or all of the light fraction of the material is entrained in the airstream. Heavy fractions meanwhile fall through the airstream to be recovered by a collecting conveyor beneath the classifier base.

The light fraction is removed from the airstream by a cyclone separator, inserted between the column and the fan. Air from the cyclone passes through the fan control damper, through the fan, and usually onto a splitting damper. There it may be proportionally diverted to a dust extraction plant and recirculated back to the classifier base.

Material infeed to the classifier is through a rotary valve, where the device is charged by a conveyor. However, where a vertical shaft granulator is providing the initial size reduction, it is normal for the column to be fitted directly onto the granulator outlet. This arrangement has considerable advantages, in that the high velocity of the horizontal discharge from the granulator allows the material to be introduced into the air stream at the ideal angle of presentation, and in a manner that creates maximum turbulence. In such an application, rotary valves are unnecessary.

While in theory mathematical modeling of a column classifier performance should be relatively straightforward, the opposite is actually the case. A number of mathematical techniques appear to confirm experimental results, but they are reliable only within a very narrow range of parameters. However, the purpose of an air classifier in an RDF process is to separate the combustible light fraction from the noncombustible heavy fraction. If that is achieved, then the behavior of the heavies in the airstream is unimportant. Modeling can concentrate upon the light fraction.

Even then there are difficulties. Conceptually, a particle of paper becomes entrained in the airstream for the same reason that it may be carried by the wind: the moving air is imparting sufficient energy to it to overcome gravitational force. In that case, the particle has weight, which the airstream must support. It also has mass which, if it is initially moving in the opposite direction to the air, creates a momentum that the airstream must overcome. Were the particle to be of a regular size and shape, of a consistent density, and to always present one of those sides to the air, then its reaction to the forces created by the air could be calculated precisely. Unfortunately, that is not so. Paper from a granulator is delivered in small and irregular flakes. Its density and therefore weight varies considerably according to its moisture content, and the characteristic behavior of the flakes in a moving airstream is to continuously rotate. Very little air speed is needed to move a sheet of paper when the air is presented to the flat face, but moving it by applying air flow to the edge requires very much more. In spite of that, air classifiers do give surprisingly predictable results and acceptable separation at air speeds of between 21.3 ft/s and 24.6 ft/s (6.5 and 7.5 m/s), with a consistent infeed.

They are, however, not sophisticated machines, and so while they will separate metals and minerals from granulated paper efficiently, they do not eliminate organics and dense plastics with any reliability at all. This and other characteristics of a typical column classifier are shown in Figure 9.17a–h. The separation graphs are a compilation of data from several devices, and are reasonable representatives of the performance of any of a similar design. They reveal that at air speeds above 22 ft/s (7 m/s), for example, although paper and card recovery is maximized, there is an increasing contamination from organic materials. They also suggest that at almost any air speed there is likely to be contamination by nonferrous metals, particularly in the particle size range between $^3/_4$ to $1^1/_2$ in. (20 to 40 mm). In that range, the metals will almost invariably be aluminum foils.

A further observation arising from examination of the data is that column classifiers operate mainly in the air speed range from 6 m/s to 8 m/s. In that

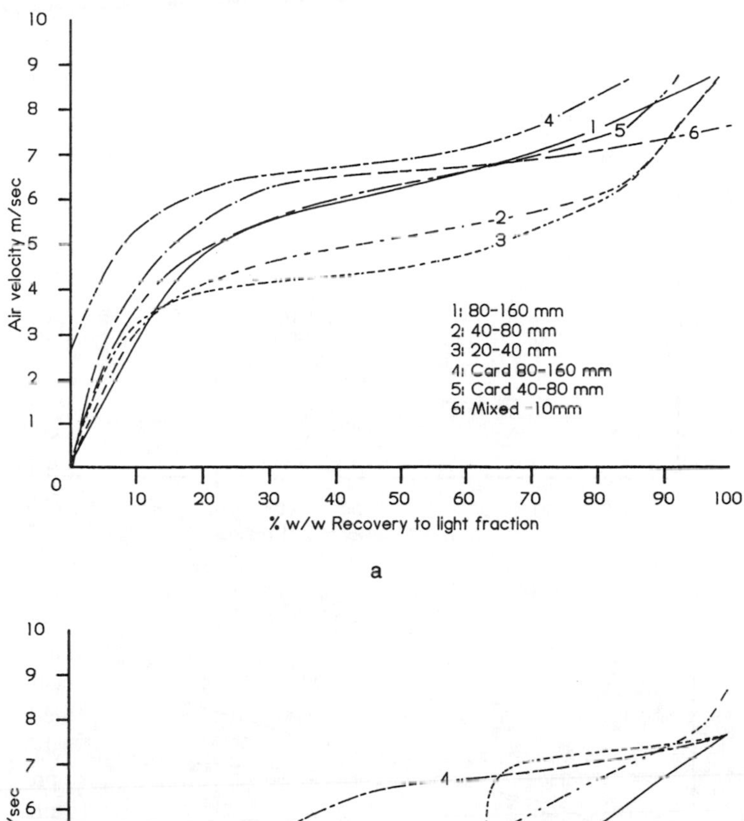

a

b

Figure 9.17 The behavior of various materials in the column classifier. a. Paper and card. b. Plastic film. c. Dense plastic. d. Textiles. e. Organics. f. Ferrous metals. g. Nonferrous metals. h. Glass fines.

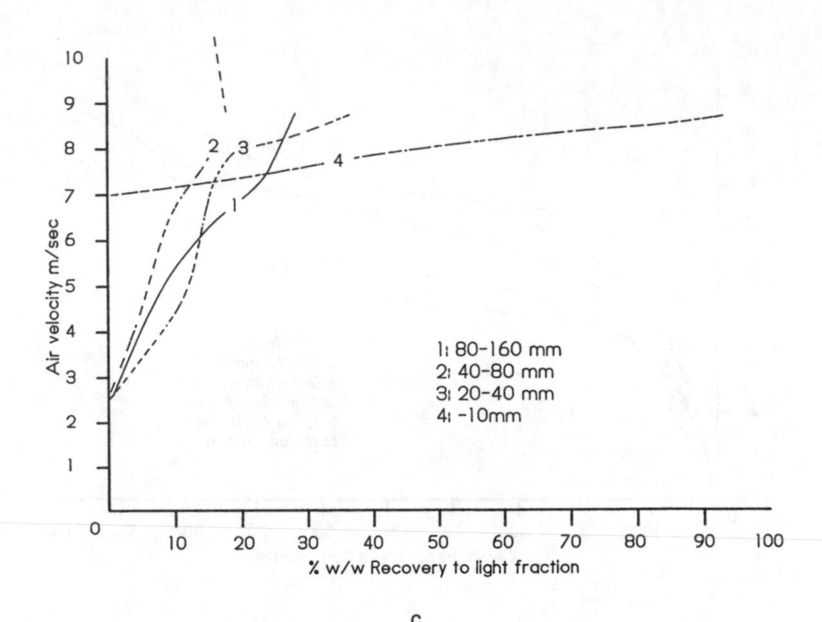

c

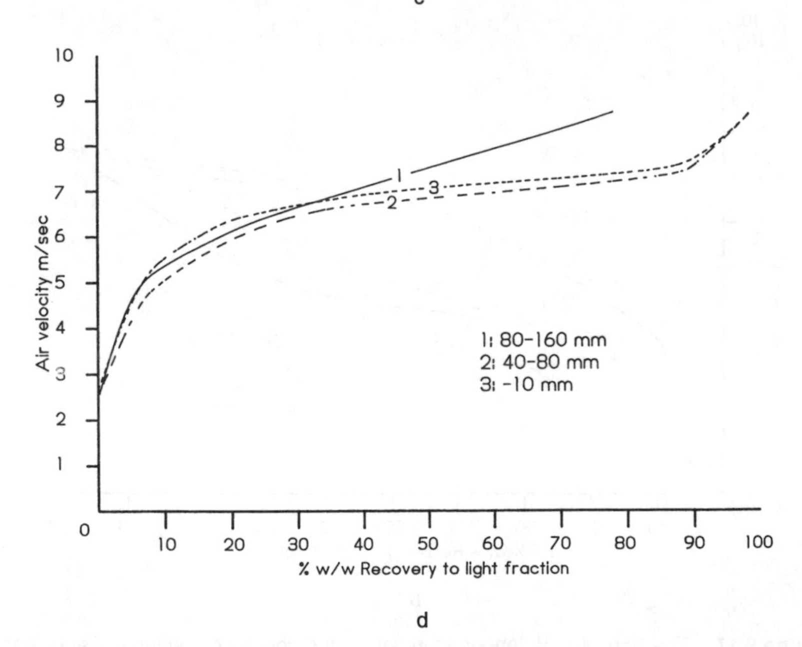

d

Figure 9.17(*Continued*).

range, a very small increase in air speed results in a correspondingly large increase in recovery to the light fraction. This is particularly noticeable in respect to the waste fractions of interest in an RDF process. Paper and card milled to >$^1/_2$ in. (10 mm) is recovered at 15% at an air speed of 19 ft/s (6 m/s), and at over 60% at 22.5 ft/s (7 m/s).

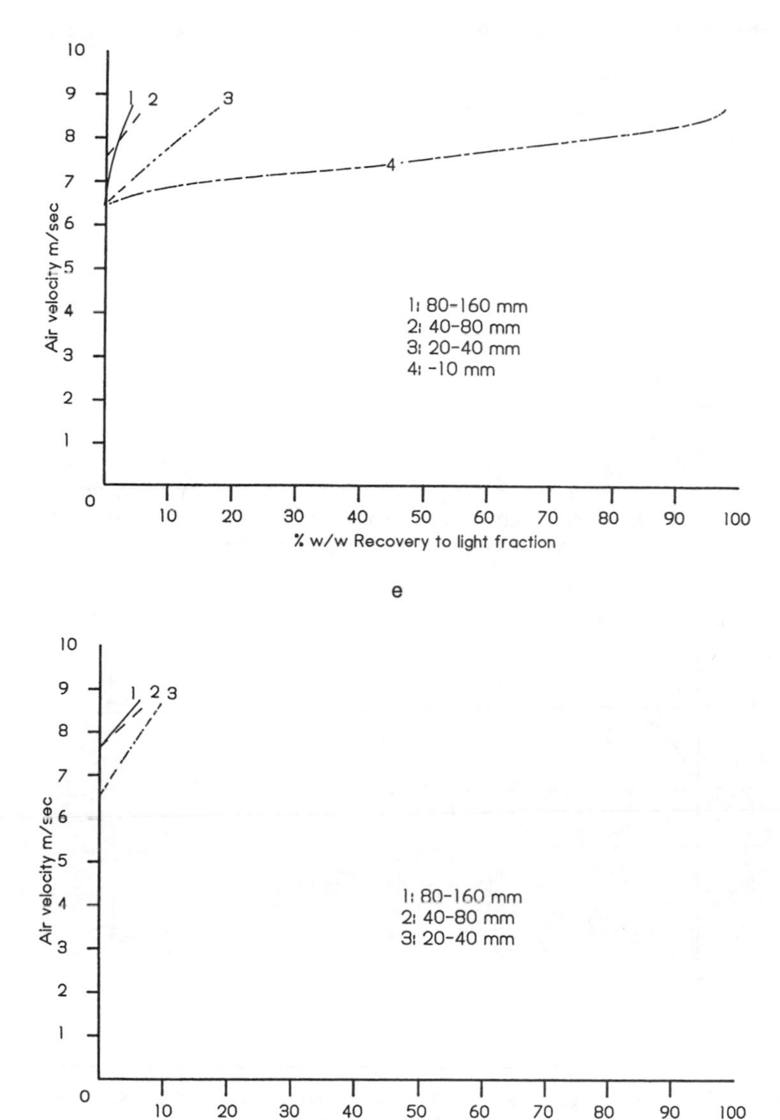

Figure 9.17(*Continued*).

Paper and card that has been granulated to that degree also exhibits a behavior that appears inconsistent with the general trend in reaction to granularity. Figure 9.17a shows that the air speed necessary for recovery to the light fraction is reduced in direct relationship to the particle size, but the $>^1/_2$ in. (10 mm) fraction appears to require a higher speed than any of the others over most of the speed range. The explanation for this probably lies in the

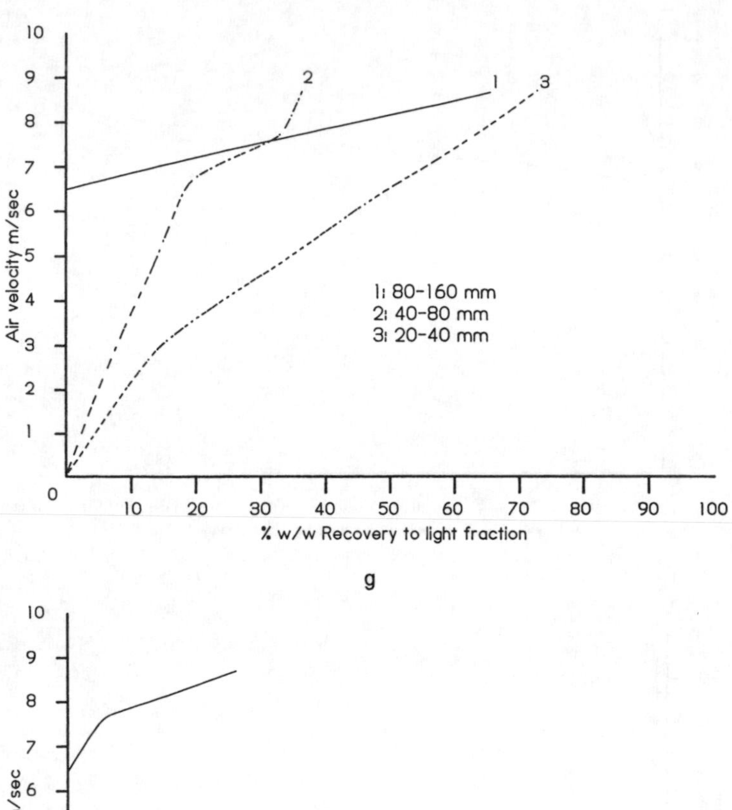

h

Figure 9.17(*Continued*).

change in nature of the material when it is granulated to very small particle sizes. In that event, there is some loss of cohesion, and particles are much more fibrous. In that form they coagulate into lumps that do not easily break up when released into the classifier, with the result that they behave as if their mass were more similar to much larger particles.

9.2.4.2 Zigzag Classifiers

Figure 9.16 also shows the general principles of a zigzag classifier, so called because of the configuration of steeply inclined ledges inside the separating chamber. In operation, material is introduced through a rotary valve in the upper casing in such a way that it falls upon the highest of the ledges, down which it slides until it passes over the throat onto the lower ledge. The lower ledge then releases the material over a second throat into the heavies collecting chamber in the classifier base.

Air is drawn from the top of the classifier, via a cyclone to a fan as before, and again a proportion of the air is recirculated back to the base of the machine. In traveling through the classifier casing, the air is forced to follow the path of the inclined plates, passing upward through each of the throats. Each throat then acts in a manner rather similar to that of an air knife classifier, but at much lower air pressure and flow.

Zigzag machines are rather more efficient than column classifiers, since friction against the inclined plates slows the feedstock down considerably. The slower the rate of fall of the material, the lower the momentum it contains, and the lower the air flow required to overcome it. Additionally, the passage of the material over each of the throats creates considerable turbulence within its mass, and this has the effect of disrupting any lumps that may have been formed by earlier mechanical handling.

Although the inclined plates improve efficiency, they add yet another variable to the list that would require data before a mathematical model could be created. The feedstock to the classifier may no longer be crude waste, but it still is a sufficiently inconsistent material for any accurate assessment to be made of its behavior. For that reason, manufacturers again rely upon test results to produce performance data, which is the only recourse of the process designer.

9.2.4.3 Drum Classifiers

While somewhat uncommon, machines of this design may still be found occasionally in waste treatment processes. In general principle they are a rotating drum screen through which air is passed into a collecting chamber beyond. They are therefore a combination of both screening and air classifying technology. Figure 9.18 shows the general principles.

The design theory is intuitively sound. When mixed waste is screened in a drum, as discussed earlier in the section on trommels, the rotary action of the drum causes it to be elevated to a considerable height and then dropped against the screen plates. Normal air resistance causes some separation of lights from heavies quite naturally, and the possibility of capturing the lights in an airstream is therefore attractive. It appears to make a drum screen capable of three-way separation, since once the lights are entrained in the air, they are

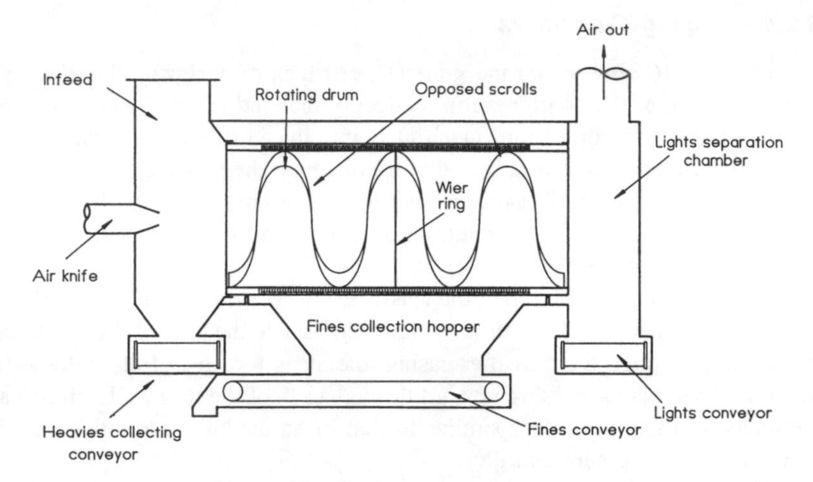

Figure 9.18 Drum classifier.

likely to stay so until recovered by a cyclone or separation chamber. Organic materials, glass, and soils would pass through the screen plates. Oversized heavies such as cans and heavy plastics would be discharged from the end of the drum onto a collecting conveyor. Oversized lights in the form of paper and plastic film would be carried onward by the air flow.

The behavior of the material in the drum conforms generally to the model derived for a conventional rotary screen, and the entrainment of the light fraction changes that very little. In fact, maintaining the light fraction in suspension could be expected to improve the breaking of the heavy residues, since there is less possibility of blinding of the screen plates. Removal of the light fraction is efficient, using air flows slightly lower than those expected in a column classifier. The light materials are already airborne when influenced by the classifying air, and their initial momentum is being overcome only to the extent of changing their direction through 90°, rather than the 180° that would be necessary in a conventional air classifier.

Opposed scrolls are fitted to the inner circumference of the drum, arranged to transport material toward the drum ends. As a result, any large, heavy items that have been forced into the drum by the action of the air knife stage are returned to the heavies collecting conveyor. The light, combustible fraction meanwhile passes through the central weir ring and across the second stage of the screening drum. There any remaining fine contaminants are screened out, and the combustibles are finally released from the air flow in the separation chamber.

While the machines are attractive in the facilities they offer, they have the major disadvantages that they cannot easily be fed directly from a granulator, and sealing is more complex than that of a conventional classifier. Since one of the main purposes is the removal of fine heavies by screening, pregranulation would be counterproductive. It would mix the materials excessively and would increase the loss of the light, combustible fraction beyond acceptable limits. Therefore, the machines must attempt to air classify particles much larger than would be experienced in a more conventional device. Cardboard is likely to escape separation, being too heavy in large pieces, and in consequence it preferentially reports to the heavy oversize stream.

Simplified versions of the machines, without scrolls or inlet end separation chambers, are not unusual for product refining. They have occasionally been employed for the removal of plastic films from municipal waste compost, for example, although in that application it is more a case of providing a through-draft in a conventional drum screen than it is of designing a true classifier.

While an air knife is used at the infeed end to achieve the initial heavies/lights separation, it is a rather more efficient arrangement than would normally be expected. Generally air knives are crude, inefficient devices with little control over the behavior of particles in the airstream. However, where they are used in drum classifiers, they are discharging in the direction of a large-diameter drum where there is adequate space for expansion. Furthermore, since the drum is subsequently vented by a separate suction fan, the pressure differential balancing effect upon the air flow greatly assists in achieving streamlining.

9.2.4.4 Air Knife Classifiers

Air knives are hardly machines in any recognized sense. In their most common form they are simply an air jet directed across a vertical chute, usually into a branch or hopper on the opposite face. Material is dropped down the casing, which may be no more than a transfer chute between two conveyors, and in passing through the air jet the light fraction is blown off. They are particularly inefficient devices, requiring much greater air pressures than any other air classifier (two or three times the air movement of a column classifier), and because they operate by positive rather than negative pressure, control of the light fraction is difficult.

The design of a pressurized air jet that will provide a streamlined flow is extremely complex, and is not usually successful where material is passing the nozzle vertically. As a result, excessive turbulence is created in the separation zone. While that is an advantage with a negative pressure classifier, it is the reverse in this case, since there is then no means or space to restore the streamlining that would be necessary for good product control.

The devices have some application in RDF plants, but they are generally only used as a last defense against metals entering into pelletizers and other sensitive machines. They are insufficiently sophisticated to form part of the main process flowline, and are mentioned here only to complete the discussions on air classifier types.

9.2.4.5 Ballistic Classifiers

While air classifiers are simple devices, their supporting installations of fans, cyclones, and dust extraction plants are large and expensive. Ballistic classifiers were first introduced into the RDF industry in an attempt to establish a cheaper alternative, and they became quite common in the 1970s and 1980s. Again, there are several designs available, ranging from little more than a high-speed belt conveyor that throws material across a partitioned discharge chute, to a complex arrangement of oscillating, perforated decks.

In a sense, some ballistic classifiers are air classifiers in reverse. An air classifier operates by passing air through a mixture of materials in such a way that the lighter fractions become entrained. Belt-type ballistic classifiers, however, attempt to move the material mixture at speed through a region of stationary air. Theoretically, the effect is the same, in that the heavier items are less impeded by the air and so are able to pass on to a distant collecting chamber. Light fractions, meanwhile, are arrested by air resistance, and are caused to fall separately. The benefits offered with this design are that, since there is no air movement, there is no requirement for fans, cyclones, ductwork, or dust extraction plants.

In spite of the apparent attractions, however, belt-type ballistic classifiers have not made a great impact upon the RDF industry. Although theoretically they may be adjusted to achieve separation efficiency by regulating the belt speed, obtaining the desired result is rarely straightforward. They are much more sensitive than air classifiers to variations in feedstock quality and quantity, and at high loads the trajectory of heavies is increasingly impeded by the light fraction.

Occasionally, the machines are described as being capable of separating not only heavies from lights, but also, at the expense of a multi-partitioned discharge chute, of segregating various grades of heavies. Consideration of physics, however, suggests that such a characteristic could not exist, since the trajectory of a projectile is determined solely by the acceleration due to gravity and by air resistance. The mass of the projectile is irrelevant as far as gravitational attraction is concerned, and with heavy materials, air resistance has a minimal effect. A machine may be capable of separating ferrous metal pieces from aluminum foil by entirely ballistic methods, but some of the more exotic claims made cannot be true.

If an air classifier has to provide air speeds in the order of 24 ft/s (7.5 m/s), then to achieve separation of the heavy and light fractions the materials must be projected into stationary air at least at that speed. The requirement is therefore for a belt conveyor capable of an operating speed in the order of 1476.5 ft/min (450 m/min). Such a conveyor is extremely fast by any standards, and is too fast for the practicalities of conveyor belt design. Slower speeds allow the design to work to some extent, but separation reliability drops accordingly.

There is a limit to how slowly the belt can run before the device becomes impractical, although there is no clear demarcation point. The separation efficiency is to a very large extent dependent upon the nature of the materials being treated and the response of the fractions to air resistance. However, the low-speed limit is still extremely fast for a belt conveyor, and it introduces considerable design difficulties. Not least of those is the matter of belt tracking, where any tendency of the belt to move sideways is likely to be catastrophic. Tracking idlers are not an ideal solution because at high travel speeds they are inclined to cause excessive wear to the edges. Domed head and tail drums, where the center diameter is greater than that of the ends, are a better solution, but only at the expense of some stressing of the belt plies.

Cleaning and edge sealing are, if anything, an even more difficult matter to deal with, since at high speeds the frictional resistance between the casing and the scrapers or skirting seals is likely to generate excessive heat and wear. There is no ideal solution in this case, and sealing particularly is essential. The consequences of any material falling from the carrying strand and lodging inside the return strand would, again, be catastrophic.

The third difficulty with the design is encountered in the way in which material is introduced to the device. Some form of transfer chute is inescapable, and however large it may be, the speed of the product when it meets the belt is determined entirely by the acceleration due to gravity. If the chute is installed at a right angle to the belt, then the material must change direction through 90° and must accelerate to the belt speed both at the same time. The initial velocity as a result of falling in the chute has no benefits, and in respect to the momentum gained, may even be harmful. Some assistance is possible by inclining the transfer chute so that it is discharging as nearly as possible in the direction of the belt travel, but municipal waste possesses a steep angle of repose. It will not reliably slide down an incline of less than 60° to the horizontal, and even at that angle it gains insufficient speed to have any significant effect upon the acceleration required.

The consequences of a comparatively slow-moving mass falling upon a high-speed conveyor are several. The mass contains inertia, and not enough friction exists than is necessary to achieve acceleration without slip. There is an immediate potential for excessive wear of the cover ply of the belt, and the faster the machine runs, the worse the wear is likely to be. In addition, as the particles meet the conveyor, quite considerable impact forces will be experienced. The resultant force of the momentum of the particle and the horizontal velocity of the belt is, in the case of an inclined chute, at some angle to the horizontal. The effect of the force is to project the more inelastic particles upward away from the belt with some velocity. Where the particles are metallic or mineral, there is a good possibility of belt surface scuffing.

It is the projection of heavier, inelastic items as a result of the initial impact with the belt that considerably reduces the efficiency of a ballistic belt classifier. Materials are only likely to be accelerated to the optimum speed for separation if they stay in contact with the belt surface. As soon as they have left it, air resistance begins to reduce their speed. Their trajectory then returns them to the belt, where further impacts and projections occur, and so the items travel toward the discharge end in a series of bounces. The only opportunity for acceleration to belt speed occurs if they become trapped by light materials, and that, in a sense, is exactly what the machine is supposed to prevent their doing.

9.2.4.6 Oscillating Ballistic Classifiers

Figure 9.19 shows a general arrangement of a fairly common oscillating ballistic classifier design. Here a group of long, narrow, perforated decks are mounted individually upon the cranks of two parallel crankshafts, and are caused to oscillate in a circular path at considerable speed. The decks are inclined to the horizontal, and their throw causes any material placed upon them to travel toward the high end.

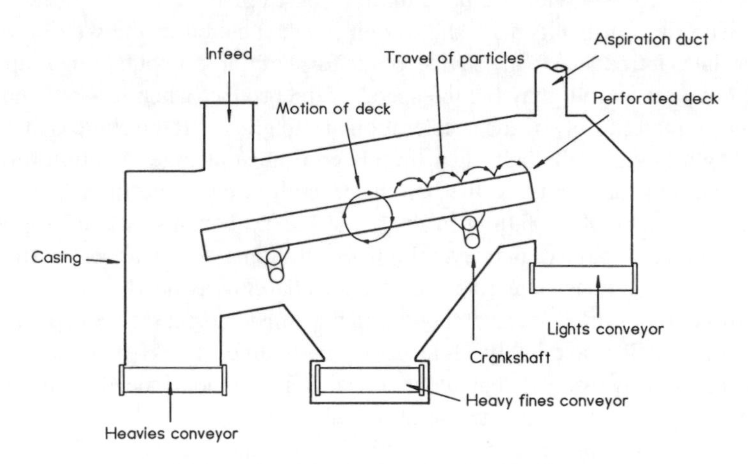

Figure 9.19 Oscillating ballistic classifier.

Since the individual cranks on each shaft are offset in relation to each other, the decks oscillate independently and out of phase. This results in the load upon them being continuously turned in addition to being transported. Fine heavy materials drop through the deck perforations into the collection hopper beneath. Larger heavy items are intended to ricochet from the decks immediately below the infeed chute, and to fall rearward to the heavies collecting conveyor. The machine is therefore intended to create a three-way separation of infeed into a fine heavy fraction, an oversized heavy fraction, and a light fraction.

As a mechanism for crude separation and classification of crude waste, the oscillating ballistic classifier is reasonably efficient, but it is somewhat less so when dealing with a granulated fraction that has already experienced some degree of refining. In those circumstances, a considerable volume of light combustibles also passes through the deck perforations and is lost to the reject lines. Large heavy items, meanwhile, do not always ricochet from the tail end of the decks as intended, because the decks are blinded by a layer of soft and yielding light fraction.

Although the individual decks are fabricated from sheet materials 0.118 in. (3 mm) thick, their physical size and the rapid, high amplitude oscillations they experience impose very large stresses upon their structure. This is particularly apparent at the mountings for the crankshaft bearings, and so maintenance costs are usually somewhat high. Stresses are further increased by the absence of any drive connection between the two crankshafts. They are held in the correct relationship to each other solely by the decks, which effectively reduces the high end shaft to an idling role. All of the forces and stresses involved in driving the machine are therefore experienced by the low end crankshaft alone.

In any densified RDF process, there must be some positive method of preventing any metals from remaining in the combustible fraction by the time it reaches the densifiers, otherwise severe damage to the machines becomes likely. In any form of RDF, metals must not reach the fuel customer's premises, or the manufacturer may find himself facing claims for plant damage and loss of income. Ballistic classifiers in general do not and cannot offer guaranteed metals removal, and that failing is exacerbated by their being just as likely to pass large and heavy metallic items as small ones.

9.2.5 Cyclone Separators

Cyclones are so widely used throughout industry that it almost appears superfluous to discuss them here. Their principles are even used in many makes of domestic vacuum cleaners. However, they are fundamental to the operation of any RDF process, throughout which they appear in many forms.

A cyclone separator is an entirely static device without any moving parts whatever. The rotary valve in the hopper bottom is there only because the devices are generally installed in a process line rather than at the end of it, and the air they handle is being passed further down the line to other machines. There are, however, many applications for cyclones where even the valve is absent.

The principle of operation, as shown in Figure 9.20, is solely one of the application of centrifugal force, created by the high-speed circulation of air within the cylindrical casing. Air and the product entrained within it is introduced tangentially into the chamber, from which the only unobstructed outlet is the central exhaust tube. As the air–product mixture rotates in the casing, the centrifugal forces that are created affect both the air and its burden. However, since the density of air is likely to always be the lower by a considerable margin, it is the material which experiences the greatest influence.

As a result, it is forced to the periphery of the casing while the air, having no other escape route, is constrained to departure through the center tube. At the periphery of the casing, friction between the air–material mixture and the casing itself creates a slower-moving layer where the forces are much reduced. Released from entrainment by the slower speed, the material falls to the hopper bottom, from where it may be extracted.

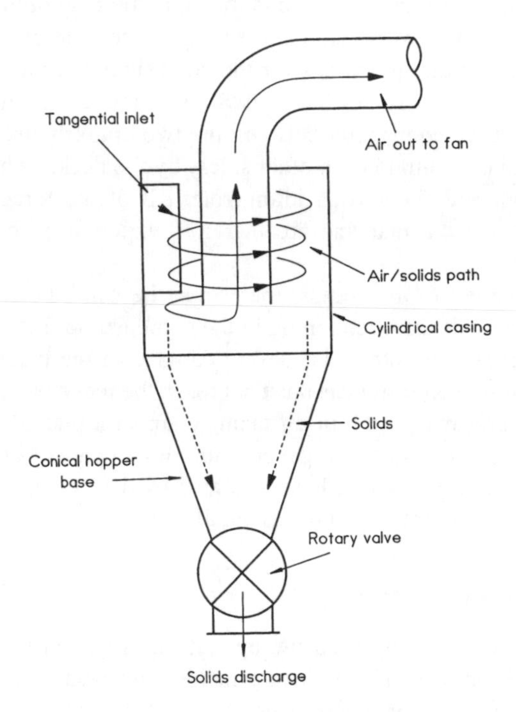

Figure 9.20 Cyclone separator.

In high-efficiency cyclones, it is usual to install a "vortex cone." This is a sheet metal cone placed beneath the inlet of the air exhaust pipe, immediately above the hopper bottom. Its purpose is to stabilize the vortex formed by the passage of the air into the exhaust pipe, and to prevent its influencing the material that has fallen into the hopper. It also has the effect of creating a localized, very high-speed rotation in the air at its tip, leading to intense centrifugal forces there and virtually ensuring that no solids can enter the exhaust pipe.

The formula for the calculation of centrifugal force, or more properly for the *centripedal* force, which is the equal and opposite reaction to centrifugal force, is

$$C_f = wV^2/gr \qquad (14)$$

where w is the weight of a particle, V is its velocity, g is the acceleration due to gravity, and r is the radius of the circular path.

If that concept is applied to a mass of paper entrained in an airstream moving through a cyclone, a force will be applied to both. Since air has an accepted density for most purposes of 0.08 lb/ft^3 (1.3 kg/m^3), and paper has a true bulk density in the range 44 to 68.6 lb/ft^3 (700 to 1100 kg/m^3), then it follows that the force applied to the paper is approximately 500 to 700 times greater than that applied to the air.

The explanation is however somewhat simplistic, since there are a number of other factors influencing the behavior of the mixture. The density of air, for example, varies according to temperature and pressure. A cubic foot of hot air weighs less than a cubic foot of cold air, which is the reason why hot air rises. If the air is constrained at pressures greater than atmospheric, then again there is a density change, in that case an increase. Therefore, although the centrifugal force applied to a solid particle remains dependent only upon its weight and its velocity in the circular path, when applied to the air it may vary considerably.

It follows from those observations that before the cyclone, the air density should be as high as possible in order that it may transport the material better. In the cyclone chamber, however, it should be as low as possible in order to increase the centrifugal force difference. It is for that reason that cyclone efficiency and design depends upon a degree of pressure drop across the device. Specialist manufacturers of cyclones usually specify the pressure drop range, within which a suitable efficiency may be expected, for a wide range of materials and granularities. They also provide charts showing performance characteristics based upon test work, and those are more likely to be relevant than any mathematical model to the process designer. The inclusion of a cyclone in the process flowline is more a case of selection from a range available, than of calculation and detailed design.

9.2.6 Driers

Two types of driers are most common in RDF processes — thermopneumatic and rotary cascade. The two designs differ fundamentally from each other, and their relative merits have been hotly contested for many years. Rotary cascade driers are supported by a wealth of experience in other industries, and as such are seen in many circles to be at least well understood and capable of evaluation. Thermopneumatic driers are less common elsewhere, and their simplicity appears to make them somewhat suspect. That is unfortunate, since they are by far the most effective devices for processing the finely granulated combustibles that provide the feedstock for RDF.

Both types, and in fact all driers, work by heating water to a temperature sufficient for it to change state into vapor, and by providing a means by which the vapor may be conducted away from the dried material. Heat must be provided in a medium also capable of transporting the vapor once the heat is lost, and it must be delivered at a temperature above the boiling point of water.

The method used to produce the heat for drying is also a matter of some dispute. In some circles it is considered appropriate to do so using solid fuel combustion systems, preferably burning the fuel which the plant creates, or even by incinerating the low-calorific value residues. Unfortunately for that philosophy, solid fuel combustors respond too slowly to be able to deal with the almost continual changes in feedstock moisture content. As a result, in one design that was prominent in the early 1980s, the heat output was maintained at a steady state and it was the feedstock flow that was varied. Large rotary cascade driers were used, and the feed rate was modulated according to the gas exhaust temperature. Since the residence time of the material in the drier was several min, it follows that the response time to any control command was also several min, with the result that the feedstock was always either too wet or too dry. It was rarely exactly correct.

The alternative to solid fuel combustors is gas or oil burners, and those are always used in thermopneumatic systems. Although they are using primary fuels, the burners have very rapid response times measured in seconds. Accordingly, they can be modulated to accommodate repeated changes in moisture content and to produce a much more consistent product.

Cost–benefit analyses do not show such a design to be as disadvantageous as might be expected. Certainly, the primary fuels both cost more than the "homemade" RDF, but they are also very much more efficient. The correct evaluation of the alternatives is made by comparing the cost per usable heat unit for each fuel, where the cost of RDF is the value at which it is being sold in the open market. If the market value of RDF is one half that of oil, for example, but twice as much is required to provide the same drier output, then the overall costs per heat unit are the same. In such a case, it is better to sell the RDF to a solid fuel user, and to manufacture it using primary fuels.

Drying does not of course necessarily involve high temperatures. If it did, then the normal domestic activity of drying the laundry outdoors would be impossible. Water begins to evaporate at a temperature only slightly above freezing point, but below 212°F (100°C) the excitation of its molecules is slight. Low-temperature drying is a lengthy process, and is inappropriate in a production facility where drying has to be achieved in seconds or fractions of a second. Therefore, the minimum practical temperature for an industrial drier is 212°F (100°C).

In RDF processes there is also a maximum permissible temperature. Lignocellulosics have an ignition temperature of between 284°F and 320°F (140°C and 160°C) at atmospheric pressure. They also contain between 65% and 75% volatile matter, which is essential for the maintenance of a good calorific value in the fuel product, and that volatile matter begins to be liberated at temperatures very little higher than the boiling point of water. It would therefore appear that a drier in an RDF process must operate at precisely 212°F

(100°C) if it is to achieve drying without ignition or elimination of volatile matter.

Fortunately that is not strictly so. Insofar as the ignition temperature is concerned, it is the temperature of the material which matters, not that of the heat source. Heat gain by the material is controlled by its thermal conductivity and by its moisture content. Given a sufficiently low thermal conductivity, a material is unlikely to gain sufficient heat to ignite in the time allowed. Also, since water has a very high thermal conductivity, all of the available heat will be used in changing its state before the material itself can begin to gain temperature. Water cannot be heated to more than 212°F (100°C) at normal atmospheric pressure, since that is the vaporization temperature. Therefore, as long as the material containing it has some residue of water, it also cannot be heated to much above 212°F (100°C).

The term "much above" is used with some caution. In theory, the material could not be heated to greater than the boiling point of water as long as it has a detectable moisture content. However, no material is so consistent in terms of thermal conductivity as to make theory and practice coincident. The outside of a particle may well reach a considerably higher temperature before its core has become hot enough to evaporate water. The temperature experienced in practice, where granulated lignocellulosics are concerned, is between 248°F and 284°F (120°C and 140°C). As long as they contain moisture, that is the highest attainable temperature. It follows from those considerations that a dryer in an RDF process must be able to provide heat at sufficient temperature to eliminate a proportion of the moisture rapidly, but must then discharge the dried product at a temperature below the ignition point. This heat/temperature balance and the physics supporting it is fundamental to the design of a drier.

It is convenient to imagine the heat input to the device being in three distinct tranches. There is a heat demand to (1) increase the temperature of the transporting medium to conduct heat to the product, (2) increase the temperature of the water to its vaporization point, and (3) replace the heat losses due to conduction and radiation through the drier structure. In this conceptual representation, it should be noted that the heat initially contained by the transporting medium is a combination of all three. However, (2) and (3) are given up to the product and to the drier structure, respectively. Only (1) remains in the medium.

There are two measures of heat most commonly used in the design of driers: The joule and the British thermal unit. The joule is the universal SI unit of energy, and it therefore applies equally well to mechanical work as to heat. For the purposes of drier design, it is equivalent to 0.2388 calories, where a calorie is the amount of heat necessary to heat 1 g of air-free water from 14.5°C to 15.5°C at a constant pressure of 1 atm. In other words 4.188 J are necessary to achieve that effect. More properly, the calorie unit in question is known as the "15°C calorie" (cal_{15}). There are also International Table and thermochemical calories, each with a joule equivalent. However, the difference between them in practical terms is slight, and insufficient to have any effect in drier calculations. The cal_{15} and its corresponding joule unit is generally used. The unit is based upon heating water 1°C from 14.5°C because the specific heat of water varies with temperature. Again, the amounts involved are insignificant to drier calculations, and so a heat requirement of 4.18 J per degree Celsius is adequate.

The British thermal unit, meanwhile, is defined as the amount of heat necessary to raise the temperature of 1 lb of water 1°F. It is therefore equivalent to 1055.06 J, or more usually 1 Btu/lb is equal to 2.326 kJ/kg.

Given these concepts, to heat 1 g of water from 14.5°C to its transformation temperature of 100°C requires 357.4 J. Therefore, if a drier feedstock contains m_1% of water by weight when m_2% is required, and it is delivered at a rate of 1 t/h, then the heat requirement is

$$H_w = \frac{4.18W(m_1 - m_2)(100 - t_{in})}{100} \tag{15}$$

where H_w is the heat required in joules, W is the weight of material, m_1 and m_2 are the initial and final moisture contents in percentages by weight, and t_{in} is the temperature of the infeed.

Supposing, as an example, an RDF process were to treat 37.5 t/h of waste, of which 35% was the total combustible fraction. The initial moisture content of the combustible fraction is 30%, and the final moisture content is required to be 15%. Ambient temperature is 15°C. Then the equation yields

1. 35% of the total input per h is 13,125 kg/h.
2. From equation (15) above, the heat required to increase the water temperature to boiling point is

$$H_w = \frac{4.180 \times 13{,}125 \times (30 - 15) \times (100 - 15)}{100} \text{ kJ/h}$$

Therefore, H_w = 699,496.88 kJ/h (663,030.21 Btu/h).

Published steam tables are available to provide data that defines the thermal behavior of water and steam, and reference to them is essential to proceed to the next point in the mathematical model of a drying process. The amount of heat required to raise the temperature of the water contained within the material to boiling point at atmospheric pressure is relatively small. The tables however reveal that to change the state of the water into vapor requires several orders of magnitude more, 2256.7 kJ/kg at 1 atm and 100°C. The total heat of the steam (H_s) at that point and in the example then becomes

$$699{,}496.88 + (2258 \times 1968.75) = 5{,}142{,}375 \text{ kJ/h}$$

$$= (4{,}874{,}289.1 \text{ Btu/h}).$$

The calculation has so far assumed that the drying process is 100% efficient, which of course it cannot be. One aspect at least of the efficiency loss is that usable heat must remain in the exhaust gas from the system, otherwise the water driven off will simply recondense on the product. Supposing then that it was decided to introduce hot gas at 400°C (which is the average gas temperature used in RDF processes), and to exhaust it at 120°C. The system thermal efficiency could be simplified to

$$\text{Efficiency} = \frac{(t_1 - t_2)}{t_1} \times 100\% \tag{16}$$

where t_1 and t_2 are the gas inlet and exhaust temperatures. The total heat input then required becomes

$$H_g = \frac{H_s}{E}\ \text{kJ/h} \tag{17}$$

where H_g is the heat input to the gas, and E is the efficiency calculated from equation (16) above and used as a decimal.

In the example used, therefore, the new total heat required becomes

$$H_g = \frac{5{,}142{,}375}{0.07}\ \text{kJ/h}$$

$$H_g = 7{,}346{,}250\ \text{kJ}\ (6{,}963{,}270\ \text{Btu}).$$

This has established the heat required to drive off the moisture and to overcome exhaust gas losses. All that remains is to calculate the heat loss due to conduction and radiation through the drier structure, and here some assumptions must be made. While there are published tables defining the thermal conductivity of various materials, the data is always expressed in terms of square unit area of transmitting surface. Therefore, the estimation of the losses requires knowledge of the dimensions of the drier casing. Before those can be identified, it is necessary to determine the air flows which will be involved in conducting the heat to the product.

In considering the air flows there are two essential physical characteristics to be taken into account. First, the density of air or gas varies according to temperature at constant pressure. Second, air or gas has a specific heat capacity. At 400°C, the density of air is 0.032 lb/ft^3 (0.52 kg/m^3). Conveniently, 1 kJ of heat increases the temperature of 1 kg of air 1°C.

There are some inaccuracies involved when using these figures. For example, the hot gas is not air alone. It is a mixture of carbon dioxide, nitrogen, nitrogen compounds, and depending upon the fuel source, it may also contain sulfur dioxide, carbon monoxide, and complex hydrocarbons as well as air.

There are no simple means of accounting for all of those variables, but air has the lowest density. Therefore, if the calculations are based upon air they will err on the high side. In addition, the effect of heat energy upon air is only true up to a point, in that it assumes a constant pressure situation. That cannot be guaranteed. Heating at either constant pressure or constant volume only happens under laboratory conditions. In a fullscale process, both are likely to vary, and will be influenced by, among other factors, the infeed rate of the feedstock.

Any attempt to consider all of the variables within the model would create a need for very sophisticated mathematics, which would produce a far greater accuracy than is necessary. The purpose of the mathematical model is simply to determine the global parameters to which the system will be designed. The designer will then need to form a judgment upon the extent to which he should overspecify, in order to ensure that the system performs in circumstances outside the original assumptions.

From equation (17) it is determined that H_g kJ/h of heat will be supplied to the gas in order to achieve the required product drying. Using a heating capacity of 1 kJ/kg/°C for the gas, and a density of 0.52 kg/m³, the calculation becomes as follows: At the heating capacity stated, heating the gas to its initial temperature will require 1 kJ/kg/°C, or in this case 400 kJ/kg. Therefore the gas *volume* required will be:

$$V_g = \frac{H_g}{t_1 \times 1 \times 0.52} \text{m}^3/\text{h} \ (\times 35.31 \text{ ft}^3/\text{h}) \tag{18}$$

where V_g is the gas volume, H_g is the total heat input to the gas, and t_1 is the initial temperature to which it is heated.

In the example, this results in a demand for

$$V_g = \frac{7,346,250}{400 \times 1 \times 0.52} \text{m}^3 \ (\times 35.31 \text{ ft}^3).$$

Therefore, $V_g = 35{,}318.5$ m³/h (× 35.31 ft³/h).

Given the necessary gas volume, it now becomes possible to determine the dimensions of the drier casing, but a decision must be made upon the velocity that will be applied. If the device is to be thermopneumatic, then an air speed in the order of 15 m/s (49.2 ft/s) will be needed. If it is a cascade drier, then any speed greater than 1 m/s (3.3 ft/s) should be avoided.

Assuming first therefore that the device is to be thermopneumatic, V_g m³/h of gas will be carried at a velocity of 15 m/s. The duct diameter is therefore

$$D = \frac{4V_g}{15\pi} \text{m}. \tag{19}$$

Again, in the example this provides

$$D = \frac{4 \times 9.811}{47.13}\,\text{m}.$$

Therefore, D = 0.9126 m diameter.

Now that the duct diameter is established, it becomes possible to return to the question of heat losses through the structure. If a decision is taken upon the overall length of the duct, which as will be discussed later is a matter of convenience rather than of thermodynamic necessity, then the radiating surface area can be calculated as

$$A = \frac{\pi D^2}{4} \times L\ \text{m}^2 \tag{20}$$

where A is the area, D is the diameter calculated from equation (19), and L is the duct length.

At this stage it is necessary to consider what steps will be taken to minimize heat losses through the structure. If the duct were to be left uncovered, then it would radiate according to the thermal conductivity of steel, which is very high. A great deal of heat would be lost, as would the revenue that paid for it. Insulation is clearly necessary, and in that case the important figure is the thermal conductivity of the insulation rather than of the drier casing. If that is Kilojoules per square meter per second per degree Celsius, a figure provided by data tables from manufacturers, then the heat loss for the duct becomes

$$H_1 = \frac{3{,}600 K_i A\left(\left[\frac{t_1 + t_2}{2}\right] - t_a\right)}{1000}\ \text{kJ/h} \tag{21}$$

where t_1 and t_2 are the inlet and exhaust temperatures, and t_a is the ambient temperature of the surrounding air.

In this formula t_1 and t_2 are combined to estimate the average temperature of the duct. Once again, this is not strictly accurate, since there is a nonlinear temperature gradient of between 400°C and 120°C throughout the system. The thermodynamic characteristics calculated by the methods outlined will show an erroneously low heat loss at the inlet end of the line, and a correspondingly high one at the exhaust end. Applying the calculation to the example system and using a conductivity of 0.42 J/m²/s/°C (for 100 mm thickness of glass fiber insulation) clarifies the limited importance of that:

$$H_1 = \frac{3{,}600 \times 0.42 \times A \times \left(\frac{400 + 120}{2}\right) - 15}{1000}\,\text{kJ/h}.$$

Therefore, H_l = 12,744 kJ/h.

The model predicts a conductive heat loss of 12,744 kJ/h in a system absorbing 7,346,250 kJ/h, which is the reason for the limited importance of it in the overall assessment. However, had the duct not been insulated, then the heat loss in this respect would be higher than the total heat input established thus far. That is perhaps an indication of the necessity of high-quality insulation.

Finally, the conductive heat losses can be added to the previous gas heating load to establish the total heat for the system. At that point, strict accuracy would demand that the air flows are recalculated to account for the extra heat to make up conductive losses, but again the actual amounts are small in relation to the whole. Such recalculation is hardly worthwhile.

The same model can of course be applied to a rotary cascade drier, substituting the appropriate air flows and dimensions to arrive at a heat load.

Throughout the logic train here, it has been assumed that if the original waste contained 30% by weight of water, then so must the fraction at the drier. This is not strictly valid, since the moisture remaining depends upon the moisture distribution between categories of materials. Clearly paper and card can retain more water than plastic film, and organic materials, having considerable quantities within their cellular structure, can retain more than paper. This concept is explored in more detail in Section 9.3, where the concept of moisture distribution between categories is developed.

9.2.6.1 Thermopneumatic Driers

Figure 9.21 shows the general design of a conventional thermopneumatic drier. As the name suggests, it is a device which handles product pneumatically while applying heat. In operation it is effectively a pneumatic conveyor which uses hot gas as the transporting medium. It is a duct with a heat source at one end and a cyclone separator and suction fan at the other.

The hot gas is conventionally provided by either a gas or oil burner located at the inlet end of the duct, and exhausting directly into it. There is a free air inlet around the burner flame tube to provide the extra air necessary for temperature control and flame tube cooling, and often, where space is at a premium, there is a spark arrester immediately beyond the burner. Material is introduced into the duct at a point beyond the flame tip — about 3 m away in the most usual installations. Introduction may be through a rotary valve, although that is not always necessary. It depends somewhat upon how the product is delivered to the drier area.

The temperature created by the burner at the point where the product is introduced is essentially between 660°F and 750°F (350°C and 400°C), which in view of the earlier comments upon ignition temperatures may be surprising. However, at that point there is little risk of ignition, since the material is too moist. The heat is very rapidly consumed in the dissipation of the water, to

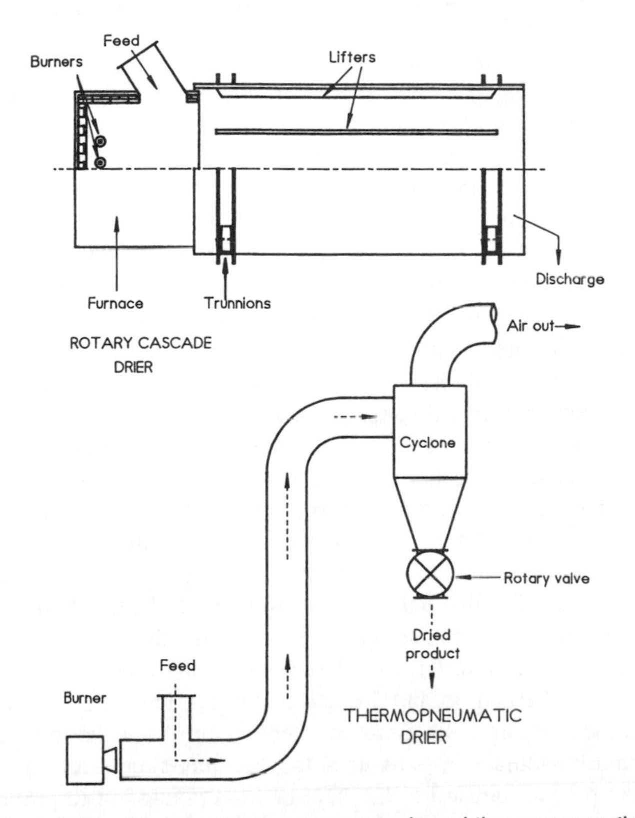

Figure 9.21 Common driers: rotary cascade and thermopneumatic.

the extent that some 9 ft to 10 ft (3 m) further along the duct the gas temperature will have reduced to less than 320°F (160°C).

A second factor constraining the possibility of ignition is of course that the transportation medium is not oxygen-rich air. Rather, it is mainly combustion gas from the burner unit. Oxygen levels are low, while carbon dioxide is predominant. The material is being conveyed in what is mainly very hot, inert gas. Even so, thermopneumatic driers can and often do ignite the material. Granulated RDF feedstock is always somewhat unpredictable in terms of moisture content, and it is not impossible for a very dry fraction to pass briefly through the process. It is when this happens that the major advantage of such a drier becomes apparent.

The air speed through the duct is necessarily considerable. Air (or hot gas and air mixed) has a low density, which may be as little as half that of air at ambient temperatures. To maintain the product in suspension, much higher speeds are required than, for example, in an air classifier, and conventionally they range between 48 and 64 m/s (15 and 20 m/s). It is rare for building space availability to permit a single-run thermopneumatic line to be more than about 38 to 40 ft (12 m) in effective length, and so the residence time within it is on average less than 1 second. The contents of the duct are moving too rapidly for flame propagation to take place upstream, while downstream there is only a sealed cyclone separator and a scrubber for dealing with the final exhaust gas. Therefore, product ignition simply results in a very transient fireball which passes through the system and into the scrubber, often without its occurrence having ever been noticed.

The evidence of ignition is usually first observed at the cyclone rotary valve, where charred material may appear briefly. Continued combustion at that point is very rare, since there is normally insufficient oxygen in the cyclone hopper bottom to support it.

These comments in respect to operating temperatures and the possibility of product ignition reveal the essential nature of a thermopneumatic drier. It is a "flash" drier, where the moisture contained within the material is removed in a small fraction of a second. Once the material has accelerated to the same speed as the gas stream, very little further drying takes place. A speed differential is necessary, otherwise the water vapor is able to remain with the particles of product, and to recondense upon them when the temperature conditions permit. For that reason, it is customary to install thermopneumatic driers so that the duct becomes vertical in as short a distance as possible from the product introduction. Gravity then restricts the ascending speed of the particles, while thermal lift and the suction fan continue to accelerate the gas.

Even then, drying is likely to be effective only over the first 13 ft (4 m) of duct, and the remainder is reduced to becoming simply a means of transporting the product onward to the next process point, and to provide access to the cyclone separator and exhaust scrubber. For that reason, in addition to building height limitations, the devices are rarely more than 38 to 40 ft (12 m) long. In fact, any length greater than 13 ft (4 m) is of limited value.

Occasionally process plants are designed to deal with waste in which it is expected that the moisture content will be excessively high. Those dealing with transfer station waste are particularly likely to experience such a phenomenon. In such a situation it may be that a single thermopneumatic drier cannot alone provide sufficient heat capacity to eliminate the moisture in the time available. Then it is common to install two driers in series, where the first discharges directly into the second.

An essential but often unrecognized design requirement in relation the thermopneumatic driers is a means of dealing with erosion in the duct. Hot, dry paper and card granules traveling at high speeds are surprisingly abrasive, and steel at 750°F (400°C) is somewhat less resistant to wear than is cold metal. The bends where the duct inclines to the vertical, and later where it returns to the horizontal to discharge into the cyclone, are particularly vulnerable. Thin-walled, circular cross section pipe in those areas erodes very rapidly.

For reasons not clearly understood, bends fabricated from rectangular-section pipe with a wall thickness greater than $^3/_8$ in. (8 mm) offer much better wear resistance. Of the possible explanations for this, it may be that a circular-section pipe has the effect of concentrating the particle impacts at one point as a result of centrifugal forces, while in a rectangular pipe the impacts are spread over the full available area. Certainly circular bend thermopneumatic ducts always wear in a narrow strip around the outside circumference of the bend.

A second possible explanation, or at least a contributing factor to wear reduction, may be that the continual impacts of the product particles initially remove metal, but progressively harden that which remains. If the duct wall is of thin metal $^1/_4$ in. (5 mm) or less, it may be that most of the metal is lost before the hardening can develop.

In common with all other hot gas driers, thermopneumatic devices have to be heated to operating temperature before effective product treatment can begin. If material is introduced immediately upon start-up, much of the heat contained in the combustion gas will be lost in the drier casing. Steel is a considerably better conductor of heat than paper. It is in this respect that the minimum $^3/_8$ in. (8 mm) wall ducting is advantageous, since it retains heat for much longer between operating shifts. Where that characteristic has been further enhanced by heavy insulation, driers have been shown to lose only 200 to 210°F (100°C) even in overnight shutdowns.

9.2.6.2 Rotary Cascade Driers

As also shown in Figure 9.21, a rotary cascade drier consists of a large, rotating cylindrical drum in which material and hot gas are introduced at one end and exhaust from the other. The inside circumference of the drum is fitted with longitudinal "lifters," which are simply extended ledges intended to raise the material as the drum rotates and release it toward the highest point of the rotation. The drums used are necessarily large in order that they may contain the volumes of hot gas necessary. Diameters of 16 ft (5 m) and lengths of 32 ft (10 m) are common, and in such a large structure, operation at high percentages of the critical speed are impractical. Therefore, product distribution cannot safely be a matter of rotational speed as it is with a trommel screen, for example.

The physical size and weight of such a drier also prevents inclination being used as a means of ensuring a steady product flow. The stresses imposed upon a tracking wheel would be too large even with an inclination of only 3°. Instead, the machine has to rely upon the gas flow to ensure that the product moves consistently, and the more granulated the product, the less likely that is to occur.

Rotary cascade driers achieved some impact upon the European RDF industry, mainly because they were readily available from a number of manufacturers and had a lengthy track record in other quite dissimilar industries. It was perhaps felt that their operating and design characteristics were well understood, and that they therefore were a more secure technology than, for example, thermopneumatic driers. That theory founders, however, upon the fact that rotary cascade driers are principally designed to deal with heavy particulates that do not readily become suspended in a gas flow. Gas volumes can then be comparatively low, and residence times within the drum can be long. The material can be tumbled through the hot atmosphere many times until the desired degree of dryness is achieved.

However, where the product is both of a very low apparent bulk density 1.9 to 3.7 lb/ft^3 (30 to 60 kg/m^3), and is finely granulated to $>^1/_2$ in. (10 mm) as is common in RDF processes, then the theory fails. The particles readily become entrained in even a very low gas flow, and even at flow speeds of less than $6^1/_2$ ft/s (2 m/s) will advance a considerable distance along the drum before falling out of suspension. A machine with a very large enclosed volume, which should offer long residence times, in fact delivers exactly the opposite.

There are unfortunately few viable options for dealing with that situation, since the behavior of individual particles can never be accurately predicted. A reduction in gas flow is not a solution, since it correspondingly reduces the heat input to the material. Scrolls on the circumference of the drum cannot be used, since they would have an adverse effect upon the gas flow, and would tend to exclude much of the product from it. The solution for the designers of some European plants was to first densify the material, making it sufficiently heavy to avoid becoming entrained in the gas flow. This resulted in the "3D" system of densifying, drying, and then final densification. It also resulted in impurities and contaminants becoming locked into the feedstock beyond the reach of final screening, and consequently created a fuel with an excessive ash content.

A further difficulty with this type of drier is that hot gas is not easily distributed throughout the full volume of the drum. In such a large volume, the flow tends to be laminar, with clearly defined high-temperature streams penetrating comparatively cool, static regions. Depending upon where the high-temperature streams are located, and they do vary their position continuously, much of the product may escape them entirely; it may never be heated to sufficient temperature to ensure consistent drying. Meanwhile, material that does encounter a gas stream is likely to be ejected from the drier more rapidly, since the streams are invariably faster moving than the surrounding gases.

Even distribution of the temperature is difficult if not impossible. The cooling effect of the product is, among other factors, likely to be a major cause of irregular laminar flow, and diffusers or similar devices at the combustion chamber outlet only have an effect for a short distance downstream. Where a drier is handling a heavy product, that limitation is of less importance, since residence times can be established to accommodate any imbalance. However, as discussed earlier, very low bulk density products are strongly influenced by even the lowest gas flows. The facility to control residence times is strictly limited.

9.2.7 Predensifiers

Where fully densified RDF (dRDF) is to be manufactured, the final densification stage, in European plants is almost always achieved by means of pelletizers. In U.S. plants cubes are more common. These machines will be discussed later, and so for the moment it is sufficient to observe that while they offer extremely high fuel densities, they are somewhat difficult to feed with very low density feedstock. The sheer volumes involved in any realistic throughput are on the limit of their standard-design charging capacity. Their design is derived from machines intended for animal feed production, etc., where initial bulk densities of over 50 lb/ft^3 (800 kg/m^3) are common. Dried, undensified RDF feedstock may have a bulk density of no more than 1.9 lb/ft^3 (30 kg/m^3). Any means of increasing that bulk density offers better pelletizer performance.

9.2.7.1 Screw Predensifiers

Screw predensifiers are generally intended by their manufacturers to be final densifiers in their own right. They are not widely used as such in their adaptation to the RDF industry, because they do not create a product in the form or at the bulk density necessary for a consistent fuel. They are, however, much better able to deal with low-density materials and to compact them sufficiently to prepare them for pelletization or cubing.

Figure 9.22 shows the general principles of a screw predensifier. In this case, only one screw is shown in order to preserve clarity, although most machines have two. The screw is a very heavy alloy steel casting, keyed to a large-diameter shaft 4 in. (100 mm) and greater, and running in a casing lined with manganese–steel liners. Drive power is supplied through a gearbox from a three-phase electric motor, and distributed at high torque and low speed to each shaft.

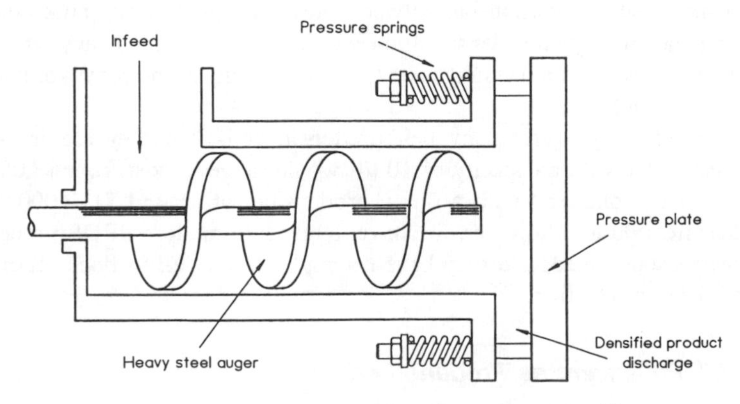

Figure 9.22 General principles of a screw predensifier.

A spring-loaded pressure plate is clamped between compression springs at the discharge end of the machine, and they restrict the exit of material as it is delivered by the screws. As a result, the material can only escape from the machine with a significant volume reduction and corresponding increase in density.

The drive power to the more common machines ranges from about 65 hp to over 100 hp, and the very high friction in the compression chamber and against the pressure plate results in the generation of considerable heat. Thus, although the machines are intended as predensifiers, they are also very efficient driers.

Depending upon the setting of the pressure plate, discharge bulk densities of up to 19 lb/ft^3 (300 kg/m^3) are obtainable. The granularity is extremely inconsistent, generally ranging from less than $^1/_4$ in. (5 mm) to over 2 in. (50 mm), which apart from the density is a reason why the machines are unsuitable for the direct production of fuel. However, the aggressive action of the compaction screws permits the machines to accept coarsely granulated feedstock in the size range 2 to 4 in. (50 to 100 mm). Such a facility is valuable in an RDF process, since it reduces the dependence upon fine granulation earlier in the flowline. It makes possible the use of rotary shears for waste pretreatment instead of the much more expensive flail granulators.

9.2.7.2 Belt Predensifiers

Figure 9.23 describes a typical belt predensifier, a device which is no longer commonly used in dRDF flowlines due to its low densification performance. In operation, feedstock is delivered to the load belt and is conveyed through the pinch point between the load and press belts, where it is compressed in one plane only. The two belts are powered by a common drive to ensure that their speeds are compatible, and the belt carcasses are impressed with a dimple pattern to ensure a firm grip upon the material.

The gap at the pinch point is adjustable, although there is a minimum acceptable clearance. Compaction to two or three times the original volume is possible, but the relationship between pressure and resulting bulk density is nonlinear. Any greater densification would require a very heavy structure and motive power, which would be beyond the capacity of the conveyor belting used or available.

The advantage offered by belt predensifiers is that they are low-cost machines. One with a capacity of 10 t/h would range in cost from $3,000 to $5,000 U.S., compared with a screw predensifier at around $150,000. Belt predensifiers do not, however, fit conveniently into modern dRDF processes for the reasons stated, and they have no application at all in flock RDF and crumb RDF production.

9.2.7.3 Pelletizers as Predensifiers

Some European process plants in the last two decades have utilized modified pelletizers as predensifiers, and the most notable to do so were those

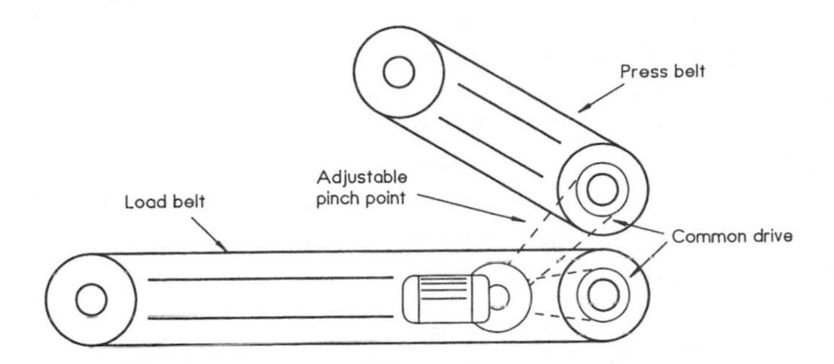

Figure 9.23 Principles of a belt predensifier.

given the title of "3D" (densify, dry, densify) processes. The machines used for the purpose were in most ways identical to the final pelletizers, but with the exception that their dies holes were bored to $1^3/_8$ in. (35 mm) rather than $^3/_4$ in. (18 to 20 mm), and force-feeders were used to charge them. In that application, they treated undried, granulated combustibles, and so were unable to achieve densities much greater than 12.5 lb/ft^3 (200 kg/m^3).

In many ways the use of pelletizers in this application was in defiance of logic. If a feedstock is of too low a density to be delivered reliably into the die of a conventional machine, then one modified simply by enlarging the extrusion holes is unlikely to be an improvement. The problem to be faced is how to introduce the feedstock in the first place, not how to extrude it when it has reached the die. Equally, if a force-feeder can overcome the delivery problem, then logic would place it upon the final pelletizer, to the exclusion of any form of predensifier.

In fact, the application had more to do with the use of rotary cascade driers and their limited ability to handle very low density feedstock. It developed at a time when screw predensifiers were not readily available, and since pelletizers were often somewhat cheaper than the early screw-based machines, there was often little choice in the matter. Unfortunately, the modified pelletizers never completely overcame the drier difficulties, since with $1^3/_8$ in. (35 mm) extrusion bores, plastic film escaped densification almost entirely. In fact to some extent a degree of classification took place, where the film was released from the bulk of the feedstock and subsequently was carried from the drier by the gas flow, in preference to the paper and card fraction.

While there are still some plants in Europe using the principle, it is no longer considered to be appropriate for new plant designs.

9.2.8 Pelletizers and Cubers

9.2.8.1 Pelletizers

The pelletizers used in the dRDF industry are derivatives of, and remain very similar to, those intended originally for the production of animal feed.

Typically there are two basic designs available, differing from each other by the die mechanism. The general principle of both, however, is that a cylindrical die of hardened steel, perforated with holes slightly tapered at their inlets, encloses a frame upon which smaller-diameter rollers are mounted. In one of the two designs, the die is rotated around the fixed rollers (the "roll shells"). Such a machine is revealed in Figure 9.24. In the other design, the roll shells are driven inside the fixed die. Predictably, the machines are classified by the generic titles of "fixed die" and "rotary die."

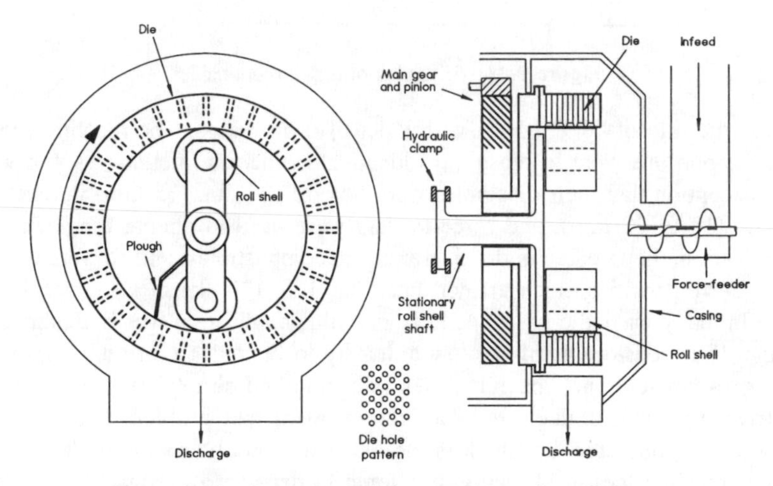

Figure 9.24 General principles of a rotary die pelletizer.

The dies are usually manufactured from case-hardened alloy steel, which is suitable for the original purpose of treating animal feeds, but is quite unsuitable for dRDF production. In the latter industry, it is common for through-hardened dies to be used. Granulated paper is quite abrasive, and wear rates in the dies are significant. Once the surface hardening has worn away from a case-hardened die, the wear rate accelerates rapidly to the point where densification is no longer possible. Through-hardening ensures a consistent wear rate throughout the life of the die, extending the service availability substantially.

In the rotary die machines, which have largely replaced fixed dies in current applications, drive power is provided by a pair of electric motors through pinion gears to a large diameter main gearwheel. The shaft between the gearwheel and the die is hollow, and the shaft holding the roll shells passes through it to an external hydraulic clamp or shear pin assembly. Both are protective systems, intended to permit the roll shell shaft to rotate with the die in the event of jamming or overload. The hydraulic clamp system is usually further interlinked with other sensors, such that if a fault occurs anywhere in the mechanism, hydraulic pressure is relieved and the shaft is again permitted to rotate. The result of this is to immediately reduce the load upon the machine, and an electrical connection between the hydraulic system and the force-feeder stops the infeed.

The shaft disc, which is clamped hydraulically, is usually of a similar profile to a sprocket wheel around its periphery, and electrical limit switches are positioned upon it. If a severe overload occurs to the extent that the roll shell shaft is forced to rotate against the hydraulic clamps, perhaps as a result of foreign material entering the die, then the limit switches detect the movement and again stop the machine.

In a fixed-die pelletizer, the die is rigidly clamped into the machine casing. The roll shells are once again mounted in a frame upon a shaft, but in this case there is no outer shaft. Drive power is usually provided by a single electric motor, through drive belts and a large-diameter pulley wheel. Overload protection in this case is by means of a shear pin assembly between the pulley wheel and the shaft. If an overload occurs, the shear pin fails and the pulley wheel becomes free to rotate upon the shaft without transmitting any drive.

The principle of operation of both types of pelletizer is the same. Granulated material is delivered to the bore of the die by a force-feeder, which is simply an independent, short-screw conveyor installed in the casing on the longitudinal center line of the mainshaft. When the feedstock enters the die bore, it becomes trapped between the roll shells and the inside face of the die, where it is forced into the die holes. Considerable frictional resistance is experienced against the hole sides, creating considerable heat. The temperature is sufficient to produce thermal fusion of the lignin and, to some extent, the plastics fraction in the feedstock, resulting in a hard, dense, and firmly bonded pellet.

The design of the roll shells is as fundamental to the efficiency of the machine as that of the die. Shells are offered with a variety of surface finishes, all of which are intended to limit slipping in the material as it is introduced into the nip point against the die. The patterns most commonly available are "fluted," where the shell periphery is deeply grooved parallel to its longitudinal axis, "dimpled," where the surface is entirely covered by small, hemispherical protrusions, and carborundum coated. In dRDF manufacturing, the fluted patterns are generally the most successful.

Dies are offered in a range of hole diameters and inlet tapers; which diameter is chosen largely depends upon the intended use of the product. However, small diameters $^1/_2$ to $^5/_8$ in. (10 to 15 mm) are uncommon. Dies of those hole diameters absorb more power, and they produce fuel pellets that are somewhat small for conventional boilers' fuel handling systems. Diameters of $^3/_4$ in. (18 to 20 mm) are the most widely used, although upon occasion $^7/_8$ in. (22 mm) dies have found applications. A section through a typical die is shown in Figure 9.25.

While pelletizer design is well understood, the conditions that prevail inside the die holes are not. It is known that friction against the hole wall is essential for the production of good-quality pellets, and logic suggests that this plays a part both in ensuring high extrusion pressures and the creation of

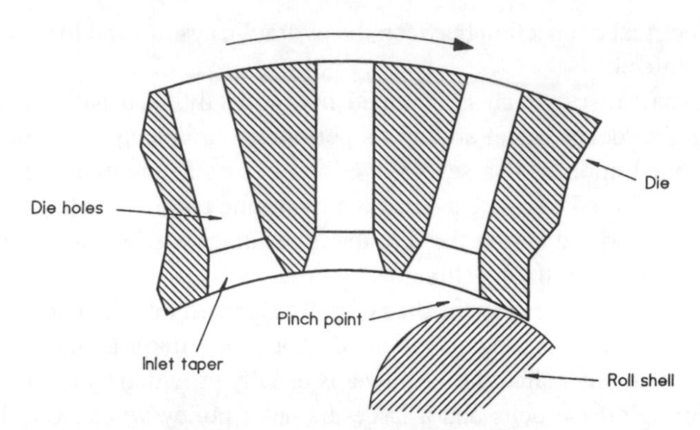

Figure 9.25 Detail of pelletizer die.

heat. High temperatures [about 285°F (140°C)]. are clearly important in producing thermal fusion, and high pressures ensure the required bulk density of the product. However, moisture content in the feedstock also has a fundamental role. Operating experience shows that if there is insufficient moisture, pellets do not bond effectively: they tend to disintegrate upon leaving the die. Meanwhile, if the moisture content is too high, then excessive steam is produced inside the pellet, sufficient to burst it apart upon exiting the die hole. Each make and type of pelletizer demands a quite specific feedstock moisture content, and the range in which it will produce good-quality fuel pellets is limited to only one or two percentage points either side of the optimum. Thus for example, one widely used design produces the best results at 16% moisture content. At either 14% or 18%, performance is noticeably affected, and beyond those limits it hardly produces recognizable pellets at all.

It is extremely difficult to monitor the moisture content of the feedstock at the die entry, and it is impossible to modify it once it has reached the roll shells. Steam injection into the transfer chute immediately before the force feeder has been used with some success occasionally, but control of such a system depends upon the final pellet quality, which cannot be measured automatically. Generally, the only solution is to ensure that the correct, and exact, moisture content is achieved before the material reaches the pelletizer. For that reason, the driers in a dRDF flowline must have very precise control systems.

A considerable element of operational control comes from experience. An astute plant operator is able to assess the quality of the feedstock by examining the finished pellets, and by listening to the machine operating. He will be aware that if the moisture content is somewhat low, then the resulting pellets will be very hard and hot. Their surfaces will be dark or black, and very shiny. They will be quite easy to break between the fingers. Alternatively, in the event of a high moisture content, the pellets will be light gray in color, will show quite clear axial laminations, and will exude a smell reminiscent of hot biscuits. Again they may be broken easily between the fingers. At the precisely correct moisture content, the pellets will be very hot, dark but not black in color, and almost impossible to break apart.

A final indication of correct operating conditions can be gained simply by listening to the machine as it runs. When the feedstock is of exactly the right quality, the die emits a very characteristic metallic "cracking" noise, which can only be described as the sound of a ruler being slapped against a metal table. In any other circumstances, the machines make little noise other than a soft rumble.

Die hole geometry is also important, although again it is not always clear why one performs better than another. There are many schools of thought concerning the angle and length of the inlet taper, although it appears that a steep, shallow taper works best. This is taken almost to the extreme in "well-head" dies, where it is reduced to a depth of only a fraction of an inch, with an entry bore of approximately $1^1/_2$ times the hole diameter. One effect of such a design is to reduce the "land" area — the area of metal between each of the die holes. Compression of the material between the roll shells and the land area is in a sense counterproductive, in that it only succeeds in increasing the power demand without creating pellets. The smaller the land area is, the more of the drive power is used to compress material into the die holes.

Residence times in the dies varies on average between 2 and 6 seconds.[1] Short residence times are the result of high levels of infeed, and vice versa. The quality of the finished pellets is very much influenced by the residence time, in that where it is long, the pellets produced are very hard, dense, and homogeneous. If the time is sufficiently long they may even become so hard as to be incapable of use in conventional combustion plant (see Section 11.3 on product markets).

This effect also introduces operational procedures which, if omitted, seriously risk plant availability. If a machine is to be shut down for any length of time, for example overnight from the end of an operating shift, the last few min of running will inevitably be at very low loads. If the resulting die contents are allowed to remain in the holes as the machine cools, they will harden to the extent that it will be found impossible to restart the pelletizer some h later. For that reason, pelletizers should never be shut down after any period of low-load working particularly without clearing the dies.

There are several ways of achieving a clear die at shutdown, and they mostly depend upon the opinions and convenience of the plant operators. However, the two most common ways involve either introducing some material that will not easily pelletize, or simply pouring some water into the feed chute. Among the materials that will not pelletize, and which therefore push the normal feedstock out of the die, are sawdust, wet granulated paper, or even the dry residues from a pellet screening process. The addition of water is a somewhat more spectacular approach involving clouds of steam, but it does ensure that the residues left in the die are too wet to form solid pellets.

Starting pelletizers from cold also involves some ingenuity, even when they have been shut down judiciously. Most operators have discovered that it is advisable to retain a supply of finished pellets beside each machine, and to charge the die with them when it is started. Because these pellets have already been compressed to the maximum density of which the pelletizer is capable, they will not compress further. They therefore force the residues out of the die holes and prepare them for feedstock introduction.

Pelletizers are again a type of machine that defies any realistic attempt at mathematical modeling. The conditions in the die are not well enough understood, and are sufficiently imprecise, that it would be difficult to develop a mathematical expression of their characteristics. It is possible to predict product bulk densities, given a level and known quality of infeed and knowledge of the die specification, but only if a residence time is assumed. Since the residence time is determined entirely by frictional resistance, any such assumption has little validity. It is therefore difficult to rely upon anything more than experimental data resulting from manufacturers' test work. The only reliable measure of pelletizer performance is to accept that production rates for dRDF will be one half to two thirds of those when making animal feeds. The latter rate is most commonly used by the manufacturers as their basis for performance.

9.2.8.2 Cubers

As suggested by Figure 9.26, a typical cuber is not entirely dissimilar to a pelletizer. Here again, material is delivered to a pinch point between a roller and a cylindrical die, through which it is forced by mechanical pressure. There are, however, several fundamental differences.

Cubers generally do not produce pellets with a circular cross section. Rather, as their name suggests, they produce square or rectangular blocks, and common production dimensions are $1^1/_2$ to 2 in. square (38 to 50 mm). They are not, however, cubes, but instead are upon average about $1^1/_2$ times as long as the length of their sides. The actual length is usually dependent upon the position of an inclined plate enclosing the outside diameter of the die, against which the compressed materials are forced. Deflection against the plate causes them to break off level with the die hole opening. The position of the deflector plate is not adjustable, since it is conical in section and circular in face. Therefore, the final lengths of the product are set during the machine design and cannot easily be changed later.

The dies also differ from pelletizer designs in that they are normally a single, radial row of individual hollow castings, clamped into place and held firmly inside the casing. Each casting is in the form of a square "U" section, and they are clamped in inverted pairs to create a square-section bore.

Usually there is a single rotating press ring instead of a pair of roll shells. The ring is mounted on a crank box on the machine mainshaft, which also carries a series of scrolls to deliver feedstock directly to the die. The configuration permits the use of a ring both narrower and of a greater diameter than the roll shell of a pelletizer. Of the many advantages claimed for this design,

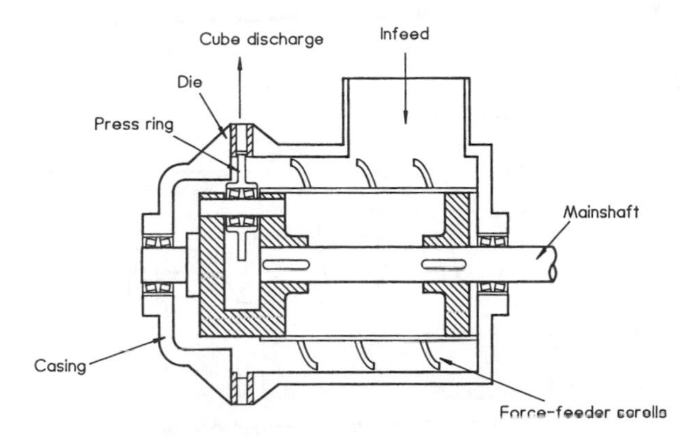

Figure 9.26 Typical cuber head.

two are readily apparent. First, since the ring is of a larger diamcter, the angle of the pinch point between it and the die is much shallower. Greater pressures can be applied to the material without risk of slipping. Second, as the ring is no wider than the single row of the die, all of the drive power of the machine is concentrated over a width of $1^1/_2$ to 2 in. (38 to 50 mm). The surface area of contact is less than half that of a single roll shell, and so the pressure exerted is potentially four times as great as for a pair of roll shells, given the same drive installation.

Comparisons of die pressures are however somewhat academic, since the die holes in cubers are of considerably greater enclosed volume than those of pelletizers. Frictional resistances and flow characteristics are quite different. The bulk density of the final product is limited to the maximum possible for the material, which in the case of paper and card is in the order of 44 lb/ft^3 (700 kg/m^3). When that density has been achieved, the cube will be discharged from the die whatever the press ring pressure happens to be.

While common in the U.S., cubers have made very little or no impact upon the European dRDF industry to date, mainly because the combustion processes that might use the fuel have demanded granularities of between $^1/_2$ to $^3/_4$ in. (15 and 20 mm). However, cubers do offer some advantages of comparative mechanical simplicity and lower cost. There is a growing indication of a willingness to build dedicated combustion plant in order to accommodate cuber outputs, and as interest in dRDF develops it is likely that the machines will assume the importance their design warrants.

9.2.8.3 Screw Extruders

Screw extruders, as shown in Figure 9.27, are little more than a screw predensifier with an extrusion tube in place of the pressure plate. In fact, most screw predensifiers are designed to accept an extrusion head, and many were originally intended for use as extruders.

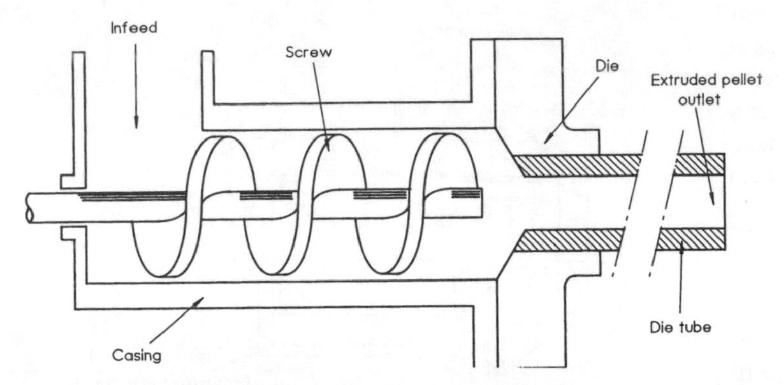

Figure 9.27 Principles of screw extruders.

Machines of this type have made no impact at all on the dRDF industry, mainly because their single extrusion tube limits their output capacity to a level well below that which would be acceptable in a full-scale process plant. They have however found a niche in manufacturing artificial logs from sawdust and wood chip for the domestic fuel market.

While the pressures and temperatures they achieve are somewhat less than those experienced within pelletizers and cubers, extruders have the benefit of extrusion tubes which can be of almost any length. Residence times in the equivalent of a die can therefore be predetermined in a way that is impossible in the other machines, and as a result almost any material can be successfully processed.

9.2.9 Coolers

While there are some differences of opinion in the matter, it is commonly believed that dRDF pellets have to be cooled immediately after extrusion from the presses in order to make them hard enough to be handled without breakage. The opinion has some validity, in that the fusion of lignin and plastic within the pellet creates the density and cohesion required, and that fusion requires heat. Therefore, it follows that it will not be complete until the pellet has cooled to some temperature below the melting point of both materials. Until then, the product is likely to remain somewhat ductile at least, and may be broken into unacceptably short lengths quite easily. The disputed matter is whether a completed pellet can actually be cooled rapidly enough for cooling to have any significance for product quality. The material from which it is made has a very low thermal conductivity, and is in fact a very good insulator. Therefore although the outer surface may be chilled, the core remains largely unaffected. In a short time after cooling, dRDF pellets return close to their original temperature, as the core heat slowly escapes.

It is the low thermal conductivity and its attendant slow release of heat that is the key to the argument. The purpose of attempting to cool the product is simply to make it durable enough for subsequent mechanical handling. Once it has reached the place of storage, it will no longer be disturbed for some period of time and its durability will become somewhat irrelevant. In those circumstances, if the outside of the pellet can be chilled sufficiently to prevent breakage at the exposed ends, and if it can be transported to store before the temperature is restored, then cooling is worthwhile. It is an inescapable fact that production of dRDF without a cooler in the circuit leads to a higher level of particulates within the final product, although the extent of the importance of that characteristic is determined by the market.

The most common design of cooler used in the industry is shown in Figure 9.28. Here, a series of short conveyor belts of lightweight, perforated metal sheet run inside an enclosed casing, the hood of which is aspirated by a remote fan. Air from the plant room is drawn through the base of the unit so that it rises through the perforated belts, which overlap at their ends. Hot fuel pellets direct from the pelletizers are introduced onto the highest of the belts, and are transported slowly through the machine, falling from one belt to the next until they reach the discharge. Residence times are usually several min, but the relevance of that must be considered in light of the fact that it takes dRDF up to 48 h to cool naturally when stored in bulk.

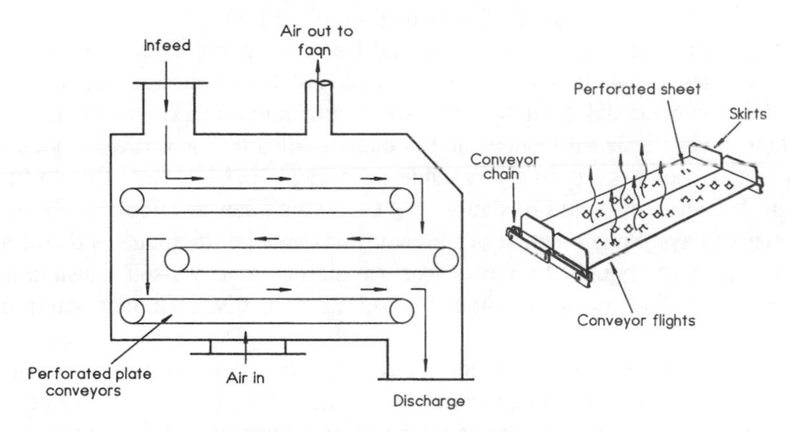

Figure 9.28 Perforated belt cooler.

The degree of cooling achieved is governed by the depth of material upon the belts and by the length of time in which it is subjected to the air flow. Ambient air temperatures may have some effect, although it is hardly detectable in most cases. Residence time is the most significant influence upon pellet quality, and that depends upon the production rate in effect at any time. As a result, in a plant where the handling system beyond the cooler is extensive, final product quality deteriorates with increased production rates.

Where a plant can be designed with the product storage very close to the final densifiers, there are some merits in excluding coolers. In spite of the claims made for them, the machines can often actually contribute to pellet degradation. They are after all mechanical handling devices, and if mechanical handling damages the product then so must they. Conveying does not harm the product in any way. It is the transfer between one handling machine and the next which causes damage. Badly designed transfer chutes between conveyors, unnecessarily steep angles of elevation, and excessive discharge heights are all major contributors. If they can be excluded, then in many ways a process is better without coolers at all. In that event, however, it is important that at least the pellet collecting conveyor serving the densifiers is equipped with aspiration. Otherwise, condensation of moisture is likely to create product degradation which exclusion of the cooler is seeking to avoid.

9.2.10 Pellet Classifiers

9.2.10.1 Vibrating Screens

Some degree of product degradation is inescapable. No pelletizer, cuber, or extrusion press ever produces only intact pellets. There is always some level of particulates in the product immediately following densification, and whatever the type of press used, the pellets always have some exposed edges that can be broken. An excessive particulate fraction is both unacceptable to most fuel customers, and is potentially damaging to the fuel.

Densified RDF, where it is clean and free from particulates, is very free-flowing, has a shallow angle of repose, and is stable in storage. All of those qualities can be devalued by excessive contamination. Then the product becomes very resistant to flow, and is capable of forming vertical angles of repose. It becomes extremely unstable in storage, and is susceptible to biological deterioration and oxidation, both of which can develop rapidly in a matter of days. Some method of removing unwanted particulates is therefore essential, even though the mechanism for doing so may itself cause some degradation. The most common method used involves sloping vibrating screens.

Generally, these machines consist of an inclined deck of perforated sheet or wedge-wire, clamped in a rectangular casing, with a particulates collecting chamber beneath. They are driven by vibrating motors in such a way that their throw is almost horizontal, so as to minimize the disturbance of the product. Pellets from the coolers are introduced to the high end of the deck, passing down the incline to be discharged at the bottom. Particulates fall through the screen plate and are removed through a separate port.

While vibrating screens are by far the most common devices for dRDF improvement, they are among the least effective, and the reason for this lies in the considerable difference between the *apparent* bulk densities of the respective materials. The particulates are, in comparison with the fuel product, extremely light, and as a result of the influence of the coolers, are likely also to be much drier. When the mass of dRDF and particulates is aggressively vibrated, the heavy pellets sink to the bottom of the material bed, and the particulates rise to the top. The pellets therefore effectively blind the screen and prevent the particulates from passing through. Wedge-wire decks, where the wires are arranged in parallel with the longitudinal axis of the machine, are some improvement, but even then there is a predisposition to blinding.

There is also, it could be argued, some failure of logic in subjecting a supposedly delicate product to the violent vibrations of a screen of this type. Even where the screen has been designed with great care and a clear understanding of the nature of the product, it is still likely to create as much degradation as it removes.

9.2.10.2 Classifiers

Air classifiers are an alternative to vibrating screens that are very worthy of consideration. Here the difference in densities between the product and the particulates works in favor of the design. It is sufficiently great that the conventional air classification method of dropping the product through a rising air stream or air jet is unnecessary. It is possible to draw the particulates clear by a method similar to the action of a domestic vacuum cleaner upon carpets.

Figure 9.29 shows a design concept worthy of consideration. It involves a simple modification to a drag link conveyor used for elevating the product to storage, and a logical extension of the principle leads to its replacing both a screen and a cooler. Drag link conveyors are well suited to transporting completed fuel, since they are capable of steep angles of inclination without load roll-back. As the product is being pushed along a deck plate which very quickly becomes highly polished, they create no detectable degradation, and the installation of air-operated gate valves in their decks (see Figure 9.30) allows them to discharge in a number of selectable positions. The modification which allows such a machine to also classify the product involves no more than replacing a section of the deck plate with perforated sheet, and in the absence of any discharge valves, inserting an air inlet in the side. A conveyor skirting rubber flap at each side of the classification position restricts air flow to the deck, while yielding to any abnormal depth of load-bed. Provided that the deck plate perforations are not larger than 25% of the pellet diameter, there is no noticeable "cheese grater" effect. The airflow is created by a remote fan, the design of which depends upon the final destiny of the particulates recovered. For simplicity, a chopper fan, where the material actually passes through the fan with the air, is worth considering, since it dispenses with cyclone separators.

The same principle, without the conveyor skirting rubber flaps, performs surprisingly well as a pellet cooler. The structure of the conveyor, together with its chain and flights, is likely to be of much thicker metal than a conventional cooler, and in this application the airflow across it has more significance in cooling the metal than the pellets. Product cooling results largely from conduction to the cool metal than to convection in the air, and since steel has a significantly greater thermal conductivity than air, higher efficiencies are the result.

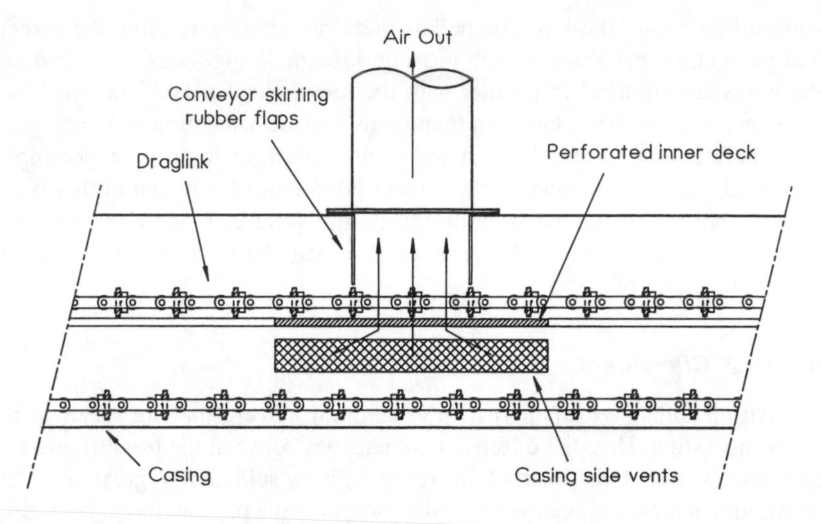

Figure 9.29 Simple fuel product classifier

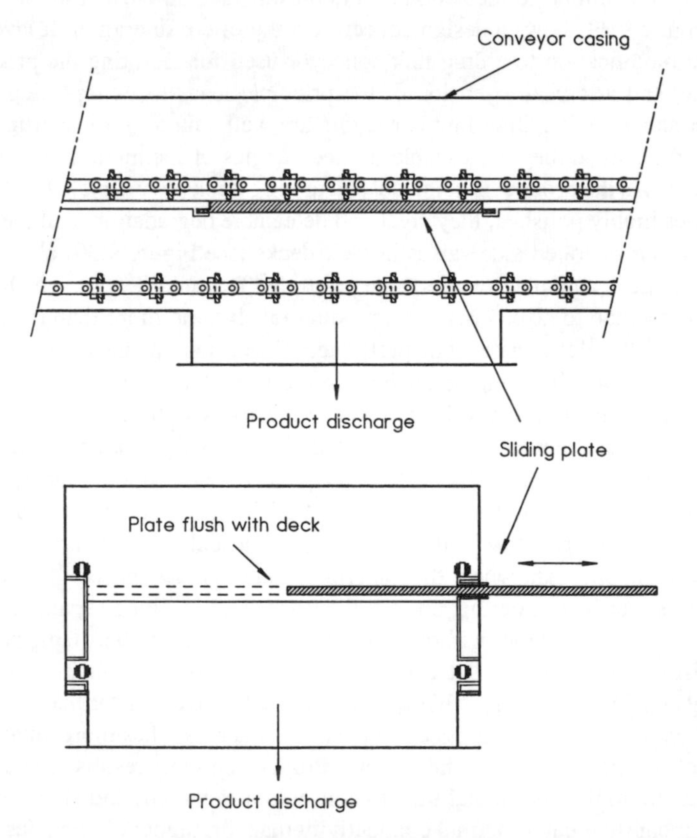

Figure 9.30 Drag link conveyor discharge valves.

9.2.11 Buffer Hoppers

As a result of the very low feedstock bulk densities, and the wide variations within them, that occur in RDF processes, ensuring a steady and consistent infeed rate to individual processing machines is always a problem. It is for example quite normal for a load of abnormally high combustibles content waste to be introduced to the process line. Such a load has a significant impact as it passes through the trommel and the primary granulator, causing a volumetric surge at the classifier stage. Even if the material is of normal moisture content, the consequences for the drier will be a sudden demand for considerably more heat energy, and the drier may simply be unable to respond sufficiently rapidly. Even if it does respond and succeeds in maintaining the desired level of drying, the surge volume will continue to the pelletizers or densifiers, with a very high possibility of an overload situation developing. It is therefore necessary in all RDF processes, with the possible exception of flock manufacturing, to introduce some form of metering into the system.

Surprisingly there are still some process designers who offer "storage" conveyors for this purpose, under the impression that a horizontal conveyor with abnormally high side plating will absorb surges and maintain a consistent flow rate. A moment of reflection should reveal that the concept is flawed. Conveyors do not store material, they transport it. A horizontal conveyor must necessarily discharge at exactly the same rate as that at which it is loaded. Therefore, if a surge load is applied to the charging end, then in due course the same surge will be released at the discharge end. A device that will deliver at a consistent rate irrespective of its loading is needed.

There are two most common methods used to achieve such a situation, buffer storage hoppers and live bottom hoppers, of which the former are generally the more reliable. Both work upon the principle of a mechanical handling device being called upon to remove material from the bottom of a storage volume, the first doing so by means of a heavy-duty plate feeder conveyor, and the second using a group of parallel screw feeders. In both, since the conveying mechanism is operating at a steady speed, feedstock will be withdrawn from the hopper above at the same rate irrespective of the level of infeed to it. Any surges therefore only increase the level in the hopper temporarily; they do not pass further down the flowline.

Figure 9.31 shows a example of a buffer hopper. The design concept is simple, and it relies upon a plate-feeder conveyor, a section of which is inclined at 45° to the horizontal and served by a charging chute of considerably greater volume than would be normal for a machine intended for conveying alone. The steep incline of the conveyor, together with lifters welded to the flights (see Section 3.2 on mechanical handling), causes the conveyor to unload the charging hopper at a consistent rate. Any depth of material beyond a predetermined level upon the belt simply rolls back on the incline.

Devices of this type are rarely included in a process solely as surge arrestors. More commonly, the plate-feeder conveyor is used as a conventional

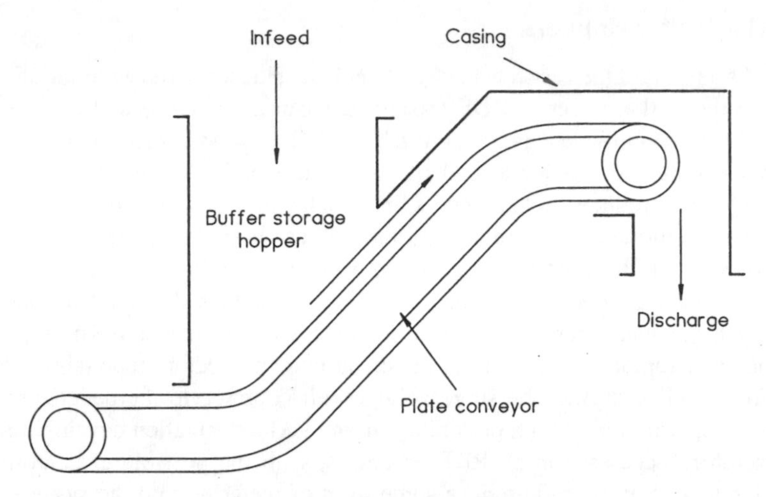

Figure 9.31 Buffer storage hopper.

mechanical handling device in every sense, for example to transfer material from a classifier to a drier. In such a situation, some degree of elevation is already necessary to collect the material from the base of the classifier and to raise it to the inlet hopper of the drier. The buffer hopper, therefore, becomes almost an extension of the classifier, rather than a device in its own right.

A live bottom hopper, meanwhile, is essentially a stand-alone machine, since the screw feeders that form the discharge mechanism cannot conveniently be used to transport the material elsewhere. The name given to the principle arises from the number of screws contained in the base of the hopper. Commonly, they are installed in parallel with each other across the base, and are driven by a single drive gearbox with multiple transmission chain transfer to each shaft. The space between each screw is minimal, being usually no more that 1 or 2 in., and so the bottom of the hopper is "live" as the name suggests.

The screws discharge material through the casing opening, as shown in Figure 9.32, and may deliver onto a belt conveyor, pneumatic system, drier, or even directly into the charging device of a densifier. In order to ensure reliable discharge and freedom from blockages at the unloading end, it is customary to fit a plate deflector to each shaft. This is no more than a piece of flat steel plate, welded in place, and rotating with the shaft in such a way as to push feedstock clear of the end of the screw and into the discharge chute.

Both devices need careful design or they can create very serious blockages. They are handling material of a very low bulk density but capable of very high rates of compaction indeed. As a result of the way in which they must work, both machines cause a degree of recirculation of material against the hopper platework at its discharge end, and screw feeders do so to a greater extent than do plate-feeder conveyors. In screw feeders this is due to their collecting material across the whole of the base of the hopper, where the full weight of the stored mass is applied. In a buffer storage hopper, however, the

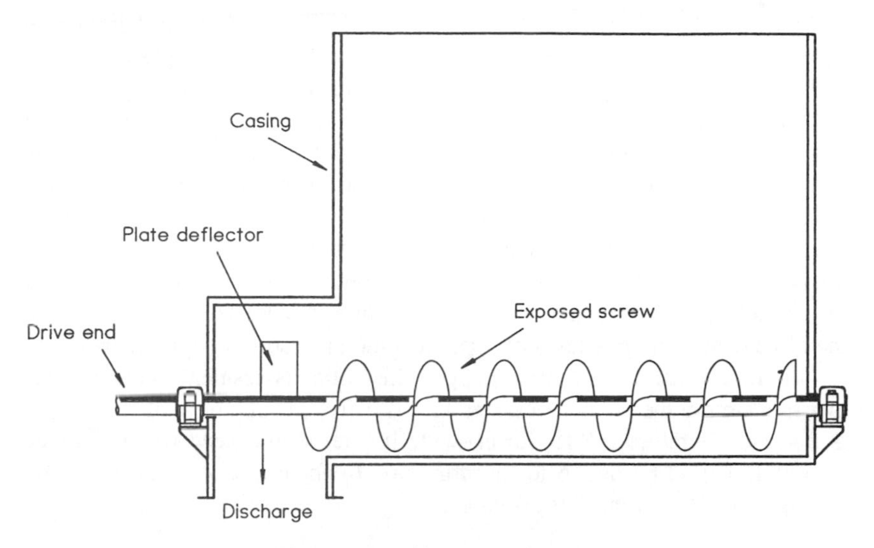

Figure 9.32 Live bottom hopper.

plate-feeder conveyor is called upon to remove material at an oblique angle. Compaction forces due to the depth of the load are reduced and are never applied at right angles to the direction of transport.

Determination of the storage volume to be installed is more a question of choice based upon experience than upon mathematical modeling. It requires a decision upon the extent of any surge that may be expected, and then a simple calculation upon the volumetric consequences of that surge within the flowline. The basis for that assessment and calculation should be that the surge hopper facility should, under normal circumstances, be running empty or almost so. It should be acting simply as a somewhat oversized transfer chute, otherwise difficulties with recirculation of material at the discharge, and with compaction inside the hopper, are almost inevitable.

As a guide to the allowance for surges, a practical surge capacity is 10% of the hourly throughput of waste at the chosen position of the hopper, in volumetric terms. Therefore, if a process is treating 37.5 t of waste an h in two flowlines, then with an average waste analysis it is likely that some 35% of that, or 6.56 t/h per line, will appear immediately after the classifier. At that point, the apparent bulk density will be within the range 1.87 to 3.74 lb/ft^3 (30 to 60 kg/m^3). In the worst case of the lowest density, the normal process volume will then be 285 yd^3 (218 m^3/h), of which 10% is 28.5 yd^3 (2.18 m^3). A hopper to contain that volume, mounted upon a 5 ft (1.5-m)-wide plate conveyor for example, would need sides of area 154 ft^2 (14.3 m^2).

Where a conveyor is used, the base of the hopper will be angled at 45° to the horizontal, and so the product of the length of the top and the vertical height of the backplate must be twice the area, or 308 ft^2 (28.6 m^2). The square root of that figure is 17.5 ft (5.3 m), and so a hopper 17.5 ft (5.3 m) long and of equal height can contain the expected surge. Some freeboard is clearly necessary, since the material does not possess a horizontal angle of repose, and so perhaps a backplate height of 20 ft (6.5 m) would be appropriate.

While upon first sight the result of the assessment appears to be a somewhat small hopper in relation to the flowline capacity, it should be recalled that the conveyor will, in normal circumstances, be transferring waste at a rate of 285 yd^3/h (218 m^3/h). The surge will almost inevitably be followed immediately by a reduction in throughput as the primary granulator reacts to the overload which it has experienced. A 10% *reduction* in throughput following a surge will result in the conveyor clearing the original excess in approximately 6 min. It could therefore deal with up to ten surges per h, which is a somewhat unlikely event.

In the case of a live bottom hopper, the initial assessment is unchanged, and it is simply the hopper side area that is different. The base of the hopper is flat, and therefore a 5 ft (1.5-m)-wide hopper would need to have a side area of 154 ft^2 (14.3 m^2). Allowing for some freeboard, it could be 15 ft (4.7 m) high and 12 ft 6 in. (3.9 m) long.

9.3 RDF MANUFACTURING PROCESSES

In this section our purpose is not to discuss every possible RDF process design that has been or could be built, since that would require a volume of much greater size. Instead, our purpose is to explore a process for manufacturing each of flock, coarse or crumb, and fully densified RDF to a quality that compares adequately with the conventional fuels that would normally be used. Processes that simply granulate waste and then "refine" the product by extracting the ferrous metals fraction are mass incineration processes given an over-optimistic title. They are not RDF, and judging from recent legislation in many countries, it is only a matter of time before they are prohibited from claiming otherwise.

Later in this section, Figures 9.33, 9.34, and 9.35 describe the processes that will be discussed, and examination of them reveals that each contains elements of the others. A fully densified fuel (dRDF) is achieved by first manufacturing flock (fRDF), and then crumb (cRDF), before final treatment and densification. At each stage, the quality of the product increases, so that fRDF is in many ways similar to low-grade pulverized fuels used in power plants. "Brown" coal with a high lignin content, or even peat, is used in many countries in large industrial boiler plant, and in that application cRDF compares reasonably favorably. Higher-quality solid fuels are normally washed and graded to a regular granularity, and are intended for use in installations with sophisticated fuel handling systems and high-efficiency combustion plant. In those applications, dRDF can generally find a ready market.

Given that each successive design is a product of the processes that precede it, the discussion of the technology can begin with fRDF and be completed with dRDF. At each stage, a portion of the essential mass and volume balances can be constructed, and for the purposes of example, the hypothetical plant used throughout this book, with a throughput of 37.5 t/h, will be considered.

To assist in that, Table 9.3 lists the apparent bulk densities that might be expected at each stage of the process, and summarizes them to an average bulk density for the whole waste there. The resulting whole waste bulk density is derived from the mass and volume balance, so that while for example the classifier lights and heavies component densities are identical, the masses reporting to each stream are not. There will be little paper and card in the heavy fraction, but most of the heavy plastic, resulting in a relatively high apparent density for the whole of the heavies fraction. The data has been derived from experimental work and statistical records from operating process plants over a number of years, and corrected for an average as-received waste moisture content of 30% w/w. It relates to a waste analysis by weight of

Paper and card (PACD) — 25%
Plastic film (PLF) — 5%
Dense plastic (DP) — 6%
Textiles (Tx) — 6%
Wood — 3%
Steel/iron (Fe) — 7%
Nonferrous metal (Nfe) — 1%
Organic/putrescible (OP) — 35%
Soil — 2%
Glass (Gl) — 8%
Other (–10 mm) — 2%
Total water — 30%

The expected apparent densities of the individual fractions are listed in such detail in a sense because of the inevitable inaccuracies that occur when dealing with comingled waste. One can never be sure, for example, where the paper fraction all falls within the true bulk density range for paper of between 44 lb/ft^3 and 69 lb/ft^3 (700 and 1100 kg/m^3). By treating the waste as a large number of components, and by identifying their individual response to processing stages, at least the designer avoids the possibility of compounding errors. Therefore, while a particular apparent density may not be precisely the same in every application, the final result is realistic for most processes.

9.3.1 Flock Refuse-Derived Fuel (fRDF)

As has been shown repeatedly in the preceding sections of this book, the reprocessing of municipal waste into new consumer products must begin with the removal of fractions unsuitable for such products. In a fuel process, the first step must therefore be to broadly separate the combustible fraction from that which is either noncombustible or only slightly so. The highly combustible fractions of municipal waste are paper, cardboard, and plastic film.

Table 9.3 Fraction Apparent Bulk Densities

fRDF Stages

		Apparent bulk densities lb/ft³ (kg/m³) after:				
		Trommel[a]			Classifier[b]	
Fraction	Waste	O/s	U/s	Granulator	L	H
PACD	6.24 (100)	3.7 (60)	7.5 (120)	1.9 (30)	1.8 (28)	1.8 (28)
PLF	5.0 (80)	4.4 (70)	5.0 (80)	2.5 (40)	2.5 (40)	2.5 (40)
DP	12.5 (200)	12.5 (200)	13.7 (220)	17.5 (280)	17.5 (280)	17.5 (280)
Tx	7.5 (120)	7.5 (120)	11.2 (180)	11.2 (180)	11.2 (180)	11.2 (180)
Wood	18.7 (300)	21.8 (350)	18.7 (300)	25.0 (400)	25.0 (400)	25.0 (400)
Fe	93.6 (1500)	93.6 (1500)	187.2 (3000)	249.6 (4000)		
Nfe	49.9 (800)	49.9 (800)	124.8 (2000)	156.0 (2500)		
O/P	15.6 (250)	15.6 (250)	17.2 (275)	25.0 (400)	25.0 (400)	25.0 (400)
Soil	62.4 (1000)	62.4 (1000)	62.4 (1000)	62.4 (1000)		
GL	49.9 (800)	49.9 (800)	74.9 (1200)	74.9 (1200)		
Other	31.2 (500)	31.2 (500)	34.3 (550)	34.3 (550)	34.3 (550)	34.3 (550)
Mixed	11.0 (177)	5.0 (81)	20.2 (324)	2.8 (45)	2.1 (33)	21.5 (344)

cRDF Stages

	Apparent bulk densities lb/ft³ (kg/m³) after:				
			Secondary screen[a]		
Fraction	Buffer	Drier	O/s	U/s	Predensifier
PACD	5.0 (80)	1.6 (25)	1.6 (25)	1.3 (20)	18.7 (300)
PLF	2.5 (40)	1.9 (30)	1.9 (30)	1.6 (25)	12.5 (200)
DP	17.5 (280)	21.8 (350)	21.8 (350)	23.7 (380)	25.0 (400)
Tx	11.2 (180)	7.5 (120)	7.5 (120)	3.7 (60)	12.5 (200)
Wood	25.0 (400)	18.7 (300)	18.7 (300)	12.5 (200)	25.0 (400)

cRDF Stages

Fraction	Apparent bulk densities lb/ft³ (kg/m³) after:				
			Secondary screen[a]		
	Buffer	Drier	O/s	U/s	Predensifier
O/P	25.0 (400)	6.2 (100)	6.2 (100)	4.4 (70)	25.0 (400)
Other	34.3 (550)	18.7 (300)	18.7 (300)	12.5 (200)	31.2 (500)
Mixed	4.8 (77)	1.8 (29)	1.75 (28)	2.9 (46)	17.35 (278)

dRDF Stages

Fraction	Apparent bulk densities lb/ft³ (kg/m³) after:
	Pelletizer/cuber
PACD	37.44 (600)
PLF	31.2 (500)
DP	31.2 (500)
Tx	25.0 (400)
Wood	31.2 (500)
Mixed	37.3 (598)

Note: Paper and card (PACD); plastic film (PLF); dense plastic (DP); textiles (Tx); steel iron (Fe); nonferrous metal (NFe); organic putrescible (O/P); glass (Gl).

[a] Oversized (O/s); undersized (U/s).

[b] Lights (L); heavies (H).

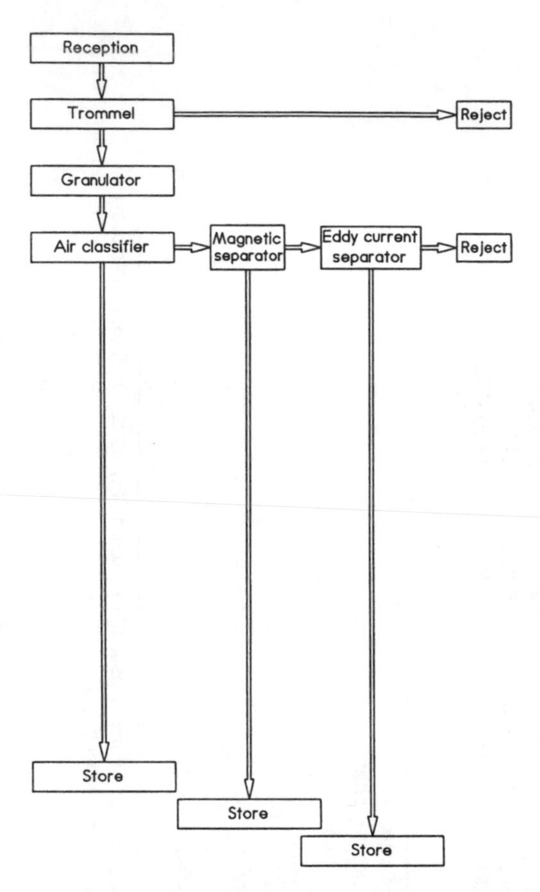

Figure 9.33 Flock refuse-derived fuel (fRDF) flowline.

Those fractions are ductile, and they arise mainly in sheet form, and so they are suitable for removal from the bulk by means of a trommel screen. The machine will not deliver an entirely uncontaminated product, since metal cans and plastic containers at least will remain with the oversized, combustible fraction. So, to an extent, will textiles and some residues of organic wastes. Glass and ceramics will be broken, but the oversized materials will be moist, and so a certain amount of glass in fine shards may adhere to it. However, operating experience of trommels indicates what to expect for each category of oversize. It is therefore possible to derive a mass balance from which a volume balance may be calculated, using the apparent densities provided in Table 9.4.

9.3.1.1 Trommel Stage

The balances show that at the trommel oversize discharge there will be 9.6 t/h (8.8 te/h) per line, with a volume of 3879 ft^3 (108 m^3) and a total bulk

Table 9.4 Trommel Oversize Stage Mass and Volume Balance

Fraction	Separation efficiency %	Mass		Volume	
		t/h	(te/h)	ft^3/h	(m^3/h)
PACD	10	4.651	(4.219)	2514.2	(63.9)
PLF	13	0.899	(0.816)	408.8	(10.5)
DP	10	1.116	(1.013)	178.6	(4.6)
Tx	20	0.992	(0.900)	264.6	(6.8)
Wood	8	0.571	(0.518)	52.3	(1.3)
Fe	20	1.158	(1.050)	24.7	(0.63)
Nfe	7	0.192	(0.174)	7.7	(0.20)
O/P	99	0.072	(0.066)	9.3	(2.40)
Soil	98	0.008	(0.008)	0.3	(0.007)
GL	98	0.033	(0.030)	1.3	(0.034)
Other	99	0.004	(0.004)	0.27	(0.007)
Total		9.697	(8.784)	3879	(108.4)

Note: Paper and card (PACD); plastic film (PLF); dense plastic (DP); textiles (Tx); steel/iron (Fe); nonferrous metal (NFe); organic/putrescible (OP); glass (GL).
Separation efficiency is a measure of the ability of the trommel to remove fractions from the waste stream as undersize.

density of 4.9 lb/ft³ (81.5 kg/m³). Since the infeed to the trommel is one half of 37.5 t/h, then 10.7 t have been rejected in the undersized fraction. At this point it is interesting to note that the apparent bulk density of the original waste was 11 lb/ft³ (177 kg/m³), resulting in a volumetric infeed of 105.8 m³/h. The volume of the trommel oversize fraction is 13% greater than that, so although 48% of the mass has been lost, the volume has increased slightly. This is an entirely normal result of trommel operations, even though it may be somewhat surprising. It is also the reason why processes designed upon mass balances and true densities rarely achieve more than 50% of the capacity for which they were intended.

9.3.1.2 Transfer to Granulator

The immediate relevance of the figures derived for the trommel oversized fraction is that a conveyor can be designed to move it to the next stage of the process. The conveyor must be able to carry 9.7 t and 3879 ft³/h at a bulk density of 4.9 lb/ft³ (81.5 kg/m³). A suitable conveyor can then be specified using the formulas established in Section 3.2

9.3.1.3 Granulating Stage

The next part of the process must be to reduce the material mix to a condition where further separation and refining becomes possible, and so granulating is necessary. More screening is pointless, since it will not perform significantly better than the trommel, but will reject more of the valuable combustible fraction. It is the heavy, noncombustible residues that need to be removed, and air classification is the most appropriate method. The granulating stage must therefore seek to create a consistent particle size in order that the classifier may achieve the highest efficiency.

Table 9.5 examines the product of the granulator, where the weight delivered is exactly that which was discharged from the trommel. Here it is assumed that the degree of granulation achieved is 95% below $^3/_4$ in. (20 mm), and the immediate point of note is that while the fraction masses have remained unchanged, the total volume has nearly doubled. This reinforces the observation made in Section 9.2, that granulators do not necessarily create a volume reduction as is commonly supposed. In this model, the machine has indeed reduced the volume of some of the components, but it has also substantially *increased* that of others, particularly the paper and card.

Table 9.5 Granulator Stage Mass and Volume Balance

Fraction	Mass		Volume	
	t/h	(te/h)	ft³/h	(m³/h)
PACD	4.651	4.213	4896.1	140.44
PLF	0.899	0.815	719.4	20.36
DP	1.116	1.011	127.6	3.61
Tx	0.992	0.899	177.2	4.99
Wood	0.571	0.517	45.6	1.29
Fe	1.158	1.049	9.3	0.26
Nfe	0.192	0.174	2.5	0.07
O/P	0.072	0.066	5.8	0.06
Soil	0.008	0.007	0.27	0.007
GL	0.033	0.030	0.88	0.023
Other	0.004	0.004	0.24	0.007
Total	9.697	8.784	6927	195.2

Note: Paper and card (PACD); plastic film (PLF); dense plastic (DP); textiles (Tx); steel/iron (Fe); nonferrous metal (NFe); organic/putrescible (O/P); glass (GL).

The change of state has occurred because the machine has altered not just the granularity of the fraction, but also the shape of the individual particles. No longer are they in sheet form, but instead they have been converted to fibrous nuggets. They cannot now interlock together to the exclusion of the air space between them, and so while the mass is unchanged, the volume occupied by that mass is increased.

Metal cans, meanwhile, will also have been shredded to nuggets, with the result that the original mass now occupies *less* volume. This is also the case with textiles, heavy plastics, and wood, but the effect upon the metals is the most significant from the point of view of recycling. In being reduced in a mixture of paper and card, the metal is very effectively cleaned of organic residues, and is discharged in a condition that makes it attractive to the metal salvage industry. It becomes weight-limiting in transport, where the payload carried is that of the maximum load capacity of the vehicle without exceeding its volumetric capacity.

9.3.1.4 Transfer to Classifier

In the flowline proposed in Figure 9.35, the material from the granulator is passed on to the air classifier, and the method by which that is achieved depends upon the type of granulator installed. A vertical shaft machine can discharge directly into a column classifier, while a horizontal shaft machine must employ a transfer conveyor of some type. With the former there is no calculation to be made in relation to the transfer mechanism. In the latter, a plate feeder conveyor is necessary to withstand the impacts of items being discharged at high velocities. A conveyor that can accommodate 9.7 t/h at a volume of 6927 ft^3/h is required.

In this application, the width of the conveyor is determined by the discharge throat opening of the granulator. Installing a conveyor that is narrower than the throat risks almost continual blockages, as the very low-density product is dropped into a narrowing transfer chute. Equally, there is little point in making the conveyor much wider than the throat, since in that case the load would be concentrated along its center line. Therefore, in this case the matters requiring mathematical modeling are constrained to the bed depth on the belt and the resulting traveling speed.

9.3.1.5 Classifier Stage

Table 9.6 extends the mass/volume calculations to the classifier stage, where again a volumetric change occurs. The lights volume discharged from the classifier is marginally greater than the total volume from the granulator, even though a significant proportion of the mass has been rejected into the heavy stream. Once again it is the paper and card which is largely responsible for this, the effect having been caused by the aeration of the mass. The air stream has distributed the particles more widely, and their original isolation from each other is not completely restored afterwards.

At the classifier, 6.59 t/h and 5621 ft^3/h are discharged as light fraction, while 3.11 t and 635.5 ft^3 are rejected into the heavy fraction. Where the purpose of the process is to create a low-grade, low-cost fuel for use with pulverized fuel (PF) combustion plant, the light fraction would now be passed on to a storage facility. In such an event, the very low overall bulk density of the material, 2.3 lb/ft^3 (17.3 kg/m^3) makes any form of stockpiling somewhat impractical. A 100 t stock, which is insignificant by PF boiler standards, would occupy 86,960 ft^3, and although there would in such cases be a considerable degree of compaction under its own weight, the stockpile would still demand an unacceptable amount of spacc.

For that reason, it is customary for the classifier cyclone to discharge directly into an electrohydraulic packer, which compacts the material into

Table 9.6 Classifier Stage Mass and Volume Balance

Fraction	Separation efficiency %[a]	Mass te/h (t/h)[b]		Volume ft³/h (m³/h)[b]	
		H	L	H	L
PACD	92	0.37 (0.34)	4.28 (3.88)	413 (12)	4755 (138)
PLF	92	0.07 (0.07)	0.83 (0.75)	58 (1.6)	662 (19)
DP	23	0.86 (0.78)	0.26 (0.23)	98 (2.8)	29 (1)
Tx	90	0.10 (0.09)	0.89 (0.81)	18 (0.5)	159 (5)
Wood	25	0.43 (0.39)	0.14 (0.13)	34 (1.0)	11 (0.3)
Fe	3	1.12 (1.02)	0.03 (0.03)	9 (0.3)	0.3 (0.01)
Nfe	60	0.08 (0.07)	0.12 (0.10)	1 (0.03)	1.5 (0.04)
O/P	48	0.04 (0.03)	0.03 (0.03)	3 (0.1)	2.8 (0.08)
Soil	5	0.01 (0.01)	4×10^{-4} (4×10^{-4})	0.3 (0.01)	0.01 (4×10^{-4})
GL	5	0.03 (0.028)	17×10^{-4} (15×10^{-4})	0.8 (0.02)	0.04 (1×10^{-3})
Other	10	0.004 (0.003)	4×10^{-4} (4×10^{-4})	0.2 (0.01)	0.02 (7×10^{-4})
Total		3.11 (2.82)	6.59 (5.96)	635.5 (18.3)	5621 (163)

Note: Paper and card (PACD); plastic film (PLF); dense plastic (DP); textiles (Tx); steel/iron (Fe); nonferrous metal (NFe); organic/putrescible (O/P); glass (GL).

closed containers both for storage and transport. The question of conveyor transfer does not therefore arise. Only the heavy fraction has to be conveyed onward, usually to magnetic and eddy-current separators for the recovery of the metals. A conveyor suitable for the purpose again is prescribed in terms of width by the heavies discharge aperture of the classifier which, in an 18.75 t/h flowline, is unlikely to be narrower than 5 ft (1.5 m). A conventional rubber belt conveyor is suitable, and one of that width designed to carry 3.11 t/h and 635.5 ft^3/h will be a very lightly loaded conveyor indeed. However, light loading at this stage is a valuable attribute, since it will encourage the highest efficiency from the metals separators.

9.3.2 Crumb Refuse-Derived Fuel (cRDF)

9.3.2.1 Buffer Hopper

Where the process is intended to manufacture cRDF or dRDF, the light fraction from the classifier will have to be passed through a buffer system to eliminate any volume surges. Normally the classifier cyclone would discharge directly into the hopper of the surge arrestor, and in the flowline under consideration it would do so at the rate of 6.4 t/h (5.8 te/h) and 2580 ft^3/h (72.9 m^3/h). In Section 9.2 it was suggested that a suitable buffer hopper capacity would be 10% of the volumetric throughput per h, and so in this case a capacity of 258 ft^3 (7.3 m^3) is required.

The plate feeder conveyor that forms the basis of the buffer design could be almost any width, since it is independent of the classifier casing. However, it will have to discharge into the feed hopper of the drier, and that in turn will be prescribed by the dimensions of the drier itself. It is most unlikely that the drier would be able to accept a direct discharge width of greater than 5 ft (1.5 m), and so that must be the width limit of the buffer belt. According to the formula proposed in Section 9.2, the buffer hopper would need to be at least 1.6 m high and 1.6 m long. Alternatively, if a live-bottom hopper is to be used, it would be of 1.1 m^2 sides and 1.5 m wide, excluding the necessary freeboard.

The discharge from the buffer is likely to be as represented by Table 9.7, where 6.43 t and 2580 ft^3/h will be discharged. Some 3041 ft^3 (90 m^3) of volume reduction has occurred, which is consistent with the operation of a buffer system. Both a plate-feeder conveyor and a live bottom will create some degree of compaction, which is counterproductive as far as the drier is concerned, but equally is inescapable. Ideally, the material should be introduced to the drier with as little compaction as possible, since the drier must work best by creating a volume of hot gas between each of the particles. However, the extent of volume reduction observed is not likely to form an insurmountable barrier to drying efficiency.

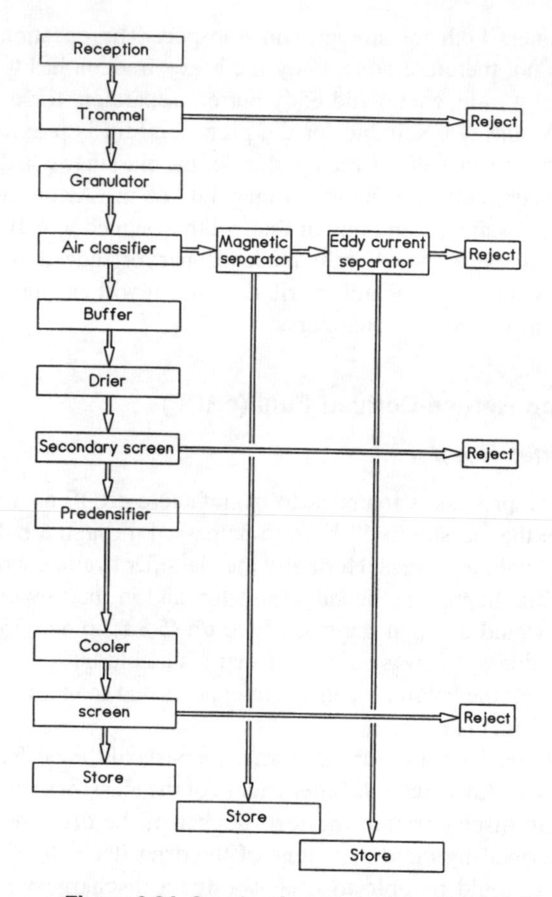

Figure 9.34 Crumb refuse-derived fuel (cRDF) flowline.

Table 9.7 Buffer Stage Mass and Volume Balance

Fraction	Mass		Volume	
	t/h	te/h	ft³/h	m³/h
PACD	4.28	3.88	1714	48.45
PLF	0.83	0.75	663	18.73
DP	0.26	0.23	29	0.83
Tx	0.89	0.81	159	4.49
Wood	0.14	0.13	11	0.32
O/P	0.035	0.031	3	0.08
Other	0.0004	0.0004	0.02	0.001
Total	6.43	5.83	2580	72.91

Note: Paper and card (PACD); plastic film (PLF); dense plastic (DP); textiles (Tx); steel/iron (Fe); nonferrous metal (NFe); organic/putrescible (O/P); glass (GL).

9.3.2.2 Drier

If the drier is to reduce the moisture content of the combustible fraction to the 16% w/w, which is suitable for a densifier or pelletizer feedstock, then with the waste analysis proposed in this example it must remove 14% w/w. Most of the water in the original waste will have existed in two forms, and be distributed mainly between two fractions. As discussed in Section 2.2, it will exist partly as free moisture, and partly as organic moisture. Most of the free water will be contained in the paper and card fraction, while most of the organic will be within the organic and putrescible fraction. The organics have been almost completely removed well before the drier stage, and only a small residue is left adhering to the combustibles. Therefore, while the total moisture content of the original waste was assessed at 30% w/w, the same will not be true of the fraction entering the drier. It will instead depend upon the moisture distribution between categories, and a working average distribution is shown in Table 9.8.

Table 9.8 Moisture Content Distribution

Fraction	Index	Moisture % w/w
PACD	1.15	34.50
PLF	0.11	3.30
DP	0.11	3.30
Tx	0.39	11.40
Wood	0.30	9.00
Fe	0.12	3.60
Nfe	0.12	3.60
O/P	1.75	52.50
Soil	0.95	28.50
Gl	0.12	3.60
Other	0.90	27.00

Note: Paper and card (PACD); plastic film (PLF); dense plastic (DP); textiles (Tx); steel/iron (Fe); nonferrous metal (NFe); organic/putrescible (O/P); glass (Gl).

The data recorded in Table 9.8 have again been established by laboratory analysis over a number of samples, and are sufficiently representative for a process plant design. They are derived by the application of an "absorbency index" to each fraction, where the index is a numerical measure of the ability of a component to retain moisture in comparison with the remaining components.

It will be noted that the moisture contents listed do not total 100%. This is because they are individually expressed as percentages of the *total moisture in the waste*, not of the waste itself. Thus, 34.5% of the moisture is contained in the paper and card fraction, 52.5% in the organics and putrescibles, and so on. Equally, if the indices are applied to the waste analysis given, the resulting moisture content revealed is not always identical to the original estimate. If the estimate has been based upon analytical work, then it is almost inevitable that the recorded total moisture content will have been influenced by some loss of organic matter in addition to water.

In this case, the drier is called upon to remove 0.78 t/h, and having done so will discharge a mainly combustible fraction according to the mass and volume balance shown in Table 9.9. Although only water has been removed, the waste volumetric change is greater than can be explained by the loss of that alone. It is in fact perfectly possible, depending upon the initial waste analysis, to obtain either an increase or a decrease in volume at this stage. The effect is a consequence of the way in which the temperature in the drier affects the dry solids.

Table 9.9 Drier Stage Mass and Volume Balance

Fraction	Mass		Volume	
	t/h	te/h	ft³/h	m³/h
PACD	3.39	3.07	4236	140.62
PLF	0.81	0.73	848	22.66
DP	0.26	0.23	24	0.60
Tx	0.48	0.43	127	6.11
Wood	0.13	0.12	14	0.39
O/P	0.12	0.105	37	0.29
Other	1×10^{-3}	1×10^{-3}	0.11	3×10^{-3}
Total	5.17	4.69	5287	153.04

Note: Paper and card (PACD); plastic film (PLF); dense plastic (DP); textiles (Tx); steel/iron (Fe); nonferrous metal (NFe); organic/putrescible (OP); glass (GL).

Paper and card, for example, have reversed the previous trend and lost some volume, compared with the classifer discharge and the extent of that loss is greater than the volume of water driven off. The figure is derived from the application of the average apparent bulk density of paper and card at a drier discharge listed in Table 9.3, and as discussed in earlier sections, apparent bulk densities are notoriously difficult to establish. The lower they are, the harder it is to measure them accurately, and a point is reached where all that can be said is that the material must have a density of approximately "*x*" in order for it to behave in the way that it does. This is true of the paper and card component of the drier discharge, and so while a figure of 1.6 lb/ft³ (25 kg/m³) is applied, it must be regarded with some caution. A difference of 0.06 lb/ft³ (1 kg/m³) incurs an apparent volume change of 230 ft³ (6.5 m³).

There is some intuitive validity in the balances shown, however. The masses can be relied upon, since in simplistic terms, if one removes a kilogram of water from a material, then the remainder must weigh one kilogram less. If damp paper and card granulated to a highly fibrous state is dried rapidly at very high temperatures, then predictably there would be some element of shrinkage. Plastic film meanwhile, would be more likely to distort and lose its flake-like structure. The cellular structure of any remaining organics would be disrupted, causing a transient increase in particle size under the internal pressure of steam.

In Section 9.2, the relevance of drier temperature to the volatile matter content of the combustible fraction was discussed. It was suggested that the matter begins to be liberated by temperatures above 212°F (100°C), but that the heat required to change the state of water from liquid to steam generally constrains the solids to that temperature as a maximum. Such a condition cannot of course be guaranteed, since to be so it would be necessary for every particle in the feedstock to be of exactly identical moisture content. As that is not so, some particles will already be at a moisture content close to the desired level, and in a drier designed to deal with the *average* rather than the individual, those will be overheated. Some volatile matter will always be driven off in any drier, however critically it is controlled, and most of the easily accessible volatiles are contained in the paper and card. The volatile matter is mainly detected in the form of the complex hydrocarbons limonene, methanol, and ethanol, all of which are of low density. Therefore, their release may have no measurable impact upon the mass of the feedstock, but over a period of an h of operation, it may be revealed in the volume balance.

It is the emission of the volatile gases which creates the characteristic, and occasionally intrusive, odor from RDF processes. This is the single most significant aspect of the whole production operation that is likely to lead to public complaint, and in view of the nature of the odor-creating materials, water scrubbers are clearly ineffective. Light, volatile oils are not water-soluble or even easily contained in suspension. Biological filters are capable of dealing with these oils, as are afterburners in the emission duct. Where the exhaust heat is to be used in a later process, in drying a plastic fraction for example, then some form of burner would be necessary in any case. Efficiency dictates that it be used for a dual purpose wherever possible.

The result of the drying process in the example has been for the mass of the combustible fraction to be reduced to 5.17 t/h (4.69 te/h), while the volume has decreased slightly to 5287 ft^3/h (153 m^3/h). The apparent density at that point is accordingly 1.96 lb/ft^3 (30.65 kg/m^3). The next stage in the route to cRDF or dRDF is screening to remove the organic and incombustible residues, now that the combustible fraction has become too dry to retain them. The mechanism of transfer to the screen depends upon the type of drier used.

A thermopneumatic drier has the particulates in suspension in air, and when it has exhausted its heating capacity, it becomes simply a pneumatic conveyor for its remaining length. Logic would suggest that in such a case the drier cyclone separator should be mounted directly above the screen feed chute, and should discharge directly into it. No conveyors would then be required. A rotary cascade drier, however, would need the services of a conveyor; considerable care will need to be taken over the design of such a drier.

A product with an apparent density of 19.6 lb/ft^3 (30 kg/m^3) is less than ideal for conventional mechanical handling, since it is very prone indeed to bridging in transfer chutes and conveyor casings. In addition, although it has left the drier, it is still hot and it retains some moisture. It will continue to emit water vapor for some time after leaving the drier, and so in an enclosed conveyor, condensation will inevitably create difficulties. The inside of the conveyor structure will become moist, and the very light feedstock will adhere to it with surprising rigidity, again leading to rapid blockages.

Section 3.2 considered the various types of conveyor that could be used effectively to handle granulated product, which is mainly paper and card, and plate feeders, sidewall belts, and folding belts were shown to be appropriate. In dealing with the discharge from a drier, that remains so in purely mechanical handling terms. However, another factor should be taken into account in this particular application.

Rotary cascade driers are just as capable of initiating combustion as are thermopneumatic systems, but they are much less well-equipped to deal with it. Generally, they are fitted with water spray bars over their discharge apertures, controlled by temperature sensors in the discharge hood. Although such an installation is an improvement over no action at all, it is likely to be of limited effectiveness in the event of a major fire in the drier. The very low density of the feedstock limits its ability to absorb water in quantity, since the first to reach the mass simply causes surface compaction. The water never reaches the material beneath the surface, and so combustion continues there.

The nature of the combustion is however changed fundamentally, since the water in the surface layers tends to exclude sufficient oxygen for flame to be apparent. The fire smolders instead, and at very high temperatures. If the feedstock is discharged onto a conveyor in that condition, oxygen will be admitted and the conflagration will erupt with some violence. Clearly, if that occurs upon a synthetic rubber sidewall or folding conveyor belt, plant damage is likely to be catastrophic.

If water sprays in the drier hood cannot be relied upon to control fires, then a conveyor that can withstand them is necessary. Once again, the ubiquitous plate feeder is the obvious choice. It would be untrue to suggest that such a machine cannot be damaged by fire, but it is much less sensitive to it than any other type. In addition, being highly water-resistant but inherently self-draining, it lends itself to the possibility of further water sprays above it being used to control any residual combustion.

For a plate feeder to be effective in such a circumstance, it must be of sufficient width and traveling speed to minimize the depth of material upon it. The shallower the bed is, the more efficient water sprays will be in dealing with a fire. Therefore, in applying the above concepts to the drier used as an example, the plate feeder must be able to accommodate 5.17 t/h and 5287 ft^3/h at an overall apparent bulk density of 1.96 lb/ft^3.

It may be considered that a suitable bed depth of material upon the conveyor would be 4 in. (100 mm), in which case formula (5) for the calculation of belt speed defined in Section 3.1 can be applied.

$$\text{Belt speed} = \frac{\text{Volume per h}}{\text{Bed depth} \times \text{belt width} \times 1},$$

At this point, a degree of trial and error is introduced, in the sense that either a belt width or a traveling speed must be assumed. The choice is somewhat arbitrary and depends upon which is the more critical to the overall process design. If a belt width of 5 ft is selected, then the equation produces

$$\text{Belt speed} = \frac{5287}{(^4/_{12}) \times 5 \times 1} \ \text{m/h}$$

$$= 3172 \text{ ft/h or } 52.9 \text{ ft/min.}$$

Such a speed is not excessive for a plate feeder conveyor, and the selected belt width is consistent with the width of the discharge aperture of most rotary driers. The conveyor indicated by the calculation would therefore be adequate for the duty.

9.3.2.3 Secondary Screen

Before the effect of the screen can be established in mass and volume terms, it is necessary to determine what its efficiency is likely to be in respect to each of the remaining fractions of the feedstock. There is no exact method for doing so, but there is a useful logic train to give some guidance.

The machines used in this application are always drum screens. Vibrating deck screens become blinded instantly. It is known that the granularity of the combustible fraction is 95% below 1 in. (25 mm), and it can safely be assumed that at least the same proportion is greater than $^1/_4$ in. (5 mm). The organic and noncombustible residues, however, are those which adhered to the paper and card fraction due to its moisture content. The residues could be expected with equal confidence to be of a granularity very much smaller than $^1/_4$ in. Very fine perforations in the screen plates will lead to blinding, while large ones will pass too much product, and an acceptable compromise would be $^1/_4$ in. Some of the paper and card fraction will be rejected by a drum screen with screen plates of that perforation, but that is the cost of the removal of the components that would otherwise seriously compromise the quality of the product.

Table 9.10 provides a realistic assessment of drum screen efficiency with the feedstock in question. Here, the efficiency is expressed as the extent of the screen's ability to pass a fraction to undersize. It is expected to eliminate only 2% of the paper and card, but 99% of any glass residues, which is consistent with the measured performance of an experimental $^1/_4$ in. mesh screen. The efficiencies are then applied to the material discharged from the drier with the result shown in Table 9.11.

Table 9.10 Secondary Screen Efficiency

Fraction	Screen Efficiency %
PACD	2.0
PLF	2.0
DP	30.0
Tx	1.0
Wood	10.0
Fe	40.0
Nfe	40.0
O/P	99.0
Soil	99.0
Gl	99.0
Other	99.0

Note: Paper and card (PACD); plastic film (PLF); dense plastic (DP); textiles (Tx); steel/iron (Fe); nonferrous metal (NFe); organic/putrescible (O/P); glass (Gl).

Table 9.11 Secondary Screen Stage Mass and Volume Balance

Fraction	Separation efficiency %	Mass t/h O/s	Mass (te/h) U/s	Volume ft³/h O/s	Volume (m³/hr) U/s
PACD	2	3.32 (3.01)	0.07 (0.06)	4152 (120.5)	108.5 (3.07)
PLF	2	0.79 (0.72)	0.02 (0.01)	831 (23.9)	20.6 (0.58)
DP	30	0.18 (0.16)	0.08 (0.07)	16 (0.5)	6.5 (0.18)
Tx	1	0.47 (0.43)	— —	126 (3.6)	2.6 (0.07)
Wood	10	0.12 (0.10)	0.01 (0.01)	12 (0.4)	2.1 (0.06)
O/P	99	— —	0.11 (0.10)	— —	52.5 (1.49)
Other	99	— —	0.001 (0.001)	— —	0.2 (0.01)
Total		4.88 (4.43)	0.29 (0.27)	5138 (148.8)	192.8 (5.46)

Note: Paper and card (PACD); plastic film (PLF); dense plastic (DP); textiles (Tx); steel/iron (Fe); nonferrous metal (NFe); organic/putrescible (OP); glass (GL).

Separation efficiency is a measure of the ability of the screen to remove fractions of the waste stream to undersize.

The screen delivers an oversized fraction of 4.88 t/h (4.43 te/h) and 5138 ft^3/h (148.8 m^3/h) at an overall apparent bulk density of 1.9 lb/ft^3 (29.8 kg/m^3). It also produces an undersized fraction of 0.29 t/h (0.27 te/h) and 192.8 ft^3/h (5.46 m^3/h) at an overall density of 3 lb/ft^3 (49.5 kg/m^3). The undersized fraction, in spite of its low mass, contains materials that would, if left in the feedstock, increase the chlorine content from less than 0.5% to over 1.5%. It would change the total ash from about 8% to as high as 18%, and it would increase the heavy metals emissions significantly. Analysis of that fraction shows that it contains substantial amounts of lead, zinc, copper, and iron, and that there is chlorine present in large quantities in a number of forms and chemical combinations. It is therefore a potential source of dioxins and furans, of metallic fume emissions, and of ash and chlorine levels that would make the fuel unusable in most conventional boilers.

The sources of the contaminants in the undersized fraction are many, and are the subject of some conjecture. However, it is reasonable to imagine that most of the lead originates in vacuum cleaner dusts, as a result of lead from street dirt being introduced to domestic residences on the occupants' shoes. The hypothesis is supported to some extent by microscopic examination of the materials, where large quantities of the small, bead-like grains of proprietary carpet fresheners are always in evidence. The presence of zinc is more difficult to explain, as the quantities are always higher than the most obvious sources in food wastes, etc. would support. Copper is again probably a major component in vacuum cleaner dusts, while the many chlorine compounds arise mainly from food residues, common salt, and residues of heavy, chlorinated plastics.

At the discharge from the screen, the oversized fraction has two possible destinies. It is still in fRDF form, but is a much cleaner and higher energy content product than that which would have been supplied direct from the classifier. Its calorific value should be in the order of 8600 Btu/lb (20,000 kJ/kg), but it has a very low energy density. It would be suitable for a PF combustion plant, or it could now be densified into cRDF.

Table 9.12 suggests the mass and volume balance expected at the completion of a screw-type predensifying stage. The product has been reduced in volume to 567.4 ft^3/h (16.06 m^3/h), and at a mass of 4.76 t/h (4.31 te/h) it has a bulk density of 16.78 lb/ft^3 (268.4 kg/m^3). Here it should be noted that the density achieved is a *true* value, not an apparent one. The material will now remain at that density virtually irrespective of how it is further handled. It exhibits very little tendency to compact further under its own weight, and disturbing it does not cause a corresponding increase in volume.

Table 9.12 Predensifier Stage Mass and Volume Balance

	Mass		Volume	
Fraction	t/h	te/h	ft^3/h	m^3/h
PACD	3.21	2.91	342.9	9.7
PLF	0.79	0.71	126.2	3.6
DP	0.18	0.16	14.3	0.4
Tx	0.47	0.42	74.8	2.1
Wood	0.11	0.10	9.1	0.3
Total	4.76	4.31	567.4	16.1

Note: Paper and card (PACD); plastic film (PLF); dense plastic (DP); textiles (Tx).

As a result of the very low feedstock density, screw predensifiers necessarily have low throughputs. A capacity of between 3 and 4 t/h should be regarded as the maximum, and in the case of the example, the mass of feedstock exceeds those figures. Therefore, two predensifiers will be required, although for most of their duty cycle they will be operating at lower than maximum load. This is inescapable, since the process design is limited by the range of proprietary machines available, not by the calculated demand.

The energy needed to reduce the volume from 5138 ft^3 to 567 ft^3/h is necessarily substantial, since the volumetric compaction ratio is slightly more than 9:1. This compares with, for example, electrohydraulic static packers with 40 t hydraulic rams, where the equivalent ratio is rarely more than about 5:1 with the same material. Since the energy delivered and that released as losses in the form of heat are directly related, it follows that the high-powered screw predensifiers at least must reach very considerable temperatures. When they do so, they must inevitably cause some further loss of moisture in the feedstock. It is impossible to predict mathematically what that loss may be, since there are too many variables involved, but in the light of experience it is reasonable to consider a reduction of 20% w/w in the moisture contained in each fraction remaining. This realizes a moisture loss of 0.12 t/h as a result of predensification to 16.78 lb/ft^3 (268.4 kg/m^3).

At this point, the mechanical handling system needed to remove the product to storage ceases to be constrained by the volumes involved. Both the weight and the volume of the cRDF are such that it is well within the capacity of even a small belt conveyor. The question of design therefore becomes more of one based upon the minimum practical width of a belt than of a calculated delivery rate. A belt conveyor of less than 20 in. (500 mm) width is difficult if not impossible to trough sufficiently to retain a bulk material correctly. If the predensifier is assumed to produce particles of an average granularity of $^3/_4$ in. (20 mm), then such a conveyor with a bed depth of one particle would have to travel at slightly less than 90.72 ft/min (290 m/min), which is slow by belt conveyor standards.

9.3.3 Densified Refuse-Derived Fuel (dRDF)

The final stage of processing to achieve the most highly refined fuel product possible involves progressing to dRDF. In that case, the predensifiers would normally be installed in such a position that they discharge directly into the feed apertures of the pelletizers or cubers without any intermediate mechanical handling. Such an installation design is fundamental to RDF production in general, and it encompasses two principles. First, at the classifier and drier stages, the feedstock is being elevated to some considerable height, and mechanical infrastructures can be minimized by allowing gravity to deal with transfers between machines. So for example, where a thermopneumatic drier is used, it would be sensible to install the secondary screen at a high level with the predensifiers immediately beneath it. The pelletizers or cubers could then be installed beneath the predensifiers, with the result that the feedstock can pass through a total of four process stages without encountering any conveyors at all.

The second fundamental principle where dRDF is manufactured is that wherever possible the feedstock temperature should be maintained. Pelletizers produce a better product if their dies are running at temperatures in excess of 230°F (110°C). It is wasteful in the extreme if, having expended a great deal

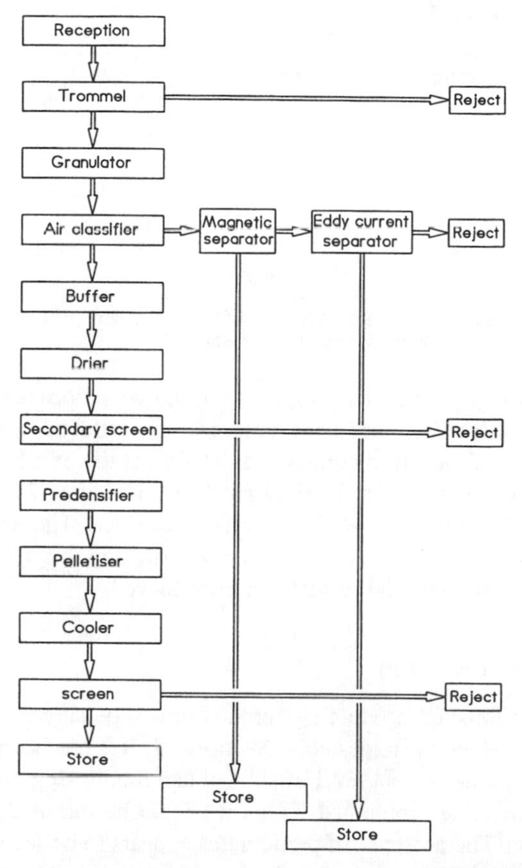

Figure 9.3 Densified refuse-derived fuel (dRDF) flowline.

of energy at the predensifiers, the feedstock is allowed to cool again before being introduced to them.

At this point, it is worthy of note that although the die temperatures of pelletizers and cubers exceed those at which volatile matter would normally be driven off, such losses do not in fact occur to any significant extent. While there is no direct scientific evidence for such an assertion, it is probable that the very high pressures experienced in the dies have some effect upon the vapor pressure of the volatiles. While that may or may not be the case, it is certainly true of the remaining moisture in the feedstock, since the temperatures should be sufficient to remove all of it. In fact, it is normal for a reduction overall of little more than about 30% of the total moisture content to be liberated.

When the material is introduced to a pelletizer or cuber, the results are likely to be somewhat as summarized in Table 9.13. A calculation based upon the apparent density data in Table 9.3 suggests that 4.72 t/h (4.28 te/h) and 275.9 ft^3/h (7.81 m^3/h) will be produced as fuel pellets, with an average bulk

Table 9.13 Pelletizer Stage Mass and Volume Balance

Fraction	Mass		Volume	
	t/h	te/h	ft^3/h	m^3/h
PACD	3.17	2.88	169.5	4.8
PLF	0.79	0.71	50.5	1.4
DP	0.18	0.16	11.4	0.3
Tx	0.47	0.42	37.3	1.1
Wood	0.11	0.10	7.3	0.2
Total	4.72	4.28	275.9	7.8

Note: Paper and card (PACD); plastic film (PLF); dense plastic (DP); textiles (Tx).

density of 34.19 lb/ft^3 (547.9 kg/m^3). During the whole process, the moisture contained in the fractions has been progressively reduced. At the drier discharge, the residual moisture content was 16.6%, at the exit from the predensifiers it was 14.3%, while in the final product it becomes 7.1%.

In Table 9.14, a typical dRDF analysis is recorded. The data was derived from a process plant of the flowline design described in this section for dRDF, and it indicates the fuel quality that can be achieved.

9.3.3.1 Final Screening

Before the finished product is stored it has to be screened or classified, using one of the devices suggested in Section 9.2. If it has been manufactured in a process with the correct level of control and monitoring, then the amount of flake or particulates contained within it should be minimal, but there will always be some. The presence of particulates appears to be seen as a challenge by some process designers, who feel a need to recycle them to some earlier part of the process so that they may finally be turned into product. The urge should be resisted.

The particulates that arise at the final screen are extremely dry, and may be of somewhat higher bulk density than the original feedstock. If they are reintroduced into the pelletizers, which appears to be the favored option, then they have the effect of reducing the overall moisture content below the minimum limit for effective densification. The result is that the finished pellets are more prone to breakage, which in turn leads to more particulates being screened off. The exercise therefore rapidly becomes self-sustaining.

Recycling the particulates to any point downstream of the drier is likely to have a similar effect, since the drier has been designed and tuned to provide the moisture content required by the pelletizers, not to account for varying degrees of drying caused by the addition of flake. However, it does not follow that the material could therefore be reintroduced at the drier, even though doing so would appear to reduce the heat energy demand. The more probable outcome would be to initiate repeated drier fires, since the flake would already

Table 9.14 Typical Third Generation Refuse Derived Fuel Analysis

Physical:	
Calorific value	8600 Btu/lb (20,000 kJ/kg)
Total sulfur	0.16% w/w
Total chlorine	0.2% w/w
Moisture content	9% w/w
Ash content	9%
Bulk density	37.44 lb/ft^3 (600 kg/m^3)
Chemical:[a]	
Si	1.5
P	—[b]
S	0.14
Cl	0.099
K	0.299
Ca	1.7
Ti	0.18
Cr	0.02
Mn	0.01
Fe	0.22
Ni	0.002
Cu	0.007
Zn	0.04
Ga	0.0005
As	—[b]
Se	0.0003
Br	0.0023
Rb	0.0018
Sr	0.0043
Y	—[b]
Zr	0.0023
Nb	—[b]
Mo	—[b]
Sn	—[b]
Ba	0.0112
Pb	0.0048

Note: This specimen analysis defines third-generation refuse-derived fuel from a modern process. It should not be confused with pre-1985 products, which were of a much lower quality.

[a] Method: particle-induced X-ray emission (PIXE). Concentrations in % w/w.

[b] Not detected.

be so dry that it would have insufficient water to resist the temperatures involved.

Transferring the material to the buffer hopper is possible, since there it would at least have sufficient time to absorb some moisture and to have some effect upon the overall moisture content of the feedstock. The effect is likely to be minimal in view of the small quantities, and intimate mixing would be difficult to ensure. Any small benefit would be gained at the expense of another conveyor or pneumatic system, the capital and running costs of which may make the particulates the most expensive product in the whole process.

A strong case can be made for treating particulates from the final screen as a reject, to be disposed of together with the scrubber residues and the undersize from the secondary screen. Freed of the compounding effect of screenings reintroduction, the pelletizers or cubers will create a much higher-quality product from which the reject levels should be minimal, certainly undetectable within the overall economics of the facility.

9.3.4 Product Storage

9.3.4.1 Flock Refuse-Derived Fuel

Flock fuel is extremely difficult if not impossible to store successfully for any more than a few hours at most. Where it is extracted directly after classification, it contains significant amounts of putrescible material that is mainly in liquid form at that stage, but the results of its presence are potentially extremely damaging. Biological activity begins long before the material is processed, and nothing has been done to change that or to minimize it. The product will therefore self-heat quite rapidly, and its high moisture content (in the example used, over 34% w/w) increases the possibility of oxidation of the lignocellulosic fraction. Even in a matter of h, it will begin to form homogenous masses which do not break up easily, and it begins to transform into a material similar to leaf mold. Fungal activity will become established in cool zones near the surface, although that will not be apparent for some days.

For those reasons, fRDF direct from the classifier can only be regarded as a very low-grade and unstable product that only has a value if it is used immediately after production.

Where the material has been further treated by drying and by secondary screening, it becomes somewhat more stable. The putrescible contamination is largely removed, and that which is left is sterilized. However, that is achieved at the expense of even more difficulty in handling, since the product density is reduced. Its storage life possibly becomes months rather than h, although eventually the final result is the same. It will develop biological activity at some time, because even a dried and sterilized putrescible contamination cannot be protected indefinitely.

Although the mass and volume balances offered as an example in this section suggest that putrescible contamination has been eliminated at the drier and secondary screen stages, that is not strictly so. There are putrescible residues in all qualities of RDF, although a point is reached in the process flowline where their mass and volume impact is so small as to be insignificant. At that point, there is some digression between that which is significant to the process designer as against that to the operator. The designer is indifferent to anything that ceases to affect the mathematical model of the process; he is not overly concerned with the durability of the product in storage.

The difference in perspective is perhaps not extreme, since even from the design point of view, storage of fRDF for any length of time is limited by the sheer volumes involved if not by the durability. The product can be pressure-packed into closed containers, and as a result it will occupy somewhat less space, but even then a volumetric problem remains. A 46 yd^3 (35 m^3) compaction container may only hold 5 t of material even with a compaction ratio of 5:1, and so storage of even so small a quantity in fuel terms of 100 t would require 20 containers. Apart from the space consideration, the capital investment in the containers and in the handling plant for them would be extreme.

While fRDF has a place in the solid fuel market, it is a highly specialized product only suitable for immediate use in a major combustion plant. Classifying and drying only improves its combustion characteristics. It does little to change its attendant handling and storage inconvenience.

9.3.4.2 Crumb Refuse-Derived Fuel

There is no history of cRDF being used directly as a fuel as far as the writers are aware, although there is no obvious reason why it should not be. Possibly it is because it does not conveniently meet any of the more common characteristics demanded of a solid fuel. It does not have a regular granularity, and its energy density is low, and so it cannot conveniently be used in combustion plant designed for graded coal. Its bulk density is however sufficiently high for it to be extremely resistant to milling down to the grain size of pulverized solid fuel. As an energy source for gasification, it would appear to have many merits, since in such an application the low bulk density works in its favor. Unfortunately, applications of that type have yet to recover from the demise of the manufacture town gas industry.

In terms of storage, it is placed between fRDF and dRDF. Its density is still insufficient for conventional fuel stockpiling to be cost effective, and its potentially heavy contamination by fine particulates would render such a principle hazardous. Deterioration in a matter of weeks could be expected to result from the development of biological activity, and oxidation is again likely to be a problem. Therefore, the most obvious application for the product is one where it could be consumed almost immediately after production.

9.3.4.3 Densified Refuse-Derived Fuel

When RDF has been densified to 37 lb/ft^3 (600 kg/m^3), it becomes extremely stable if provided with suitable storage conditions. In those circumstances, it can be stockpiled for some years with confidence. The conditions necessary are however somewhat critical.

It must not be stored for longer than 1 month in the open air, and even then, in adverse climatic conditions, deterioration can be expected. Direct contact with water either from rainfall or even from high atmospheric humidity causes absorption in the surface layer of the stockpile, down to a depth of about 8 in. (200 mm). The fuel pellets in the surface layer swell and their bulk density reduces by up to one half of the original. However, the increase in pellet size creates a semi-impervious cover over the remainder, limiting further water penetration. Where the product is stored for use in large-scale combustion plant, it is quite normal for it to be stored outdoors for many months, since in that application the deterioration of a small surface layer is not usually significant.

For more discerning applications, storage under cover is essential, and then some means of ventilation must be provided. When dRDF first leaves the production line it may, as a result of its passage through the coolers, appear to have nearly reached ambient temperatures. That, however, is a characteristic only of the surface of each pellet or cube. The very low thermal conductivity of the material ensures that the core remains hot for some time, with the result that dRDF appears to reheat quite rapidly when first placed into storage. Commonly, its temperature can be expected to rise to over 175°F (80°C) within a few h, reaching a peak in about 1 day. Depending upon the size of the stockpile, it may then remain at that temperature for a further 2 days, before beginning to cool again very slowly. It is not unusual for the product to take 1 week to cool to ambient temperatures.

The effect of the reheating and subsequent cooling is to cause the residual moisture contained within the product to be driven off. RDF with an as-produced moisture content of 8% w/w will dry to 4% or so by the time its temperature has stabilized. In an unventilated store, condensation is likely to create severe corrosion upon metal structures, and in falling back into the stockpile, will have a similar effect to rainfall in outdoors storage.

Extraction fans alone are hardly adequate, since the condensation of moisture can occur equally well in the upper surface layers of the stockpile. Again, the low thermal conductivity ensures that the core temperature of the stockpile remains at a considerably higher temperature than the surface. In an enclosed building it is possible for such condensation to interact with the heat in the pile and to establish biological activity and oxidation.

The occurrence of oxidation is easily recognizable, as it always occurs in clearly defined strata about 12 in. (400 mm) down from the surface. The color of the product there changes from dark gray or black to dark brown, and a characteristic odor of charred paper may be present. The material within the strata forms into brittle plates, revealing an appearance best described as shale-like. Severe degradation of the product both in physical and chemical terms occurs. Individual pellets again swell and become fused together. They lose their original cohesion and break up into fine particulates easily. Increases in particulate contamination enhance the moisture absorbency of the layer, and deterioration becomes compounded.

RDF is much less sensitive to oxidation than is coal, and there is no accepted maximum storage depth. Its lack of sensitivity to oxidation ensures that spontaneous combustion is unlikely, and as far as is known there has never been a recorded case of its occurrence in a dRDF stockpile. However, oxidation is invariably initiated by moisture and temperature, where the heat may be as a consequence of biological activity or simply a residue of the production process. Once a sufficient temperature has been achieved, oxidation becomes self-sustaining, creating more heat and moisture.

The very regular granularity of dRDF appears to work against oxidation reaching the point of sustainability, probably because, since the effect must necessarily take place where oxygen is present in copious quantities, only the material near to the top of the stockpile has access to it. There the moisture liberated by the increase in temperature can escape fairly easily, and so the hotter the material becomes, the more water is driven off. Eventually it becomes too dry to support either biological or chemical decomposition, and the process ceases. For that reason, while the pile temperature may reach 175°F (80°C), it never remains at that level or rises significantly above it. In the intervening period, however, deterioration of some of the product is possible, leading to the reduction in quality described.

Fortunately, the nature of the industrial solid fuel market is such that the conditions are rarely encountered, since the stockpile is turned over too rapidly. If a plant is manufacturing at a rate of 150 t a day, then it must sell at the same rate; otherwise, it will gradually accumulate product for which there is no outlet, or which must be sold in bulk in the "spot" market.

Such a production level also imposes some transport constraints that have an impact upon the storage facilities. In a state where the maximum permitted gross vehicle weight is, for example, 44 t, the heaviest payload possible may be 30 t. To move the products at a rate of 150 t per day therefore involves the permanent commitment of five vehicles. If on any day one load less is removed, then the next day six vehicles will be required to regain the steady state. Clearly, the consequences of any shortfall can escalate rapidly, to the point where it becomes impossible to obtain the numbers of vehicles needed. Then the duration of product storage may change suddenly from a matter of h to an indefinite period.

The balance the plant operator must seek to achieve is therefore to have only sufficient product in stock to ensure that all delivery vehicles can be loaded immediately upon their arrival at the plant. Such a policy limits the stockpile to little more than 1 day of storage capacity, and careful stock rotation ensures that no material remains longer than that on site. Where that is achievable, the process designer need make no special provision other than a covered area. However, if longer periods of storage appear likely, then some form of under-floor air injection of the design used in grain stores is worth consideration.

Dry and stable dRDF with a bulk density of 37 lb/ft^3 (600 kg/m^3) has a very low angle of repose. The sides of a stockpile upon a flat floor adopt an angle of about 30° to the horizontal, and will spread easily to an even shallower angle if disturbed. While contamination by particulates increases the possible angle very significantly, the designer can only assume that the plant will produce fuel to the quality standard intended, and that particulate inclusions will be minimal. Some advantage can be gained from the absence of risk of spontaneous combustion by seeking to store the product to the maximum pile height, and 19$^{1}/_{2}$ ft (6 m) would be common for a distribution conveyor installation. Elementary trigonometry then suggests that at a 30° angle of repose, the length of the side of the stockpile would be

$$l = \frac{2 \times \text{Height}}{\tan 30°}. \tag{22}$$

Therefore, l = 68 ft 2 in. (20.78 m).

If, as suggested, the plant is to store 150 t for 1 day at a time, then the length of the face of the pile is determined by

$$w = \frac{2 \times \text{Volume}}{hl} \tag{23}$$

where h is the storage height at the apex of the pile.

In the example used, storage of 150 t for a day in these conditions would therefore require a bay width of 13 ft (4 m). Such a bay would be an extremely convenient size for a shovel loader to deal with, but a store building 67 ft (21 m) deep and 13 ft (4 m) wide is somewhat impractical. An area of three or four times that front width is rather more practical, and it could be included in the main process plant building without requiring separate facilities. The design would then permit storage of product for 3 or 4 days, but with each day's production separated from the others' by bay walls. Stock rotation would be easily accommodated, and no product would remain on site for any longer periods. Some additional benefit would be gained by the facility to isolate any product that fails to meet the quality standards required, and by the material having time to reach its peak temperature and begin to cool again. Figure 9.36 describes a storage facility that meets the conditions discussed.

The second of the potential benefits is of more value to the fuel customer than it is to the plant operator. Inevitably, the customer will be an established user of coal or other conventional solid fuels, and he will almost certainly be aware of the consequences of oxidation and spontaneous combustion. Fuel that is visibly emitting steam when it is delivered may cause some alarm, while a stable, cool product will be received with equanimity.

9.4 RDF PROCESS CONTROL SYSTEMS

9.4.1 Control System Design Principles

Irrespective of the mechanism of control used in a process, the electrical design will almost always employ the principles of ladder logic. As the name implies, this is no more than a method of preparing a graphic representation

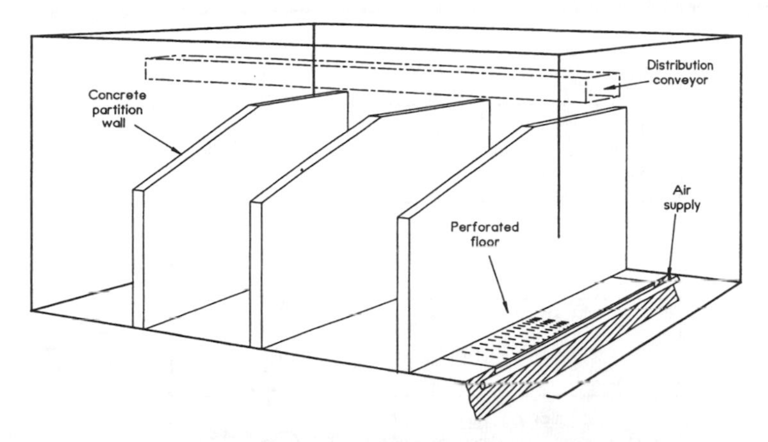

Figure 9.36 Bay storage system for dRDF.

of a control system in a way similar in appearance to a ladder. A simple example is shown in Figure 9.37, where the initiation controls with sequence-starting are shown for three electric motors.

In this diagram, relay coils operate a series of electrical contacts, all of which are actually contained within each relay casing, although the logic diagram does not show them to be. Each contact is designated by a reference that relates to the relay of which it is part. So, for example, relay R1 has contacts R1C1 to R1C4 inclusive. Where a contact is further designated "NO" it is "open," or in other words it is breaking the electrical circuit when the relay coil is deenergized. An "NC" contact meanwhile is "closed" when the coil is deenergized, and is therefore maintaining a circuit.

This arrangement of open and closed contacts and relay coils has been the basis of control systems for some generations, and the two alternative functions provide for an almost unlimited number of combinations in control philosophy. Further NO and NC contacts may be included in a circuit by the use of limit switches, temperature and rotation sensors, relays in which the coils are current or voltage-sensitive, and devices that respond to the proximity of metals. All of those are "feedback" devices, where they respond to a change in circumstances and initiate a control system response.

Ladder diagrams are relatively easy to read and the logic path is demonstrated clearly. In Figure 9.37, for example, relay coil R1 is energized by the operation of starting switch S1. As it becomes so, the coil closes contact R1C1. Switch S1 is spring loaded in the off open position, and so it will disconnect the circuit to the coil immediately as it is released. R1C1, however, maintains the electrical supply to the coil and holds it in its energized state. It also provides a supply to the "running" indicator, while the NC contact R1C2 opens to disconnect the supply to the "off" indicator. R1C3 meanwhile closes to establish a circuit to the motor supply contactor MCC1, and the first electric motor in the series is started.

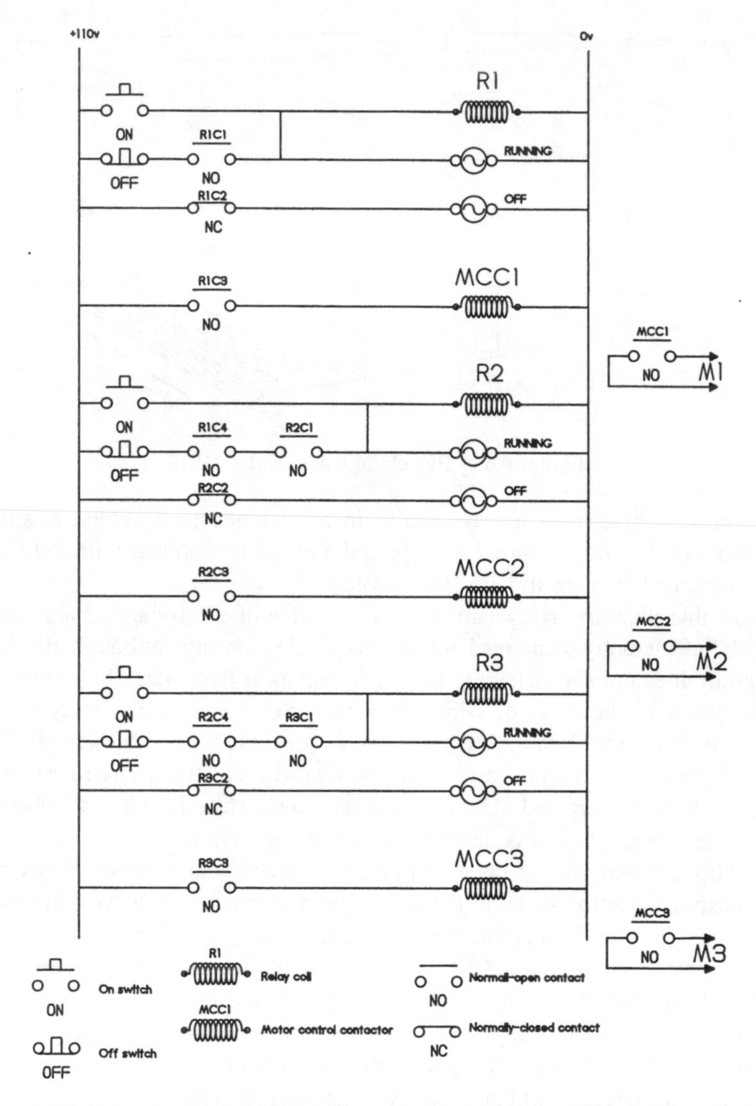

Figure 9.37 Example of a ladder logic diagram with sequence starting for three motors.

The starter for the second motor is identical with the exception that it has an NO contact from relay coil R1 in its maintenance circuit. Relay R2 cannot therefore be held energized unless motor M1 has been started, and since the starter for motor M3 has a similar R2 contact in its circuit, it cannot run without M2 being initiated. The circuit therefore permits motor M1 to operate alone, for M1 and M2 to operate together, or for all three to run. It does not, however, permit M2 to operate without M1, or M3 without both M1 and M2.

The logic diagram shown is somewhat oversimplified, and not likely to be found in a modern control system. There is no feedback to prove that the selected motor is running, and no device to disconnect the circuits if a motor experiences an overload. Nothing prevents motors M1 and M2 from continuing to run if M3 fails for any reason. These characteristics used to be common in early control installations, and they created significant difficulties in RDF plant operations. There, the failure of a drive without the immediate disconnection of all drives upstream of it had catastrophic results when dealing with high volumes of low-density materials. Overcoming the problem was relatively straightforward, but it involved "cascades" of relays in such numbers that, although they are extremely reliable devices, a failure at some point was almost inevitable. Among the most common causes of failures were loose wiring connections, of which in the average system of the day there may be many thousands.

The appearance of programmable logic controllers (PLCs) largely eliminated the problems of complex control systems and made multiple feedback possible. PLCs almost invariably retain the ladder logic concept, simply replacing electromechanical hardware with a software program run on a computer board. They do so even to the extent of being able to display their ladder logic upon a video monitor, and to print it on wide-platen dot-matrix printer. They are usually programmable via a special keyboard with electrical symbol keys in addition to letters and numbers. Using them, it is possible to portray the logic diagram on the monitor and carry out amendments at the keyboard. Most importantly, there is almost no internal hard wiring, and so connection problems are unlikely to be encountered.

A conventional PLC for use in a process plant is usually a series of integrated circuits inside plug-in cards, each of which is mounted in a chassis with external input and output terminals (I/O ports). The device is therefore capable of being extended according to the control requirement, and numbers of I/O ports running into hundreds is common. As a result, the ladder logic diagram for a large RDF plant may fill several hundred sheets of computer paper.

The controller output ports deliver electrical signals to operating devices according to the commands of the software. For example, instead of a pushbutton switch energizing a relay coil in order to provide a circuit for a motor contactor, the PLC simply completes a circuit direct to it. Manual initiation of the motor may be from a pushbutton switch that delivers a signal to the PLC via an input port, again without any relays or mechanical contactors being involved. It is even becoming common for the pushbutton switch to be an entirely solid-state device, which operates upon the change in capacitance it experiences when touched. The touch screen principle takes that concept to the extreme, by allowing the operator to make changes or to start and stop machines simply by touching symbols on the monitor screen.

Feedback into the PLC is still largely dependent upon electromechanical devices, since they are needed to translate a mechanical occurrence into an electrical signal. However, a growing number of entirely electrical detectors have appeared with the development of solid-state electronics.

Optical devices are now available to detect shaft rotation, as are inductive proximity switches. Product flows can now be monitored by equipment that transmits and receives low-power radiation and is unaffected by dust and moisture. Capacitive devices can monitor the chemical analysis of a circulating fluid, and can respond to contamination beyond a very precise and predetermined level. The opportunities for extremely efficient process control have become almost boundless.

Feedback devices may be either analog or digital. An analog device produces an electrical voltage or current that varies in direct proportion to some external stimulus. A digital device produces either a signal or no signal. So for example, a pressure gauge is analog in operation while a switch is digital.

Some feedback has to be in analog form to be meaningful. A temperature measurement is useless if the only information it provides is that something either is or is not hot. Equally, some data has to be in digital form, for instance where an electric motor running indicator is concerned. The motor is either running or it is stopped. As a result, a PLC must be capable of receiving, processing, and reacting to both digital and analog signals, and so it becomes a hybrid.

In control systems, a digital microprocessor is capable of producing graphic representations of process flows, of printing reports and operating data loggers, and of initiating control actions. However, it can only accept input or feedback signals which are either "1" or "0" — on or off. An analog unit, meanwhile, receives data in infinitely variable form, but is difficult to program and operate. The combination of both in a hybrid system, with digital/analog/digital converter interfaces offers the advantages of each while escaping the disadvantages.

The types of feedback devices most commonly used in RDF flowlines are as follows.

1. Optical Rotation Sensors: solid-state devices that transmit a signal in visible light or infrared to a reflector disc or bands on a rotating shaft and then respond to the reflected signal. Microcircuitry counts the signals emitted and compares them with those reflected, and responds to any failure of them to match.
2. Inductive Proximity Rotation Sensors: solid state devices that respond to electromagnetic induction in a nearby conductor. Generally, they operate as rotation sensors by detecting the regular passage of a slotted disc across their path of influence. The disc is mounted upon the rotating shaft. In effect, an electronic relay is held either open or closed by a time delay that requires an inductive signal in order to reset. In the absence of such a signal, the relay "times out" and either connects or disconnects a circuit to the relevant input port on the PLC.
3. Optical Speed Sensors: these devices are virtually identical to their equivalent rotation sensors. Instead of simply monitoring that a reflected signal is received for every one emitted, they deliver a continuous light beam and

count the frequency of the reflected signals. Commonly, a single reflective target is fixed to the rotating shaft, and the detector measures the number of times a min that the target passes its receiver. Very slow shaft speeds can be accommodated by the use of a number of reflective strips equally spaced around the shaft or on a shaft-mounted disc, and the PLC can be programmed to integrate the resulting signals.

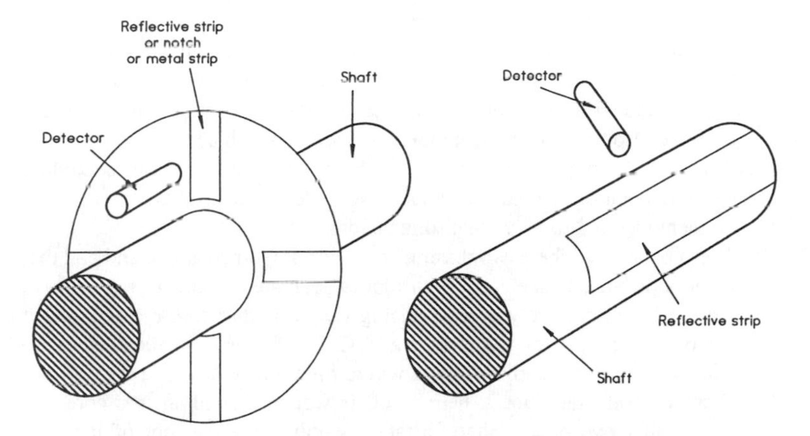

Figure 9.38 Typical shaft rotation/speed detectors.

4. Inductive Proximity Speed Sensors: again, it is common for the detector to be a simple inductive proximity switch, and for its target to be a slotted disc fixed to the shaft. Where speed monitoring is required, the PLC is programmed to count and integrate the input signals to its port, translating the frequency of the signals into a speed indication. Figure 9.38 shows the two more conventional methods of rotation and speed detection.
5. Strain Gauges: the electrical resistance of some metals varies according to the tension or compression to which they are subjected, and the changes in current flow that occur as a result of either may be detected and amplified. Very small increments of tension or compression, resulting in increases or decreases in gauge length as small as thousandths of an inch, are possible. As a result, strain gauges are used in a number of movement detectors and monitors, but are most common in loadcells.
6. Loadcells: devices where an increase or decrease of tension, compression, or pressure affects the gauge length of a strain gauge, resulting in a change of current flow through the gauge. The current flow is amplified from milliamperes.
7. Inductive Proximity Switches: these are again solid-state devices and they are identical to those used in rotation and speed sensors. When used as switches, their internal electrical current flow is influenced by the change in inductance caused by adjacent metallic objects.
8. Electromechanical Limit Switches: entirely mechanical devices where an electrical circuit is connected or disconnected by a switch. The switch is operated by a linkage moved by some external influence.

9. Pressure Switches: usually designed around strain gauges, although some of the more primitive units employ bellows or diaphragms linked to a mechanical limit switch. The strain gauge types are preferable in RDF flowlines, as they are entirely solid-state and are largely impervious to dust and moisture.
10. Ultrasonic Devices: these emit high-frequency sound waves from a tuned crystal and detect the rebound signal from any target object. Given the provision to measure the time difference between emission and reception, the precise distance of the target from the emitter can be established. Ultrasonics are therefore quite widely used to monitor quantities of materials contained in tanks and silos, measure the thickness of metals or the distance of objects, or indicate the presence of an object.
11. Thermistors: an electrical resistor in which the resistance varies according the temperature. The output signal is in analog form, and the devices can be manufactured to very small dimensions.
12. Thermocouples: where two dissimilar metals are joined at one end and that junction is heated, an electrical current is generated in direct proportion to the temperature. A very precise analog output is therefore created, and it can be used for process control via a PLC. In RDF process, the devices are widely used, particularly in driers where high temperatures are involved.
13. Displacement Transducers: here a coil is wound around an iron core that is free to move over a short distance within it. Movement of the core generates a very small electric current in the coil, and the current can be detected, amplified, and calibrated to suggest the amount of physical movement involved.

In the following section, a full-scale process plant manufacturing dRDF will be reviewed. Its flowline is that shown in Section 9.3, Figure 9.35. Each processing stage will be examined in order to define the control requirement and the inputs and outputs to and from a PLC. A general representation of the control loop logic, Figure 9.39, shows the interaction between flowline components.

9.4.2 Process Stage 1 — Reception Conveyor

PLC outputs:

Drive motor starting and stopping
Failure and running alarms and indicators
Running speed variation

PLC inputs:

Drive motor overload detection
Drive rotation detection
Running speed monitors
Blocked chute monitors
Conveyor chain stretch/failure detection
Primary granulator load sensing
Infeed rate weight sensing

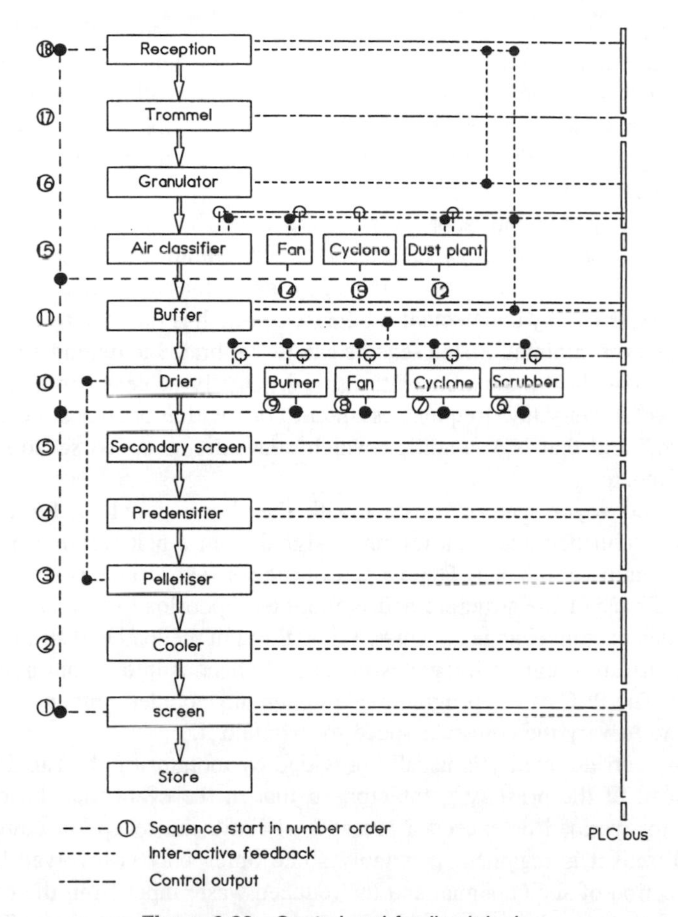

Figure 9.39 Control and feedback logic.

In terms of direct control, there must clearly be a means of starting and stopping the conveyor, and of monitoring that it is actually running and not just delivering a signal that it is. The latter requirement is satisfied by the use of optical or proximity detectors positioned against the drive motor shaft, although that does not of course ensure that the conveyor is actually running. It is possible that the gearbox could fail, for example, in which case the control system would show the conveyor in motion when in fact only the motor was turning. That, however, is a somewhat unlikely event not generally addressed in design.

In addition to starting and stopping, the conveyor must be capable of speed variation according to the demands of the process flowlines. It is necessary for a speed to be set in accordance with the chosen plant throughput, and to then allow that to be modified according to the load conditions in respect to downstream machines. The initial speed control may be either by means of a variable-speed gearbox, where the variation is achieved by an electromechanical servomotor that progressively changes the gear ratio, or by a frequency modulator. In the latter case, the PLC circuitry is capable of changing the frequency of the supply current to the motor, with lower frequencies resulting in slower running speeds. The lowest possible frequency is zero, when the power supply becomes direct rather than alternating current, at which point the drive motor is braked.

Frequency modulators, or inverters, are somewhat prone to motor stalling at very low speeds. In those circumstances, the motor may still be drawing electrical current, but at so low a frequency that it does not actually turn. Such a condition is largely undetectable at the PLC, which is satisfied by any current flow, however minimal. Speed sensors can be calibrated to defend against that situation, but their indication of the speed actually achieved is likely to be unreliable. At very low frequencies, the motor rotation becomes increasingly "stepped," and that is a condition which conventional speed sensors cannot accommodate.

The conveyor speed consistent with the chosen flowline throughput is usually established against a reference signal from a belt weigher on a convenient section of the belt. There a part of the conveyor frame is isolated from the remainder of the structure and is mounted upon load cells. An amplifier circuit detects min changes in current flow through the load cell strain gauges, and a software program integrates the signal changes into a continuous load readout. The PLC can be programmed to maintain a predetermined continuous load, and to vary the conveyor speed to maintain it.

An override facility is usually provided by monitoring the running electrical load of the primary granulator, so that in the event that the machine begins to overload or exceed a preset load limit, the reception conveyor is slowed until it is regained. Inevitably some time delay is involved between the detection of such a signal and the reduced waste input from the conveyor reaching the granulator, since the trommel screen is between them. The time delay varies according to the residence time in the trommel, and may be several min. It is an undesirable situation, but one not easy to correct, since the trommel output cannot be restricted without risk of major blockages developing. Generally, the only recourse is to make the response to granulator load very sensitive, so that the conveyor speed is continuously varying slightly.

Where transfer chutes are included in the reception conveyor design, some form of blocked chute detector is useful. The most simple and reliable of these is a cover plate hinged such that a buildup of material behind or beneath it will cause it to open. The movement is then detected by either an inductive proximity switch or a limit switch. The signal from the detector is used to instruct the PLC to stop the conveyor and sound an alarm.

Conveyor chain stretch/breakage monitors are not widely used, mainly because reception conveyors are of such mechanical strength that failures are uncommon. However, inductive proximity switches are sometimes used against target plates on each of the conveyor chains, or on the tail sprockets where both are free to rotate independently. As long as all switches are signaling, then the chains must be intact and running, while the failure of one signal suggests that the chain on that side has failed. Stretching can be detected by requiring both signals to be not only present but also synchronized.

9.4.3 Process Stage 2 — Trommel Screen

PLC outputs
- Drive motor starting and stopping
- Failure and running alarms and indicators
- Running speed variation

PLC inputs
- Drive motor overload detection
- Drive rotation detection
- Running speed monitors
- Blocked chute monitors
- Drum position monitors

Drive motor controls and failure alarms are common to those upon all other machines including the reception conveyor, as the running speed control may also be. While efficient trommel design precludes the need for variable-speed drive, there may be some circumstances where it is felt that some flexibility may be of advantage. The ability to rotate the machine very slowly is sometimes of value during maintenance periods, for example, but to achieve that, limited inverter control is all that is necessary. However, any speed control facility should be strictly limited to only a small variation close to 70% of the critical speed of the drum when under automatic control. As was discussed in Section 9.2, the performance of a trommel is very sensitive to its rotational speed. A reduction to 60% of critical, for example, may be sufficient to cause sudden and substantial carry-over of putrescible and other materials into the combustible fractions.

Drive rotation detection is somewhat more simple in respect to a trommel screen, since the large-diameter drum is rotating sufficiently rapidly for it to be monitored. The most convenient method for achieving that is usually with proximity or optical sensors on the trunnion wheels, rather than on the electric motors.

Drum position indicators or monitors are an additional safety factor, the purpose of which is to disconnect the drive in the event that the drum moves from its correct running position. While extremely unusual, it is possible for a structural failure to release a trommel drum. In the rare circumstances when it occurs, the fault usually lies with the tracking idler, which holds the drum against its 3° inclination. The idler is a small wheel, mounted upon a vertical stub shaft between the track ring flanges at the low end of the machine, and its small diameter limits the mechanical strength that can be accommodated. A failure of the roller does not imply that the drum is about to ride off of its trunnions completely, since the track ring flanges prevent that. However, if the flanges are permitted to contact the sides of the trunnions, then some damage is to be expected.

The most effective way of detecting such an event is by means of an inductive proximity switch, mounted by a track ring flange in such a way that a failure would permit the flange to move away from it. Inductive proximity switches have a very limited and precise detection range, usually considerably less than 1 in. (25 mm), and so the drum is not permitted much movement before the switch can no longer detect its presence. In those circumstances, a normally open switch, held closed by the presence of a continuous signal, breaks the control circuit and initiates a motor shutdown. The switch location described is preferable to one where the switch detects the *approach* of a track ring, since in that case any movement of the drum would be likely to destroy the switch, and damaged switch cannot be relied upon to respond in the manner required.

9.4.4 Process Stage 3 — Primary Granulator

PLC outputs:

- Drive motor starting and stopping
- Oil pump starting and stopping
- Failure and running alarms and indicators
- Direction of rotation selection

PLC inputs:

- Drive motor overload detection
- Drive rotation detection
- Running speed monitors
- Blocked chute monitors
- Vibration monitors
- Oil pressure monitors
- Explosion detectors

The primary granulator is invariably the highest-powered single machine in the process line. Where flail mills are employed, drive motors of 350 hp are common, and in many machines, including all vertical shaft types, the bearings are pressure-lubricated by a separate oil pump. Those machines with oil pumps must not be started without full oil pressure having been developed if catastrophic bearing failure is to be avoided. For that reason, the oil pump control circuit is always interconnected with the main motor control center in such a way that the main contactors cannot engage until that condition has been obtained. Consequently, the main motor power supply only remains connected as long as the oil pressure is within the permitted range.

Feedback from the pressure system is therefore fundamental to the operation of the primary machine, and good design practice requires that there are several ways in which the feedback is provided. All should be interdependent, and so the main motor should not be capable of being started or run unless load sensors indicate that the oil pump motor is running, that the oil reservoir level is adequate, and that full oil pressure exists. A sequence contact in the pump control circuit, similar to that shown as part of the sequence start in the illustration, is all that is required to ensure that its motor is energized. A limit switch linked to a float determines that the oil reservoir is filled, and a pressure switch in the oil circuit ensures satisfactory pressure.

Vertical shaft granulators can be driven in either direction of rotation, and it is common practice to extend flail arm life by reversing the rotation on a regular basis. In that case, feedback is necessary to indicate the selected direction of rotation. That is usually no more than a set of contacts in the main motor control center. The contacts used are mechanically linked to those energizing the main contactors for the appropriate rotation.

Motor overload detection takes two forms. At the first level, current-sensitive relays in the motor control center detect an increase in load current, using toroidal transformers that enclose each of the main conductors. Alternatively, the output from the transformers may be detected by the PLC, which is itself programmed to react to a preset level of load. Whichever method is chosen, the purpose is to provide for speed variation of the reception conveyor as described earlier.

The second level of motor overload is common to all motor control centers throughout the flowline, and involves thermal overload relays. In these devices, the electrical supply to a relay coil is delivered through a contact maintained by a bimetallic strip. The strip is manufactured as a lamination of two dissimilar metals which, when heated by the passage of an electric current, expand at dissimilar rates. If the current exceeds a predetermined value, the strip overheats and the difference in expansion rates causes it to bend, disconnecting the circuit. Overload relays of this type have been in use for many years, and they are fail-safe. In other words, any mechanical damage or failure of the relay brings about a condition similar to an overload, and the main motor circuit is disconnected.

Generally, very high-powered motors also have thermistor protection. Here the heat-sensitive resistor principle is used to construct a number of small heat-sensitive detectors included in the motor field windings. Since overloading and overheating of field windings are directly related, an excessive overload results in one or more of the thermistors issuing an analog signal in proportion to the temperature. The signal may be conditioned through a voltage-sensitive relay and caused to interrupt the power supply to the motor, or it may be passed to a PLC for load reduction to be initiated.

Vibration monitors are usually forms of strain gauge load cells bolted to parts of the machine casing nearest the mainshaft. Increases in vibration amplitude have a similar effect to a continuous load upon the cell, and again amplifiers can detect the change in current flow and pass it on to the PLC. The system can therefore respond to a level of vibration in excess of one that has been judged safe. The devices are useful in primary granulator operation, because they are able to detect flail arm breakage. In a machine with a 350 hp motor, running at 800 rpm and driving a rotor carrying 16 flail arms each with an average weight of 45 lb (20 kg), the out-of-balance forces created by the loss of one flail are very severe. Flail damage or breakage is not uncommon, and is most often created by heavy metallic items finding their way into the machine with the trommel oversized fraction.

Explosion detection is only of significance in high-speed flail granulators. There, even cans of paint thinners can explode when struck by a flail arm, and many serious fires in waste processing plants have been caused by them. In many ways, the detectors are somewhat superfluous, in that any explosion of sufficient severity to cause damage should be immediately apparent to operating staff without their needing an indicator to tell them of its occurrence. However, granulators should always be installed in a separate room that has pressure relief openings in the roof and cannot be entered when the machine is running. In those circumstances, it is not so much the explosion that needs to be indicated as the possibility of a flash fire that usually attends it.

Fire detection equipment, however, either senses the presence of heat by means of either fusible links in a tension wire to a spring-loaded switch, or by infrared radiation. Alternatively, devices with a mildly radioactive sensor detect the presence of smoke. All of them may take a few moments to react, and flash fires initiated by a granulator explosion travel very rapidly indeed. It is quite possible for the flame to spread throughout a conveyor system before fire detectors react to it. An explosion detector, meanwhile, reacts instantly to the pressure wave rather than the heat or radiation.

The design of explosion detector that is both the most simple and the most reliable is nothing more than a pressure relief door held closed by a spring catch. Doors are usually fitted above the granulator in the infeed hood in vertical shaft machines, and in both the infeed hood and the discharge casing in horizontal shaft machines. The extent of movement of the doors is constrained by a surrounding steel framework, which is designed to ensure that the door can only open sufficiently to vent the pressure vertically. Inductive proximity switches mounted beside the door can be held energized as long as it is closed, and in that position they react to the door target plate moving away from them in the event of an explosion.

Any loss of signal from the proximity switches should then be used by the PLC to initiate high-volume water sprays in the granulator infeed hood and conveyor, and in horizontal shaft machines on the discharge conveyor also.

Vertical shaft granulators that discharge directly into column air classifiers do not present the same risks of fire spread, since the product is entirely enclosed until it is released by a remote cyclone separator. In that case, the PLC should be programmed to immediately stop the cyclone rotary valve and to shut down the classifier fan and heavy rejects conveyor. That action alone should then be sufficient to contain any fire downstream of the granulator, and water fog atomizers in the classifier casing will prove effective in extinguishing it.

Where granulator discharge conveyors are unavoidable, the water sprays alone cannot be relied upon to completely extinguish any fires. The product at that stage of the process is of such a low density that the initial explosion may distribute it over a wide area beyond the reach of the sprays. Even where it remains upon the conveyor, it is possible for the water deluge to only wet the surface, which then compacts sufficiently to shed the remaining water before it can penetrate. Although the PLC should immediately stop the conveyor when an explosion is detected, operating staff still require some provision for dealing with the situation rather than being forced to wait for water sprays to work. It is therefore good design practice to provide for a manual override at the PLC such that the conveyor can be started with its discharge transferred to a dump chute, unloading into a fire bay where any residual conflagration can be dealt with.

Blocked chute detectors are not normally associated with flail granulators, often because there is no obvious place in which to site them. The discharge casings of horizontal shaft machines do not lend themselves to the devices, since the only outlet from them is occupied by a conveyor rather than a chute. Vertical shaft machines discharge their product in a high-speed airstream, and so again there is no chute in which to install a detector. However, there is some merit in placing a device in the highest part of the infeed conveyor casing, since that is where blockages are most likely to occur. If the design of the detector is as described earlier, then it can form a part of the conveyor casing as easily as the cover of a chute.

Where a device is fitted, there is little point in its initiating a shutdown of the granulator. An infeed blockage will already have unloaded the machine, and securing it only ensures that there is no outlet for the material if the blockage is cleared. It is more important that the PLC reacts by stopping the trommel screen, reception conveyor, and the conveyor upon which the blockage has occurred, leaving the flowline downstream running normally.

9.4.5 Process Stage 4 — Classifiers

PLC outputs:

- Fan motor starting and stopping
- Cyclone rotary valve starting and stopping
- Dust extraction plant starting and stopping
- Failure and running alarms and indicators
- Fan damper controls

PLC inputs:

- Drive motor overload detection
- Ductwork air pressure switches
- Dust extraction plant differential pressure monitors
- Fan damper position monitors
- Cyclone differential pressure monitors

The control system for an air classifier should be designed for internal sequence starting, where the dust plant is initiated first, followed by the cyclone rotary valve. The fan control damper should be closed and proved so, and finally the fan itself should be started. As with the whole process flowline, the logic is that all equipment should be started in the reverse direction to the product flow, in order that every component is in operation before any material reaches it. The PLC should be programmed to always close the fan control damper upon shut-down, and to test that it is closed before start-up. Classifier fans are necessarily high-power devices, and 100 hp drive motors are common. Starting against a closed damper is therefore essential to minimize the initial current surge.

When the system is running, the control system has to maintain classifier efficiency against varying loads by modifying the air flow through the system, and to do so it requires a feedback of the classifier air pressure. Air flow devices would be attractive, but most conventional types are unable to tolerate the heavy particulate burden. Differential pressure monitoring is however sufficient for most purposes, since an increase in throughput with an unchanged air flow results in a pressure drop in the classifier casing and ductwork. A strain-gauge pressure transducer in a straight section of duct between the classifier and the fan is the most effective, but care has to be taken to ensure that it is in an area of nonturbulent air flow. The PLC can then be programmed to maintain the pressure within set limits by regulating the fan control damper.

Differential air pressure transducers in the dust extraction plant are necessarily interactive with those in the classifier duct, since any increase in the differential across the dust plant indicates that air flows cannot be maintained. However, dust plant sensors identify where a fault lies in that event. In a bag filter plant, an increase suggests that the filters are becoming blocked, and that the reverse jets are for some reason ineffective. A wet scrubber only experiences differential pressure increases if either the baffles or water tank are becoming choked with particulates, or if the water level is too high. Since in most machines of this type the water level is controlled by an overflow weir, there is little opportunity for the latter occurrence.

In some bag filter systems, the PLC determines the operation of the reverse jet cleaning mechanism, rather than any internal control system in the bag house. In that case, the feedback is operational. Otherwise, it is purely a monitoring and alarm signal.

The fan damper position monitor signals are both operational and informative feedbacks. The PLC can position the damper correctly without a feedback from it, but only by regulating on the air pressure transducers in the classifier duct. Since that latter signal is never steady, the result of doing so is a fan damper that is continually cycling between open and closed. The better principle therefore is one where the PLC compares a damper position signal with one indicating air pressure, and is programmed to only move the damper when the pressure moves away from a preset range.

The cyclone differential pressure signals are also interactive with those from the classifier, since the cyclone is at the discharge end of the air duct. Here again, the transducers perform more of an alarm function than an operational one, in that they only indicate when a cyclone blockage is developing. Since cyclone separator efficiency is determined by the pressure drop across the unit, it is important that the PLC is programmed to react to any departure from the operating range. Any movement approaching the maximum permitted differential pressure should result in an alarm signal, and further movement beyond it should initiate a plant shutdown. Blockages in cyclones, although fortunately a rare occurrence, are notoriously difficult to clear, and so a temporary loss of production immediately may prevent a major one later.

9.4.6 Process Stage 5 — Buffer Hoppers

PLC outputs:
- Drive motor starting and stopping
- Running speed variation

PLC Inputs:
- Drive motor overload detection
- Rotation sensors
- Running speed monitors
- Chain failure sensors
- Hopper level sensors

Since the purpose of the buffer is to eliminate surges in the product flow before they reach the drier and densifiers, it must be able to react to changes at the air classifier. The feedback and control from and to both units must therefore be interactive to the extent that an increasing buffer hopper level should, as it approaches its maximum limit, cause a signal to the PLC to slow down the whole upstream process.

In that event, two actions may be taken. A throughput reduction which starts with the reception conveyor will take some min before it begins to have any effect upon the buffer, but in situations of fine tuning, that will normally be sufficient to deal with routine process variations. It is however possible, if fairly rare, that a surge could be of such volume that it is likely to overwhelm the hopper if allowed to continue. In that case, the feedback to the PLC may show a ramp (the rate of change) that is steeper than one expected of a normal process variation, and although the hopper may be able to accommodate the surge, it may not have sufficient space to contain the restored normal volumes immediately.

Since the PLC is a microprocessor, it has a built-in clock facility with which it can behave as if it had timing relays in its circuitry. It is possible for it to respond to a rate of change such that a signal level has been attained in less than the permitted time. In the event of an unacceptably steep ramp in the buffer hopper level signal, the only action that can achieve an almost immediate result is the closing in of the classifier fan damper. The consequence of that action will be to increase the volumes of rejects and reduce those of the light fraction. Some combustible material will be lost, but that is preferable to the buffer blockages and drier overloading that might otherwise occur.

A speed variation control and feedback in respect to the buffer is of use only in those circumstances where the overall processing rate is being changed. The purpose of the buffer is to smooth out the flow rates, and it does that by running at a steady and predetermined speed. The speed variator should not be interactive with any other part of the system, but rather it should be a part of a manual control that allows global adjustment.

The measurement of hopper levels is beyond the capability of most devices based upon light signals or ultrasonics. The materials in the hopper are likely to be sufficiently dusty as to cause obscuration of the lenses of a light-emitter, and if ultrasonics are used, loose flakes are likely to cause spurious signals. A worthwhile alternative uses a series of devices in each of which an external paddle is mounted upon a shaft driven by a slow-speed electric motor. The motor is in turn mounted upon a spring-loaded pivot inside a casing, and is mechanically linked to a microswitch. As long as the paddle, which protrudes into the hopper, is turning, then the device issues no signal. However, if the paddle is prevented from turning by a buildup of material around it, then the shaft torque moves the motor on its pivot. The microswitch becomes engaged and a signal is dispatched.

Such a device is of course entirely digital. There is either a signal or no signal, without any stages in between. An analog output would be preferable, particularly with the preference for the PLC to be able to detect a steep ramp in any level increase. However, a series of the paddle switches in a vertical orientation can provide almost an element of analog feedback. The PLC program then can detect a ramp by monitoring the time between the receipt of signals from each successive switch.

9.4.7 Process Stage 6 — Drier

PLC outputs:

- Exhaust fan drive motor starting and stopping
- Exhaust fan damper modulation
- Combustor/burner starting and stopping
- Combustor/burner modulation
- Exhaust scrubber starting and stopping
- Cyclone separator rotary valve starting and stopping
- Drier inlet rotary valve starting and stopping

PLC inputs:

- Drive motor overload detection
- Rotation sensors
- Drier gas pressure differential

Exhaust fan damper position sensing
Temperature sensors
Scrubber pressure differential

The purpose of modulation and control of the drier is solely to ensure that the product is discharged at the correct moisture content for densification, and there are two mechanisms by which that may be achieved. The first, monitoring of the product moisture content before delivery, is desirable if not strictly essential. The second is indispensable, and it involves modulating the drier heat supply in accordance with the exhaust gas temperature.

Measurement of the moisture content of the feedstock is difficult due to its physical nature, and it involves a considerable amount of computing power to calculate a measurement. For that reason, it is not a widely used principle. Control by monitoring the drier exhaust is much more common, and in fact has been the standard method for many years. The theory behind the practice is that the heat energy necessary to change the state of water to steam is a very precise quantity. If the initial heat input to a system is known and controlled, then a proportion of that heat will be absorbed by the water as it changes state. If the heat is conducted by a gas flow, then as the water is driven off the temperature of the gas will be reduced in direct proportion to the evaporation rate.

Using an example in explanation of the concept, one could imagine a drier system where the gas exhaust temperature is 250°F (120°C). The feedstock is entering the drier at 30% and leaving it at 16% moisture content. Now a transient quantity of material with a moisture content of 35% arrives. As the extra moisture begins to change state, the exhaust gas temperature drops. The amount of gas and the heat energy it contains has remained unchanged, but more heat is being taken from it. The reduction in exhaust temperature is detected by the sensors which feed back an analog signal to the PLC, which responds by increasing the firing rate of the heat source. More heat is introduced into the gas stream, and the exhaust temperature is restored. The feedstock continues to leave the drier at the required moisture content because more water is being released.

Now the original feedstock quality is regained, and the initial moisture content drops back to 30%. Immediately, the drier exhaust temperature begins to climb, as less water requires less heat to boil it off. Again the PLC responds, this time by reducing the firing rate of the heat source. There is now both less water and less heat, and stability is once more restored.

In Section 9.2, the nature of the heat source for the drier was discussed, with the recommendation that solid fuel combustors should be avoided. It is the need for rapid response and precise modulation as described above which justifies that recommendation. No solid fuel combustor can ever respond quickly enough to meet the changing conditions in an RDF process. Gas or oil burners, however, can. Their modulation is achieved by the simple expedient of a servo-operated fuel supply valve, and because the fuel is extremely consistent, the PLC can be programmed to make exactly sufficient adjustment to achieve the desired effect. The principle works equally well with thermopneumatic and with rotary cascade driers, although the response time of the latter is somewhat longer.

Some drier designs also employ exhaust fan damper positioning, where in the event of a change in heat source firing rate there is an equivalent change in the gas flow, modulated by the damper. Here the theory is that a change in heat demand might be a result of a change in throughput rather than in unit moisture content, and a lower throughput requires a lower hot gas flow because it has less water in it. There are some merits in the theory, although not for the reason usually given. There is a direct relationship between the temperature of a gas and its density, and there is also a clear if not easily established relationship between its density and its ability to hold particulates in suspension. In a system with exhaust gas temperature sensing, a reduction in moisture content for whatever reason causes a reduction in heat input to the gas. The gas density therefore becomes marginally higher, and its ability to hold particulates in suspension increases, doing so at a time when there are less particulates to be suspended. Modulation of the exhaust fan damper therefore seeks to maintain the relationship between gas temperature and flow rate irrespective of changes in the quality or quantity of the feedstock.

The extent to which such a facility is worthwhile is arguable. Although an increase in the temperature of air from 32°F to 750°F (0° to 400°C). reduces its density by half, the result of a change of a few degrees is hardly detectable. It is certainly unlikely to have any detectable effect upon the suspension of the feedstock or upon the residence time in the drier. It does however give the PLC another parameter to reconcile with heat source modulation, and its attempts to do so often cause conflict. For that reason, thermopneumatic lines particularly are best left to run at a gas flow consistent with maximum throughput, and to modulate the heat input alone. Small quantities of product will not pass through the system noticeably faster than large quantities, because the gas velocity remains substantially the same.

With both types of drier, monitoring of the scrubber pressure differential is essential. The drier exhaust will always carry some level of dust burden that the cyclone separator cannot be expected to remove. It is also not unusual for local statutes to require it to be scrubbed to remove odors, hydrocarbons, and sulfur and acid emissions from the primary fuel. Any variation in the differential pressure across the scrubber will have a considerable effect upon the gas flow through the drier, and although there is normally no reason for the differential to change, it must be monitored and related to that of the drier. That would normally be achieved by establishing a set point for the overall differential across the system, to be maintained by the PLC.

Where wet bath scrubbers are used, and they are the most common for RDF drier exhaust conditioning, the scrubber fan usually runs at a constant speed without any provision for regulation by dampers or other control devices.

The only facility for modulation of the system is therefore at the drier exhaust fan, and by means of free air inlets between it and the scrubber. A free air inlet is simply an open port on the transmission duct which, when opened, permits air from the plant room to enter the scrubber and to thereby reduce the depression on the drier exhaust. When in use, the port maintains the steady air flow the wet bath scrubber needs for efficient operation, while permitting the drier to operate at low load.

The extent of modulation is not unlimited. As examined in Section 9.2, the gas flow through the drier is not simply established according to the need to deliver heat to the product. In both thermopneumatic and rotary cascade driers, the gas has to transport the product and remove the water vapor, and in the former it must actually lift the material vertically over a considerable distance. To do so reliably it must have a velocity in excess of 26 ft/s (8 m/s), and preferably not less than 32 ft/s (10 m/s). Variations of heat input, however, necessarily result in the heat source providing more or less gas, since the air required for combustion is directly related to the quantity of fuel burned. In normal operations, the waste moisture content does not vary more than about 20% at most, and so the burner modulation and resulting gas generation rate is also not unlimited. The control requirement is therefore not to regulate the flow from zero to maximum, but rather to balance the steady state of the scrubber with a potential drier differential range of about 16 ft/s (5 m/s).

The most straightforward control loop which achieves that is one where the free air inlet port is regulated by a damper operated by a servomotor under the control of the PLC. The control logic is then that the drier differential pressure is monitored and regulated to maintain the correct gas flow under all heat input rates, with the free air inlet control following it to ensure a constant pressure at the scrubber.

9.4.8 Process Stage 7 — Secondary Screen

PLC outputs:
- Drive motor starting and stopping

PLC inputs:
- Drive motor overload detection
- Rotation sensors
- Drum position sensors
- Blocked chute detectors
- Fine screenings flow detectors
- Fire detectors

The most sensitive part of secondary screen operation is made so by the nature of the feedstock being processed. It is dry, of very low bulk density, and possibly hot. It does not flow easily, and it forms blockages in transfer chutes very rapidly. For those reasons, although the secondary screen is a single-speed machine with little facility for modulation, condition monitoring is important. A blocked chute detector in the feed chute to the screen is vital, and depending upon the design of the assembly, some form of pneumatic pusher is worth consideration. A suitable device for the purpose need be no more than a rod fixed to the end of a compressed air cylinder and projecting inside the feed chute. In its retracted position the end of the rod should be flush with the chute casing, and operation of the cylinder should be initiated by the PLC in the event of the blocked chute detector registering an alarm. The control could be arranged to operate the cylinder a predetermined number of times after which, if the detector has not turned off, an operator alarm should be sounded and the process line should be shut down.

The fine screenings created by this stage of the process are even more difficult to deal with than the original feedstock. Although in this case the bulk density appears to be considerably higher than that of the feedstock, the major fraction is fibrous dust with a much lower bulk density. The small inert component makes the material appear more dense. The absence of any free-flowing characteristic is certainly true of this fraction, and in addition it is particularly sensitive to atmospheric humidity. Given damp conditions, it is quite capable of adhering to the sides of a transfer chute or hopper even though they may be inclined near to the vertical.

The dust component of the undersized fraction is so light that it is capable of remaining airborne for some considerable period, with the result that the collecting space beneath the screen is normally filled with floating particles. Any ingress of air, or even windage from the rotation of the screen drum, can present an opportunity for blockages to develop. They are likely to do so at random anywhere in the hopper, particularly where there are any obstructions or even irregularities caused by structural welds. Blocked chute detectors are in this case unreliable, since they cannot be located in the areas most likely to block.

The very fine nature of the dust, and its resistance to flow, makes some form of conditioning essential before it can be transported onward for disposal, and so for that reason it should be delivered immediately to a mixer–conditioner mounted below the screen. Here it can be intimately mixed with water to form a heavy slurry, the flow of which can be monitored. That is best achieved by the use of a spring-loaded, hinged plate in the discharge chute from the machine, installed so that it is held open by the flow of material down the chute. Absence of a steady flow allows the plate to close, and when it does so the event can be detected by proximity or limit switches.

Unfortunately, if a signal is received from the flow indicator, there is little the PLC can do in the way of remedial action. The cause of the flow failure could be anywhere in the hopper, and the methods needed to deal with it could be different for every possible location. However, the rate at which the fine screenings are produced is not excessive, and plant operators, given a warning alarm, have several min in which to deal with the situation.

Secondary screens are not normally inclined in order to ensure consistent throughput, since they are short in comparison to trommels. The drums will fill to about 15% of the available volume and then, as more feedstock enters, that which is already in the screen will begin to spill from the discharge end. However, the absence of any inclination does not preclude tracking idlers and drum position sensors, although the justification for them is different. The very dry and dusty nature of the feedstock makes effective sealing between the feed chute and the rotating drum structure extremely important, otherwise dust will be released into the plant room. Effective sealing in turn requires that the drum maintains an exact position in relation to the chute, and any departure from that needs to be the subject of an alarm condition.

The low density of the feedstock also calls for blocked chute detectors in the discharge from the screen, since it is there that blockages are also likely. Again, a hinged cover plate with limit switches is sufficient.

Fire detection is important in this application, particularly in the dust hopper beneath the drum. The machine is being loaded directly from the drier, and any ignition there will almost certainly be transferred to it. The dust is capable of reaching sufficient concentrations in air to provide the conditions necessary for a dust explosion, although that is an uncommon event. It is much more likely that it will smolder on ignition, creating very large volumes of smoke and sufficient heat to damage the hopper casing if left unattended. Conventional smoke detectors or infrared sensors are unlikely to work in the hopper environment, since the airborne dusts blind them and obscure lenses. A more effective detection method involves the securing of thermocouples around the hopper casing, hardwired in parallel so that any single probe can register a warning signal. An increase in temperature beyond a preset limit can be used to sound an alarm and, if considered necessary, to operate a water extinguishing system.

9.4.9 Process Stage 8 — Predensifiers

PLC outputs:
- Drive motor starting and stopping

PLC inputs:
- Drive motor overload detection
- Rotation sensors
- Blocked chute detectors

Predensifiers are constant-speed machines and so other than starting, stopping, and condition monitoring, there is little that a PLC control system can do to modulate them. However, since their purpose in an RDF flowline is to reduce the volume and increase the bulk density of the feedstock ready for pelletization or cubing, they deal with material in the most difficult condition in respect to mechanical handling. For that reason, any failure of throughput leads to immediate and severe blockages, and blocked chute detectors are of vital importance.

9.4.10 Process Stage 9 — Pelletizers

PLC outputs:
- Drive motor starting and stopping
- Force feeder motor starting and stopping
- Grease pump starting and stopping

PLC inputs:
- Drive motor overload detection
- Rotation sensors
- Blocked chute detectors
- Lubrication pressure sensors
- Die temperature sensors
- Shearpin failure sensors
- Clutch slip detectors
- Main drive motor load sensors
- Hydraulic oil pressure monitors

The most common pelletizers in modern RDF processes are rotary die machines, with force feeders to ensure consistent material feed rates. Starting has to be carried out in a sequence that begins with the grease pump which supplies the bearings, followed by the main motor or motors, and finally by the force feeder. Shutdown is the reverse of that sequence, and there the grease pump is not finally secured until rotation sensors detect that the pelletizer has stopped. Not all machines are fitted with grease pumps with separate drives. Occasionally, they are served by grease mains running from a central lubrication installation, in which case the starting sequence includes the operation of servo-valves in the grease manifolds. Since the machines must not be allowed to run without pressure lubrication being available for the mainshaft bearings, it is normal for the presence of a grease pressure signal to be mandatory for start-up.

On some makes of rotary die machines, the roll shell shafts are held stationary against the rotation of the die by an external hydraulic clutch. Here the clutch disc is clamped between hydraulically powered clutch pads similar to those on an automotive disc brake. Loss of hydraulic pressure allows the disc to turn, and further densification becomes impossible as the roll shells rotate with the die. Therefore, any movement of the clutch disc has to be detected, and the signal must be used to stop the force feeder. Otherwise, the die can become completely blocked by materials that will, if left, compact to a state where removal becomes very difficult indeed. The principle is identical on machines with shear pin protection instead of hydraulic clutches, and the detection of movement and the reaction to it are the same. Usually, a microswitch registers against lobes or notches on the periphery of the disc, and is activated by being engaged by a lobe or disengaged by a notch.

Main motor load sensors are a valuable if not essential addition to the control system, mainly for their operator information capability. As with all process machines, pelletizer main drives take electrical power in direct proportion to the volume of feedstock passing through them, and it is therefore useful to know what the loading rate is. In more sophisticated process flowlines, each pelletizer may be served by its own buffer hopper, the speed of which can be related to the pelletizer motor load current. In those circumstances, it is possible to install a manual load selector which establishes what the throughput of each machine shall be in terms of a percentage of full load, and to program the PLC to maintain those conditions by varying the speed of the buffer or feeder.

Die temperature sensors are also a useful addition to the feedback circuit, although machine operation is not dependent upon them. The sensors can provide plant operators with information about how the machine is performing, since the best product quality is obtained from dies that are at the correct working temperature. There does not need to be any interaction with the other control and feedback circuits, since an incorrect die temperature usually requires manual decision and intervention. Low temperatures may, for example, be caused by a low throughput, and a control system cannot be designed to react to that. If it did, then starting the machine would be impossible.

Blocked chute detectors are essential in the infeed chute, but are irrelevant to the discharge. In spite of any predensification, the infeed to the pelletizers is still of relatively low density, and the feed chutes are necessarily of small cross-sectional area. Blockages are always a real possibility, and the detection of any must result in the PLC shutting down any infeed to the affected machine. In doing so, it clearly must slow down all of the process production flow, otherwise the remaining machines will become overloaded. For that reason, all of the signals resulting from failure of a pelletizer are interactive through the PLC with the main plant control loop. Detectors are not usually considered necessary in the pellet discharge chute, since the finished product is dense, very free-flowing, and most unlikely to form blockages.

9.4.11 Process Stage 10 — Product Cooler

PLC outputs:
- Drive motor starting and stopping
- Fan motor starting and stopping

PLC inputs:
- Drive motor overload detection
- Fan motor overload detection

The cooler is a single-speed machine with which the facility or need for control is somewhat limited. There are no parameters needing continuous modulation, and the only feedback signal of immediate importance to production is one of overload or failure of the main drive motor.

9.4.12 Process Stage 11 — Product Screens

PLC outputs:
Drive motor starting and stopping

PLC inputs:
Drive motor overload detection

Other than start/stop controls and machine running feedback, little can or needs to be controlled at this stage of the process. If the installation is designed in such a way that the product continues to feed through even with the screening facility out of service, then all that is needed in addition is a failure alarm. The small amount of particulates that may carry over into the product store in such an event is negligible, and there is time for manual intervention without initiating a full-scale shutdown.

9.4.13 Alarm Systems

In addition to the control and feedback loops, operator information displays are essential, and a number of developing situations need to be advised. PLCs are not intelligent devices, and they cannot be programmed to make predictions in anything other than very simple situations. Information displays can be considered as falling into two categories from the design point of view: those that simply record what the process is doing when everything is within the set parameters, and those to which the operators' attention must be drawn. The second category involves some form of audio and visual alarm.

Alarms themselves can then be conveniently categorized into three levels in reverse order. In this philosophy, a level 3 alarm is one where a situation is developing over which the PLC has insufficient or no control, but which does not threaten production. So, for example, the failure of the extraction fan on the pellet cooler is a level 3 alarm. The cooler will continue to operate without it, albeit inefficiently, and product will still flow through the cooler.

A level 2 alarm is one where a situation is developing or has occurred which, if not dealt with in a set period of time, may threaten either production or the security of plant. In this case, the PLC should issue the warning through an annunciator and should begin counting down through the time period that has been established for a level 2 alarm. If at the end of that period the fault still exists, then flowline shutdown is initiated. It is possible to program the PLC for a different time period in respect to each individual alarm signal within the level 2 category, and where this is done the process visual display unit should warn the operator of the time elapsed and remaining before shutdown is initiated. As an example, a level 2 alarm might be issued in response to a loss of flow from the screenings hopper of a secondary screen. The loss of flow may occur simply because the screen has just been started and has no product passing through it, and so immediate shutdown is not required. Conversely, if there is a blockage but the hopper capacity is equivalent to 30 min of full-load operation, then the operator can safely be permitted 20 min in which to attend to the matter or to start the introduction of product.

A level 1 alarm is issued in response to a situation where there is an immediate danger to plant or personnel, and where the PLC has insufficient or no control. It should initiate an immediate flowline shutdown involving all main and ancillary machinery, and should provide a detailed visual display in the control room to indicate to the operator what the fault is and where it has occurred. It should not be possible to cancel the alarm until the fault has been rectified, and all process machines should only be capable of starting with their controls switched to manual. The provision for manual operation is important if the alarm has been initiated by a blockage in a transfer chute, for example. Generally, such a fault cannot be cleared without running the downstream machinery to remove the material.

Reflection upon the above alarm rankings suggests that they are interactive to some extent. A level 3 alarm could eventually become a level 2 if nothing is done about it, even though it did not initially threaten production or plant. Any abnormal condition is likely to worsen or affect other parts of the flowline sooner or later. Equally, a level 2 alarm becomes a level 1 if action upon it is not taken within the preset time limit.

It also follows that all manual emergency stop buttons, switches, and pull-wires constitute level 1 alarm situations, since they all cause immediate flowline shutdowns.

9.5 OPERATIONAL CONSIDERATIONS

As with all production processes, RDF flowlines have their own peculiarities that demand operator experience rather than monitors and controls. A competent operator develops a feel for the plant under his supervision, and is often aware of a developing situation long before an alarm warns him of it. He is also aware that with a highly variable feedstock, there are some materials or changes in overall quality that will affect production and may even threaten its continuity.

It would be impossible in a work such as this to condense many years of process operating experience into a few pages. A feel for machinery cannot be learned from a book. However, operator training is an essential prerequisite of a plant start-up, and some pointers can be offered in that respect. The essential peculiarities of an RDF process can to that extent be categorized into those relating to the feedstock or product, and those relating to the individual machines.

9.5.1 Feedstock-Related Elements

As repeatedly observed throughout this book, municipal waste is not a material. It is instead a mixture of materials, each with its own handling peculiarities. Therefore, any waste treatment process can be likened to trying to manufacture an automobile without any control over the initial stocks of metal.

In theory, the waste should be reasonably consistent. Almost all developed countries have legislation or local statutes defining the nature of municipal waste, and what a householder may deposit. In practice, few householders are aware of the content of the legislation, and even fewer make any real attempt to conform to it. They can assume with reasonable confidence that once something is sealed up within a trash bag, no one is going to open it again to check the contents. If later the item is discovered or if it causes difficulties in a process plant, no one will be able to identify its source. It is therefore not at all unusual to encounter items or materials that have the potential to cause damage to machinery, to initiate fires or explosions, and possibly to threaten the safety and health of personnel.

The first line of defense is always with the operators of the shovel loaders, cranes, or other means of delivering the material to the process lines. Their task is primarily to be observant, and to isolate anything of which they are uncertain. Drums, unmarked containers, and any large metallic items should be treated with suspicion, as should any quantity of construction material such as concrete.

Their secondary task is to be selective in the way in which they load the waste. The maximizing of throughput and the quality of the product depends largely upon their selecting a reasonably consistent mix of waste at the front end. They should certainly not just load the waste in as it comes, but instead should aim for the same proportion of paper and card in every load. That is not always as simple an operation as one might suspect, since it is quite normal for several waste deliveries to the facility to have quite dissimilar combustible fraction contents. For example, loads that have arisen in areas that are mainly commercial are likely to contain much more paper and card than those from entirely residential areas. The facility may even accept commercial waste from special deliveries, in which case the loads will be almost entirely paper and card. Infeeding such a delivery without mixing it with the other municipal waste will result in a very significant increase in the volumetric loading in the early stages of the process lines, and may even overwhelm individual machines.

Another important aspect of the task is that the operators should attempt to load the waste as steadily as possible. Large quantities dumped onto the reception conveyors with gaps in between will simply cause the plant control system to continually adjust for changes in feedstock flow. Although the reception conveyors are designed to even out the feed, they cannot be expected to deal with variations ranging from overload to no load at all. Steady operating conditions will never be reached in those circumstances, and it is inevitable that the average throughput will be much lower than the design capacity.

Once the material has entered the process lines, the plant operators should be aware of the mass and volume changes occurring at the various stages. The indicators showing the flow rate on the reception conveyor are only a guide to how the plant is performing in treating the day's waste input. They give no guidance to what is likely to happen further down the flowline, since they are incapable of assessing the volumetric impact of changes of quality. Therefore, the temptation to increase the throughput because the front end indicators show less than optimum should be resisted, at least until the reason for it has been identified. It may be, for example, that in spite of the diligence of the loader drivers, there is a temporary surplus of paper in the mix on the reception conveyors. The volumes involved will increase, but the recorded weight per min will reduce in accordance with the lower bulk density. If at that stage the throughput is increased, then again the volumetric impact upon the machinery may be too great to be sustained.

Operators in RDF plants should be encouraged to become very sensitive to unusual odors. Fresh municipal waste does not have a significant or particularly strong smell, and once the process is under way, all trace of its original smell are lost. It is replaced by an odor the nearest description of which is hot biscuits. It does not smell resinous, or like paint, and there should be no charred paper odors; any of those suggest that something either has or is about to go seriously wrong.

Possibly the odors to which operators should be most sensitive are those suggesting aromatic liquids, since paint thinners and gasoline do upon occasion find their way into the waste stream. When any indications of their presence are received, the operators should begin to prepare themselves for dealing with fires.

In Section 9.3 a process model was considered, and that reveals the very significant volume changes that occur throughout the process. Often those changes are not what logic would seem to suggest, and it is quite normal for more material in volume terms to leave a machine than entered it. These changes have more effect upon the process operation than any other single factor, and they arise from the progressive removal of fractions until only one, the combustible component, is left. Therefore, as demonstrated, the behavior of the feedstock will change repeatedly, to the extent that the material arriving at the pelletizers is quite different in every sense from that which entered the trommels. A competent operator should be able to identify the process stage from which it arose by a cursory visual examination of a small quantity of material.

To be able to do so requires a clear understanding of the difference between the mass and volume balances. The mass balance will show a steady reduction throughout the process, from a maximum at the infeed end to a minimum at the product output. The volume balance, however, will reveal continual increases and decreases up to the predensifiers, and the greatest volumes are likely to occur immediately after the primary granulators. Since a volume surge creates the blockages to which all waste treatment plants are prone, operators should be most vigilant in stages of the flowline with the greatest volume increases.

To some extent, as control systems become more sophisticated, they encourage excessive confidence in their facility to run the plant with a minimum of human intervention. To a large extent they do so successfully, and as many failures are caused by unnecessary intervention as by control errors. However, even the most complex system cannot detect an event before it occurs, while a trained operator often can. By understanding the materials with which he is dealing, he can see a trend developing and can predict its outcome if it is left unattended.

9.5.2 Machine-Related Elements

9.5.2.1 Reception Conveyors

Since they receive the waste in its most aggressive state, the reception conveyors are among the parts of the process that are the most susceptible to problems. They are very heavy-duty machines, with reserves of power sufficient to overcome most difficulties, but there are some conditions they cannot be expected to accommodate. The waste is always likely to contain "tramp" items, such as large metallic objects, and it is possible that they may cause some damage to the conveyor flights. Therefore, regular inspection of the conveyor deck is essential, and any bent or damaged flights should be replaced. Otherwise, spillage at least will threaten damage to the under-structure and running gear of the machine, and could cause premature chain failure.

Regular inspection of the deck is also important in monitoring any buildup of fatty deposits. Generally, those are self-limiting, but in some circumstances seasonal variations in waste quality may cause them to accumulate, particularly if there is any quantity of ash present. In those cases, the deposits are sure to be very abrasive, and may be sufficiently so to increase the wear rates on the flight overlaps. Steam cleaning then appears to be an attractive way of dealing with the situation, but it should be avoided: it only succeeds in releasing the accumulations from the deck and depositing them upon the running gear beneath, where their potential for causing damage is much greater. Loosening the accumulations with hand scrapers is not a pleasant undertaking, but it is the best way of dealing with them, since they will then be extracted by the trommel screen during the next periods of operation.

9.5.2.2 Trommel Screens

The most significant of the few problems to which trommels are subject is the buildup of wire and metal objects on the screen plates. It is for example rare for an operator to enter a trommel drum without finding some areas of plate completely blinded by accumulations of metal, particularly the lids of cans. Usually at the center of the accumulation there will be the remains of a radio loudspeaker, the magnetic core of which has secured itself to the screen plate and then attracted the lighter pieces of metal. Wire also is often present in municipal waste in surprising quantities, although it only becomes apparent when concentrated in a process flowline. Again it is the trommels that are likely to first experience difficulties with wire, as it binds in the screen plate perforations and then causes a buildup of light materials and other wire around it. One of the most consistent sources of wire is the remains of mattresses, where the textile covering is rapidly destroyed by the action of the trommel, releasing the spring interior.

Both of these conditions can only be dealt with by regular maintenance and observation, and so trommel drum interiors should be inspected at least weekly, and any accumulations of metal or wire should be removed.

9.5.2.3 Primary Granulators

The high-speed rotating shafts of primary granulators make them also highly susceptible to wire from mattresses, etc. In addition, they do not deal easily with lengths of man-made textiles, or with rope or strips of carpet. All of those materials wrap around the rotors and bind the flail arms. Horizontal shaft machines are more sensitive in this respect than are vertical shaft machines. There is no solution to the problem other than to exclude those items wherever possible, although in small quantities they are unlikely to cause trouble.

Flail arm breakage can occur if a tramp item that is both large and hard is introduced. Scrap components from automobile engines are a common cause, although the problem appears to have reduced in frequency over the last 10 years. Breakage causes very severe out-of-balance forces in the rotor, to the extent that the fault is usually immediately identifiable by the noise level alone. Replacement of the damaged or broken flail arm is rarely sufficient by itself, since the remaining flails on the rotor will have sustained some degree of wear during the time they have been in service. If a new and unworn component is installed, then it will be heavier that its nearest companions, and the out-of-balance forces will continue in existence. The correct solution is to always replace flail arms in diametrically opposed pairs, and to match them by weighing first.

Progressive wear of flail arms is a routine matter, and it is rarely uniform along or down the rotor. Invariably there is one area where the wear is greatest, and in horizontal shaft machines it is predominant toward the rotor center. On vertical shaft machines, it is usually predisposed toward the top of the rotor. The immediate consequence of wear is a gradual increase in granularity, and if left unattended the product particle size may become too large or too inconsistent for the process line to maintain full production levels. Primary granulators are not difficult to open for inspection, and there are many merits in doing so at the end of every day's operations. Any excessively worn flails can then be changed, again always in matched pairs, and doing so will be rewarded by production levels that remain as designed throughout.

Vertical shaft flail granulators offer some advantages where flail replacement is necessary. The conical shape of their casings make it possible to reuse worn flails at a lower position on the rotor, by cutting off the worn face and thereby obtaining a shorter arm. Where a worn flail would fit in a rotor position without any cutting, the temptation to do so should be resisted, since the worn edges will be rounded and blunt. Cutting them off restores the chopping ability of the machine. Again, weight matching any flails so treated is of the utmost importance.

9.5.2.4 Air Classifiers

Since air classifiers have no moving parts, there is little that can go wrong with them. However, the fundamental principle of their operation is that there is a steady and smooth air flow through them. Streamlining of the air column caused by obstructions or casing damage can create areas of reduced or even reversed air flow, with a corresponding loss of light fraction material. Where column classifiers are used with vertical shaft granulators, it is not unknown for casing damage to occur as a result of impacts of heavy items ejected from the granulator discharge. The usual solution is to install a sacrificial lining plate immediately opposite the discharge aperture, but that must then be inspected regularly. If it is allowed to wear excessively, it will eventually become distorted, at which point it will begin to interfere with the air flow through the machine.

The transfer duct between the classifier and its cyclone separator also requires regular inspection, since wear rates on bends can become critical. Perforation can occur quite rapidly once a level of erosion has occurred, and when it does, air ingress will begin to reduce the classifier performance. Inspection of the ducts is made easier by painting them in a light color, preferably white, since thinning of the metal beyond a certain point is always accompanied by localized heat generated by friction. When the metal is becoming thinned, white paint turns brown, clearly indicating where the wastage is worst.

Wet bath scrubbers are often used to treat all or part of the classifier exhaust air, and although they are highly efficient in doing so, they are sensitive to very fine paper granules and fibers, and to small particles of plastic film. The classifier cyclones cannot always be relied upon to remove all entrained solids from the airstream, and some will always escape to the scrubber. The most immediate consequence is a buildup of floating debris on the surface of the water bath, to the extent that it eventually begins to interfere with the pressure differential upon which the machine depends. It is also possible for fibers to remain in suspension in the water, and to create a situation where the water appears to become viscous. Again, the result is interference with the pressure differential and a loss of scrubbing efficiency.

Monitoring of the pressure differential is essential if the problem is to be identified before it becomes critical. The evidence of it is always a very gradual increase in differential, and a PLC control system with a graphic display capability is well-equipped to create statistical reports that show it clearly. Dealing with the contamination of the bath is a fairly simple maintenance task, involving pumping out into a road tanker for disposal as trade effluent.

Wet bath scrubbers are also prone to biological contamination of the water. The air classifier processes waste in which there is a residue of organic and putrescible matter, some of which is both biologically active and in sufficiently small particles to be carried over into the scrubber. It is also likely that the intense air flows through the system will liberate bacteria, and there is no possibility of the cyclone separator intercepting them. The water bath is in many ways an ideal breeding ground for bacteria, in that the water is kept oxygenated by the turbulence upon which the efficiency of the design depends. It is also likely to be rich in suitable nutrients, and to be warm enough to encourage growth.

Colonization by bacteria may at best create unpleasant odors that will be liberated into the surrounding atmosphere, and may also cause some deterioration in parts of the scrubber casing. At worst, it may release infectious organisms, and although there have been no recorded case to substantiate the concern, there appears no reason to suppose that legionella could not survive in the water bath.

Chemical dosing is the preferred treatment for dealing with contamination, but there are limits to what it can achieve. Dosing concentrations must of necessity be maintained at low levels, or atmospheric emissions may become intrusive. In consequence, bacteria can become immune, and can continue to develop. Therefore, continuous dosing must be supplemented by periodic shock treatment using massive doses, preferably of a different chemical formulation. This should be undertaken during shutdown periods, when emission to atmosphere is prevented. Also, since bacteria can survive even shock dosing if they are insulated within a solids residue in the base of the machine, the treatment should wherever possible be preceded by draining the bath down, washing out residues, and refilling with fresh water.

9.5.2.5 Buffer Hoppers and Conveyors

Machines of this type are both simple in design and robust in construction, and so there are few difficulties associated with their operation. The most important matter requiring attention and vigilance concerns conscientious housekeeping.

The feedstock at the buffer stage is of a very low bulk density, and it is highly fibrous. Some spillage is inevitable, and if allowed to remain in contact with the plate feeder conveyor chains, it will become compacted and may eventually overstress drive components. Regular inspection and cleaning is of the utmost importance.

9.5.2.6 Thermopneumatic Driers

The two problems most commonly associated with the operation of thermopneumatic driers are wear and product ignition, and the former is often the cause of the latter. Hot, dry paper, moving at speed in a gas flow that may be corrosive, is surprisingly abrasive. Abrasion is always concentrated at bends, or where any turbulence is being created, and severe damage often disrupts the gas flow sufficiently to cause a localized buildup of solid matter. When that occurs, the solid matter will rapidly ignite, since almost the whole of the drier is at a temperature above the ignition point of dry paper.

The design of a drier that is resistant to wear was discussed in Section 9.2, but no design can exclude wear completely. Therefore, monitoring of the casing thickness at the outer circumference of bends should be a part of the process maintenance strategy. Ultrasonic thickness meters are adequate for the purpose, and are both cheap and simple to operate. A log of readings over a period of time identifies both a developing wear problem and any inconsistency in the gas flow.

The second most common cause of product ignition is sparks from the burner unit. Fouled nozzles on oil burners may cause burning droplets of fuel to pass into the feedstock as a result of inefficient atomization. Dust and debris may, if allowed to accumulate near the free air inlet around the burner flame tube, be drawn in and ignited by the flame. Some protection is offered by fitting spark arrestors beyond the flame tip position, but they can often become a maintenance liability in their own right. Regular burner servicing and good housekeeping in the surrounding area is more productive.

9.5.2.7 Rotary Cascade Driers

The most significant difficulties experienced with rotary cascade driers are associated with the heat source and the relatively slow movement of product. Generally, there is little that can be done about either, and operational practice is of necessity concerned with accommodating rather than correcting them.

The machines are invariably of large diameter, and so obtaining a gas flow that is evenly distributed across the drum is difficult if not impossible. At any time there will always be streams of very hot gas interspersed with strata that are considerably cooler, and the positions of both change continually. Some improvement can be obtained by designing the combustion chamber to create as much turbulence in the gas as possible, but even then the gas flows are inclined to streamline at some point along the drier casing. It is possible that selective heat loss as product falls through the gas stream is a contributor to the problem, since the particles never become evenly distributed throughout the drum.

In spite of the influence of the gas stream, the behavior of materials is similar to that in a trommel or drum screen. They are lifted in the direction of rotation and allowed to fall clear at some point above the horizontal. In a drier, that point is established by longitudinal lifters rather than by rotational speed, but the result is the same. The sector of the drum on the opposite side has very little product in it for most of the time, and the gas streams there may remain at a higher temperature. Differences in gas density, temperature, and humidity caused by intermittent contact with the product are probably the main influences upon the stratification of the gas.

The operation of a cascade drier is fundamentally different from that of a thermopneumatic system in that in a cascade drier the product is in intimate contact with the casing for much of the time it spends inside the drum. With a heat input at temperatures of between 660°F and 750°F (350°C and 400°C), the casing becomes very hot, and thermoplastics become softened by contact with it. A progressive buildup of charred plastic material develops on the lifters, and paper and card particles adhere to it, gradually charring themselves until a carbonaceous, soft, and sticky coating is established. Although the layer could be expected to constitute a serious fire risk, it is rarely the cause of fires. The more common cause is light fraction drifting back from the feed chute into the combustion chamber, under the influence of the turbulence created to minimize the gas streamlining. However, when a fire is initiated, the layers of charred plastics rapidly become contributors to it, and they ensure that the resulting conflagration will be difficult to extinguish.

The buildup can and should be removed periodically. If it is left to develop, it will at least reduce drier efficiency, quite apart from its potential to enhance fires. Removal is however less than easy. There is no obvious way of doing so while the drier is in service and the deposits are soft, but when the drier is cool enough for personnel to enter, the deposits harden and can become bonded quite firmly to the casing metal. One procedure that appears to assist in their removal is to continue to rotate the drier with the burners off but with the exhaust fan running, for some time after shut down. Fairly rapid cooling of the deposits, combined with differences in contraction rates between the metal and the plastic residues, limits the degree of adhesion achieved and makes removal possible with the assistance of pneumatic chisels. It is a lengthy task, but not a particularly arduous one, and the effort is rapidly rewarded by improvements in the drier performance.

9.5.2.8 Drier Exhaust Gas Treatment

The exhausts from all RDF process driers produce a characteristic odor, which many people find intrusive. The odors are caused mainly by volatile matter containing limonene, methanol, and ethanol. In addition, and depending upon the nature of the heat source, there may also be components of sulfur dioxide, carbon dioxide and monoxide, and hydrogen chloride. In many countries, emissions abatement legislation imposes controls over what may be discharged to atmosphere, and so invariably RDF drier exhausts have to be treated.

The conventional wisdom is to treat the gases in a wet scrubber, possibly with alkali dosing of the scrubber liquor to eliminate the acids. Unfortunately, doing so has little effect upon the volatile compounds that create the odors, since in the main they are insoluble. Biological filters could, in theory, deal with them, but before passing through the scrubber the gases are too hot for that form of treatment; after the scrubber they are too wet. Thermal oxidation by means of afterburners is a proven solution, but it is expensive since primary fuel is being consumed without any production reward.

In many plants the compromise solution is to extend the exhaust stack height to obtain the maximum thermal lift, and to fit an accelerator cone on the top. Here the reasoning is that if the gases can be discharged at sufficient height, then dilution will ensure the reduction of odors to vanishing point by the time they return to ground level. For much of the time the policy works, but plant operators should be warned that there will be occasions when it does not. Atmospheric conditions that temporarily eliminate much of the thermal lift and gas dilution are not uncommon, and when they occur complaints of odors are likely to be received.

9.5.2.9 Predensifiers and Pelletizers

Perhaps the greatest of the historical problems with predensifiers have been mechanical failures. The machines operate by applying large amounts of energy to materials that are abrasive and, beyond a certain point, almost incompressible. In some of the early machines, product became forced into joints in lining plates and exerted sufficient pressure to dislodge them. Shaft failures were not uncommon, and were often a result of distorted lining plates. However, most of those causes of failure have been designed out of the machines, and so there are no longer any particular barriers to their use.

Pelletizers are in many ways more sensitive machines. While there is little operator input to the predensifying process, there is much to that of pelletizing. The design of the die holes, the material from which the die is made, how wear is dealt with, the effects of feed rate and feedstock quality and temperature, are all significant contributors to a good-quality product.

The first three parameters, those of die hole configuration, die material, and wear rates, are all interactive. The pelletization (or cubing) process depends upon material being compressed under very high pressures into an aperture though a tapered entrance. As the material is gradually forced through the hole to exit from the die, heat created by friction causes fusion of the lignin and thermoplastic components, and the longer it remains in the die the greater the degree of fusion. The pressure exerted upon the material is dependent upon the "pinch angle" between the roll shell and the die, the die–roll-shell clearance, and the state of wear of the inlet taper. The pinch angle is predetermined by the diameter of the roll shell in relation to that of the die, and is incapable of being changed during operation. The clearance between the roll shell and the die, however, is entirely dependent upon the wear of the die "roadway" — the inside circumference against which the roll shell rotates. Although the roll shells do not actually touch the die roadway, the continual introduction of hot, dry feedstock between them creates wear upon both. The die roadway wears down relatively evenly, thereby increasing the clearance against the roll shells and reducing the pressures imposed upon the feedstock. Production rates slowly decline, since with incorrect clearances material is forced out sideways rather than through the die holes.

Product quality also declines progressively with die roadway wear. As the feedstock is trapped between the die and the roll shells, it is crushed and reduced against the lands between the holes, and the more significant the crushing action, the higher the finished pellet density. As the clearance increases, that response decreases, and pellet density suffers. Dealing with the matter requires regular adjustment of the roll shells in order to maintain the correct clearance, and immediate replacement of the dies as soon as the limit of roll shell adjustment is reached.

Wear in the tapered bores of the die is not a matter that can be dealt with by adjustment. It can only be corrected by replacing the die completely, but it can be alleviated by the use of dies that are "through hardened." These are manufactured from alloy steel that has been heat-treated to a consistent hardness throughout, rather than just in a surface layer. They are more resistant to roadway wear, and the bores are likely to remain efficient throughout the full range of roll shell adjustment.

A considerable element of operational control comes from experience. An astute plant operator is able to assess the quality of the feedstock by examination of the finished pellets, and by listening to the machine operating. He will be aware that if the moisture content is somewhat low, then the resulting pellets will be very hard and hot. Their surfaces will be dark or black, and very shiny. They will be quite easy to break between the fingers. Alternatively, in the event of a high moisture content, the pellets will be light gray in color, will show quite clear axial laminations, and will exude a smell reminiscent of hot biscuits. Again they may be broken easily between the fingers. At the precisely correct moisture content, the pellets will be very hot, dark but not black in color, and will be almost impossible to break apart.

A final indication of correct operating conditions can be gained simply by listening to the machine as it runs. When the feedstock is of exactly the right quality, the die emits a very characteristic metallic cracking noise, which can only be described as the sound of a ruler being slapped against a metal table. In any other circumstances, the machines make little noise other than a soft rumble.

The various operational adjustments that are possible lead to the temptation to manufacture very hard and dense fuel pellets. Such a product has an attractive appearance similar in many ways to that of conventional solid fuels, and its bulk density is directly related to energy density. The more firmly each pellet is compressed, the higher is its energy content per unit volume. There is, however, a point beyond which the pellet density becomes counterproductive.

dRDF pellets contain volatile matter and carbon in an approximate ratio of 65% to 25% on an ash-free basis. When they are ignited, they begin to break up in laminations across their diameters as the volatile gases are released, and it is not until that process has been completed that the carbon is consumed. If the pellet density is higher than about 40 lb/ft^3 (650 kg/m^3), then the time taken by the release of volatiles is extended quite substantially, and combustion plant is designed to take in fuel and discharge ash at a consistent rate. Stoking equipment delivers fuel in quantities determined by the heat demand of the boiler and by the calorific value of the fuel. Low calorific values demand higher supply rates, and so an inverse relationship exists between the calorific value and the running speed of the stoker. In those circumstances, excessively dense dRDF does not have sufficient time to burn out completely, and carbon residues are discharged with the ash. In that event a considerable proportion of the original calorific value is lost, requiring more fuel to replace it. The situation rapidly escalates until the maximum speed of the stoker is achieved, at which point boiler output begins to decline.

9.5.2.10 Coolers

Product coolers are such primitive devices that there is little in their operation that requires specialist knowledge. Other than ensuring that the air flow is maintained and that the running speed of the transporting belts is consistent with production rates, there is nothing requiring operator attention.

9.5.2.11 Product Screens and Classifiers

The only operator attention from which vibrating product screens benefit is the maintenance of a clear deck. Particulates can upon occasion block the deck perforations, although where wedge-wire decks are used that is a rare occurrence. Clearing usually involves nothing more than a brief sweeping with a stiff-bristle brush while the machine is in operation.

Product classifiers are even less likely to need any attention provided that a steady air flow is maintained. There is no need to regulate the air speed, since the difference in density between the product and any flake contamination is large. The air flow necessary to draw in intact pellets is many times greater than that for flake, and so the device can be operated at a set flow rate irrespective of the production level.

9.6 HEALTH AND SAFETY

RDF manufacturing processes are not particularly hazardous, and most of the threats to operator health are those related to waste handling. Heavy rotating machinery is employed in the flowlines and that of course brings its own attendant risks, but no more so than in any other heavy industrial process. There are some hazards that are not always immediately apparent, however, and the purpose of this section is to concentrate upon them.

9.6.1 Product-Related Risks

Any process plant that is manufacturing fuel must necessarily encounter fire hazards, and an RDF plant is no exception. In some ways, the risks are greater than those of a conventional solid fuel process because at some stages the feedstock is dry, finely granulated, and in intimate contact with air. Fires in driers are a common occurrence, while fire spread along conveyor systems from primary granulators is infrequent but is an ever-present risk.

When RDF flock ignites, it does not generally burn with any great vigor, but it does create very large volumes of dense gray smoke. As well as the more usual particles of unburned carbon, the smoke is likely to contain the products of plastics combustion, and it may therefore be to some extent toxic. The smoke has an extremely pungent smell, and even in low concentrations it causes severe respiratory discomfort.

Once a fire becomes established, it is difficult to extinguish. Chemical fire suppressants are generally valueless, since they do not reduce the temperature of the burning material. When their application is ceased, the material simply reignites. Water is more effective, but flock feedstock quickly absorbs it into its surface layer and then sheds the remainder, leaving deep-rooted fires still smoldering. Dealing with fires can therefore become a lengthy process.

Where that is so, operators cannot be expected to remain in an environment of dense and choking smoke for very long, and so provision must be made for their protection. The smoke cannot be eliminated until the fire is out, and breathing apparatus needs special training. The smoke can however be removed, and although there is an attendant risk of introducing more oxygen to sustain the fire, at least it becomes possible to see it. Plant ventilation systems provide the only solution, and they should be at high level so that they lift the smoke above the operators.

Dust is also an ever-present part of RDF process operations, and it can never be eliminated entirely. It is by nature fibrous and often biologically active, and breathing it can cause respiratory irritation in predisposed people. There is some evidence of continuous exposure causing increased sensitivity, although health records derived from existing process plants do not suggest that it is a major problem. The use of simple dust masks is usually sufficient to deal with the risk when working in the more affected areas, while adequate ventilation elsewhere reduces it to an acceptable level.

Physical handling of any part of the feedstock between the primary granulator and the drier should be discouraged, as the granulation tends to obscure hazardous items or materials. Hypodermic syringes are unfortunately becoming increasingly present in waste, and after the granulator they are no longer sufficiently intact to be recognizable. However, they are still capable of skin penetration, and of transmitting infections.

One effect of the very low density of RDF feedstock is its inability to bear any weight, and that, together with its propensity for creating blockages, makes it less innocuous than it seems. Where blockages occur in hoppers or storages, entering to clear them must be undertaken with considerable caution. While the material may be highly compacted in the vicinity of the blockage, it is unlikely to be so elsewhere, and anyone stepping into it may find himself instantly submerged. In those circumstances, suffocation is a very real possibility as the fine particulates enter the respiratory passages.

9.6.2 Machinery-Related Risks

Much of the machinery used in RDF processes is large and heavy, and rotating at some speed. It is unlikely to be affected by the presence of personnel trapped inside it, and its response to emergency situations may be injurious to anyone nearby. While in general most of the machines are totally enclosed by virtue of their operation, there are some hazards associated with them that are not always apparent. It is those less apparent hazards that are considered in the following sections.

9.6.2.1 Reception Conveyors

The plate feeder conveyors used for waste reception are very slow-moving, and they are not in themselves particularly threatening. Apart from the more obvious precautions of guarding on drives and moving parts, where statutory controls are likely to apply, there would seem little that needs to be done to protect operators. However, there is some record of personnel injuries sustained from contact with the machines as a result of actions involving the waste. The most common scenario leading to injuries begins when an operator observes something on the belt that he believes may be unsuitable for the process lines. The instinctive human reaction is always to reach for it and attempt to remove it.

Unfortunately, the human brain is not a good judge of the weight of an item purely upon the basis of its appearance. Often the suspect item is much heavier than the operator expected it to be, and it may be held quite firmly in place by the surrounding wastes. At the point of attempting to lift it, the operator is almost certainly unbalanced, and also is probably overreaching. Muscular strain is a very real possibility, as is overbalancing and falling into the conveyor. In that event, there is a possibility of his being carried into the trommel screen before he can recover, and such an event would be invariably fatal.

For that reason, access to the conveyor deck anywhere but in the loading area should be made as difficult as possible, and where walkways pass close to it, platework should extend to a height of at least $3^1/_4$ ft (1 m). Trip wires parallel to the sides and level with the top of the platework guards are a valuable safety precaution, since any overbalancing causes the wires to be disturbed, stopping the machine instantly.

9.6.2.2 Trommel Screens

Trommels, particularly when fitted with bag splitting spikes, are aggressive machines. It would clearly be impossible for a human being to survive being inside one while it is operating, and total enclosure is essential to prevent that ever occurring. There are, however, circumstances under which a machine can begin to rotate even though the power is off, and so entry for cleaning and maintenance is more hazardous than it appears. Trunnion-mounted machines rotate very easily upon their bearings, although their physical size suggests that to be unlikely. It is quite possible for one person, standing inside the drum to one side of the lowest point, to cause the machine to turn solely under the influence of his own weight. In that event, the inertia of the drum is sufficient to keep it turning for some part of a revolution, and certainly enough to throw the operator off balance. Falling against the screen plates or lifters may easily cause limb fractures, since the structure is designed for the purpose of breaking organic matter. Falling against the bag-splitting spikes has consequences that hardly need elaborating.

Avoidance of the risk of course involves ensuring that the drum cannot turn when anyone is inside, and the most appropriate way of achieving that is to clamp the drive shaft before the trunnions. The more basic approach that has been used upon occasion — using wooden wedges between the track rings and the trunnions — should be avoided, since it is not unknown for the locking mechanism to be forgotten upon restarting. A clamp on the drive shaft will simply prevent the start-up and cause the drive motor to register an overload. Wedges between the track rings and the trunnions will cause the drum to lift, with potentially serious structural consequences.

9.6.2.3 Primary Granulators

Although the machines are always totally enclosed in very thick lining plates, they are the single most likely origin of fires and explosions. The detonation of some explosive material or item is unlikely to cause any damage to the machine, but the blast may be much less sustainable to nearby personnel. In this case, the solution is to install primary granulators in separate rooms, with blast-resistant walls and an access door secured by interlocks with the control system. The interlocks should prevent the door being opened while the machine is running, and should be incapable of opening for some min after it has been switched off, in order to allow time for the rotor to become stationary.

9.6.3 Maintenance Procedures

In a complex mechanical process, maintenance procedures are likely to involve some element of risk, and many accidents are caused by machinery being started unwittingly by operating staff who were unaware of the presence of a maintenance engineer. There are many procedures for dealing with these situations, and many of them are inadequate. The most secure system involves the use of "permits to work."

The permit to work system is remarkably simple, yet it excludes any possibility of any plant or machinery being started inadvertently. In its general principles, a maintenance engineer applies to the senior operator for a permit to undertake certain agreed works upon a machine or flowline. The senior operator establishes what electrical and mechanical isolations are necessary to ensure that nothing that could affect the work or the person can be started, and he padlocks each isolator.

The keys from the isolators are then placed in a key safe, which is simply a cabinet containing a number of small compartments, each of which can be individually locked. With the keys inside, the keysafe is itself padlocked, and the key to it is stapled to the copy of the permit-to-work, which is handed to the maintenance engineer.

The permit should state to whom it has been issued, by whom, upon what date, and for what duration. It should clearly define the work to be undertaken, and list all of the isolations and closures that have been used to ensure safety. It should also list all adjacent plant and machinery that is not isolated, and should carry warnings of any aspects of the work that the senior operator considers may present risks. A copy of the permit is kept in an "open permit" file by the senior operator, and a further copy is locked in the keysafe with the isolator keys.

In these circumstances, the plant operators cannot start any of the affected machinery, since the keys to the isolators are not available to them. Equally, the maintenance engineer cannot do so either, since all he has is the key to the keysafe cabinet. The maintenance engineer has absolute control of his own safety in the sense that only he can decide when the work is complete, but even then he cannot start any of the machinery without the knowledge and consent of the senior operator.

When all works are completed, the maintenance engineer must return his copy of the permit to the operator, having signed it in declaration of completion and stating clearly what has been done where it differs in any way from the original details. The operator must then check the machine and the work location and sign the appropriate declaration upon the permit. He may then release the isolator keys and reinstate the plant.

9.7 CAPITAL AND RUNNING COSTS

As is usual in capital plant costings, those of RDF facilities depend very much upon where they are, the nature of the site upon which they are to be built, and the standard of product they are intended to manufacture. In the following costs of machinery installations, site and civil engineering and project management and design are all combined into single figures that assume a plant in an industrial building upon a level site with good soil conditions. Land costs are excluded, as are the costs of providing road access and mains services to the site: they relate to the manufacturing of RDF in the three grades discussed in Section 9.1.

The figures listed are independent of any national considerations, and so they bear no charges for planning approvals or other statutory requirements. Running costs assume a level of staffing determined solely by the demands of the process, and without accounting for any national or local labor agreements or legislation.

The derivation of the figures is from "bench marking" of historical costs of existing plants, amended according to latest designs, and reconciled with prices quoted in Great Britain in 1995. They are quoted for plant waste input capacities of 50,000, 100,000, and 150,000 t per year, respectively.

Table 9.15 lists the financial analysis for each of the main processes in terms of construction cost, construction cost in terms of annual and daily waste input, operating costs, and power consumption in terms of electrical units. The operating costs include a notional electricity supply cost of UK£0.04 (US$0.064) per unit, including any maximum demand charges.

As is usual with process plant designs, the costs are not directly related to the throughput of waste. A 100,000 t plant does not cost twice as much as a 50,000 t plant, and it is always financially sensible to build the largest that the waste arisings can accommodate. In terms of production costs per ton of waste input, it is more cost-effective to build a 100,000 t dRDF facility than a 50,000 t fRDF one, because the smaller plant takes up almost the same size of building as the large one. The machinery used is manufactured in standard sizes, not specifically those which any individual plant may require, and is therefore likely to be working under capacity in the smaller plant. Labor establishments are much the same for both plants, since control of machinery takes a specific number of hands irrespective of its size or capacity.

Nothing in the figures provided gives any indication of the values of the final products. It would be unrealistic to attempt to do so, since prices vary considerably between regions and countries. Equally, circumstances often arise where there are markets for one grade of product, but none at all for the others. For example, in Great Britain in 1995 there are thriving markets for dRDF, but none for cRDF or fRDF. However, where outlets for all three exist, it would be reasonable to expect value ratios of 1:2:6 for fRDF, cRDF, and dRDF. It therefore appears that in every case where there is a market for it, dRDF should be the production target.

Generally that is true. dRDF has much to commend it, being easy to store in quantity, economical to transport, and efficient in use. There are, however, other considerations. A large plant needs a large combustion facility to supply, and most large combustion facilities are designed around water-tube boilers

Table 9.15 Process Capital and Running Costs

Product	Waste Input (tons per year) 50,000	100,000	150,000
	Constructon Costs $m (£m)		
fRDF	7.84 (4.9)	8.48 (5.3)	9.92 (6.2)
cRDF	9.92 (6.2)	10.72 (6.7)	11.68 (7.3)
dRDF	10.72 (6.7)	12.32 (7.7)	13.60 (8.5)
	Construction Cost per ton per year $ (£)		
fRDF	156.8 (98)	84.8 (53)	65.6 (41)
cRDF	198.4 (124)	91.2 (57)	67.2 (42)
dRDF	182.4 (114)	123.2 (77)	91.2 (57)
	Construction Cost per ton per day, 350 day year £000 ($000)		
fRDF	54.9 (34.3)	29.6 (18.5)	23.0 (14.4)
cRDF	69.7 (43.4)	37.5 (23.4)	27.1 (16.9)
dRDF	75.0 (46.9)	43.0 (26.9)	31.7 (19.8)
	Power Consumption (MW)		
fRDF	0.5	0.6	0.8
cRDF	0.6	0.8	1.0
dRDF	0.9	1.1	1.4
	Running Cost per year $m (£m)		
fRDF	1.12 (0.7)	1.28 (0.8)	1.6 (1.0)
cRDF	1.44 (0.9)	1.76 (1.1)	1.92 (1.2)
dRDF	1.92 (1.2)	2.72 (1.7)	3.04 (1.9)
	Running Cost per ton per year $ (£)		
fRDF	22.4 (14)	12.8 (8)	10.72 (6.7)
cRDF	28.8 (18)	17.6 (11)	12.8 (8)
dRDF	38.4 (24)	27.2 (17)	20.16 (12.6)

burning pulverized fuel. dRDF is not easily pulverized, and even if it were there is little logic in densifying it only to mill it down again. Small process plants meanwhile produce small quantities of fuel, and those quantities are often insufficient to be of any interest to the large combustion plant operators. Small combustion plants are usually designed to burn solid fuels, using chain grates or other similar fuel handling devices.

In consequence, it is often inescapable that dRDF is the only viable route for small process plants, and fRDF for the large ones. The economic balance is irrelevant to the market for the product.

REFERENCES

1. Burlace, C. J., *The Pelletization of Refuse Derived Fuel,* Warren Spring Laboratory, Stevenage, Hertfordshire, England, 1987.

10 COMBINED RDF/COMPOST/RECYCLING PLANTS

10.1 INITIAL SORTING PROCESSES

The first stages of the treatment process for both RDF and compost are identical. Each process simply makes use of those fractions which, in the main, are of no use to the other. In addition, since each of the first-stage processes are designed with the sole purpose of separating, from the fractions required for the main products, recycling becomes a possibility. Glass, metals, plastics, and textiles have to be removed from both product lines, and with comparatively little extra effort they can be further refined.

In Chapter 6, more sophisticated mechanical composting systems were reviewed, while in Chapter 9, RDF processes were examined. The first stage of the composting lines was reception, conveying, and tromelling to recover the organic fraction mixed with some contaminants. The first stage of RDF production was also reception, conveying, and tromelling, and the machines used in both cases were identical. It therefore follows that having split the two major fractions at the trommel screens, the one front end can supply both processes.

This concept is shown in Figure 10.1. Here the solid connecting lines denote the main flowlines, while the open lines denote recycling systems. Although the main lines separate at the trommel, there continues to be a number of material transfers between them at later stages. There are also material transfers from them to common recycling stages. The figure is intended simply to show what can be achieved. It is not an indication that there is necessarily any value in the materials recovered, and neither does it suggest further processing to convert them into new consumer products. The discussion in the previous sections concentrated upon recovering clean feedstock for the composting and RDF lines, and there is no point in repeating them. In this case, the subject under consideration is the secondary products.

Both flowlines will produce metals, some separation of which begins at the trommel screen. Here the classification is mainly into cans, which pass on into the first stage of the RDF line, and heavy metallic items, which start down the composting line. The cans and larger pieces that escape the trommel will be removed by the air classifier, while the metals that set off toward the composting line are first attended to by a magnetic separator immediately downstream of the trommel.

Recovery at this stage is not essential, but it does have some advantages in respect to the glass separator that immediately follows it. Quite long lengths

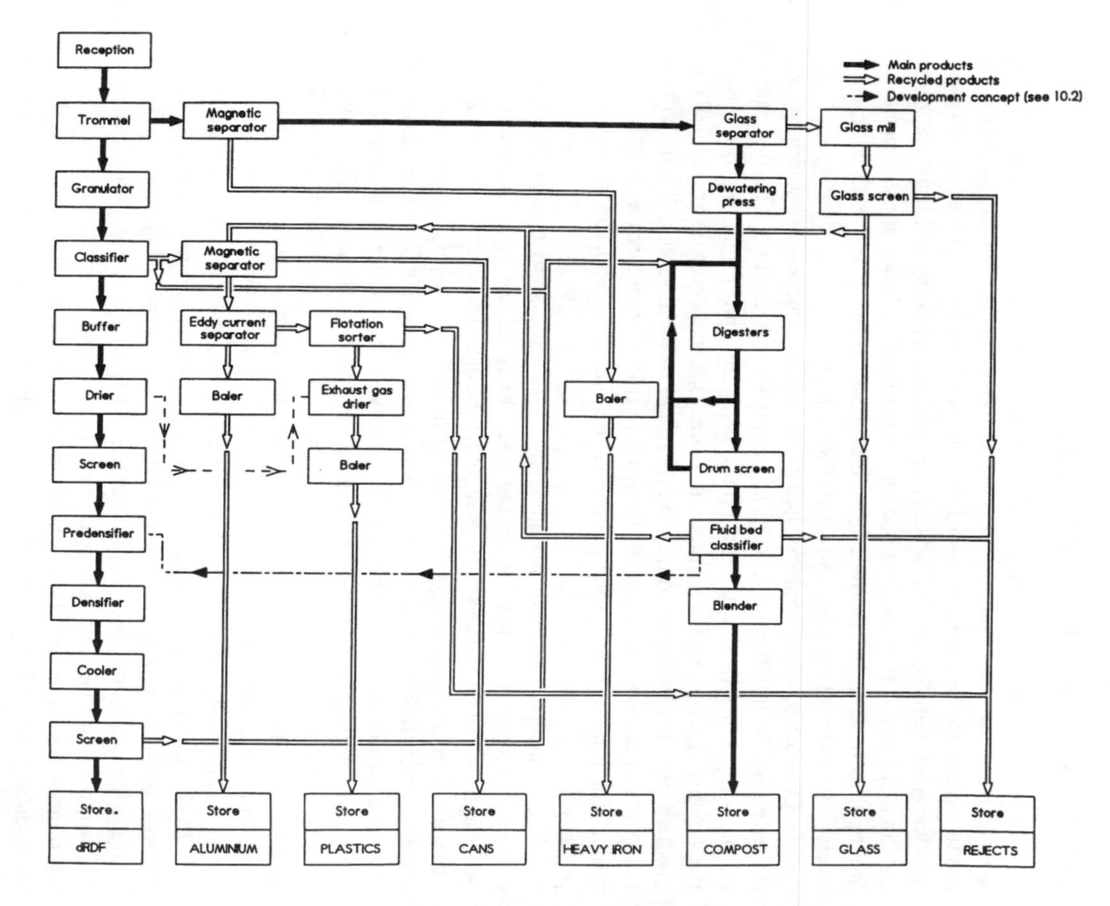

Figure 10.1 Combined dRDF/compost/recycling flowline.

of metal are not unusual occupants of the trommel undersize fraction, since if they are presented to the screen perforations end-on, they will pass through. Where they are nonferrous in nature, that is not particularly important, since they are too ductile to affect the glass separator. Lengths of steel pipe, etc. are more potentially hazardous in that they are extremely likely to cause jamming of the separator drag link conveyor. Magnetic separation on the mixed undersize fraction transfer conveyor is not highly efficient, but anything removed at that point is something less to cause trouble later.

Figure 10.1 shows the metals removed as being delivered to a baling machine. Generally, baling ferrous scrap is not recommended, particularly where it consists mainly of cans. Doing so makes the work of detinning plants considerably more difficult, and it therefore reduces the value of the material. However, the scrap initially removed from the trommel undersize is a mixture of a wide range of ferrous materials and alloys, with very little high-value metal in it. Cutlery, kitchen tools, scrap iron, and pieces of domestic appliances are all very common, and their only value is as low-grade scrap for remelting at steel works. When remelted and treated, all of it is reduced to mild steel whatever its original alloys may have been, and mild steel has the lowest value of all ferrous products. It is not therefore worth increasing transport costs by leaving the mixture as it was recovered.

When at least some of the ferrous metals have been removed, the trommel undersize fraction passes on to the glass separator in preparation for composting. In spite of its name, the glass separator removes by fluid classification anything that is not organic, with the exception of some textiles and plastics. In the heavy fraction that is removed there will be glass, ceramics, stones, metals that are mainly nonferrous, heavy plastics, currency, and bones. Such a diverse mixture of materials has no value, and in that condition is suitable only for landfill. There are, however, valuable materials within it, and so the glass separator delivers it onward to the glass mill.

At the glass mill, the glass and any ceramics will be reduced to shard-free grains, the granularity of which is determined by the ball-fill of the machine, and by the feed rate. It is possible to reduce both components to dust if that were ever required, but more normally they would be ground to about the same particle size as sand. Doing so permits the use of the mixture as aggregate and filler for concrete products, etc. Ball mills of the type used to mill the glass separator heavies do not have any significant effect upon plastics, metals, or textiles, and so if the milled mixture is screened through a drum screen with plates perforated to the desired glass granularity, clean glass is recovered as an undersize fraction. The oversize at this stage contains plastics, traces of textiles, and metals.

Since most of the metals are likely to be nonferrous, there is a worthwhile income to be gained from their recovery through an eddy-current separator. However, eddy-current separators are at risk if ferrous metals are allowed to enter them, since the metals are retained against the rotating drum and not ejected. For that reason, the oversized fraction from the glass mill screen is delivered first to the magnetic separator which serves the classifier. Some tin-free ferrous metal will be introduced into the cans from the classifier, but not enough to make any detectable difference to the quality of the scrap recovered.

Once any ferrous metals have been removed, the remaining fraction can be passed on to the eddy-current separator, which also extracts the quite substantial quantities of nonferrous metals from the air classifier. The recovered nonferrous metals are still mixed to some extent, although partitioning of the discharge chute of the eddy-current separator can classify them to some extent. However, once that point has been reached, little more can be done with them in the process plant. Without the addition of considerably more specialized plant and equipment, it is impossible to manufacture any consumer products from them on the same site, and so the most obvious market for them is with secondary recyclers.

The residues from the eddy-current separator are plastics, bones, and a mixture generally described as "unclassified," containing small quantities of a wide range of materials such as leather, rubber, ceramics, etc. The unclassified fraction has insufficient of any specific material to make further separation worthwhile, even if the technology were available to do so. However, some of the plastic can be recovered in a flotation process, and then dried using waste heat from the RDF drier before baling. Most of the mixture is likely to be thermoplastics, which can either be sold to secondary users or can be processed on site into low-grade plastic products. None of the plastics fractions recovered in a plant of the design shown are suitable for reintroduction into food packaging applications, since they cannot be cleaned sufficiently.

Upon completion of the organics digestion process, the digestates still contain some plastics, glass, and metals. The glass separators and magnetic separators can achieve a degree of purity such that the organic fraction before composting contains less than 3% of glass, for example, but they cannot be expected to extract the pieces of glass or fragments of metal that have become impacted into some organic matter. The digestion process liberates such contaminants, and they may then be removed by a combination of screening and fluid-bed separation. Once more, a residue exists that is largely composed of nonferrous metal, which can be transferred to the air classifier magnetic separator and then on to the eddy-current separator for recovery. The final residues after that stage are no longer of any value and are rejected to landfill.

At the product screening stage of the RDF flowline, the final transfer of materials takes place. In Chapter 9, when describing RDF processing, it was stated that reintroduction of the screenings from the fuel product into the pelletizers or cubers was counterproductive; in an RDF-only plant, they should be treated as rejects. However, where a composting process is also installed they have some value, since the organic fraction requires a bulking agent to be composted efficiently. The screenings from the RDF line are unlikely to be in sufficient quantity to provide all of the bulking agent that the composting process needs, but at least they are put to use rather than becoming a charge against operations.

10.2 CONFLICT BETWEEN RDF AND COMPOSTING LINES

While the composting and RDF processes in general each make use of materials that are unsuitable for the other, there are some exceptions where conflict occurs. The most significant area of conflict arises in the need of the composting process for bulking agent, when the only agent readily available is also the primary feedstock for RDF. Therefore, in a combined plant there always has to be a compromise between maximizing the production of RDF and of compost.

In some ways there are benefits to be gained from the conflict. In winter, there is almost no market for compost, but there is potentially an increased demand for RDF. The reverse of that situation arises in summer. It is sensible economics to seek the highest possible RDF production in winter, and the highest compost production in summer. Compost production demands a minimum level of bulking agent, but there is no maximum. It is quite possible to manufacture compost from paper and card in which there is only a trace of organic material. Therefore, at the appropriate times when RDF demand is low, a major proportion of the feedstock for it can be diverted to the compost line, where it will provide for a substantial increase in the volumes produced. It is for that purpose that Figure 10.1 shows a path from the RDF line air classifier to the recirculation system of the digesters.

While the procedure is convenient for tuning production to market demands, there is one factor that may influence the decision. Compost is normally sold by volume rather than by weight, and nutrient levels are expressed in that form. A significant increase in inert bulking agents will therefore change the chemical analysis of the product, leading to a position where the compost manufactured throughout the winter is different from that manufactured in the summer.

The extent to which that factor is of importance depends largely upon the markets. If the material is all being sold for low-grade land reclamation applications, then the variation is unlikely to be of any particular significance. However, if the market is an agricultural or landscaping one, it may be more important. In those applications, the buyers are more likely to specify the levels of nutrients they require, and to expect consistency between batches. Where that is the case, there are merits in considering the production of two grades of compost. One, made during the winter months, could be sold as a high-nutrient product for soil amelioration. The other, manufactured in summer, could be marketed as a soil conditioner.

In recent years there has been some examination of yet a further alternative, and some work has been carried out to explore the opportunities it offers. When the organic fraction has been reduced by aerobic digestion, it exhibits a substantial increase in calorific value approaching that of RDF. It is then finely granular, relatively dry, and biologically stable. It has been proved to be capable of pelletization, and it appears that it could be used as a fuel either on its own or as a part of a conventional RDF product.

This observation is made with some caution, since although there is a small amount of experience of the use of the material as a fuel, very little published data exists on emissions and the effects upon combustion systems. The chemical nature of digested organic materials hints that there could be problems with the creation of dioxins and furans, and that other complex chemicals could be created at high temperatures. At least it would appear likely that any fuel made from it may contain very high levels of chlorine compounds, which would seriously derate any boiler plant upon which it was used. More research appears to be necessary, but there are sufficient justifications for it to be undertaken, since it would introduce a further degree of flexibility to a process that is extremely market-sensitive.

10.3 CAPITAL AND RUNNING COSTS

As a process becomes more complex, so any attempt to provide accurate cost estimates becomes more difficult. All capital plant costs are sensitive to geographical location, and to local geology and topography. This is especially so of combined RDF/compost/recycling facilities, where difficult site conditions alone can easily increase construction costs by over 10%.

The nature and quality of the products to be manufactured also has an inevitable impact upon costs. It is a fundamental principle of any manufacturing process that the more sophisticated or diverse the products are to be, the higher the cost of construction. In the case of a combined waste reprocessing plant, there are three potential grades of RDF, and within them almost any number of quality standards. There are at least the same number of potential grades of compost, further complicated by the possibility of varying nutrient levels by modifying digestion periods.

In this situation, any attempt to define cost ranges would be unrealistic, and so the best that can be done is to suggest the *minimum* likely costs for a process of a specific design. Therefore, in Table 10.1 the cost structure is developed upon the following basis:

1. The process plant is assumed to produce landscaping-quality compost, without any subsequent blending of nutrients or improvers.
2. The municipal waste analysis is assumed to be such that it contains 28% w/w of paper and card, and 35% of organics and putrescibles.
3. The RDF installation is expected to manufacture a product to each of the three main grades, and to achieve and maintain the quality standards defined in Chapter 9 on RDF processes.
4. Each of the secondary fractions recovered for recycling are assumed to be sold to other processors, rather than receiving any treatment on site other than recovery.
5. The initial process stages of reception, reception conveyors, and trommel screens are common to both main flowlines.
6. Both main processes are housed in a single industrial-quality building.

Table 10.1 Process Capital and Running Costs

Product	Waste Input (tons per year)		
	50,000	100,000	150,000
	Constructon Costs $m (£m)		
fRDF/RDC	9.62 (6.0)	10.2 (7.2)	13.8 (8.7)
cRDF/RDC	11.7 (7.3)	13.7 (8.6)	15.5 (9.8)
dRDF/RDC	12.5 (7.8)	15.3 (9.6)	17.5 (11.0)
	Construction Cost per ton per year $ (£)		
fRDF/RDC	192.5 (120)	114.6 (72)	91.2 (57)
cRDF/RDC	234.1 (146)	121.0 (76)	92.8 (58)
dRDF/RDC	218.1 (136)	153.0 (96)	116.8 (73)
	Construction Cost per ton per day, 350 day year £000 ($000)		
fRDF/RDC	67.4 (42.1)	40.1 (25.1)	32.0 (20.1)
cRDF/RDC	81.9 (51.2)	47.9 (29.9)	36.1 (22.5)
dRDF/RDC	87.5 (54.7)	53.4 (33.4)	40.7 (25.1)
	Power Consumption (MW)		
fRDF/RDC	0.65	0.85	1.15
cRDF/RDC	0.75	1.05	1.35
dRDF/RDC	0.9	1.35	1.75
	Running Cost per year $m (£m)		
fRDF/RDC	1.44 (0.9)	1.76 (1.1)	2.24 (1.4)
cRDF/RDC	1.76 (1.1)	2.24 (1.4)	2.56 (1.6)
dRDF/RDC	2.24 (1.4)	3.20 (2.0)	3.68 (2.1)
	Running Cost per ton per year $ (£)		
fRDF/RDC	29.0 (18)	17.6 (11)	15.0 (9.4)
cRDF/RDC	35.2 (22)	22.4 (14)	17.1 (10.7)
dRDF/RDC	44.8 (28)	32.0 (20)	24.5 (15.3)

7. The composting process is based upon four aerobic static digesters, operating in series.
8. No provision is made for access roads, site works such as parking areas and interconnecting roads, or overall site drainage.
9. No provision is made for the connection of mains services, all of which are assumed to be available at the site boundary.
10. Electrical power demand is assumed to be consistent throughout the operating year, and its cost is assumed to be stable over the same period. No allowance is made for seasonal maximum demand charges.

The data provided in Table 10.1 are established by "bench marking" from historical costs of existing process plants of a similar design where possible. Current international heavy machinery prices are then used to correct the data to 1995 conditions. In view of the variability of the conditions listed above, while any of the processes defined are unlikely to cost less than suggested, a price range to at least 10% above the minimum is possible.

11 MARKETS FOR RECYCLED PRODUCTS

11.1 SECONDARY PRODUCTS

For the purposes of this text, secondary products are defined as glass, ferrous and nonferrous metals, plastics, and textiles. There are of course many other components of municipal waste, but none of the remainder occur in sufficient quantities to make recycling worthwhile.

A reprocessing or recycling plant is entirely a manufacturing venture, with little to do with waste disposal. This is a fundamental position that must never be departed from. Waste is the feedstock, and nothing more than that. Waste disposal is a fortunate benefit arising from the development of sound markets for products. If the products are of unacceptable quality, consistency, or repeatability, then they will not sell and there will be no waste disposal stream. There is sufficient proof of that statement throughout the Western hemisphere, in abandoned facilities built upon the premise that the marketplace was both simple to enter and imperceptive with respect to quality. Whatever the avowed interest in and concern for the environment, no one will buy anything simply because it is recycled. They will only do so if the product suits their needs of the moment.

This is never more true than for the secondary products from the facility. The marketplace for them is not the public, where packaging and emotional appeal may have some effect. It is instead entirely industrial, where quality, consistency, repeatability, and price are the primary considerations. The purpose of this section is not to identify the markets and name the potential buyers, since that would be an impossible task. Markets are firmly based upon regional and national circumstances. Instead, the purpose is to identify what any industrial buyer is likely to expect of each product, and to attempt to emphasize the severity of the competition that may be encountered.

11.1.1 Glass

Glass recovered from materials recovery facilities (MRFs) and street-corner bottle banks has some value provided that it is both color-segregated and free from contamination by ceramics. The slightest detectable trace of any ceramic material is likely to result in the rejection of the entire load in which it occurs. Even any noticeable contamination of one color by another is likely to lead to rejection.

Products of those qualities cannot be derived from waste processing plants, where the glass is essentially of mixed color and almost inevitably contaminated by some ceramics. None of the separation machinery available within existing technology is capable of distinguishing between any of those criteria. Markets that are not discerning in any of those areas must therefore be discovered.

In seeking to develop the markets, it is worth recognizing what glass actually is. It is primarily sand with some chemical additives. Sand may not have any high value, but at least it is a product for which there is a market. To meet the requirements of those markets, it must be relatively clean, free from metallic contamination, and free of shards. Sand is a product that is likely to be handled, and skin penetration by sharp fragments will not be regarded lightly.

If the glass recovered from a reprocessing plant is milled as described in earlier sections, then it is free of shards and metals. It is in mixed colors, but that can be used as a marketing attribute rather than a disadvantage. For example, if it is used as the aggregate in the manufacturing of paving slabs, and if the slabs are etched upon completion, then some of the glass particles are left in relief upon the surface. Mixed glass has an overall green color, and paving made from it therefore has a faint green coloration.

There are also somewhat more limited markets for decorative sand and for construction products made from them. Aquarium sand is one such possibility, while facing blockwork for buildings is another. Neither attract large sales as yet, but the opportunities are there. Milled, shard-free glass is potentially an excellent product that deserves imaginative marketing.

One of the more imaginative applications developed in the United States some years ago, where mixed, granulated cullet was used in the surfacing of public highways as a replacement for flint or granite chips. It was reported that the surfaces were hard-wearing, and that they possessed a high skid-resistance, with the added advantage that the particles reflected headlight beams to some extent. Night driving, particularly out-of-town, was assisted by the way in which the whole road surface showed up clearly. In such an application the technology has approached the ideal, in moving from a means of disposing of a waste product to one of taking advantage of that product's special attributes.

There are other potential markets that are largely indifferent to the color of the material, and the most prominent among them involves the production of glass fiber insulation. Those markets are, however, extremely dependent upon the location of competing industries, and those manufacturing products from newly created glass are the most competitive. They always produce a considerable proportion of glass scrap, and for many of them any attempt to recycle it within their own process is counterproductive. They often see glass fiber manufacturing as a way of ridding themselves of their clean waste, without the risk that their formulations may become available to a competitor. As a result, in many countries, glass fiber insulation manufacturers requirements are satisfied by nearby glassmaking industries. Competition for the market is difficult if not pointless, since a reprocessing plant cannot compete with another who sees the material he is offering as a waste, which he would pay to rid himself of if necessary.

11.1.2 Ferrous Metals

Ferrous metals occur in municipal waste in the form of tinned containers, and as diverse heavy scrap of a wide range of alloys and qualities. Cans generally have a ready market for detinning, but viability depends upon there being a suitable detinning works within a reasonably economic transport distance. Otherwise, they have no value other than that of their steel content, which is minimal. They are considered to be very low-grade scrap, usually with a low recovery rate, and in the scrap market their tin and lead content is a positive disadvantage. By the time the tin and lead has been removed together with the residues of paper and organic wastes always associated with them, metal recovery rates can be as low as 70%. There are always some losses due to oxidation upon remelting, and so the final yield from scrap cans may reduce to no more than 50%.

The heavy scrap fraction has somewhat more value as a material for recycling into new steel products, provided that it can be separated from both cans and nonferrous metals. Irrespective of the alloys recovered, the result of remelting at steel works is to reduce all of the scrap to mild steel. There is therefore no benefit to be gained from attempting to segregate individual alloys such as low-grade stainless steel from cutlery, even if it were possible to do so mechanically. Much of household cutlery, although "stainless," is still magnetic, and so magnetic separators cannot be relied upon to distinguish between it and any other ferrous material.

Generally, steelmakers have targets for their recycled metals requirements, and increasingly they are failing to meet them. In theory, the result should be an increased demand and a corresponding increase in prices paid for scrap. However, steelmaking worldwide is increasingly becoming centralized into a small number of very large facilities, which are prepared to support their needs by import where necessary. Centralization may reduce the value of scrap to the producer if not to the buyer, simply because the nearest user is too far away for transport to be economical.

11.1.3 Nonferrous Metals

Aluminum in the form of beverage cans is the most prevalent form of nonferrous metal in municipal waste, and there is a significant demand for it. For that reason, it is a very popular target for voluntary recycling groups, and much of the scrap aluminum arisings therefore never reach the recovery facility end of the waste stream. Generally, the proportion of aluminum in the waste at that point is less than 1% by weight, and although the scrap value is high, there may be insufficient to justify the recovery costs.

While the market is fairly consistent in terms of quantities, it is much less so in value. Aluminum is subject to international market forces and is almost a commodity to be traded, with the result that large price variations are not unusual. That perhaps is unimportant to a voluntary group, for whom the price is always attractive, even when it is at its lowest. They, after all, do not bear labor or fixed plant costs, and almost all of the income is profit. The same does not of course apply to a processing facility, and even an MRF may find that at some times its aluminum recovery lines are uneconomic.

The remainder of nonferrous metals recovered from waste treatment facilities are mainly copper or copper-based alloys, and coinage is one of the more prominent among them. A large, combined reprocessing plant, with a waste throughput of 150,000 t per year, will quite commonly recover up to 17 ft^3 (0.5 m^3) of coins a week, and the value may of course be almost any amount. As metal scrap, the coinage is not particularly valuable, and in many countries defacing the coinage is an offense. Cleaning and sorting the large number of coins involved to remove foreign currency is extremely labor intensive, and so in spite of the potential value of the result, it is a task usually left to voluntary groups to whom the mixed coins are given freely.

Where the nonferrous fraction has been recovered by the use of eddy-current separators, refining the scrap into various types of metals is possible if not precise. Some improvement of quality may be achievable, but the worth of doing so depends upon the market outlets. More commonly, the scrap is sold in mixed form to secondary materials merchants, who may themselves further separate it using their own specialist plant. In those cases, careful refining and separation at the recovery facility may add no value whatever to the material.

11.1.4 Plastics

Plastics may be categorized for waste recycling purposes as thermoplastics and thermosets. Thermoplastics, among the most prominent of which are polyethylenes, polypropylene, ABS, polycarbonate, polystyrene, acrylics, PVC, nylon, and polyacetal, are transformed by heat into a viscous liquid that can in theory be reprocessed any number of times. In thermosetting plastics such as phenolics, ureas, melamines, epoxides, and polyesters, heating creates strong chemical bonds between polymer molecules, and so recycling is impossible. For those reasons, plastics of interest to the waste recycling industry are the thermoplastics.

Unfortunately, with some exceptions it is difficult if not impossible to distinguish between the various polymers without expert knowledge. In a waste recycling plant the only reliable method is by identification of the product into which they are made, rather than by identification of the polymers themselves. Even then there will be many materials that defy classification by that means. For example, it is fairly certain that a container made of thin, clear plastic will be polyethylene terephthalate (PET), but one made from colored, opaque plastic might be melamine formaldehyde, which is a thermosetting plastic.

Generally, over 50% of the plastics found in the municipal waste stream are polyethylene film, and are therefore reasonably easily identifiable. However, a certain amount of clear film used in packaging is cellulose, which is not a plastic at all but is derived from wood. There the only difference leading to immediate product verification is that cellulose film does not stretch, and it tears easily from an initial penetration. Polyethylene films, meanwhile, are capable of stretching to several times their original length, and they possess a high tear resistance.

Even where the plastics are identified, their reuse is limited by contamination. PET that has been in prolonged contact with comingled waste, and particularly that which has been granulated in a flail mill with paper, metals, and some organic residues, is very difficult to clean to a standard that would be acceptable to food hygiene agencies. For that reason, most of the plastics recycling occurs as a result of "bring" systems of source separation, where the materials never enter the municipal waste stream in the first place. There the segregation of material types depends upon the limited knowledge of the general public, and so it is restricted to PET, polyvinyl chloride (PVC), and polyethylene (PE).

Some mechanical classification can be achieved by density-based methods. Most plastics have densities within the range 56 to 94 lb/ft^3 (900 to 1500 kg/m^3), and those of thermosetting materials are generally toward the higher. Flotation methods can therefore provide crude separation, but frequently several stages are necessary to obtain a clean single product. The plant and technology involved is complex and expensive, and so it is not widely found in waste recycling and reprocessing plants. It is more within the province of secondary recyclers, who may accept mixed plastics for recovery of useful fractions elsewhere.

Of the more common products made from recycled plastics, all are used in lower-value applications than they were in their virgin state. Thus, PET is increasingly used for fiber fillings for domestic goods and clothing, and for industrial applications such as filter media. Scrap PVC is quite widely used to manufacture flooring materials and drainage pipes. PE is used for nonfood packaging, and for the manufacturing of trash bags and garden products.

As the availability of recycled thermoplastics increases, new applications for them are developing. Among those are "synthetic timber" fence posts, trashcans, and road cones, for example. It is therefore worth considering, when contemplating the development of a waste processing facility, the possibility of encouraging a start-up industry in one of those fields or even including one or more of the processes in the main plant.

Otherwise, unless a waste treatment facility is fortunate enough to discover a major manufacturer of any of those products close by, transportation over some considerable distance is likely. Then the very low bulk densities of plastics become apparent, and baling is inescapable if viable payloads are to be achieved. In that case, there is some merit in seeking intermediate dealers for all plastics recovered by the facility.

11.1.5 Textiles

The days when textiles were recovered in considerable quantities for use as wiping rags in industry are effectively over. Much of the modern textile fraction in municipal waste is synthetic fiber, which is largely nonabsorbent, or wool, which is equally unsuitable for that purpose. Synthetic fibers cannot generally be reprocessed into anything else, and so they have little or no value in their original form. Wool can be reprocessed into fiber for other purposes, although again all of the applications involve "down-cycling" into lower-grade products. The most prominent of the uses for wool are as fillers for padding, and for underlays for carpets, etc.

A growing application for clean and undamaged textile items exists in the underdeveloped countries, where there is a shortage of suitable clothing for impoverished people. Most of the trade is through dealers and voluntary organizations, and prices obtained are unlikely to be high. The markets are difficult or even nonviable for direct penetration by the operators of a waste processing facility, since the small quantities of textiles now found in waste make the effort hardly worth the trouble. There is also of course the political viewpoint, which has been known to argue that supplying discarded clothing in the municipal waste stream to poor countries is patronizing in the extreme.

11.2 REFUSE-DERIVED COMPOSTS

11.2.1 Green (Yard) Waste Compost

Generally, the only market preparation likely to be needed in a green waste composting facility is bagging, palletizing, and shrink-wrapping. Whether or not bagging plant is actually required at all depends entirely upon the nature of the target product market. Even where it is considered necessary, it may be that there is a reasonably local contract bagging facility that could undertake the task of handling the product under contract.

The markets for green waste compost are such that bagging is essential to realize a good income from the product. There are a number of opportunities for bulk supplies, but they are frequently difficult to realize, and they return quite low values. The whole business of bagging is, however, quite complex, and so it is worth considering the general constraints.

First, the type and size of the bags must be determined with some care. To store compost successfully, the bags should be manufactured from 120 μm polyethylene film of at least two-ply, where the inner film is black and the outer is of a color determined by the brand image. The film should either be microperforated, or should have perforations of not greater than $^1/_4$ in. (5 mm) diameter along the edges; microperforations are best. The black inner film is essential to exclude light from the product, while the perforations allow it to "breathe," thus avoiding anaerobicity.

The best film for compost bags is actually three-ply, and it is possible to have the outer film prepared with a slightly matte surface finish, which renders it slightly less slippery. A three-ply film offers good strength and resistance to tearing, while the matte finish makes stacking on pallets somewhat easier.

In terms of outer film color, there is actually not as much choice as one would expect. Pale green, white, and ivory are the most common colors, and which is selected is entirely a matter of which will best suit the chosen artwork for the printing. Simply offering the product in unmarked black bags is unprofessional, presents a poor image to a potential customer, and leads him to expect a low-quality product at a low price. A facility handling 12,000 t of green waste a year could generate about 275 yd^3 (210 m^3) of compost a week, and if the price per bag were to be reduced by say, $1.50, as a result of a poor brand image, the result would be a loss of income of up to $8,000 (£5,000) a week.

The artwork for the bag design needs to identify the manufacturer, and to make the potential customer clearly understand what the product is and its potential for him. It must be sufficiently eye-catching to be readily identifiable among a stockpile of competing brands in a garden store. Each bag must bear printing along at least one side, identifying the product when stacked on a pallet.

In terms of bag size, there are three primary considerations: The bags must be convenient for the purchaser to handle but not so small that the quantity purchased is inadequate. People resist buying several bags at once, even if the price is the same as one bag containing an equivalent quantity. Composts compact to some extent, and it is important to select a bag dimension such that one filled initially with, say 1.4 ft^3 (40 l), does not compact on a pallet to such an extent that when opened it appears to be only half full. And the bag dimensions must be such that, when filled, the bags can be stacked in an interlocking pattern of possibly 60 bags on a standard disposable pallet.

Composts are in fact sold in a very considerable range of bag sizes, the most common, general-purpose ones being offered in packaging ranging from less than 1 ft^3 to $3^1/_2$ ft^3 (20 to 100 l). Green waste compost is, however, very heavy, and a $3^1/_2$ ft^3 (100 l) bag is much more than the average person could lift. Meanwhile, $^3/_4$ ft^3 (20 l) is likely to be too small a quantity for average domestic use. It is a volume more consistent with special-purpose composts for hanging baskets, etc. In fact, a good market launch size for green waste compost is 1.5 ft^3 (40 l), since it contains sufficient for most domestic purposes without being too heavy to lift. Such a bag may retail at less than £5.00 ($8.00), and so the investment required of the customer interested in trying it for the first time is not excessive. A convenient 1.5 ft^3 bag is of 18 × 33 in. (460 × 840 mm) dimensions when flat.

In considering the arrangement of a supply of bags, it is important to realize that once the artwork for the printing and bag design has been done and the printing plates have been made, changing to another bag supplier will be expensive. Bag costs from the original supplier could be expected to reduce as the demand increases and the initial print preparation costs have been absorbed. There are a large number of plastic bag manufacturers, but by no means all of them have experience of composts, and since compost bagging is still in some senses a "black art," it is wise to ensure that the prospective supplier has such experience!

The plant necessary to achieve packaging meeting these requirements is volumetric. Composts are not sold by weight, since changes of moisture levels in production and storage mean that the final density could never be defined accurately. In some circumstances, a bag of a given volume would only be half full at a given weight, while at other times it might be impossible to fill it to that weight. The bagging plant must therefore be based upon a volumetric metering system, not a weigher.

There are two ways of achieving volumetric metering. The simplest system is one where a small hopper is fitted with a discharge spout encircled by a bag-holding clamp. Each bag in turn is fixed to the spout and filled to capacity, which, given accurate bag design, is then assumed to be as many liters as are declared in the artwork. The bag itself determines the quantity.

The alternative, and much more expensive system, is one based upon a metering system within the bagging machine. In this case, it is the machine that determines the quantity, not the bag, and it does this by initially filling a metering chamber to a known volume, then discharging into a clamped bag through a vibrating spout.

Each system has advantages and disadvantages. Perhaps the only advantage of the simple device in fact is that it is cheap and primitive. It is, however, quite impractical when large quantities have to be processed, since it is likely to be entirely manual in operation. In the case given as an example above, 275 yd^3 (210 m^3) of compost a week equates to 5250 bags of 1.5ft^3 (40 l) a week, or a rate of 130 bags an h working without a pause. That is rather more than an operator could be expected to handle. The second system can be automated to almost any extent, although the more so it is, the higher the capital cost. It does, however, make handling almost any quantity possible, and most such machines can handle a range of bag sizes from $^3/_4$ ft^3 to $3^1/_2$ ft^3 (20 to 100 l) or longer, permitting changes in response to market demands.

All bagging plants of whatever type need a heat sealer to close the bags after filling, and the automated machines need a bag opener also. The heat sealer is in fact a separate machine, in the sense that it can be purchased separately, but it must be selected to match the bagger in terms of capacity.

The interconnection between the bagger and the heat sealer is generally a "vee" conveyor, which is two narrow belt conveyors installed such that their top surfaces are close together and at an angle to each other. Often they are lined with wooden slats to ensure a positive grip on the bags. By that means, the bags are presented to the closing rails of the heat sealer in an upright position.

Once the bags have been filled and sealed, they must be stacked upon a pallet. It is most unlikely that all will be sold directly off-site as fast as they are made, and so both storage and transportation are best achieved by shrink-wrapping pallet loads of 1 t each. It is possible to buy palletizing machines that will carry out the stacking automatically, but they are expensive and not really justified for the average green waste operation. All that is needed, in fact, is a stand upon which to place each pallet, and either a turntable or a swinging frame to hold the shrink-wrapping (or stretch-wrapping) film.

Placing the bags on the pallets in the first place can become somewhat laborious as the quantities build up, and it is always worthwhile to consider a sack lifter. There are relatively simple devices available that operate upon a vacuum system, delivered by a counterbalanced clamp plate suspended from a cantilever frame. The clamp plate is fixed to the end of a large-bore flexible pipe, in which a partial vacuum is created by a suction-blower. The size of the clamp plate is selected to be consistent with that of the bags to be lifted, and when it is applied to one, the partial vacuum holds the bag firmly to the plate. The adjustable counterweight then allows the bag to be lifted without effort, and the pivoted cantilever permits it to be placed in any location within the radius of the arm.

The application of the retaining film depends upon which type it is. Shrink-wrapping requires a heat-shrinking, hand-held tool that can be powered either electrically or by propane gas. Stretch-wrapping uses a film with built-in "memory": it tries to return to its original dimensions when released from tension, but it adheres to itself, so causing the whole package to grip the load tightly.

For reasonably small levels of production of about 60 pallets a week, a hand-operated stretch-wrapping system is normally sufficient. Such a system is not a machine in the true sense, being rather a place upon which a loaded pallet can be stood while an operator walks around it applying the film. Once the material has been palletized, it has to be handled to storage, clearly an application for a fork lift device. Assuming that the bagging operation were to take place inside a site building with a hard floor, then a hand-operated fork lift would be sufficient for the relatively small number of pallets that would be involved initially.

The final consideration in palletizing bagged compost is that the pallet load should be "capped" on completion. This is in fact simply a matter of placing a black film on top of the load and shrink- or stretch-wrapping it into place. The cap serves the purpose of improving the weather resistance of the package, and keeping the sunlight off the top bags, where it might otherwise cause the printing to fade.

An alternative strategy to a wholly owned bagging operation is to negotiate a contract with a local bagging firm. The disadvantage of such a step is of course that the product has to be double-handled, once taking the bulk material to the plant, and once taking the palletized products away to the distribution center. However, it is possible in many cases to negotiate an arrangement whereby the bagging plant operator will store the completed pallets and arrange the distribution upon receipt of individual consignment instructions. That facility is of great benefit if the product is to be sold regionally or nationally, rather than just locally.

In cost terms, one should expect a contract bagging service to charge between 13% and 15% of the retail value per bag, which will include the cost of wrapping and of the disposable pallets. The client must supply the actual bags, which are generally supplied in minimum print runs of 5000 at about 5% of the final retail value.

The most important market for green waste compost is in domestic retail sales, hence the emphasis upon packaging. It is not a product that is as yet received with enthusiasm by the professional market, where it is seen as being of inconsistent quality and performance. A professional grower operates to extremely close tolerances in the products he uses for his business. For example, a grower of potted plants for multiple stores may require every one of several thousand plants to all start to bud at exactly the same time. He needs a growing medium so consistent between each bag and each batch that he can rely upon that not just for 1 month but for every month of every year. Many growers achieve that by purchasing a material that has no nutrients at all, and then by adding their own formulations to it to create the growing medium they require. To assist them, there are a number of standard, traditional formulations requiring them to purchase additive chemicals of guaranteed purity and concentration. Green waste compost, and in fact any organic compost, cannot possibly achieve such a level of consistency, since the manufacturer has no means of controlling the biological activity to such a fine tolerance.

Some professional horticultural operations are a little less sensitive, and may be prepared to experiment. Most significant among these are the tree and shrub growers. Their need is for comparatively large quantities of growing media for every individual plant, and the exact control of nutrients is of secondary importance. As long as the trees or shrubs they grow are of an approximately similar size, neither they nor the public who buy the products will be overly concerned. Unfortunately, in most regions there are many more growers of potted plants and vegetables than there are those who specialize in trees and shrubs.

Agriculture may eventually become a market, but in most Western countries it has not yet done so. Modern farming is intensive and operators are inclined to rely upon chemical fertilizers, which are easier to apply by mechanical methods. If a farmer were to use organic composts, he would need very large quantities indeed and all at the same time of year. As a bulk buyer, he would not expect to pay premium prices, and the demand of one farm alone could overwhelm most green waste composting facilities.

In dealing with the only readily penetrable market, that of domestic retail sales, the product will be competing with others manufactured by large and highly professional companies, whose business is often international. They sell on the strength of established brand names recognized by almost all home gardeners, and they spend considerable sums of money upon product development. Competition between them is intense, to the extent that in many countries there is a surplus of compost and a resulting depression of price. That position is often made worse by the habit of multiple stores of offering composts as "loss leaders": products sold for less than their production cost to attract buyers into the stores.

It is into that market that the local green waste organic product must be introduced, and hence the importance of packaging and the establishment of a brand name. Competition with the large producers can only succeed by putting the product forward as a local initiative, seeking the natural human tribal response. The term "recycled" is best avoided, since the bad experiences with such materials over the last few years simply place ammunition in the hands of the competitors. "Multipurpose organic compost" is a more acceptable term, especially if it is supported by artwork suggesting a cozy, almost old-fashioned rural environment.

Cozy impressions and cleverly designed artwork, however, are only a part of the marketing strategy. Professional marketing demands more effort than that. The product must be supported by independently undertaken growing trials by an organization whose reputation is beyond reproach. Nutrient analyses that clearly state the range of concentrations of each component must be available, and based upon those analyses the manufacturer must be honest enough to state clearly what his product is unsuitable for. Table 11.1 shows a typical green waste compost analyses in the form in which a retailer may be interested to see it.

Green waste compost is not, for example, a seed compost. Its conductivity, alkalinity, and nutrient levels are likely to be too high for that. Neither is it suitable as a general rule for plants that grow best in acid soils. It is an excellent soil ameliorant, but it is not a good soil conditioner, since although its high nutrient levels will improve poor soils, its general lack of inert material will not improve soil texture.

11.2.2 Municipal Waste Compost

In addressing the marketing of municipal waste composts, it is necessary to consider the matter at various levels. Compost is a relatively bulk low-value product in whatever form it is produced. It is therefore logical to minimize distribution costs by seeking as much local utilization as possible. However, as the subject of waste recycling has now become such an important political issue, it is also necessary to consider those issues affecting the marketability of the products. There are many ways in which this is manifested, but the most significant is through the choice of collection systems. The type of collection system affects the level of contaminants in the feedstock, and this is reflected in end-product quality. The choice of collection systems also effects the financial and legal arrangements for the process plant and is reflected in end-product selling price. Markets are, of course, influenced by both price and quality.

Table 11.1 Typical Green Waste Multipurpose Compost Analysis

Property	Typical analysis or specification
Physical Criteria:	
Texture	Peat-like
Water-holding capacity	70%
Air capacity (AFP)	15–30%
Moisture	40–50%
Bulk density	63 lb/ft^3 (0.33 kg/l)
Chemical criteria:	
Organic matter	56%
Total nitrogen	1.7%
Total phosphorous	0.5%
Total potassium	0.8%
Total magnesium	0.2%
Carbon–nitrogen ratio	16.6:1
pH	7.9
Conductivity	<1600 mS
Cation exchange capacity	0.006 oz/lb (434 mg/kg)
Total metals % w/w:	
Zinc	0.0374
Copper	0.027
Nickel	0.002
Cadmium	0.0002
Lead	0.0244
Mercury	6.5×10^{-5}
Chromium	0.00425
General criteria:	
Weeds	Nil
Seeds	Nil
Roots of perennial weeds	Nil
Sticks	Nil
Handling	Free-flowing
Color	Light brown
Hygiene	Sanitized by natural heat treatment

The politics of waste are now firmly established on the international agenda, and the two recent examples that illustrate this are the European ban on the dumping of sewage sludge to sea and the setting of recycling targets throughout Europe. The area of solid waste management is in a state of considerable flux. Recycling and resource utilization have joined pollution as established political vote-winning issues. The European Community has responded to these changes along similar lines to those adopted earlier by United States bodies, and national organizations and legislative bodies have been established to stimulate recycling through a variety of initiatives, including direct support for investment in recycling technologies, legislation encouraging investment in recycling by large companies, and legislative standards for recycled materials.

The following list provides an overview of the situation in most European countries. It is compiled from a comprehensive survey of European compost standards.[1]

Austria: The present policy is to move away from mixed waste composting toward separately collected biowaste and thus overcome some of the quality problems previously experienced.

Belgium: The Flanders region has 26 municipalities running collection schemes for source-separated domestic waste. The scheme covers approximately 87,000 people in rural areas. The waste strategy required municipalities to have established collection schemes for domestic biowaste by the end of 1992.

Denmark: There are five biowaste collection systems and recycling plants in Denmark. Currently 100,000 households sort waste, and this number is expected to increase so that by the year 2000, 40 to 50% of mixed household waste will be recycled.

France: France is considered to be the country with the most expansive composting operation, with 90 plants processing 1,300,000 t of household refuse per year, resulting in the production of 650,000 t of compost (1989 report). Legislative standards for compost quality are also being revised and tightened.

Germany: A government-funded program is assisting in the development of organic waste composting plants. The authorities are currently drafting a new administration order that will act as "best practice" guidance. The guidelines are likely to include the requirement for separation at source of organic waste, for materials to be treated by composting. Composting of mixed waste is unlikely to be acceptable.

Greece: There are approximately ten companies involved in composting. A full-scale plant is being built with assistance from the EC.

Italy: There are 21 plants in Italy today with a compost production level of 825 t of compost per day (equivalent to 7.5% of total urban refuse), but many of the plants produce a poor quality end-product. A recent new decree was considered by the Ministry for the Environment. In this, it was recognized that a higher-quality product can be generated through the use of presorted waste. A national strategy for sampling and certification was recommended.

Netherlands: Legislation in the Netherlands has the aim of reducing in phases the pollution of the soil by heavy metals and arsenic from fertilizers (and this includes compost). The goal is that by the year 2000 the inputs to the soil will be equivalent to the outputs removed by the crops. They have therefore prepared extremely restrictive standards for composts. These constraints not only affect municipal composts: 26% of sewage sludge currently disposed of to land will have to be disposed of by other means.

Spain: There are 30 to 40 composting companies in Spain. A recent order has been introduced to protect the environment and the consumer from poor products, but the requirements are not onerous.

Sweden: There are 11 composting plants in Sweden. The aim of the Ministry of the Environment is to develop a decentralized system for composting with local control. They also wish to stop incineration and landfilling of waste as soon as it is feasible to do so, although after a period of cessation, incineration has once more been permitted.

Switzerland: There is a requirement in law that the cantons compost agricultural, garden, and domestic wastes. The Technical Ordinance on Waste imposes a duty of decentralized composting, separation of wastes at source and the utilization of compostable refuse. The overall strategy of the legislation is to enforce product quality, to promote waste separation at source, and to reduce the amount of mixed waste used as a compost feedstock. There are 117 composting plants in Switzerland with an annual capacity over 100 t. Of the 117 plants, 38 have an annual throughput in excess of 1000 t. Additionally, 30 more composting plants are being planned, 6 of which are in the private sector. Existing plants process 230,000 t of refuse per year (63% of which is agricultural and garden waste), which results in 140,000 t of compost. Of this output, 80,000 t are used in agriculture, 45,000 t in professional horticulture, and 15,000 t in amateur horticulture.

United States: State composting initiatives for both yard waste and crude municipal wastes are too numerous to list here, but composting in general is a technology which has been received with considerable enthusiasm throughout the U.S. Many states and municipalities have legislation or voluntary programs for the recycling and use of organic materials.

Appendix B shows the current French standards for composts by way of an example.

The above discussion indicates the important role that composting is beginning to play in the development of organic waste treatment systems, and there are a number of influences shaping the future of composting.

Co-mingled waste composting has been extensively undertaken particularly in the Southern European countries over many years. The quality of separation engineering in these plants was not high, and the resultant compost was of a poor quality. However, an increase in environmental awareness has given rise to concern about pollution when such composts are applied to soils, promoting a backlash against mixed-waste composting.

Meanwhile, there has also been a significant rise in popular pressure for the recycling of waste materials. As organic waste is a substantial component in the waste streams, there is a vociferous lobby demanding the utilization of municipal waste compost, as a natural process for recycling those materials. Those potentially conflicting pressures (to recycle but not to pollute) are being reconciled through the introduction of segregation at source of the organic waste stream. There are cost and logistical implications to these strategies, so governments are using a variety of legislative instruments to encourage them. Some are planning to make source separation a legal requirement, while for others it is a stated "best practice," and the finance to support the strategy is being provided. Whatever national strategies are evolved, there are some statistics which point strongly to the long-term pressure to consider municipal waste compost as an essential resource. The World Resources Institute, in Washington D.C., suggests that more than 3 billion acres of land worldwide have been seriously degraded in the past 50 years, and that over 20 million acres are no longer able to grow crops. Long term projections indicate that at the present rate of soil erosion, caused mainly by the failure to maintain humus levels, the plains states of the U.S. will contain no topsoil at all in 60 years time.

The U.S. produces approximately 192.5 million tons of municipal waste a year, of which some 3.9 million tons are composted. At average waste analyses, and after accounting for process losses, the potential however is for 33.7 million tons of compost. At an application rate of 45 tons per acre per year therefore, some 750,000 acres of land could be provided with and ongoing supply of essential humus. In terms of continental land mass 750,000 acres is small, leading to the conclusion that the potential market is huge and largely insatiable, provided that a sufficiently high quality is achieved. Spain and Portugal, for example, compost 17% and 15% of their municipal waste. They still find ready markets, in spite of a higher population density and an agricultural land bank which is smaller as a consequence.

11.2.3 Compost Standards

For some time it has been recognized that compost standards are important. In an attempt to discourage poor-quality composts that have a pollution potential and to encourage confidence in the "quality" composts, many countries have established standards. Unfortunately, there is little or no uniformity or consistency between them. The result is a somewhat confused picture. In Europe, 8 out of 12 countries have legally defined compost standards, and within them there can be as many as four grades and as few as one.

Some standards define analytical and quality control requirements and others do not. There are as many sets of parameters and values as there are compost grades; for example, there are 15 different sets of heavy metal limits. Some of the controls apply to marketing labels, while others are for environmental protection. This unsatisfactory position is further complicated by the recent involvements by international organizations. For instance, the European Committee for Standardization (CEN) has recently established a technical committee to standardize soil conditioners and growing media. CEN has the power to require member states to adopt as its national standard whatever European CEN standard is finally agreed. Composting is also included in other CEN committees such as packaging, eco-labeling, and classification of waste.

It appears probable that in its attempt to bring waste transport under control, the EC may well also include some definitions and standards for composting and compost in order to be able to define when waste has become a compost product. One reason for such an initiative is the need to provide acceptable standards for trade in the free market. The conclusions to be drawn from this analysis are that the recycling of organic waste using composting will be widely adopted throughout all regions of Europe, with the process being prescribed in legislation in many countries.

To ensure environmental protection, compost standards are being compiled, and steps have already been taken to move toward European standardization in this field. To support this, the development of segregated waste as a policy is growing, and is being fostered through national and regional legislative initiatives. The impact of both quality standards and the source segregation issue is difficult to predict, and many countries are not likely to lead but will reluctantly follow either by being shamed (as with recycling targets) or forced to do so (as with sewage sludge dumping at sea or on agricultural land).

11.2.4 Compost Utilization

Historically, compost utilization, involving product specification, performance, and marketing, has been given a lower priority than waste processing and disposal issues. To some extent that is understandable, as it is difficult to address the question of utilization until there is a product to sell. However, unless a balance is struck between the pressure for waste recycling and the demand from the market that will use the compost, then composting operations will become constrained by surplus production. This situation is already occurring in much of Europe.

There is a significant strategic difference between the issues that influence waste disposal and compost utilization. The control and regulation of waste disposal and the initiatives for waste recycling are influenced by governments. By contrast, governments at present have almost no influence on the utilization of compost. The private sector, as represented by farmers, horticulturalists, and landscape architects, is the prime influence on decisions affecting that aspect. They are not subject to the same political and popular pressures as government organizations, and they base their compost utilization decisions on commercial criteria. Consequently, the markets for composted materials can be considered to fall into three categories based upon the use to which the compost is to be put, namely

- Soil improvers/conditioners
- Growing media component
- Organic fertilizers.

These markets can then be further divided depending upon the users. For example, soil conditioners are purchased by domestic gardeners, grounds contractors maintaining municipal flower beds, and landscape contractors reclaiming mining tips. The objective is identical in the improvement of the soil through the addition of organic matter, but the scale and nature of the task determines the materials used and manner in which sales and distribution are undertaken. The main user groups interested in the purchase of composted materials are

- Landscape contractors responsible for the reclamation of derelict land
- Landscape and grounds maintenance organizations responsible for amenity and public areas
- Professional growers of plants in containers
- The retail sales outlets that sell to the home gardener and the peat/compost blenders who produce for the retail trade
- "Organic" farmers who use compost as a source of plant nutrients.

Major changes are, however, occurring in the markets for composted materials and the horticultural peat market which has long been the hurdle over which organic composts must climb.

11.2.4.1 The Peat Market

The growing material that has dominated the markets in recent times has been peat. Some years ago, a European campaign was launched by a number of conservation organizations to lobby the market to reduce its peat usage. There were a number of justifications put forward for such a view, but the principal motive appears to have been the loss of lowland peat habitats. The conservationists have argued that for many horticultural applications, other materials can be used instead of the nonrenewable peat products. While there is still no single material that can perform in the way that peat does, nevertheless there has been a dramatic shift in attitude away from unquestioned use of peat. One example of this change of attitude has been the proliferation of "peat-free" municipalities. New products have been introduced based on other materials to satisfy the perceived public appetite for an alternative, and the single biggest share of that alternative market has been secured by coir. This is a product derived from the outer husk of coconuts and is the waste "dust" that would otherwise not be used. It is available from the major coconut-producing areas, including Sri Lanka, Phillipines, and India. This material is, however, comparatively expensive to import, and there are growing lobbies that argue that it would be more environmentally friendly to reprocess national waste materials rather than move materials halfway around the world.

A number of organizations have commissioned research into the alternatives to peat, and they probably form the most accurate assessment of the potential for many (but not all) of the composted materials markets. Peat usage in professional and amateur horticulture is a development that has occurred in the last 40 years, and the markets in which organic materials may compete are

- Retail horticulture
- Amenity/landscape
- Professional horticulture
- Vegetables propagation and transplanting
- Bulb forcing

- Mushroom casing
- Hardy ornamental nursery stock containers
- Pot plants
- Bedding plants
- Glasshouse soils

The quantities demanded by these markets are not usually sufficiently large to absorb all the potentially available compostable materials. Neither are they so small that they fail to represent substantial business opportunities for the still small and embryonic compost business. Which of these particular market sectors ought to be the target of a marketing strategy depends on local consumption of peat and other products, and on the growers and retail users in the immediate area of the composting operation.

In order to attempt to develop a marketing strategy, a limited market survey can be conducted quite simply. From the trade sections of local telephone directories, a selection of landscape architects, contractors, and nurseries can be contacted and questioned about their current compost usage and their attitude to alternatives. The results of the survey indicate fairly quickly what the local consumption of peat alternatives would be, and therefore whether additional survey work would be justified. The products that should be discussed are mulches, soil conditioners, tree planting composts, top dressings, and growing media.

Landscape architects do not themselves buy composts. They do, however, have a strong influence on the materials that the contractors buy through the writing of the landscaping job specifications. In most cases, architects express interest and will specify if they can be convinced that there is an assured supply. They are often receptive to new materials, but they also require both a well-documented product and a known supplier before they can suggest to the contractors that the product should be considered. Despite the encouraging values often placed upon the compost materials used in the industry, these types of consumers often do not utilize sufficient material to provide the steady substantial demand needed by a recycling facility.

Topsoils are widely used in landscaping, but they are often variable in quality and poorly specified. Some organic topsoil ameliorants are also used in landscaping, but again they are poorly specified and availability is erratic. Leading soil scientists have advocated specified organic topsoil ameliorants, as they can be a more consistent and cost-effective method for land restoration in landscaping projects than topsoil. One example of the use of an alternative waste material is the combining of sewage sludge and subsoil materials, marketed as an alternative to topsoil. Sales of products of this type are usually substantial, indicating that an entirely organic compost made from municipal waste might achieve some penetration. The value of composted material in these applications is low, and less than $16 per ton (£10.00) is usual, but the volume consumed is substantial. The difficulty in supplying to this type of market is that it is mainly operated on a spot basis, where materials are sold project by project, as the requirement arises. The price that the material fetches, whether it is soil or organic material, is very strongly influenced by the level of other landscaping activity and building works in the vicinity.

Farmland is in many ways the natural ecological home for the recycled organic matter. The difficulty in selling to the farmers is that the unit cost for plant nutrients, when supplied by chemical fertilizers, is very low in comparison with organic materials. There are known advantages to the application of the organic matter to the soil, but the farmer is paid by crop yield, not soil condition.

The traditional view of composts and agriculture may change in the future as a result of other pressures. Overproduction is causing changes in the way farmers operate, and in Europe set-aside is perhaps one of the most obvious examples. There, farmers are paid from a central fund to take land out of productive use. In this case it does not have any great significance to compost markets. However, "organic" farming does not permit the use of chemical fertilizers. This means that some organic farmers are already pleased to take organic wastes as a supply of plant nutrients. This in itself is of limited importance, as the number of organic farmers is generally small.

However, the numbers of organic farmers may grow, as in some countries it has been suggested that they may be eligible for support payments within a revised Common Agricultural Policy. Organic methods are now being seen as a way to cut surpluses and benefit the environment. The new trend of linking agriculture and the environment could dramatically change the opportunities for agricultural utilization of composted organic materials.

11.3 REFUSE DERIVED FUELS

As discussed earlier in this book, the refuse-derived fuel industry is encumbered with a history of low-quality or even positively harmful materials that have all been marketed optimistically under the name of RDF. In many cases, they were simply crude, comingled municipal waste from which some but not all of the metals had been removed. Burning them was mass incineration foisted upon other industries, and in the U.S. and Europe legislators reacted against the pollution that resulted. Laws were introduced to control waste incineration, and RDF was included in them. In the eyes of the European Commission, the combustion of RDF was waste burning under another name.

Potential customers also reacted against the product as a result of its history. There were many stories of severe boiler fouling or even damage, of clogged grates and burned-out fire bars, of superheater tubes that failed after 6 weeks of using "RDF." Production losses which resulted from those experiences were said to be catastrophic, although it was difficult to locate the user whose experience was being reported.

In Great Britain and in parts of Europe, the RDF industry responded by developing its technology to create products that were reliable, consistent, and nonpolluting insofar as any fuel can be. Gradually, as a result of these initiatives and the experiences of a small number of large-scale users, high-quality RDF is once more gaining acceptance.

The process designs capable of producing such a quality are limited in number, and they are necessarily virtually identical. There are only a certain number of ways in which the combustible fraction of waste can be sufficiently refined, and the principal among them are those described in Section 9.3 of this book. They are capable of producing flock, crumb, and densified RDF (fRDF, cRDF, and dRDF), with the lowest quality existing as fRDF and the highest quality as dRDF.

Crumb RDF has little application, since it is in a sense a halfway stage between two dissimilar products. It is too dense to be co-fired with coal dust in pulverized fuel plants, and yet it is insufficiently so to be burned in solid fuel systems. Therefore, in considering markets for RDF products, only fRDF and dRDF offer potential. In addition, and in view of atmospheric pollution controls now existing in most developed countries, there is little point in examining markets for low-grade products of the historical variety. It can be assumed with some confidence that where they may be used now, they will be banned in due course. Therefore, the fuels considered in the following text are those manufactured by modern processes of the designs described.

11.3.1 Flock RDF

The markets for fRDF are generally limited to major power-generating utilities and large process works that currently burn pulverized fuels (PF), and the very low density of the material restricts viable sales to local outlets only. The boilers upon which pulverized fuels are used are invariably water-tubes, operating at high pressures and superheat, and so fRDF cannot be used in isolation. Its energy density is too low for the required firing rates to be maintained.

One alternative application for RDF in general and fRDF in particular is as a supplementary fuel for sewage sludge incineration. Sewage sludges are a Western hemisphere problem, and are slowly becoming one for developing countries as well. Historically they were disposed of either by discharge into the sea or by injection into agricultural land, but the first of those options is increasingly being prevented and the latter is being overwhelmed. Incineration remains the only viable alternative within the constraints of existing technology, but the sludges require a supply of energy before they will burn — they are not self-sustaining combustible materials. When mixed with RDF, the additional calorific value is sufficient for them to then burn effectively, and that makes possible the use of several incineration designs not normally associated with sludges.

In most conventional boiler plant, which remains the principal market for fRDF, it is customary to blow the fuel into the furnace through a separate burner or burners. They are usually mounted between the higher of the PF burners where they can be ignited and kept so by the high-energy source fuel. Delivery to the pneumatic system of the burners involves an element of reshredding to break up any lumps that may have formed during handling and storage, although generally this is nothing more than a live bottom in the storage or reception hopper. An excellent example of this technology exists at the Madison plant in Wisconsin.

Sales are almost always based upon calorific value. The customer pays according to the energy he has received, rather than by weight or volume, and it is his analysis of the product upon which the supply contract is usually agreed. It is normal for the supplier to be paid a price per useful therm or megajoule, which is calculated from the tonnage supplied and the calorific value of spot samples of that load. From the suppliers point of view, that introduces some variation into his cash flow, in theory at least, since he never knows exactly how much he will be paid until the customer declares the calorific value he has received. In fact, modern fRDF processes create a very consistent fuel, and the uncertainties of calorific value determinations make it possible to negotiate payments that remain static provided that the product is within an agreed range.

11.3.2 Densified RDF

Densified RDF is a far more flexible fuel than flock, and accordingly it has a much wider market appeal. It is capable of operating successfully in combustion appliances ranging from small shell boilers to very large water-tubes, and it can be burned in most grates and fuel handling systems designed for conventional solid fuels. Before any of those markets can be approached, however, a detailed understanding of the combustion of the product is necessary. The customer is not usually concerned primarily with fuel, but is instead fully occupied with his production facilities. He does not consider himself to a be a combustion expert, and generally does not wish to become one. He expects that expertise to be offered by the fuel provider, and if the advice he is given is wrong, he reserves the right to sue.

The following sections contain sufficient details of the technology of dRDF combustion to evade most of the pitfalls.

11.3.2.1 Physical Characteristics of dRDF

Energy Density — The concept of energy density has as much impact upon combustion as almost all of the other factors combined, since it is largely that which determines whether, from the customer's point of view, it is worth using dRDF at all. Calorific values of fuels are always quoted in energy units per unit of weight, for example as kilojoules per kilogram, British thermal units per pound, or therms per ton. But the stokers and fuel handling systems used upon combustion plant are essentially volumetric devices. They are designed to deliver fuel of a particular bulk density in sufficient volumes to maintain the selected firing rate. The analysis that is important is therefore the calorific value per unit cube of the product.

The concept of energy density permits a more revealing comparison between fuels, and it suggests how a combustion system will react to varying types. For example, a coal may have a bulk density of 87 lb/ft^3 (1,400 kg/m^3) and a gross calorific value of 11200 Btu/lb (26,000 kJ/kg). Therefore, every cubic foot of coal delivered into a furnace will supply 87 × 11200, or 974,400 Btu (36,400 mJ per m^3) of energy. Meanwhile, dRDF with a density of 37.4 lb/ft^3 (600 kg/m^3) and a gross calorific value of 8600 Btu/lb (20,000 kJ/kg) provides 321,640 Btu in every cubic foot (12,000 mJ/m^3). The comparison of calorific values alone suggests that the two fuels are not vastly dissimilar, but that of the energy densities magnifies the very considerable difference that exists in reality.

In fact, the energy equation is somewhat more complex than the simple calculations suggest, since the two fuels are quite dissimilar in a number of other respects. The result of the other dissimilarities has the effect of reducing the *apparent* differences in combustion characteristic and energy yield. However, the estimation of the energy density does show that to achieve the same boiler output from dRDF as from the coal, the fuel handling system will need to deliver dRDF at a rate 1.18 times greater than that of coal in terms of weight, but 2.77 times greater in terms of volume, and that may simply not be possible. If the fuel handling system and grate has been designed for high-calorific-value coal, then it is unlikely that it has the speed capacity to do so. The best that can be hoped for in such a situation is that the dRDF could be sold as a supplementary fuel, not an alternative.

Ash Content — Fuel ash content is also a matter of concern to boiler operators, since ash handling systems have capacity limitations as well. Even if the ash content of RDF as a percentage by weight were the same as coal, the ash handling system would still have to deal with 1.18 times more of it when the plant burned dRDF. Again, the higher the calorific value (CV) the plant is designed to consume, the worse the situation becomes. A low CV dRDF may produce over 1.5 times as much ash as a high CV coal.

Moisture Content — The moisture content plays a part in limiting the amount of energy that can be delivered by a combustion system. When the fuel burns, the combustion process both produces water and vaporizes that which is already contained within the mass as free water. The conversion of water into steam absorbs a considerable amount of heat energy, and so the higher the moisture content of the fuel, the more of the gross calorific value will be used up in eliminating it. Some of that "lost" energy can in theory be recovered, if the steam is allowed to condense somewhere in the heat-exchanger, giving up the heat it has absorbed in the process. Unfortunately, the exhaust combustion gases also contain sulfur and chlorine compounds that become strong acids in the presence of water vapor. Sulfur compounds become sulfurous and then sulfuric acid, while chlorides become hydrochloric acid. Both are likely to condense at temperatures above that of the dew point of water vapor, and if they are allowed to do so upon heat-exchanger surfaces, severe corrosion is inevitable. In view of this, the energy balance holds that the higher the moisture content of the fuel, the lower the calorific value delivered.

In this case, the balance does not necessarily work to the disadvantage of dRDF. Where it is manufactured to a final moisture content of 8%, for example, its residual temperature will cause it to continue to liberate moisture in storage and transport. Within 1 or 2 days, depending upon ambient humidity, its moisture content will drop to about 4% by weight, and that compares very favorably with coal.

Chlorine and Sulfur Content — A further and extremely important measure of the suitability of a fuel is its chlorine content, since much of the chlorine will be converted to hydrochloric acid in the combustion process. Chlorine levels analyses at over 1.5% by weight have a severe effect upon the load capacity of the boiler, as they deliver very concentrated acid in the exhaust gases. The gas temperatures must be maintained artificially high to avoid the possibility of condensation occurring, and so more useful heat escapes. A similar condition applies in respect to sulfur compounds, which decompose to form sulfuric acid in quite strong concentrations.

The temperature at which condensation of either acid occurs, the dew point, is determined by a number of factors, not just the temperature at which, for example, laboratory-quality hydrochloric acid would condense. Generally the dew point appears at first sight to be low, and therefore unlikely to have any effect, since exhaust temperatures cannot be reduced easily to that extent. Temperatures down to about 140°F (60°C) are possible. However, it is important to realize that the dew point is not a characteristic of the exhaust gases, but rather of the metal surface temperatures. An economizer tube in a heat exchanger may be located in a gas stream of 390°F (200°C), but if it is handling feed water at 68°F (20°C), then its surface temperature may easily be close to the dew point of the acid vapors.

Acid emissions are again characteristics in which dRDF offers some advantages over coal. Its chlorine content, following modern manufacturing methods, is usually in the region of 0.2 to 0.3%, while sulfur levels are less than one tenth those of coal. The total acid emissions from dRDF are therefore substantially lower in spite of the higher delivery rates needed for boiler load maintenance. One major power generator in Great Britain, who consumes many thousands of tons of dRDF every year, reports that the acid emissions from his boiler plant reduce as the proportion of dRDF in the fuel mix increases.

Early Laboratory Research — For many years dRDF has suffered from a tarnished reputation as a result of successive programs of research and analysis by independent laboratories, and much of the damage has been caused by those laboratories' refusal to recognize that the industry was continually developing new technology. There were many instances where scientific reports were published, and supported by claims of authority, but were based upon analyses of fuels that had long been superseded by others of much higher quality. Irresponsibly those reports claimed that all dRDF must necessarily produce more environmental pollution than other fuels, simply because those which had been analyzed appeared to do so.

It is unfortunate that in spite of the pressure applied by the industry, certainly in Europe, for a fair hearing, pollution controls in respect to RDF in general were formulated upon the basis of the "independent" reports. Emissions abatement regulations were enacted, requiring monitoring and control far in excess of any expected for conventional fuels, in an attempt to regulate combustion of a product that no longer existed. It was regrettably a case of legislators accepting the words of those whose only knowledge was limited to some laboratory work, against the words of those who had developed the technology and who operated the processes. The experiences of some of the major users of the products will, one would hope, eventually be accepted as a salutary lesson in respect to the consequences of ill-founded environmental law.

Meanwhile, a process plant operator attempting to sell dRDF is likely to face the same questions at every sales appointment. The prospective buyer "has been told" that the fuel erodes and corrodes boiler tubes, produces very strong acids that destroy gas ducts and scrubbers, is highly malodorous and possibly infectious, and has a calorific value so low that it is of limited use even as a fuel supplement.

Table 11.2 displays a typical modern dRDF analysis from a process of the design described in this book. It is the only means by which the allegations can be refuted, and it is an indispensable marketing tool. For those reasons, marketing initiatives that have not been supported by a detailed laboratory analysis, *conducted by a laboratory experienced in fuels science*, are likely to be less than productive.

11.3.2.2 Combustion Systems

Pelletized dRDF can be used, with some constraints, upon most combustion equipment designed for solid fuel. Cubed dRDF is somewhat more specific in its application, and is suitable for use only upon equipment intended for or tolerant of large lump sizes. The more common fuel handling systems that have been used successfully with pelletized dRDF are described below.

Table 11.2 Typical Densified Refuse-Derived Fuel Analysis

Physical:	
Calorific value	8600 Btu/lb (20,000 kJ/kg)
Total sulfur	0.16% w/w
Total chlorine	0.2% w/w
Moisture content	9% w/w
Ash content	9%
Bulk density	37.44 lb/ft^3 (600 kg/m^3)
Ash fusion temperature	2100°F to 2370°F (1150 to 1300°C)
Chemical:[a]	
Si	1.5
P	—[b]
S	0.14
Cl	0.099
K	0.299
Ca	1.7
Ti	0.18
Cr	0.02
Mn	0.01
Fe	0.22
Ni	0.002
Cu	0.007
Zn	0.04
Ga	0.0005
As	—[b]
Se	0.0003
Br	0.0023
Rb	0.0018
Sr	0.0043
Y	—[b]
Zr	0.0023
Nb	—[b]
Mo	—[b]
Sn	—[b]
Ba	0.0112
Pb	0.0048

[a] Method: proton-induced X-ray emission (PIXE). Concentrations in % w/w.
[b] Not detected.

Retort ("Underfeed") Stokers — A typical retort stoker design is shown in Figure 11.1. In such a device, fuel is delivered from a hopper, along a screw tube, to a retort suspended in a division plate in a boiler furnace. The retort is assembled from hollow steel castings clamped together to form a "basket," which usually has two rows of air holes on each side and end of its inner surface. Further air ports are provided around the top of each casting, permitting combustion air to be delivered to fuel that spills over onto the furnace division plate.

Often ignition is achieved by the use of electrical hot air igniters, which deliver high-temperature air directly into one or more of the retort castings. In those cases, the igniter is controlled by a temperature sensor in the boiler

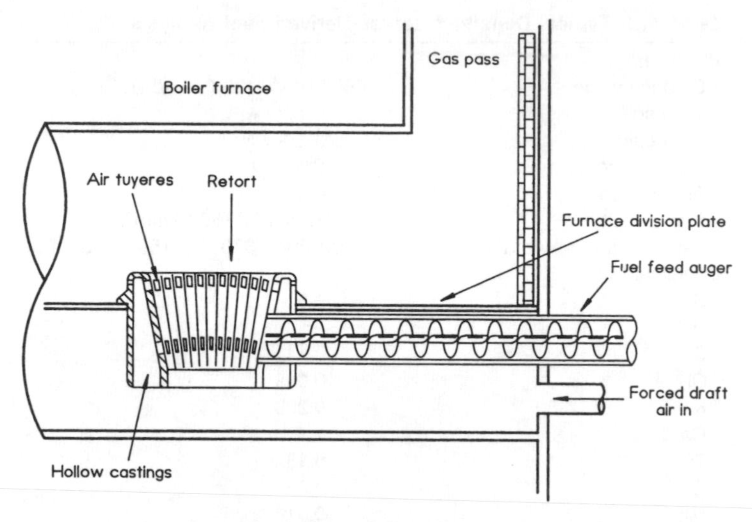

Figure 11.1 Retort Stoker.

exhaust gas duct, and will operate whenever the boiler is selected if the temperature detected is below a set point. It is also customary for the fuel feed to run only for a predetermined period when the temperature sensor registers low temperature, and not to begin continuous feed until the preset level is reached. Controls are adjusted such that sufficient fuel is delivered to the retort to permit ignition, but not to deliver more until combustion in the retort is well established. By those means, the boiler can be made automatic in the sense of entering or leaving service, and can if necessary be called upon by a time clock.

When combustion is fully established, air is delivered under pressure from a fan into the space beneath the division plate. From there it passes into the hollow castings and out through the air ports, or tuyeres (pronounced "twe-ers"). At that point, the fuel feed auger is started automatically and it begins to deliver more fuel to the retort. The rate at which it does so is controlled by the load demand upon the boiler, and it is usually intermittent. Depending upon boiler steam pressure, or hot water temperature in the case of a space heating plant, the stoker is selected to operate for a predetermined number of seconds every min.

Usually there is a limited range over which its delivery can be controlled, and so it may be able to deliver over 10, 20, 40, or 60 seconds per min. Simple electrical switching equipment is generally used for the purpose, with primary control being through a timer relay initiated by a temperature or pressure switch in the boiler riser. Low pressure or temperature initiates the relay, which in turn energizes the first of a cascade of on–off timers. The first of these starts the fuel feed auger and runs it for 10 seconds. If, at the end of that time, the primary relay is still energized by low temperature or pressure, the second timer in the cascade operates, running the stoker for a further 10 seconds, and so on.

In many installations there is also an override circuit that requires the stoker to always run at, perhaps, 10 seconds per min irrespective of the temperature or pressure in the riser. The purpose of this is to prevent the boiler controls from continually cycling between fuel feed rates.

dRDF pellets burn well upon stokers of this design, but there are some problem areas that have to be addressed. First, as discussed above, dRDF has a lower energy density than coal, and retort stokers are designed for the latter. They are usually specified in terms of their heat output capacity when burning "standard" industrial coal, with the result that any given boiler output will require a larger stoker for dRDF. For example, a boiler rated at 12 million Btu/h may need a 15 million Btu stoker rather than a 12 million Btu coal-rated unit.

Second, retort stokers are not self de-ashing, although there is a modification that can make them so. dRDF contains and delivers more ash than coal, and so manual de-ashing will have to take place more frequently. dRDF ash from retort stokers is invariably a mixture of clinker and very fine, light dust, and prevention of the latter escaping into the gas passes requires that the boiler be shut down while de-ashing is taking place.

Third, dRDF has a much higher volatile matter content than coal, and so its need for combustion air is greater above the grate than below it. Secondary air is more important than primary, yet conventional retort stokers have very little provision for the latter. Generally, some secondary air facility can be obtained by the installation of pipes in the furnace, or by slots along the edges of the division plate. The best effect, however, is achieved by means of secondary air nozzles that direct an air flow at high velocity into the area immediately above the retort. There the combustible volatile gasses are at their most turbulent, and are at their highest temperature, and so mixing with oxygen there is likely to be the most efficient method.

Finally, all retort stokers are designed with a fuel feed auger length established by the inclination of the fuel to burn back along it. When a stoker is stationary but combustion is well established, the heat from the retort will cause the fuel in the auger tube to ignite and smolder slowly. There is insufficient oxygen for vigorous combustion there, but nonetheless the fire will progress slowly along the tube toward the hopper. Provided that the tube is of sufficient length, the absence of sufficient oxygen will prevent further fire spread at some point, and that point must clearly occur before the feed hopper. dRDF, however, is capable of smoldering and liberating it volatile matter at much lower temperatures than is coal. In doing so, it is likely to draw external air in through the feed hopper and to thereby maintain an oxygen supply. It is therefore capable of burn-back over much greater distances.

During normal operation, that does not cause any problems, since the periods during which the stoker is stopped are too short for burn-back to have any significant effect. Overnight or longer shutdowns are, however, another matter, since then there is ample time for any smoldering in the screw tube to reach the hopper.

There are two potential solutions to the problem, and the first is perhaps the obvious one of simply making the tube longer in order to accommodate the more combustible nature of dRDF. In some circumstances, however, that is impossible. There may be insufficient space in the boiler room. In that case, there is one retort stoker design that has a two-stage fuel feed auger, designed in such a way that the extreme end nearest the retort always runs for longer than the main auger. By that means it creates a void space over which the combustion cannot pass, between the retort and the fuel in the feed tube.

Retort stokers are usually incapable of burning cubed dRDF, since the particles of fuel are too large to be delivered by the fuel feed auger.

Chain Grate Stokers — Chain grates of the general principles shown in Figure 11.2 are probably to most widespread of all of the solid fuel handling systems available. They have been in service for generations in combustion plants ranging from very small units of less than 12 million Btu/h capacity to large boiler installations of one hundred times that size. The principle of the design is that interlocking fire-cast fire bars are fixed to chains supported between sprockets inside the boiler furnace. The chains are driven slowly by external electric motors with variable speed transmissions, and fuel is discharged onto the top strand at the furnace front. As the grate transports the fuel along the furnace, combustion is supported by air drawn through the open furnace front by an induced-draft fan in the boiler gas exhaust. At the end of the grate travel the residual ash is deposited into an ash pit, which is usually served by a drag link conveyor immersed in a quenching water bath.

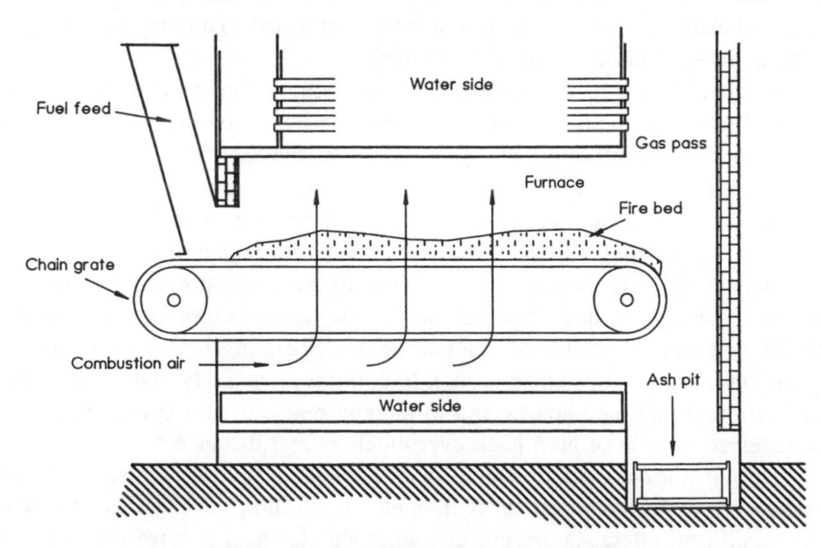

Figure 11.2 Chain grate furnace boiler.

Stokers of this type occupy the full width of the combustion chamber furnace, and where the fuel delivery to them is by means of a simple drop chute from a high-level bunker or conveyor, fuel particle size is less relevant. Chain grates are therefore usually capable of handling cubed as easily as pelletized dRDF.

Again, there are potential problems created by the unconventional nature of dRDF, and those most frequently encountered are fire bar slagging, fly-ash carry over into gas passes, and incomplete combustion of the fuel.

Fire bar slagging is a consequence of the lower ash fusion temperature of dRDF, and the need for secondary rather than primary combustion air. When a chain grate is handling coal, the air requirement is exactly reversed. Coal needs primary at the expense of secondary air, and the primary air passing between them prevents the fire bars from becoming excessively hot. Primary air also passes through the ash, from which it removes a proportion of the heat retained within it.

Where dRDF is being handled, it is possible for the limited control of the induced-draft fan to reduce the primary air supply to a point below that at which fire bar cooling is effective. The fire bars run at a higher temperature, and so does the ash which, when originating from dRDF, begins to fuse at temperatures as low as 2100°F (1150°C). Since the temperature reached in the heart of the combustion zone is likely to be at least that high, then if the fire bars are allowed to become sufficiently heated, the fused ash will begin to adhere to them. When it does so, the immediate result is a further reduction in primary air and a corresponding increase in secondary air. The situation may then worsen rapidly.

The solution to the problem is to ensure that the provision of secondary air is independent of the induced-draft fan, and is supplied by a separate forced draft unit. The control of the primary air is thereby separated from that of the secondary air, making it possible to ensure that there is always sufficient of the former to maintain the fire bars at an acceptable temperature. Some chain grate units already have that facility, but by no means all do.

Fly ash carryover into gas passes is usually a symptom of fire bar slagging. Where dRDF ash clinkers form and attach themselves to the bars, they remain highly porous. Fissures within the hot, semimolten mass become filled with very light ash. Primary air passing through the grate may then only escape into the furnace through the fissures, and it does so with a considerably increased velocity. That velocity is sufficient to suspend the light ash in the gas flow and to keep it in suspension into the gas passes. Once again, the solution is to prevent the slagging from occurring in the first place, since once it develops, prevention of ash carryover is impossible.

Incomplete combustion is a function of the density to which the dRDF is compressed and the speed of the chain grate itself. The fuel can be made to densities considerably in excess of those to which it is normally manufactured, but above 37.5 lb/ft^3 (600 kg/m^3) the results of doing so become counterproductive. Since it is a high-volatiles fuel, dRDF must first break apart to release the volatile matter before the residual carbon can burn. In a sense, it burns from the inside outward, whereas coal does exactly the reverse. If it is too dense, then the time taken to break apart increases, and may be too long for disintegration to be complete before it reaches the end of the chain grate. Carbon and even partly intact fuel pellets can be discharged from the ash pit.

The most reliable indication of such an occurrence can be gained from the color of the ash. dRDF ash should be light gray in color, perhaps with small, brittle clinkers that are visibly porous. Any black flecks in the ash indicate unburned carbon, suggesting that either the density of the fuel is too high for the combustion plant, or that the chain grate is running too quickly. In the latter event, the cause may be that the operators or the combustion controls are simply trying to fire the plant using more dRDF than it is capable of accommodating. The boiler may be calling for energy in excess of that which the fuel is capable of supplying, with the result that the grate gradually speeds up to maximum.

When that occurs, the effects become compounded rapidly. The grate accelerates to deliver more energy, but more unburned carbon remains in the ash, taking much of the heat value with it. Less energy is delivered, so the grate accelerates further, and carbon losses again increase, and so on. It has been known for chain grate automatic controls to actually run a fire out of the furnace completely, taking the boiler off line, when such a situation developed.

Fluid Bed Furnaces — Fluid bed combustors (FBCs), sometimes referred to as "bubbling beds," are furnace designs where the base of the combustion chamber is constructed as a refractory-lined steel tank, which is filled with coarse-grained sand. Air pipes are fixed in the bottom, and are fitted with air jets or tuyeres which direct air up through the sand. At a particular air pressure, which depends upon the granularity of the bed and its depth, the sand fluidizes: each grain becomes separated from its companions by air, and the mass begins to behave as a liquid. The sand will flow down shallow slopes, and while low-density items will float on the surface, high-density items will sink. Figure 11.3 indicates the design principles of a simple fluid bed combustor.

If the sand bed is fluidized and then heated to high temperatures by a supplementary fuel source, then when a solid fuel is introduced into it combustion takes place in and above the bed. The effect of the violently agitated, very hot sand is to continually erode the fuel, exposing new surfaces to heat and oxygen, and breaking the ash down into fine dust. Most of the ash and some of the degraded sand is carried away with the exhaust gases, and is collected by grit arrestors at the end of the heat exchanger. With some fuels, a certain amount of ash hardens into nodules that resist the abrasive action of the sand, and so it becomes part of the bed material. It gradually replaces the sand, and may need to be removed occasionally or performance may suffer.

While fluid bed combustors were originally designed to burn very low-grade fuels, they are sensitive in some areas. Once combustion has been

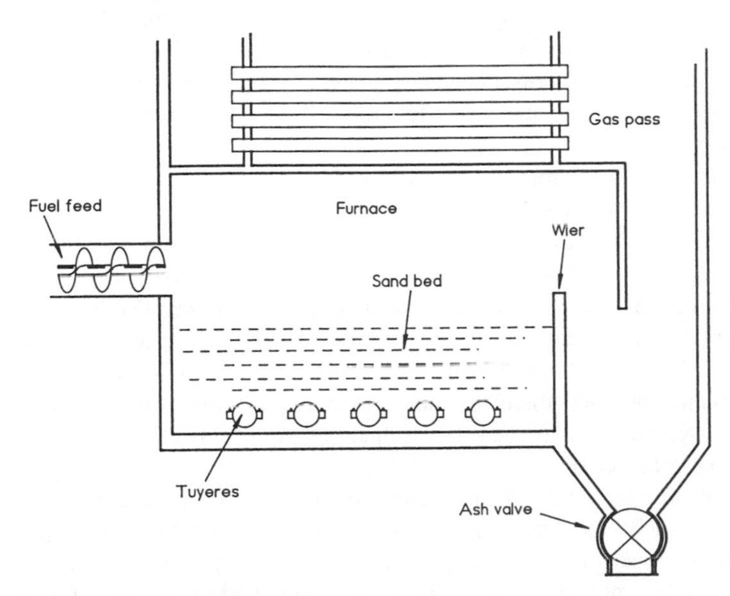

Figure 11.3 Fluid bed furnace.

established, the supplementary burners are shut off. From then on it is the combustion of the primary fuel that maintains the bed temperature. However, that is only possible if there is sufficient carbon in the fuel, since only the carbon is heavy enough to stay in the bed. Volatile matter is driven out by the fluidizing air, and it burns above the bed, supplying heat to it by radiation alone. Since radiant heat is always insufficient to maintain the bed temperature, combustion may decline and eventually cease if the volatiles content of the fuel is too high.

dRDF is a highly volatile fuel, and yet it is possible to burn it successfully in a fluid bed. To be able to do so, careful selection of the bed sand is important, since the more coarse it is, the higher the fluidizing air pressures necessary. Finely granular sands need less air, and so fluidization is less likely to drive all of the volatile matter high above the bed. The density of the dRDF pellets or cubes is important in this application also, but here is a case of seeking the maximum possible. The bed sand will abrade very dense pellets sufficiently to release the volatiles, but their weight will ensure that the carbon remains in the bed. Meanwhile the low fluidizing pressure allows the volatile gases to burn at or very close to the sand/air interface at the surface of the bed, where radiant heat can still contribute to the maintenance of the bed temperature.

Apart from the difficulties of maintaining combustion, the most frequently encountered problems with the combustion of early dRDF in fluid beds were deposits of aluminum upon back plates, and blinding of the tuyeres by wire. Before dRDF manufacturing technology was developed to its modern state, all RDF contained aluminum, mainly as scraps of foil. It also contained steel wire, which originated mainly from staples in the paper from which it was manufactured. When it was burned in a fluid bed boiler, the aluminum was vaporized and released into the combustion gases, to condense in "stalactites" of pure aluminum wherever the chamber was sufficiently cool. Such an area usually existed where the gases left the combustion chamber for the heat exchangers.

Wire from the staples, meanwhile, would sink to the bottom of the fluid bed, to accumulate around the tuyeres where the chilling of the air prevented its being burned away. Gradually, sufficient could build up there to prevent fluidizing of the sand in each local region. Eventually, fluidization of the whole bed would cease, and the plant would drop out of service. The first evidence of such a development was usually a slow but progressive increase in fluidization pressures without its being explained by a rise in bed level. Even so, the evidence was not conclusive, since an increase in fluidization pressure can equally well be caused by bed contamination by clinkers or shales, for example from low-grade coal.

Both problems have been eliminated to some extent by events. Modern dRDF is much less contaminated, and the arrival of eddy-current separation technology makes possible the production of a fuel that is substantially free of aluminum. Improved classification, shredding, and screening before densification have at least reduced the amount of wire that may be present. Simple fluid beds have been largely superseded by recirculating and fast recirculating fluid beds.

Recirculating/Fast Recirculating Fluid Bed Combustors — While recirculating fluid beds work, as shown in Figure 11.4, upon the same basic principles as a bubbling bed, namely the fluidization of very hot sand by an air blast, they differ in having no detectable sand–air interface. This condition is achieved by the use of a tall but narrow cross-section chamber, and a higher fluidizing air pressure, the two factors combining to ensure that the bed sand is entrained rather than just fluidized. To achieve this, the air flows are much faster than in a bubbling bed.

Bed sand is carried by the air (and combustion gas) flows out of the top of the combustion chamber rising shaft, and into the downcomer and grit arrestors, which return it to the bottom of the bed. When fuel is delivered into the combustion chamber, it too is entrained in the rising air–gas–sand stream, and so combustion takes place over a large part of the rising shaft volume.

Of the many advantages claimed for recirculating fluid bed combustors, the most significant is the presence of violently agitated hot sand throughout the volatiles combustion zone. The effect of this is to ensure complete mixing of volatile gases with combustion air, and equally to ensure that the bed sand remains at high temperatures without having to rely upon residual carbon to achieve them. Consequently, recirculating fluid beds are able to burn very low-grade and highly volatile fuels without difficulty. They are extremely suitable for dRDF.

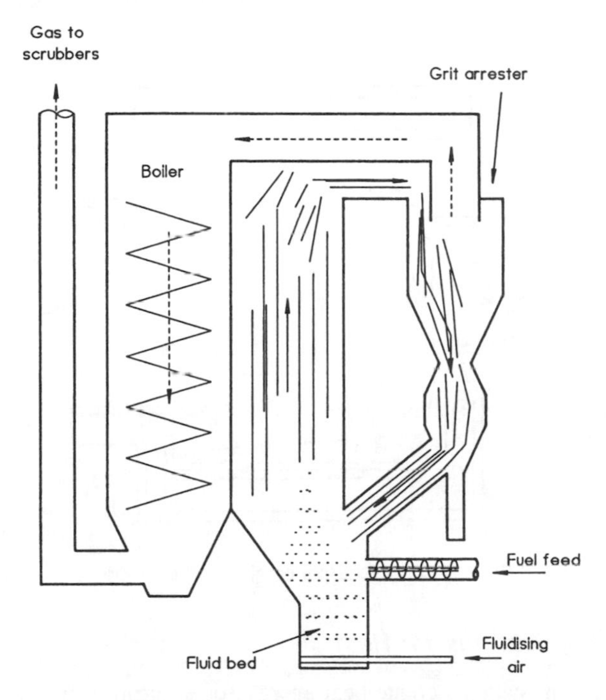

Figure 11.4 Recirculating fluid bed.

External Combustion Chambers — Although they did not achieve the wide market penetration their makers hoped for, external combustion chambers were widely advertised and researched in the late 1980s in Europe and Scandinavia. They were generally remarkably efficient, and their lack of success is surprising. Figure 11.5 shows a typical unit of the time.

The purpose of all of the models was to provide a facility for a boiler designed for oil or gas firing to benefit from solid bio-fuels. Often that would otherwise be impossible, since such a boiler would have no grate, no ash pit, and no facility for either. The external combustors were therefore intended to supply very hot combustion gas rather than fuel, and to inject it through the casing aperture normally reserved for an oil or gas burner. Although the casings of the units had to be water-cooled, that in itself was turned to advantage by being used to preheat the boiler feed water.

Gas temperatures from most of the devices were very high, usually exceeding 2370°F (1300°C) as a result of abundant combustion air and intimate combustion gas mixing in the chamber. They were designed to run sufficiently hot to fuse the ash, and so were forerunners of slagging furnaces. The ash was collected by a fixed, water-cooled scraper inside the rotating drum, and discharged usually from beneath the flame tube. Generally, external combustors were extremely efficient in burning dRDF, doing so with very low emissions.

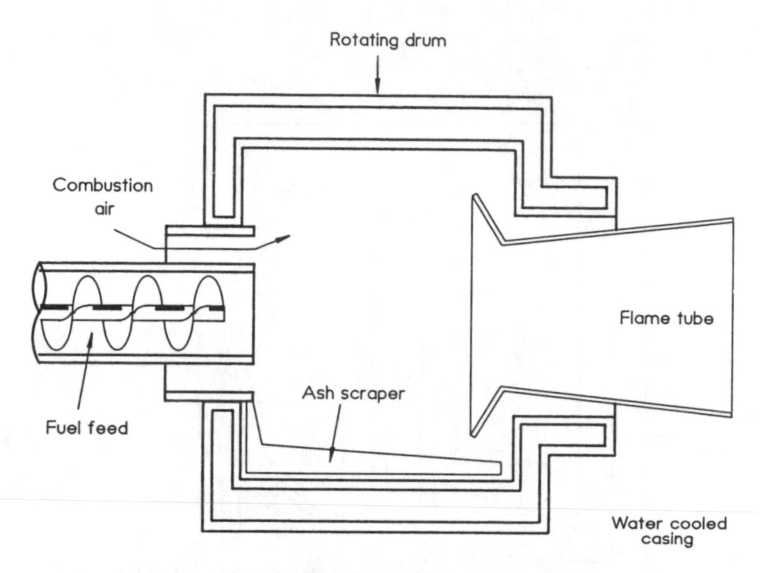

Figure 11.5 External combustion chamber.

11.3.2.3 Applications for Heat

Combustion systems create heat energy for a purpose. It may be to generate steam for industrial purposes or for power generation, to produce hot gas for a process, or to provide space heating and hot water services, and there are a limited number of ways in which it may be achieved. Steam raising uses water-tube or shell boilers, hot gas involves combustors, and hot water services are usually restricted to low-pressure shell boilers. dRDF has a reasonably successful history of use in all of those, but as with any fuel, it has peculiarities that have been known to cause trouble and must be accommodated. Some of the more common problems, and their solutions where they exist, are discussed in the following.

Shell Boilers — The general principles of a shell boiler are shown in Figure 11.6. An alternative name, and in many ways a more revealing one, used for this type of boiler is "fire-tube." The name arises from the design where combustion gases from the furnace are conducted through the fire tubes, which are immersed in the water contained within the boiler shell. The unit shown is a double-pass boiler, where the gases are reversed in direction when they reach the smoke box at the front end, and are then passed through the upper tube bank to the exhaust. Working pressures are limited by the structural strength of the shell, and so they rarely exceed about 150 psi. The figure omits any grate or combustion equipment inside the furnace in order to preserve clarity, but shell boilers are commonly fitted with oil or gas burners, chain grates, or retort stokers. Large units are occasionally fitted with ram stokers, which are simply hydraulically operated, rectangular cross-section platforms

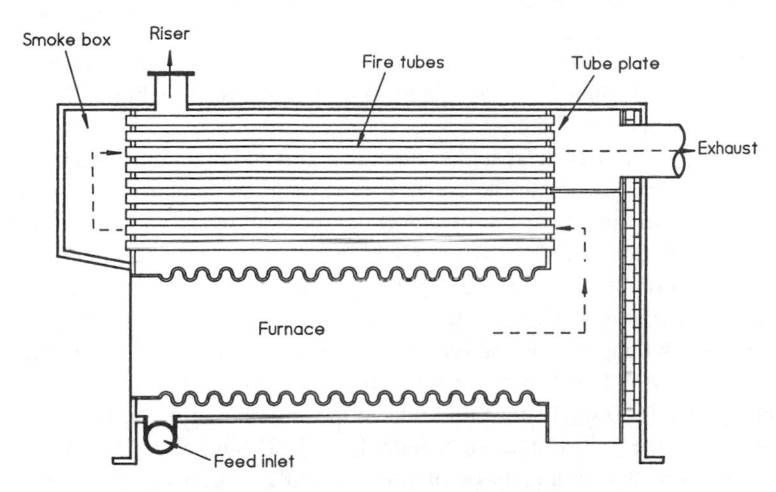

Figure 11.6 General principles of shell boiler.

in the front of the furnace. The platforms reciprocate slowly under a drop feed casing, and during each forward stroke they push fuel into the furnace, generally onto a fixed crate.

While even small shell boilers will consume dRDF with high efficiency, there are a number of considerations that have to be addressed. In Europe, the most prominent of those are the regulations controlling the combustion of RDF, contained within the EC New Plant Directive number 89/369/EEC and the Existing Plant Directive number 89/429/EEC.

The use of RDF in general is classified as a form of waste incineration, and the legislation controlling it is therefore necessarily extremely restrictive. The fact that dRDF particularly is a highly-refined product is completely ignored, with the irrationality that characterizes much of modern environmental protection regulation. Apart from the morass of controls over fuel handling, storage, and combustion monitoring, many of which are contradictory and sometimes technically impossible to meet, the most hostile to shell boiler applications is the requirement for combustion chamber gas residence times.

The legislation requires that the combustion gases created by RDF must be maintained at a temperature of 1560°F (850°C) for 2 seconds, with the intention of ensuring the destruction of dioxins and furans. That the requirement has little scientific justification or engineering reality is a point that appears to have entirely escaped the legislators, and it perhaps demonstrates clearly the consequences of permitting hysteria to overwhelm logical thought processes.

Dioxins (polychlorinated dibenzo-*p*-dioxin or PCDD) and furans (polychlorinated dibenzofurans or PCDF) are part of a very broad family of stable and long-lived chlorinated phenol compounds. Almost all of them are considered to be nontoxic, but there are a small number that may be. Toxicology evidence is as yet fairly minimal, and even massive doses of the more hazardous of the compounds do not appear to have had the damaging effect expected of them. However, those that are supposed to be toxic are capable of entering the human food chain, and as a result of their stability, they may gradually concentrate there.

A number of theoretical studies and programs of laboratory research have suggested several routes to the creation of dioxins in combustion processes. Among them are low-temperature chlorination reactions in which fly ash is a catalyst, catalytic reactions between HCL and oxygen create gaseous Cl_2, and a direct reaction between chlorides of copper and oxygen form chlorophenols without the need for a chlorine intermediary. However, irrespective of the route to their production, dioxins and furans are highly sensitive to temperature, and their formation is largely a low-temperature reaction. At temperatures above about 1200°F (650°C), they begin to decompose with increasing rapidity, and modeling by the U.S. Bureau of Standards in 1981 suggested that at normal dRDF combustion temperatures of about 2280°F (1250°C), destruction is 99.99% complete in 4 milliseconds.

The European regulations therefore require that the compounds be maintained at a temperature considerably lower than normal furnace conditions, for a length of time that is 500 times longer than the destruction times at the real furnace temperatures. They further require that those conditions are continuously monitored, although quite how the operator is to ensure and demonstrate that combustion gases stay in a furnace at 750°F (400°C) lower than their actual temperature, and that they do so for at least 2 seconds, is not made clear.

This appalling lack of logic is compounded by the fact that there has been clear scientific evidence for many years that dioxins can be formed in any zone where both the temperatures and chemical precursors are appropriate. The fact that they may be destroyed or never created in the furnace in the first place does not prevent their formation later, when the gas temperature has reduced to a suitable level; this is given somewhat less regard than it should have. The scientific research has shown that at temperatures below about 570°F (300°C), the formation of dioxins in kinetically restricted, while increases much above 570°F result in decomposition reactions. It would appear then that the critical temperature for dioxin and furan formation is 570°F (300°C), not 1560°F (850°C), and that exhaust gases should be cooled rapidly through that zone to avoid the establishment of the catalytic reactions with the fly-ash. In a shell boiler with a stack temperature of 520°F (270°C), it is reasonable to suppose that the critical 570°F will have occurred somewhere inside the final pass of fire tubes, where turbulence is pronounced and the heat loss is rapid. That condition *can* be monitored, simply by means of a thermocouple or thermometer in the gas exhaust chamber immediately after the last pass. It can also be controlled by ensuring that at that point the temperature is below 570°F.

Dioxins have been present upon the planet since long before the dawn of humanity, although in modern times their causes are prominently manufacturing processes such as steelmaking, copper smelting, motor vehicle emissions, and wood burning. Research in Canada has suggested that the most significant cause there is forest fires. It may appear somewhat strange, therefore, that RDF combustion is singled out for special treatment from all of those causes, by being included with mass incineration. This is particularly so when it has also been shown that incineration is not one of the major causes of dioxins in the first place. The legislation does, however, achieve the protection of the environment by ensuring that one of the least damaging waste disposal processes is discouraged. The "Best Environmental Option" is not a term that springs readily to the lips of politicians seeking a popularist vote.

From the point of view of the shell boiler operator who wishes to use dRDF, and the manufacturer who wishes to sell it to him, the legislation exists and has to be accommodated. Logic is not a part of the philosophy. Fortunately, something can be done in many cases.

One of the characteristics of dRDF that distinguishes it from fossil fuels is its flame speed and flame length. dRDF burns with a longer flame than does coal, and its high volatile matter content increases flame speed. The immediate effect of this, in a shell boiler designed for coal, is that more heat reaches the fire tubes and less is transferred through the furnace walls. Tube inlet ends run much hotter, and may reach temperatures sufficiently high to permit slagging to occur there. In addition to the potential for tube end blockages caused by slagging, there is an additional danger that the flame may actually impinge upon the exposed ends. Then, the reducing atmosphere will cause severe metal erosion.

Nothing can be done to reduce the flame speed, but much can be done to increase the flame path length, and Figure 11.7 suggests a way in which that may be achieved. A retort stoker is shown installed in the furnace, but the modification works equally well with most other stoker designs provided that the furnace diameter is sufficient. The modification involves the installation of a refractory arch in the upper half of the furnace exhaust end, usually suspended upon stainless steel hangers from the tube plate. The arch is assembled from perforated firebricks, through which the hangers protrude. The holes in the lowest bricks are filled with refractory concrete to trap the ends of the steel hangers and thus secure the arch.

When the boiler is in operation, the combustion gases now have to pass under the arch before they can escape from the furnace. The new length of the gas path depends upon the size of arch that can be accommodated, but in most cases it is sufficient to at least stop flame impingement and fouling on the tube ends. The modification is cheap and simple, and fairly robust, and it does have the added advantage that in forcing the gases to undertake a 180° change in direction, more particulates are ejected. The benefit in respect of the tube ends is therefore twofold, since lower particulate concentrations also lead to reduced fouling. However, whether or not the modification is sufficient to comply with the European temperature/residence time requirement is another matter. That depends entirely upon how close the furnace is to achiev-

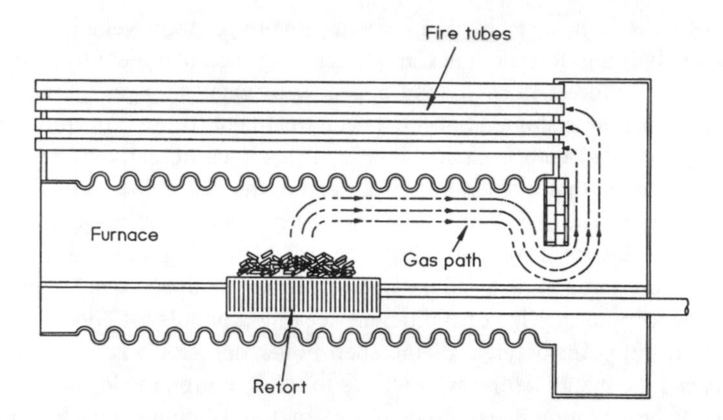

Figure 11.7 Shell boiler furnace modification.

ing it in the first place, which in turn depends upon the ratio of furnace length to firing rate. The gas behavior may be mathematically modeled using standard combustion engineering formulas and thermodynamic computer programs, but the mathematics involved are quite complex, and are beyond the scope of this book. Using them is the work of specialists in combustion engineering rather than of the sales staff of a dRDF facility.

Water-Tube Boilers — Water-tube boilers long ago replaced shell boilers for large-scale steam raising applications because of their fuel flexibility, high pressures, rapid response to load, and high thermal efficiency. In such a boiler, water is contained within the tubes rather than surrounding them, and as Figure 11.8 suggests, the tubes are connected between "headers," which are very thick-walled, water-filled drums, and the main steam drum. The design principles permit large furnaces that are completely enclosed in water-tube walls except where they are penetrated by apertures for grates or burners. Combustion gases created in the furnace pass through a "screen wall" of water tubes pitched with gaps between them, and then through the superheater to the generating tubes beyond. All of the waterwall, screen wall, and generating tubes are water-filled, while the superheater draws steam from the steam drum and is therefore steam-filled.

Figure 11.8 shows a general design concept rather than any specific make or type of water-tube boiler, since there are hundreds of both. Large boilers are erected rather than manufactured, using standard gauges and diameters of boiler tubes that have been bent ("manipulated" in trade jargon) to the profiles required. Working pressures are potentially very high because the mechanical strength of a cylindrical section under pressure is a function of its diameter. The equation which demonstrates this is

$$T = PD/2S \qquad (1)$$

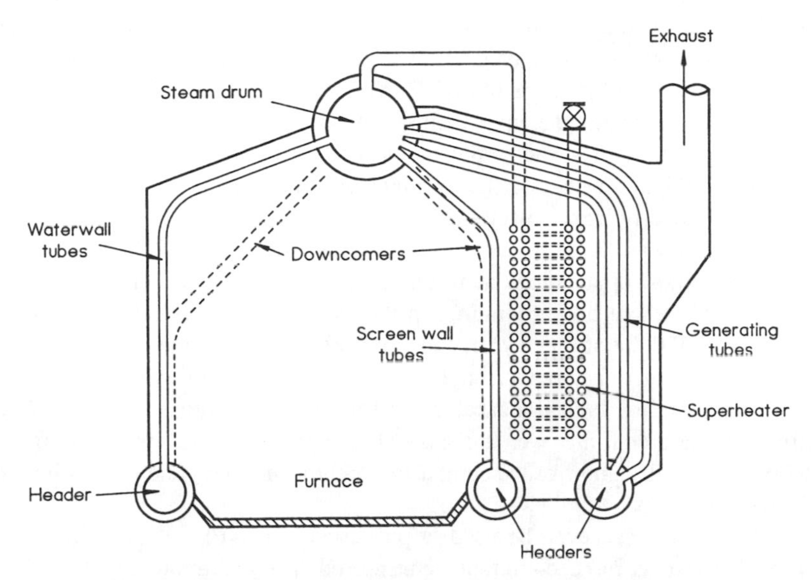

Figure 11.8 General arrangement of water-tube boiler.

where T is the bursting thickness of the cylinder wall, P is the internal pressure, D is the cylinder diameter, and S is the ultimate tensile strength of the cylinder material. It is this simple application of physics that permits water-tube boilers to be routinely designed for pressures of 2000 psi (140 bar) and above, with superheated steam temperatures limited only by the creep resistance of the superheater tube metal.

As potential applications for dRDF, large water-tube boilers are in many ways ideal, and not least of the reasons for that is the sophisticated exhaust gas scrubbing equipment with which they are commonly equipped. Electrostatic precipitators, dry and wet lime scrubbers, and bag filters are widely used, and the large furnace dimensions make gas residence times much more achievable than they are in shell boilers. As a result, there are no insurmountable difficulties in firing on dRDF or in mixing it with solid or pulverized fuels. There are, however, a few limitations.

The most significant limitation rests with the energy density of the fuel. Very high pressure boilers, with substantial degrees of superheat, require a large input of energy and a correspondingly high energy density. dRDF does not possess sufficient energy density to accommodate the demand, and so when used alone, boiler pressures are limited to around 600 psi (40 bar) and 900°F (480°C) of superheat. Even though those conditions are modest by modern standards, they still encompass a very large number of plants indeed, and of course dRDF does not have to be used in isolation.

Many water-tube installations in fact use such massive quantities of fuel that local waste arisings would not provide sufficient capacity. For example, a modern power station boiler may consume 40,000 t of coal a year, and the station may have a number of them. The total fuel demand could easily be 250,000 t a year, and in terms of dRDF, even with the most optimistic assessment of calorific value, the equivalent demand would be 300,000 t a year. At average separation rates, such a demand translates to about 1.2 million tons a year of municipal waste, and the example used would not be considered to represent an overly large power station.

As far as water-tube boilers are concerned, therefore, dRDF is a mixer fuel, and in most applications additions of up to 10% of the total fuel input are possible without boiler derating. In the example, that would still mean a demand for 30,000 t of fuel a year, and 120,000 t of municipal waste to deliver it. The conclusion to be drawn from these points is that major water-tube boiler installations are the logical market target for dRDF, and those which already burn low-grade coal are even more so. There the low acid emissions from dRDF can actually improve the combustion conditions and yield greater boiler efficiency.

The product can even be used in pulverized fuel (PF) plants with care, although fRDF is more commonly considered in those applications. dRDF cannot normally be milled to fine granules in milling plant of the type used for PF, since it is tough rather than brittle. However, conventional coal ball and roller mills will accommodate it provided that it is intimately mixed with the coal first, and that the proportions are not more than 6% dRDF in coal, measured by weight. In that case the presence of the coal is sufficient to abrade and degrade the RDF without permitting it to compact and densify further.

11.3.3 Fuel Transportation and Delivery

11.3.3.1 Flock RDF

Of all of the potential categories and subcategories of RDF, flock is the most difficult to store, handle, and transport. In addition to its propensity for compaction under its own weight, the ease with which it coagulates into a dense, fibrous mass considerably increases handling problems if it has been stored for any length of time greater than a few h. Where it is derived directly from the air classification stage immediately following milling, it contains substantial amounts of organic contamination, reducing the facility to store it still further. The presence of the organic materials encourages rapid bacteria growth, which is stimulated both by the high moisture content (about 30% w/w) and the aeration created by the granulator. This has the effect of liberating still more moisture and increasing the self-compacting capacity.

Although there are cases where fRDF has been manufactured, stored, and delivered to combustion plant over relatively short distances, the production facilities were and are small, and the amounts of product involved allow the difficulties to be overcome to some extent. Large production capacity would, however, be another matter, and the method most commonly used for transport, that of packing the material into enclosed compaction containers, would rapidly become unmanageable. Apart from any other consideration, the numbers of containers and vehicles to handle them would rapidly escalate beyond any semblance of financial viability, and the logistics involved in scheduling them would become extremely complex.

For those reasons, fRDF appears to have more application where the combustion and manufacturing facilities are on the same site, and the fuel can be delivered from the one to the other directly by conveyors or pneumatic systems. Even then, there may be some handling and storage requirements, since it is unlikely that the production facility would operate for 24 h a day. It would require routine maintenance and adjustment, and experience has demonstrated clearly that production rates can only be kept consistent if that occurs daily.

All power plants, however, are essentially continuous operations. Boilers, and particularly large water-tubes, do not react well to being taken off line at regular intervals, and turbogenerators must not be. Somehow, therefore, there must be some form of buffer storage at the interface between the two facilities, and there all of the difficulties with fRDF recur. While there are no simple solutions, the most practical is one where the fuel is stored in an enclosed building on a flat floor, with good ventilation. Recovery can then be by means of shovel loaders fitted with clamping buckets, that load the material onto conveyor systems designed to break up any lumps which may have formed.

11.3.3.2 Densified RDF

When the product has been fully densified, storage problems at least are overcome. dRDF is highly stable and is reasonably resistant to moisture. It does not absorb water from the atmosphere, and even continued rainfall takes some time to have any major effect. Where consumption is large, for example in a conventional power station, storage under cover is not usually considered worthwhile. The limited degradation likely to occur as a result of weathering is insufficient to have any detectable effect upon the combustion plant.

Where the demand or the capacity of the combustion plant is smaller than that of a large power generator, however, undercover storage is worth consideration. All fuels absorb rainwater, and the moisture content of coal can be increased substantially almost as easily as that of dRDF. Any water in the fuel reduces combustion efficiency to some extent, since the water conducts useful heat out of the boiler and into the exhaust stack. The high density of dRDF makes undercover storage economic, particularly since the storage needs to be no more than a means of keeping direct rainfall away from the product. A totally enclosed building is superfluous.

Transportation is also straightforward and reasonably economical, since dRDF is weight-limiting. In other words, a delivery vehicle can be loaded to its full weight limit before its volumetric capacity is exhausted. The fuel flows easily, and provided it is substantially free of particulates, it does not readily compact. Tipping of bulk loads presents no difficulties, and degradation resulting from vehicle vibration and movement is minimal.

Not all deliveries are necessarily in bulk, tipped loads however. Smaller combustion plants particularly are likely to store their fuel in silos, into which it is blown pneumatically by specially equipped vehicles. Silo storage plants that have their own handling and pneumatic loading facilities are something of a rarity, and so for the purposes of this text we shall restrict ourselves to bulk-blower trucks suitable for dRDF.

Although there is a long history of blown coal deliveries, and the technology involved is clearly understood, dRDF reacts rather differently in those conditions. There are two areas in which the differences are most apparent and likely to cause trouble, and the first of these concerns the rotary valve which meters the fuel into the air line from the blower. In all such vehicles, a low-pressure air blower unit driven by the truck power take-off discharges air at about 5 to 10 psi down a steel pipe to a connection for a delivery hose. Conventionally, both the pipe and the delivery hose are either 4 or 6 in. (100 or 150 mm) in bore. Between the blower and the hose connection, a hydraulically operated rotary valve in the base of the vehicle body loads fuel into the pressure pipe, forming an air seal in the process of doing so.

Almost all rotary valves used for the purpose have parallel cast steel vanes mounted upon a shaft, and the vanes are designed to fit closely to the valve casing to maintain the air seal. Inevitably some particles of fuel become trapped between the van edges and the casing as the valve rotates, but where the product is coal that has no effect. Coal is a brittle material, and any pieces so trapped simply shatter. dRDF is, however, another matter. It is tough rather than brittle, and the application of shearing forces to it is more likely to make it even harder and more resistant to pressure than to break it. Accordingly, the parallel-vane rotary valves used in conventional blower delivery vehicles cannot be relied upon to work for very long if at all. Inevitably a situation will arise where several pellets of fuel become trapped at the same time, and the valve stalls.

Usually the rotary valves have a reversing facility, designed for exactly that situation, but it is provided to deal with the infrequent occurrences of stalling when handling coal. With dRDF those conditions may occur every few min, and there are many cases on record of complete vehicle loads that should have been discharged in 20 min taking over 4 h as a result. The solution is remarkably simple, and it is to replace the parallel-vane rotor with one in which the vanes are mounted upon the shaft in the form of a widely pitched helix. Air sealing is just as efficient as it is with the parallel-vane valves, but in this design the individual vanes are not parallel to the casing. It is impossible to trap more than one pellet at a time, and even then the angle of presentation of the vane to the casing tends to push the offending pellet away rather that to trap it. On the infrequent occasions where trapping does occur, the scissor-like action of the valve cuts cleanly through the pellet in preference to crushing it.

The second condition distinguishing dRDF from coal in blown deliveries concerns its behavior in the flexible discharge hose. Its highly regular granularity maintains air spaces between the individual pellets much more consistently than is experienced with coal, and the delivery becomes dense-phase within a distance of about 16 ft (5 m). That in itself is not necessarily a matter for concern, since as long as the hose is of a smooth bore, the fuel will pass along it perfectly well. It is at the point of discharge where the problems begin to occur.

When coal reaches the end of the delivery pipe, the air propelling it can escape and expand rapidly. Coal has a high bulk density, and it falls out of the airstream almost immediately, usually within less than 20 in. (0.5 m) from the end of the hose. dRDF, however, is much lighter, and as the air expands it carries the fuel with it over much greater distances. More importantly, the air expands into an unrestricted space almost explosively, and that expansion accelerates the fuel particles violently. They may therefore be propelled across the silo or bunker with very considerable force, often sufficiently so to rapidly damage or distort the wall or casing against which they impact. The usual solution to the problem is to hang a target curtain of conveyor belting a short distance beyond the end of the hose discharge. If the belting is secured only along its top edge, then the impact of the fuel against it is absorbed by its movement, and the bunker or silo walls are protected.

11.3.4 Marketing Techniques

Fuel marketing, whatever the type of fuel being offered, is a key account business. The target is always to identify, attract, and keep substantial customers who will continue to order supplies on a regular basis over a long period of time. Doing so requires something more than the skills of a door-to-door salesman, and one of the most important attributes for both the manufacturer and fuel marketer is detailed knowledge of combustion engineering. Customers will expect their plant to be familiar to the supplier, and to be advised upon the technical, legal, and financial considerations involved in the use of the product.

Almost invariably the first step toward securing a contract involves a boiler or combustion plant survey at the customer's premises, and it is a service which the supplier must provide. Oil and gas companies will, as will solid fuel suppliers. They are able to make skilled boiler surveyors available, and to prepare detailed reports to the customer defining how their products can be used to the best advantage. The manufacturer of RDF must do likewise, and it is in that respect that so many RDF operations have failed.

Combustion plant operators are not primarily interested in recycling, however, laudable doing so may be. Neither do they see themselves as being solely responsible for the security of the environment, and they certainly will not wish to consider themselves as a waste disposal route or solution. Any attempt to sell the product by stressing those "attributes" in the absence of technical competence is doomed to failure. RDF is a manufactured, synthetic fuel, not waste. From the point of view of a marketing strategy, the fact that it is recycled from something else is immaterial and even potentially discouraging. The first point of concentration must therefore be the establishment of clear and unambiguous operating instructions based upon the combustion plant in question.

The customer will expect to be informed honestly in respect to the following considerations:

- Financial or operational advantages the product will offer. Any product must be either cheaper for the same quality or better for the same price to attract a key-account customer.
- Price basis the fuel will be offered upon. Major users buy heat, not weight, and a price per ton is irrelevant to them. Their interest is in a price per useful heat unit, such as therms, megajoules, kilowatts, British thermal units, etc. The suppliers representative will be expected to be familiar with all of those units, and to be able to convert each into another with ease.
- Extra work or commitment of resources the product's use will involve. There is no point in trying to sell a potential saving of $10,000 a year if a new fuel store and handling system costing $1 million is necessary to achieve it.
- The product's safety in respect to handling and storing in quantity, and any health risks to employees. The very name "refuse-derived fuel" suggests malodors, vermin, and rotting waste.
- Emissions are released by the product, and the extent to which they are prescribed by legislation. It is a very insular customer who is unfamiliar with the current levels of concern about dioxins and acid emissions.
- Any risks of boiler or combustion plant fouling. A boiler is a very expensive item of capital equipment, and may easily be the most expensive in the customer's works. It is probable that his whole production depends upon it, and he will be well aware of the length of time that extensive boiler repairs can take.
- Any risks of boiler or combustion plant corrosion. Most substantial owners of boiler plant will have read of experiences with burning wastes and products derived from them, simply because so much ill-informed material has been published.
- The disposal of ash residues and any controls over the methods used for doing so. Owners of solid fuel installations almost always have an ash disposal problem already, and they will not wish to exacerbate it by producing any ash even more difficult to deal with.
- The continuity of supply. The customer will not wish to go to the trouble of preparing storage facilities, retraining his employees, having his boiler combustion controls recalibrated, and negotiating a contract if the supply could decline or cease at any time.
- The calorific value of the product and what it actually represents in terms of useful heat. Most major fuel users are quite familiar with calorific value declarations that ignore excessive moisture contents. Their boilers are most

unlikely to be so efficient that they can recover the latent heat of vaporization from the fuel moisture, and they will not wish to purchase water.

- The chemical and physical specification against which the fuel will be supplied, and the extent to which that specification is guaranteed and repeatable. All discerning customers will wish to know the gross and net calorific values, moisture, ash, chlorine, and sulfur contents, and ash deformation temperatures at least, since all of those are fundamental to boiler operation. In addition, the variability must be reasonably low, otherwise the data will be useless. There is no point in suggesting a contract based upon an ash content of between 6% and 16% by weight. The customer will be comparing the specification with his existing supplies, and will not accept a variation of ash content of more than 1% to 2% at most.
- How the product will be delivered, and the freedom the customer has to select the times of day when deliveries are made. A substantial user will be accustomed to his supplies arriving infrequently in very large trucks at least, and he will be unlikely to accept almost continuous deliveries in small ones. It may also be that there are particular times of day when he is committed to moving his own production, and will already have all of his site roads occupied by his own vehicles. He may be reluctant to have suppliers' trucks attempting to move against the flow in those circumstances.
- Procedure in the event that a substandard delivery is made. All major fuel users have some experience of the odd off-specification load, and will want to be assured that the supplier will remove it and replace it with new stock, at his own expense and immediately.
- Length for which a supply contract is being offered, the price escalation formula, and "break" clauses. There is no point in seeking a 20-year contract in the fuel supply business. No customer wants to close off his options to take advantage of new technology to that extent. Equally, there is little chance of success if the escalation formula offered is based upon the retail price index. It is most unlikely that the customer's own production is controlled by that index, and he will not willingly accept it from a supplier. He is more likely to be amenable to an index used by his own industry, or one common for energy supplies. Break clauses in the contract offered are always of interest because the customer's circumstances can change. It may be, for example, that some time into the contract, a new production technique is developed in his own industry that either eliminates or considerably reduces the demand for steam. He will not then wish to be locked into a supply contract that continues to deliver fuel he cannot use, and will want to be able to give a reasonable period of notice of his intention to withdraw.
- Extent to which the supplier will support the product with technical services. If a boiler problem develops, the customer will probably blame the fuel first, and he will expect to be advised about what steps he should take to correct the situation. He will also expect to call upon the supplier to train his operators, and to visit the boiler plant regularly to ensure that optimum conditions are being maintained.

The above list is by no means a complete guide to the marketing of RDF or any other fuel. The provision of such a guide would fill several books of the size of this. The list does, however, set out to indicate the major factors involved in key-account marketing, and to demonstrate the most fundamental point of all: key-account marketing is essentially understanding the customer's business, the pressures and demands upon it, its financial sensitivities, and how the product offered can improve it. RDF marketing is difficult with those understandings; it is impossible without them.

11.4 SOME ALTERNATIVE PRODUCT MARKETS

The subject of marketing has been dealt with so far upon the premise that only the main raw materials would be converted into consumer products at the processing facility. The granulated, dried combustible fraction would be converted into an industrial solid fuel, and the digested organic fraction into a horticultural growing medium or soil conditioner.

In this book, waste reprocessing has been defined as the technology of reducing waste to clean raw materials from which entirely new consumer products can be made. Therefore, discussions upon marketing would be incomplete without some consideration of on-site manufacturing processes that might use the base products as raw materials.

11.4.1 Compost

Irrespective of whether it is derived from green (yard) waste or municipal waste, compost could be considered as a raw material rather than a product. Consumer goods result from the use of it, and that use is not constrained to any particular location. It could take place at the composting plant just as well as it could anywhere else, and such an option might be attractive if the markets for the base material are found to be limited.

If a venture were to be undertaken, then the compost would essentially have to be used as a growing medium, but whatever is grown must itself be able to reach a market sufficiently wide to absorb all or most of the compost production. That constraint alone imposes severe restrictions. There are, however, a small number of vegetable products that do command bulk markets and could use a significant amount of compost. For them the only remaining constraint is that the compost produced is suitable.

One of the most appropriate products is lawn and amenity turf, and waste-derived composts are eminently suitable for both types. The structure of the compost is such that it encourages very strong root growth, and its moisture retention capacity ensures rapid and consistent germination. It lends itself to "nursery-grown" turfs that can be produced inside tunnel-type enclosures, which are simply tubular steel frames over which ultraviolet resistant polyethylene film is stretched. This ensures that the compost remains substantially free from wind-sown weed seeds, and that the grass may become established without weed interference. Very high quality turf can be grown as a result, and the techniques for doing so are quite straightforward.

The first step is to set out long bays by laying 500 μm polyethylene membrane upon a level, prepared surface. The surface may be almost anything provided that it is sound and free from items that might penetrate the membrane. 2 in. (50 mm) of sand upon compacted, level soil is satisfactory, but tarmacadam or concrete is better if it is already available. The membrane cover should then be set out into the bays by laying upon it rows of concrete slabs, the ideal dimensions of which are 12 × 24 × 2 in. thick or similar. The slabs should be laid end to end, with a gap of at least 3 ft (1 m) between each row. The actual gap selected should be determined by the wheel spacing of the harvesting machine that will be used to recover the finished turf.

When the slabs are laid, compost should be spread between them and lightly compacted until it is level with the tops. A careful wetting with fine water sprays is worthwhile at this point, simply to ensure that the moisture content of the compost is uniform throughout the full length of the bays. If it is not, then growth rates may become inconsistent and may make harvesting difficult.

Seed of the strain or mixture required for the finished product can now be sown in accordance with the seed merchant's instructions, and when healthy growth has become established, mowing can commence. The mowing machine should run upon the paving rows rather than upon the turf, so that no damage occurs to the product. After a period, which may be of weeks or months depending upon the strains of grass, root growth becomes so dense that it is possible to roll the turf rather in the manner of strips of carpet, and when that can be done without any sign of tearing, then the turf is sufficiently robust to withstand handling. Harvesting is accomplished by simply rolling the turf up into rolls of lengths suitable for the market selected.

The method can absorb approximately 5300 yd³ per acre (10,000 m³ per hectare) before compaction into the seed bed, but it is not as economical as growing turf directly upon open land. It offers the facility to produce a value-added product, but the more traditional turf-farming approach uses more compost because the size of operation is likely to be larger.

Using a development of traditional methods, compost may be spread upon agricultural land and lightly ploughed in using quantities of 40 tons per acre (100 to 300 t/ha). The higher concentration is appropriate when the organic content of the existing soil is inadequate, or intensive growing is required. Grass yields are not directly related to the rate of compost application, and so increasing the compost addition from 40 t/acre to 120 t/acre (100 to 300 t/ha) usually results in only about 30% improvement in growth rates.

Tree and shrub growing uses much more compost than turf, particularly if the purpose is to transplant saplings for reforestation schemes. Each individual sapling may require up to 35 ft³ (1 m³) of compost, and again those composts derived from organic wastes are generally suitable.

Finally, general agriculture is worthy of consideration, although municipal waste composts should be excluded from use in growing food crops unless their heavy metals levels are shown to be within strict limits. Many countries have legislation prescribing the maximum permitted heavy metals concentrations in soils. Where the concentrations are appropriate, then grain would be a worthwhile product, since many strains have been shown to grow extremely well in derivatives of municipal waste.

Agricultural applications, and possibly new markets, could be explored using pelletized compost. It has been shown by research and experimental work in Great Britain that refuse-derived compost can be pelletized quite easily, and that at a pellet diameter of $^1/_4$ in. (5 mm) it can be delivered using conventional agricultural fertilizer spreaders or seed drills. This facility has the potential to overcome one of the primary objections of farmers to using composts, which is that it is cumbersome and inconvenient to handle in bulk, and that it is difficult to spread consistently. Again, in such an application, a spread rate of approximately 40 t/acre (100 t/ha) is generally thought suitable for most soil-conditioning and amelioration purposes. In volumetric terms of compost before pelletizing, this equates to about 8,800 ft^3 (250 m^3).

The pelletizing process is virtually identical to that employed in manufacturing dRDF, although there is usually no requirement for predrying or predensification. The base compost may be delivered directly to a pelletizing machine. The most important operational factor in the process is ensuring that at the end of a production run, the die of the pelletizer is cleared of product. Refuse-derived compost sets extremely hard if left in the die and permitted to cool, and so it must be evacuated. Feeding the machine with a small quantity of very dry compost that has been retained for the purpose is usually sufficient.

All of these applications are extremely specialized, and they require specialist expertise for success to be assured. The purpose of the above is not to suggest that the process plant operators become experts in a number of totally unrelated fields, although to some extent that is inescapable in dealing with waste. Instead, the intention is to offer some ideas that could, perhaps, form the basis of joint ventures or manufacturing licenses with local specialists.

11.4.2 Refuse-Derived Fuel

Increasingly, RDF is being seen as a fuel for electrical power generation, and there are some merits in considering generation on-site where a new facility is being constructed. Doing so eliminates the need for expensive transport at least, and the fuel and the combustion plant can be designed to operate together in a way that can never be achieved in a conventional power station. In those circumstances, the boilers can be selected to burn only RDF, without the need or provision for supplementary fuel, and the whole operation can be optimized such that the fuel is consumed as fast as it is produced, thereby eliminating any significant requirement for storage.

Generally in such applications RDF performs best with low-pressure, low-superheat boilers and low-pressure turbines, both of which are readily available. Estimates of power generation rates vary according to the efficiency of the boiler plant offered, but dRDF commonly returns boiler efficiencies of well over 80%, and so basing estimates upon 75% is realistic. The following performance has been calculated* in respect to dRDF with a calorific value of 8300 Btu/lb (19,300 kJ/kg) and 9.5% ash.

1. At boiler conditions of 17 bar and 315°C (250 psig and 600°F), with 0.1 bar (26.75 in. mercury) steam exhaust vacuum:

RDF (te/h)	5	10	15
Output (kW)	4,225	8,875	14,090

2. At boiler conditions of 40 bar and 382°C (588 psig and 652°F), with 0.1 bar (26.75 in. mercury) steam exhaust vacuum:

RDF (te/h)	5	10	15
Output (kW)	5,000	10,556	16,380

Relating these outputs to a demand for RDF, and assuming an 8000 h operating year, returns a potential consumption of 40,000 t a year at 5 t/h, 80,000 t a year at 10 t/h, and 120,000 t a year at 15 t/h. These are substantial quantities of fuel, where even the lowest of the power capacities listed would require a municipal waste availability of approximately 150,000 t per year.

The possibility leads to yet further opportunities that are increasingly being explored. A power generating plant creates its own waste in the form of heat, which could be "recycled" within the RDF plant and, even more effectively, through sales to local industries. The concept that has grown from these activities is of a combined RDF and power-generating plant in a large industrial estate, where electrical power, steam, and hot water services are provided to each of the factory units surrounding it. Markets for the RDF do not then have to be established. Only markets for the services are necessary, and the adjacent factory units would be "captive" in the sense that it would be difficult or inconvenient for them to obtain their needs elsewhere.

One extremely successful venture based on this concept has been established in Great Britain at the major industrial estate in Slough, Berkshire. There, Slough Heat & Power Ltd., a division of Slough Estates PLC, consume dRDF at a rate in excess of 30,000 t per year in large recirculating fluid bed boilers, and generate electrical power in their own power plant.

The capital costs of a power plant development are subject to some flexibility, and it is beyond the scope of this section to explore them in great detail. However, examination of a range of historical estimates suggests a total capital cost range of $10.4 to $26.5 million (£6.5 to £16.5 million) for fuel consumptions of 5 t/h to 15 t/h. Revenue estimates assuming a value of $0.065 (£0.04) per kilowatt for power sales alone then suggest a gross income of

* NEI Allen Ltd., Bedford, England 1989.

$2.56 million (£1.6 million) for the 5 t/h consumption, rising to $8.4 million (£5.25 million) for 15 t/h.

11.4.3 Glass

In the earlier sections of this book, ways of granulating and cleaning mixed glass cullet were suggested. Sales were again assumed to be to secondary recyclers for low-grade applications, possibly as construction material.

As a result of its clarity and color, glass has potential as a decorative aggregate for some consumer products, and paving and construction block are among the more obvious of those. Paving is comparatively simple to manufacture, requiring little capital investment, and since the glass has to be recovered in any case, a paving production line could be established with minimal raw materials costs. The constituents of a paving slab or construction block are, after all, only cement, sand, and aggregate, and in the development suggested, only the cement would be supplied from an external source.

This is not to suggest that a production line should be set up to produce standard items that are readily available from a number of competing manufacturers, since doing so would only transfer the marketing problem from the glass cullet to the products. Instead, a policy of value-adding would be appropriate, and such a policy might seek to take advantage of the special nature of the glass as an aggregate. For example, a wide range of paving is made with exposed aggregate, for decorative purposes and to provide a nonslip surface. Glass-aggregate slabs produced in the same manner would offer very attractive finish that has the added advantage of reflecting light at night, thereby defining very clearly to location of the pathways. Such a capability might also be appealing to architects engaged in designing office and industrial buildings, where fire escape routes prepared from glass-aggregate slabs would be clearly visible even in the event of a total failure of artificial lighting.

APPENDIX A

WEIL'S DISEASE EMPLOYEE SAFETY CARD

Precautions

In all dealings with crude domestic waste, or any materials derived from it, personal hygiene is most important. Wash hands immediately after handling refuse or anything which has been in contact with it. Treat all cuts and abrasions, and report them, however minor. Wear proper protective clothing, including suitable gloves. Do not consume food in the immediate vicinity. Avoid breathing refuse dust–wear a dust mask. Weil's disease is rare, and it is avoidable with simple precautions.

Weil's Disease

For your protection, this booklet should be kept in a safe place and shown to your doctor in the event of your illness. Ensure that he is aware of your occupation.

Weil's disease is rare but serious and can be fatal, and it is essential that it is taken into account at the earliest possible stage of illness, particularily where you have symptoms similar to influenza.

Medical Information

Leptospirosis Icterohaemorragica (Weil's Disease)

This disease normally commences as a febrile illness with varying degrees of muscular pain, tenderness, congestion of the conjunctiva, jaundice, and hemorrhages of the mucous membranes and skin. Albumin and casts may be expected in the urine, and polymorphonuclear leucocytosis is usual. Jaundice may not be encountered at all, and is unlikely during the first days.

Early in the development of the disease, symptoms are similar to those of pneumonia, tonsilitis, influenza, rheumatic fever, and nephritis. Later they are similar to those of catarrhal jaundice, and gallstones. Special laboratory tests are essential, as the disease sometimes develops rapidly after initially mild symptoms, and early hospitalisation is advised.

APPENDIX B

EXAMPLE OF NATIONAL STANDARDS LEGISLATION

Officially Recognized Organic Soil Conditioners French Standard NFU 44-051. Types and specifications, December 1981.

Definitions and Specifications

Municipal waste composts (types 13a to 13c) are divided into four types according to their granulometric composition:*

- Very fine urban compost: at least 99% of the compost must pass through a sieve with mesh 6.3 mm^2.
- Fine urban compost: at least 99% of the compost must pass through a sieve with mesh 12.5 mm^2.
- Medium urban compost: at least 99% of the compost must pass through a sieve with mesh 25 mm^2.
- Coarse urban compost: at least 99% of the compost must pass through a sieve with mesh 40 mm^2.

For types 13a to 13c, physical characteristics are as indicated in chapter 3.

Contents to be Declared and Other Markings

- For all types: organic matter, total nitrogen percentage by volume of primary product.
- For types 13a to 13c: granulometric composition.
- For types 13 to 14: declaration of the absence or presence of sharp or cutting elements**

2. Specific indications of usage, storage and handling.

* See French Standard X 11-501.
** These must obligatorily pass through a round-holed strainer (see NF E 81-061): 5 mm for coarse and medium urban composts; 2 mm for fine and very fine urban composts.

Definitions and Specifications

No.	Type	pH (H_2O)	Organic matter min. % by vol. Primary product	Organic matter min. % by vol. Dry product	Organic matter organic nitrogen max. ratio	N total max. % by vol. dry material
1	Manure	—	—	60	50	3
2	Dehydrated manure	—	—	60	50	3
3	Artificial manure	—	—	70	50	3
4	Mushroom bed manure	—	—	40	35	3
5	Mushroom bed compost	—	—	30	35	3
6	Grape marc composted >6.5	—	—	50	40	4
7	Grape marc not composted	—	—	70	50	3
8	Raw vegetal material	—	40	70	—	4
9	Non fermented vegetal conditioner	—	50	60	50	3
10	Fermented vegetal conditioner	—	35	50	55	4
11	Vegetal compost	—	20	30	55	3
12	Wood bark compost	—	30	50	80	2
13a	Fresh urban compost	—	20	—	—	2
13b	Semimature urban compost	—	20	—	60	2
13c	Mature urban compost	—	20	—	50	2
14	Town refuse	—	—	20	40	2
15.1	Acid peat <5	—	—	80	—	4
15.2	So-called alkaline peat >5	—	—	40	—	4
16	Peat compost	—	35	60	—	4

Description

Organic conditioners are described by the following:

- One of the types defined or appearing in chapter 4, possibly completed by the types specified in the individual standards.
- Origin:

    - For manure, dehydrated manure, an indication of the animal species from which it came.
    - For peat, the name of the vegetable or vegetables from which it came.
    - For wood bark composts, an indication of their origin: hardwood or resinous.
    - For raw vegetal material, nonfermented vegetal conditioners, fermented vegetal conditioners, vegetal composts and peat, their components in decreasing order of importance.
    - The degree of fineness in the case of urban composts according to one of the types defined in chapter 4.
    - Reference to this standard.

Marking

In conformity with articles 3 and 4 of Ordinance No. 80-478 of 16th June 1980, labeling or packing, as well as the accompanying documents in the case of a delivery in bulk, are to carry, with the exclusion of all others, the following markings.

Obligatory Markings

1. The term "ORGANIC CONDITIONER" in capital letters, followed by reference to this standard.
2. Type of organic conditioner such as is shown in chapter 4 of this standard, followed by the origin or, in the case of urban composts, the degree of fineness.
3. The declared organic matter and total nitrogen content, expressed in percentages, by volume of the primary product.
4. The granulometric composition, expressed in percentage of dry material, in the case of urban composts.
5. For urban composts and town refuse, declaration of the absence or presence of sharp or cutting elements.
6. The name, trade name, or registered name and the address of the party responsible for putting the product on the market, with its main office in France.
7. In the case of imported products, the name of the country of origin except for merchandise that originated in member countries of the EEC.
8. Net weight.

Optional Markings

1. Manufacturer's marking, product marking, commercial names and, if applicable, any guarantee.

2. Specific indications of usage, storage and handling.
3. Percentage by volume in primary product with minimum content of 0.5% for:
   - total dry material
   - total phosphorus, expressed in P_2O_5
   - total potassium, expressed in K_2O
   - total calcium, expressed in CaO
   - total magnesium, expressed in MgO.
4. The pH (H_2O).

INDEX

A

D

G

H

I

K

L

M

S